Biofuel Crop Sustainability

Biofuel Crop Sustainability

Edited by

Bharat P. Singh
Fort Valley State University
Fort Valley, GA, USA

WILEY Blackwell

Editorial offices: 2121 State Avenue, Ames, Iowa 50014-8300, USA
The Atrium, Southern Gate, Chichester, West Sussex, PO19 8SQ, UK
9600 Garsington Road, Oxford, OX4 2DQ, UK

For details of our global editorial offices, for customer services and for information about how to apply for permission to reuse the copyright material in this book please see our website at www.wiley.com/wiley-blackwell.

Library of Congress Cataloging-in-Publication Data

Biofuel crop sustainability / editor, Bharat P. Singh. – 1st ed.
p. cm.
Includes bibliographical references and index.
ISBN 978-0-470-96304-3 (hardback : alk. paper) – ISBN 978-1-118-63564-3 (epdf) – ISBN 978-1-118-63572-8 (epub) – ISBN 978-1-118-63578-0 (emobi) – ISBN 978-1-118-63579-7
1. Energy crops. 2. Sustainable agriculture. I. Singh, Bharat P.
SB288.B565 2013
631.5–dc23

2013002832

A catalogue record for this book is available from the British Library.

Wiley also publishes its books in a variety of electronic formats. Some content that appears in print may not be available in electronic books.

Cover design by Nicole Teut

Set in 10/11.5pt Times New Roman by Aptara® Inc., New Delhi, India
Printed and bound in Malaysia by Vivar Printing Sdn Bhd

1 2013

Contents

Contributors

Shaun D. Berry, Becker Underwood, Gillitts, South Africa

Nanthi Bolan, Centre for Environmental Risk Assessment and Remediation (CERAR), University of South Australia, Mawson Lakes, Australia

K.B. Cantrell, USDA-ARS Coastal Plains Soil, Water, and Plant Research Center, Florence, SC, USA

Dmitri Chatskikh, George Lemaître Centre for Earth and Climate Research, Earth and Life Institute, Catholic University of Louvain, Louvain-la-Neuve, Belgium

Christina Eynck, Linnaeus Plant Sciences Inc., Saskatoon Research Centre, Saskatoon, Canada

Kevin C. Falk, Agriculture and Agri-Food Canada, Saskatoon Research Centre, Saskatoon, Canada

Andrew Fieldsend, Research Institute of Agricultural Economics, Budapest, Hungary

John H. Fike, Crop and Soil Environmental Sciences, Virginia Polytechnic Institute and State University, Blacksburg, VA, USA

Wonae B. Fike, Crop and Soil Environmental Sciences, Virginia Polytechnic Institute and State University, Blacksburg, VA, USA

Thomas R. Fox, Department of Forest Resources and Environmental Conservation, Virginia Tech, Blacksburg, VA, USA

Wolfgang Friedt, Department of Plant Breeding, Justus Liebig University Giessen, Giessen, Germany

Cole Gustafson (Late Chair), Department of Agribusiness and Applied Economics, North Dakota State University, Fargo, ND, USA

A. Hastings, Institute of Biological and Environmental Sciences, School of Biological Sciences, University of Aberdeen, Aberdeen, UK

J. Hillier, Institute of Biological and Environmental Sciences, School of Biological Sciences, University of Aberdeen, Aberdeen, UK

Patrick G. Hunt, USDA-ARS Coastal Plains Soil, Water, and Plant Research Center, Florence, SC, USA

Abdullah A. Jaradat, USDA-ARS Research Lab and Department of Agronomy and Plant Genetics, University of Minnesota, Morris, MN, USA

Shailesh Joshi, South African Sugarcane Research Institute, Mount Edgecombe, South Africa

L. Chris Kiser, Department of Forest Resources, Abraham Baldwin Agricultural College, Tifton, GA, USA

Rocky Lemus, Department of Plant and Soil Sciences, Mississippi State University, MS, USA

Thein Maung, Department of Agribusiness and Applied Economics, North Dakota State University, Fargo, ND, USA

Bruce McCarl, Department of Agricultural Economics, Texas A&M University, College Station, TX, USA

Eric Obeng, Fort Valley State University, Fort Valley, GA, USA

Anna Ovchinnikova, Cargill Europe BVBA, Mechelen, Belgium

Sudhanshu S. Panda, GIS/Environmental Science Gainesville State College, Oakwood, GA, USA

David J. Parrish, Crop and Soil Environmental Sciences, Virginia Polytechnic Institute and State University, Blacksburg, VA, USA

József Popp, Faculty of Applied Economics and Rural Development, University of Debrecen, Debrecen, Hungary

David Ripplinger, Department of Agribusiness and Applied Economics, North Dakota State University, Fargo, ND, USA

K.S. Ro, USDA-ARS Coastal Plains Soil, Water, and Plant Research Center, Florence, SC, USA

David Saxowsky, Department of Agribusiness and Applied Economics, North Dakota State University, Fargo, ND, USA

Balaji Seshadri, Centre for Environmental Risk Assessment and Remediation (CERAR), University of South Australia, Mawson Lakes, Australia

Dev Shrestha, Biological and Agricultural Engineering Department, University of Idaho, Moscow, ID, USA

Bharat P. Singh, Fort Valley State University, Fort Valley, GA, USA

Hari P. Singh, Fort Valley State University, Fort Valley, GA, USA

P. Smith, Institute of Biological and Environmental Sciences, School of Biological Sciences, University of Aberdeen, Aberdeen, UK

Jeff Smithers, School of Engineering, University of KwaZulu Natal, Pietermaritzburg, South Africa

Kenneth C. Stone, USDA-ARS Coastal Plains Soil, Water, and Plant Research Center, Florence, SC, USA

Rianto van Antwerpen, South African Sugarcane Research Institute, Mount Edgecombe, Department of Soil, Crops and Climate Sciences, University of Free State, Bloemfontein, South Africa

Tania van Antwerpen, South African Sugarcane Research Institute, Mount Edgecombe, South Africa

Michael van der Laan, Department of Plant Production and Soil Science, University of Pretoria, Hatfield, Pretoria, South Africa

Johann Vollmann, Division of Plant Breeding, Department of Crop Sciences, University of Natural Resources and Life Sciences, Vienna, Tulln, Austria

J. Yeluripati, Institute of Biological and Environmental Sciences, School of Biological Sciences, University of Aberdeen, Aberdeen, UK

Preface

Agriculture by nature is an unsustainable system. Crops take more out of soil than it has the ability to replenish under normal conditions. Being aware of this fact, throughout history man has tried to supplement the difference by various means with different degrees of success. It was no accident that the location of the first agriculture-based civilization was Mesopotamia, meaning "land between two rivers." The flood water every year brought new rich alluvial soils down the river to enrich the farmland with nutrients. With increases in population, people migrated from the optimal to the best land and climate they could find, and in time were forced to settle for marginal soils and climates. However using ingenuity, mankind found ways to supplement what soil was not able to offer and used the climate to the fullest. Man's incessant desire for more, while at the same time having more mouths to feed started to take toll on the soil, the primary agricultural resource. Ancient scholars saw the development of this trend and warned against tendencies that made agricultural systems unsustainable. The evidence of such warnings is found in the literary archives of the Indus Valley, Chinese, and the Middle Eastern civilizations. In modern times, detriment to soil and climate became endemic with the large-scale use of chemicals and machineries in agriculture starting in the 1930s. Present scholars, like their ancient predecessors, raised the alarm and the "dust bowling" by mechanical agriculture created general awareness of the awaiting catastrophes from the overexploitation of agricultural resources.

The World Commission on Environment and Development of the United Nations General Assembly of 1987, also known as the Brundtland Commission provides the latest definition of sustainable agriculture. Under this definition, sustainability includes the long-term survival of agriculture as an economic enterprise benefitting not only the farmer, but the society as a whole, with due regard to the preservation of the quality of life in aesthetics, health, and culture by preserving the wholeness of the surrounding environment. It is similar to the concept followed during ancient agrarian times, components of which were lost during the Industrial Age. For example, Indian villages were a cluster of households; farming families were the nucleus and other families provided essential services to farmers, with the right of a portion of the harvest. Thus, essentially the part of the harvest a farmer could keep for his family in relation to other families in the village was fixed. Nonfarm families sold part of the harvest to exchange goods and services among themselves. This model of agrarian economy was sustainable because it created a system of exchange of goods and services that benefitted all members of the village. It also put the responsibility upon farmers to follow agricultural practices that guaranteed land to produce harvests year after year because the whole village depended on them. The farmer grew up sharing farm responsibilities from childhood and learning from his elders how to keep land productive and safe before assuming a decision-making role. People paid tribute to trees, rain, and animals and folklores were built around even the virtues of crows and vultures to ascribe their important contribution to human sustainability and to perpetuate this knowledge to future generations.

Biofuel is as old as man's discovery of how to light fire. Use of solid biofuel for cooking and the burning of plant oils for light was common until the start of the twentieth century. Using liquid biofuels for light and later as automotive fuel was not uncommon during the early 1900s. Cheap coal, kerosene, and later petroleum, however, slowly eroded plants' monopoly as energy providers and ultimately pushed them into subservient roles. Uncertainty regarding uninterrupted petroleum availability from disturbed regions of the world, which coincidentally have the greatest petroleum reserves, along with the intentions shown by petroleum-owning nations to use fuel as a political tool and fix prices outside the market domain have necessitated the shift to alternate fuel sources. Added to it was the clear evidence of detrimental impact of petro-fuels on the environment and, specifically, their connection to global warming. Thus, in the search for alternatives, there were two broad requirements: energy sources that are reliable and available year after year and secondly is environment friendly. Solar, wind, geothermal, hydro, and biofuel were perceived to meet the criteria. Biofuel is unique in the energy mix; it is the only fuel available both in solid and liquid forms and with the potential to match the multi-byproduct generation ability of petro-fuel. It is also the most suitable form of transportation fuel for the vehicles currently on the road. As the feedstock for biofuel comes from agriculture, the sustainability of feedstock production systems automatically becomes a matter of importance in consideration of this energy source. Keeping in mind that agriculture currently is mainly a food and fiber enterprise, noninfringement by biofuels of this primary function is also of paramount importance.

This book covers all aspects of sustainability as defined under the Brundtland Commission's definition, with the adage of food-over-fuel-priority underpinning all chapters. I have been fortunate to assemble the ablest authors from different countries. My sincere appreciation and thanks to all of them for graciously accepting my invitation to join in this exercise of providing a comprehensible scientific treatise on the different aspects of sustainability as it relates to biofuel crop production. The food-versus-fuel debate is highly emotional and some scientists have taken sides. I have tried my best to select authors who can provide objective deliberation and to examine each chapter carefully for science-based description. I hope this book proves useful to all concerned with agriculture, sustainability, and biofuel.

In closing, I would like to extend my sincere thanks and gratitude to my associate, Eric Obeng, for his assistance at every step of this editorial exercise. Without his help, this burden would have been lot heavier. I would like to dedicate this book to my 4-year-old grandson, Ayan—he never ceases to amaze me with his voracious appetite for reading anything with pictures and constantly attempts to discover things that are around him and which are intentionally hidden from him. What his parents call mischief, to me is just an innovative mind—the sign of a genius.

Bharat P. Singh
Fort Valley State University
Fort Valley, GA, USA

Biofuel Crop Sustainability

Chapter 1
Biofuel Crop Sustainability Paradigm

B.P. Singh

Fort Valley State University, Fort Valley, GA, USA

Introduction

Relevance of Sustainability

The topic of biological sustainability has been covered comprehensively by Morse (2010). In this review, the author contends that sustainability is more of a human centric term concerned with the survival of *Homo sapiens*. The origin of life by most accounts dates back some 3.5 billion years, to within just a billion years of Earth's own coming into existence. Living organisms evolved in many different forms and shapes (commonly referred to as species) to have multiple options of survival available for the various changes Earth may undergo over time. Sure enough, climate change is built into nature, yearly rotation with the change of season, occasional changes resulting from ocean current temperature variations, and drastic changes from gradual buildup or abrupt geological behavior such as ice age, volcanic eruptions, etc. At the same time, change through evolution is built into the constitution of living organisms, this continuous process is commonly known as mutation. Endowed with this gift of adaptability, living organisms have learned to flourish when the environmental conditions are optimal, sustain themselves when conditions become limiting, and survive when conditions turn harsh. Indeed, numerous species have disappeared in the course of time, but on the other hand, new resilient species have emerged. Sepkoski (2002) has developed a compendium of fossil marine genera, which is helpful in understanding the historical course of generation and extinction of marine species. There have been several periods of mass species extinction, one most noteworthy being Permian–Triassic event (about 250 million years ago) that killed up to 96% of marine species (Raup and Sepkoski, 1982; Rohde and Muller, 2005). As a matter of fact, though it has been stipulated that the extinction rate of living species has hovered around 99%, planet Earth remains the flourishing habitat of life. Reassuringly, there also appears to be an increase in the number of marine genera in the past 500 million years (Morse, 2010). Thus, it can be safely concluded that life has evolved a large window of survivability from catastrophic climatic events by continually transforming itself to adjust to widely different surroundings.

Biofuel Crop Sustainability, First Edition. Edited by Bharat P. Singh.

Human beings are only one among approximately 8.7 million eukaryotes inhabiting Earth. So, in nature's scheme of things, human extinction would be a mere footnote in its long history of evolution. However, for human beings, the subject of survival of *H. sapiens* is personal and of paramount importance. Creativity and innovation has been the hallmark of human existence. This human capability was first evidenced in the change from hunter/gatherer lifestyles with the constant search for food and water to being settled at a reliable water source and practicing agriculture for year-round reliable supply of food. The constant modernization since that period has brought us where mankind is today. Inventing preventions against diseases and developing shelters that provided safety from the vagaries of weather have drastically improved chances of human beings to live through the kinds of nature's episodes that resulted in the extinction of other species. These efforts have also cut the rate of mortality resulting in an exponential human population growth giving the species a better chance of being left with enough residual stock to repopulate in the event of a catastrophe. Human beings were cognizant of the fact that they were able to achieve all these feats due to their unique ability to exploit the earth to their benefit. All these successes, however, made mankind overconfident and led to the development of the notion that it was immune to nature's consequences and has the inalienable right to use Earth's resources at pleasure. However, the apparent gap between resource demand and resource availability became obvious to the wise centuries ago, and voices of concern have been raised intermittently for generations. More recently, it has become very clear that what many people and nations consider development, if not carried out more thoughtfully and better planned, will ultimately wipe out the very essential resources that man had taken for granted and the consequences could be calamitous. The book *Population Bomb* (Ehrlich, 1968), the United Nations Conference on the Human Environment (UNCHE) (UNEP, 1972), World Commission on Environment and Development (WCED) (United Nations General Assembly, 1987) (also known as the Brundtland Commission after its chairman), from which the definition for sustainable development was derived, and several subsequent worldwide forums are manifestations of concerns regarding resource availability and resource consumption. Sustainable development was defined by the WCED as "the kind of development that meets the needs of the present without compromising the ability of future generations to meet their own needs." Thus, parity in the right of the present and future generations in sharing the earth's resources was brought to clear focus. The details of the report also emphasized the importance of sharing the resources so that the poor of the world are not left behind. Thus was born the current version of the term "sustainability," which imbibes the theme of the survival and the perpetuation of high quality of life for all mankind of the present and future generations inhabiting different regions of planet Earth. The domain of sustainability born out of the environmental concern, thus, was expanded to incorporate the ingredients of sharing and social justice. Part of the reason for this change was the realization that the environment had no boundaries and all mankind must partake in its preservation, but this was only feasible if material benefits provided by resource exploitation were shared.

Sustainable Agriculture—Definition and Description

Agriculture is at the forefront of any sustainable development deliberation. This is because mankind exploits the earth most for agriculture than for any other enterprise. Agriculture, on the one hand, has the potential to provide many essentials of human life in perpetuity if harnessed appropriately, but on the other hand, if proper precautions are not exercised, can lead to the destruction of the very resources on which mankind desperately depends for survival and lifestyle support.

Many definitions of sustainable agriculture are available in the literature. The following are samples:

Allen *et al.* (1991): A sustainable agriculture is one that equitably balances concerns of environmental soundness, economic viability, and social justice among all sectors of society.

Lehman *et al.* (1993): Sustainable agriculture consists of agricultural processes, that is, involving biological activities of growth or reproduction intended to produce crops, which do not undermine our further capacity to successfully practice agriculture.

Yunlong and Smit (1994): Sustainable agriculture refers to the use of resources to produce food and fiber in such a way that the natural resource base is not damaged and that the basic needs of producers and consumers can be met over the long term.

The U.S. Code Title 7, Chapter 64, Subchapter 3103 gives the legal definition of sustainable agriculture for use by the United States Department of Agriculture. It is described as an integrated system of plant and animal production practices having a site-specific application that will over the long term:

- Satisfy human food and fiber needs
- Enhance environmental quality and the natural resource based upon which the agriculture economy depends
- Make the most efficient use of nonrenewable resources and on-farm resources and integrate, where appropriate, natural biological cycles and controls
- Sustain the economic viability of farm operations
- Enhance the quality of life for farmers and society as a whole.

The phrase "sustainable agriculture" is often used in the limited sense to describe agricultural practices that lower input requirement and preserve soil quality while maintaining economic yield. It is not unusual to associate it with organic farming. Singh *et al.* (2005) summarized the intent of sustainable agriculture into four farm-level goals: (1) to make better use of farm-based resources, (2) to minimize the needs of external inputs, (3) to prevent loss and degradation of farm soil and water resources, and (4) to maintain the quality of farm and rural life. The practice of sustainable agriculture requires the knowledge of interactions between soil and crop that result in optimum harvests at minimal economic and environmental cost. The methodologies include precision agriculture, integrated pest management, green manuring, crop residue management, soil carbon and nitrogen cycling, and other prudent farm resource managements. To be clear, sustainable agriculture does not call for going back to the farming practices that forced farmers to subsistence living and urban migration, but on the other hand, to guide them toward the right ways to adopt new agricultural innovations for maximum benefit today and times to come.

This concept of sustainability came into sharp focus during the 1980s. The need to reconsider the strategy of employing highly profitable machinery and chemical inputs at that time arose from the realization that the short-term profitability was being achieved at the expense of long-term continued fitness of the farm to produce crops and generate income on the long term. The turn-over plowing led to soil structural instability and erosion, and excessive chemical use led to pollution of the surrounding and broader environment. In addition, the constant increase in the demand for bigger machinery to plow deeper and more acres and more chemicals to control all kinds of pests rendered farming enterprise unstable, with inputs and their costs continuing to spiral upward, yields fluctuating year to year due to exacting climatic requirements, and increasing losses from pathogens and pests as they became resistant to chemicals. This created the need to develop a strategy whose aim, simply put, was to make sure that the gains made in the agricultural productivity are preserved in perpetuity. This can only happen if the soil and water resources are prudently used and rejuvenated and the soundness of farm ecology is maintained. Thus, while sustainable farmers continue to use tractors, they use new plows that disturb the soil to a minimum, thereby preventing erosion. The emphasis on the control of weeds, insects, and diseases on crop plants remains unchanged; however, new methods consist of a combination of chemicals, pest–predator control, crop rotation, and increased plant resistance, and other innovative means to prevent a toxic combination of soil, water, and crop. Farmers continue supplementing nutrients to increase the crop yield, but they use not only chemical fertilizers, but also rely on leguminous nitrogen fixation, increased availability of bound soil nutrients through enhanced microbial activity, etc. Sustainable agriculture, thus, is not a movement against industrialized agriculture, but one for an economically and environmentally viable option.

Relevance of Sustainability to Agriculture through Time

Awareness to sustainable agriculture has been shaped by the wisdom of generations starting from prehistoric times to the present, born out of experience and events of centuries. "He who plants even one tree, goes directly to Heaven and obtains *Moksha* (salvation)" (*Matsya Purana*, 59.159; period unknown, prehistoric) proclaims ancient Hindu scripture. The cutting of trees and destruction of flora were considered sinful acts. The Indian thinker Kautilya's *Arthashastra* (Aristotle's period; ~300–400 BC) prescribed various punishments for destroying trees and plants. Rapid agricultural expansion in different societies were accompanied by environmental problems. While Watson (1974, 1983) describes the "Arab Agricultural Revolution" as part of the Islamic Golden Age between the eighth and thirteenth centuries, Gari (2002) has accounted the concern expressed by several environmentalists of the period of the pollution of air, water, and soil that this revolution created as a result of wrong agricultural practices.

In the near-term historical context, the "Dust Bowl" period in American agriculture serves as a reminder to the detriment of unsustainable practices on agriculture itself, and to the environment at large. With the newly introduced farm tractors mounted with moldboard plows, farmers developed the notion that more and deeper plowing translated into better yields. They did not realize that they were endangering the most precious commodity on the farm—the soil itself. Drought is a part of nature's weather cycle and the American Great Plains went through it during the 1930s. The dry winds over the barren fields with loose soil created storm clouds stretching hundreds of miles. The dust clouds created severe health hazards and disrupted normal daily life stretching across all Great Plain States and reaching as far as the nation's capital. On Sunday, April 14, 1935, the dust cloud was so dense that the day has been remembered as Black Sunday and the whole region was referred to as "Dust Bowl" (Figure 1.1). It was estimated that 100 million acres of farm top soil were lost to the wind. This led to the passing of soil conservation legislation and the adoption of better soil management practices.

The environmental movement of the 1950s and 1960s in the United States brought to focus the need of constant vigilance to ward disaster from the well-intentioned introduction of new practices or inputs to the agricultural systems. *Silent Spring* (Carson, 1962) was a wake-up call to the increased use of pesticides, especially DDT (1,1,1-Trichloro-2,2-bis(4-chlorophenyl)ethane) post–World War II in agriculture. These pesticides were effective against crop pests, but their lethality was not targeted and DDT in particular was identified as causing

Figure 1.1. Abandoned farmstead in the "Dust Bowl" region of Oklahoma, showing the effects of wind erosion, 1937. *Source*: USDA (1937).

the thinning of bird eggs and their failure to hatch; thus the book's title was chosen to bring to attention the ultimate consequence of DDT, namely the silencing of spring because of the absence of birds. These efforts hastened the research in the development of targeted chemicals for use in agriculture and the institution of a ban on DDT in the United States at the end of 1972.

A new reminder that constant vigilance is essential to maintain the delicate balance between the agriculture and nature comes from the current near-extinction status of vultures (the Great Indian Bustard) in India. Diclofenac (2-(2,6-dichloranilino) phenylacetic acid) is an anti-inflammatory drug that was commonly prescribed to cattle before its ban for veterinary use in 2006. The vultures feeding on the carcass of recently treated dead cattle were lethally poisoned because the prescribed doses based on cattle body weights were exceptionally high for their system. Vultures have a special significance in the life chain in India, where they serve sanitary function, feeding not only on dead animals, but also on the human corpses belonging to the nature-conservationist religious group of Zoroastrians, who leave them at silent towers to be consumed by vultures and crows, who they believe were created for that purpose. The interaction of agriculture and environment is dynamic, multifaceted, and complex, and at the same time fragile, calling for constant vigilance and innovations to modify practices to fit the unique situations.

Biofuel and Biofuel Crops

Biofuel is an abbreviation for biomass fuel. Therefore, in the broad sense of the term, fossil fuels are also biofuels. Autotrophs convert a fraction of the sun's energy into the chemical energy that is held within complex organic matters. This chemical energy has two ultimate fates: much of it is eventually released (mostly as heat) while the remainder is locked within carbon-rich deposits (coal, oil, and gas). Biofuels differ from all other renewable energy sources in their unique characteristic of possessing the ability to substitute not only energy but also all fossil-fuel-based products. There are two main differences between biofuels and fossil fuels: (1) biofuel is a fuel produced from currently harvested biomass while fossil fuel was formed from biomass produced long ago and (2) biomass conversion to biofuel is in a timeframe of just a number of days while conversion to fossil fuel required millions of years. Essentially then, it is the rate of conversion that separates two similar origin fuels, one into renewable and the other into nonrenewable category.

Biofuels are currently derived from both direct and indirect biomass (animal raised on biomass) sources. The current sources of biomass are rather limited confined to selected field and tree crops, animal fat, and biological residues and wastes. However, for a futuristic outlook, the list is expansive, the geographical area for procurement vast, and may even include biomass generated from here to practically unfeasible processes like artificial photosynthesis. Odum (1971) lists the array of primary productivity (g m^{-2} yr^{-1}) of different ecosystems worldwide: deserts—3, cultivated lands—650, grasslands—600, moist forest—1290, estuaries and coral reefs—2000, continental shelf waters—350, and deep ocean—125. Although many years have passed since the quoting of these numbers by Odum, these have changed little since. The difference in the primary productivity is striking, but the reasons behind them are well understood. Deserts lack water, and continental shelf and deep-water oceans lack nutrients to support phytoplankton (Morse, 2010). Those falling in the in-between range are arranged in the order of climatic and resource optima of their ecosystems.

Many opportunities exist for creative rearrangement of ecosystems to create new optimal growth environments. Many of the vast deserts lie next to oceans, the largest water reservoir. In one scenario, algal ponds can be developed in deserts where there is plenty of sunshine and unlimited supply of water from the sea to grow salt-tolerant algal strains; otherwise these ponds can be placed in the sea itself. It is also within the realm of possibility to change the genetic makeup of selected plant species so that they can produce optimal amounts of biomass utilizing high-salt waters. Seeding of the ocean with limiting nutrients has been suggested as a means to provide suitable medium for optimal phytoplankton growth. A number of attempts have been made for seeding of the oceans with iron (Boyd, 2007), but the effectiveness of such techniques as well as their broad effects on ocean ecosystem have been much debated (Buesseler *et al.*, 2008). However, if such endeavors eventually bear fruit as a result of science and technology succeeding in creating localized phytoplankton-based ocean ecosystems, the harvests of biomass could be enormous. Some of this phytoplankton biomass may be source for biofuel feedstock. It would also lead to exponential increases in the ocean CO_2 sequestration as well as in yields of marine-based meats. The exchange of a meat-based diet with a seafood-based diet will not only be a healthy food option for consumers but will also release grasslands currently used for raising animals. Such land could then be put into producing grass-based biofuel feedstocks.

Fossil Fuel versus Biofuel

The market accepts higher prices for gasoline and diesel compared to electricity because of their unique adaptation as fuels for transportation energy. Drivers like the comfort of being able to travel for hundreds of miles after filling the tank, the presence of widespread infrastructure for refill, and the convenience of spending just a few minutes to complete the refilling process. For biofuels to compete with entrenched transportation fossil fuels, they not only need competitive pricing, but also the ability to fit into the same infrastructure built for gasoline and diesel and give out the energy output per unit volume similar to the fossil fuel so that drivers do not miss the comfort of long drives before refills. Rapid advances in research and technology have been made in recent years and proponents believe that it is just a matter of time before biofuels will be able to compete with fossil fuels on all counts. Perception is another factor favoring fossil fuel with which biofuels have to contend. Consumers do not shift from a reliable commodity that is time tested and to which they have become accustomed to another commodity new to the market, without being convinced of added benefits. In the case of biofuels, the advantage is touted in the forms of reduction of greenhouse gases (GHGs), freedom from the unreliable foreign sources of supply and distortion of price by oil cartel countries, improvement in the balance of payment and trade deficit, creation of hundreds of thousands of new jobs in rural communities, and decentralization of transportation industry. However, there is a vocal opposition that is not convinced of the environmental benefit of biofuel. This group is concerned that biofuel is detrimental to the poor because it competes for food grain, raising prices and vying for land used for producing grain, leading to food shortage. While a certain group of farmers have benefitted from the legislation-regulated biofuel boom, the farming community at large has yet to be brought on board to the enterprise by providing assured outlet at profitable prices for feedstock produced on the farm. Thus, biofuel as transportation fuel at this time can be best compared to a toddler growing into a healthy adult—whether it will grow into a mature vehicular energy source depends upon numerous conditionality, and unpredictable. However, considering the growing consciousness of the negatives attached to fossil fuel and biofuels' potential for providing the comparable substitute commodity, the current push for biofuels appear appropriate and support for research and development warranted, so that the full extent of their capability to substitute fossil fuels can be ascertained and a reasonable timeframe for achieving this potential can be set.

Biofuel Sustainability Concept

The International Energy Agency (IEA, 2011) has separated sustainability into three main categories—social, environmental, and economic—and listed indicators to aid in their evaluation (Figure 1.2).

Among social components, employment has been one of the main forces behind biofuel push at different times. Land issues will become an important factor as the biofuel industry expands and demand for volume of feedstock grows, needing large acreage. Integration of limited-resource farmers into biofuel feedstock production will be critical, especially in developing

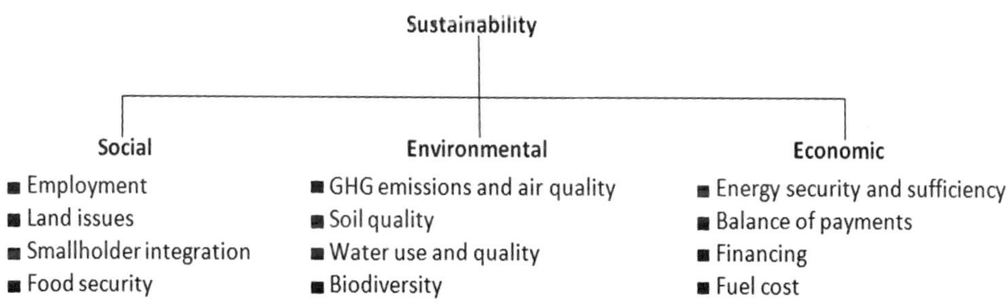

Figure 1.2. Biofuel sustainability framework. *Source*: IEA (2011).

countries dominated by small landholders. The food versus fuel debate intensified as demand of maize for ethanol drew outcry worldwide. However, the concern has been noted and there is a shift underway toward nonfood perennial crops for feedstock use but the transition will take time. In addition, to ensure that the acreage under food crops does not shrink under competition from biofuel crops, marginal lands have been targeted for future biofuel crop production, leaving aside fertile soils for food crops.

The role of plants in GHG remediation has been known for a very long time. Concerns regarding global warming caused by the burning of fossil fuels has led to the intensification of the search for nonpolluting energy sources. In this context, the historical knowledge that ethanol and plant oils can be used in place of petrol and diesel to run automobiles was the obvious reason for biofuels to come into focus. The octane-boosting property has already made biofuels a commonly found additive in gasoline. However, not all environmentalists are convinced of the GHG remediation abilities of biofuels and life-cycle analyses of different biofuel feedstock crops are in progress to determine crop-specific remediation value. Low-input systems, especially of fertilizers that are emitters of substantial GHG (carbon dioxide) during synthesis and field application (nitrous oxide) are being designed and emphasis is on crops that are perennial in nature. Perennial crops should be more efficient in nutrient recycling and threat of erosion from yearly cultivation for planting would be avoided. Perennials catch nutrient runoff to surrounding surface water and their deep root systems act as control against nutrient leaching. Alarm has been raised in recent years at the rapid rate of destruction of tropical forests, mostly in Southeast Asia. These forests harbor great diversity of flora and fauna among them, with some species designated endangered. However, the blame for forest destruction rests first on logging for timber, followed by new plantations for oil palm to meet world market demand for cooking oil. Burgeoning populations in developing countries, where most of these forests are situated, adds to the pressure from a number of new people looking for farming land. No doubt initial zeal for biofuel crop production under incentives led to some deforestation, but fortunately, many regulations and certifications have been recently put in place to safeguard against the encroachment upon virgin land that created threat to biodiversity.

Economic sustainability for biofuel industry is going to be challenging. The United States, Brazil, and European Union together with several developing countries have set definitive goals for biofuel. However to this date, price supports, incentives, import restrictions, mandates, or a combination of these have been needed to keep the biofuel industry going. Biofuels have just not been able to compete with fossil fuel on their own. Even when they manage to find a foothold in the high petroleum price environment, they will be at the mercy of the Organization of the Petroleum Exporting Countries (OPEC), which possesses the ability to flood the market at will and drive out competition, as their cost of production is extremely low. Energy security and sufficiency concerns have been raised by importing nations repeatedly at the time of high prices, but as soon as prices become affordable, these concerns evaporate. Balance of payment in trade due to the petroleum import burden is a perpetual problem for a large number of countries, but deficit financing has become the norm with their governments, notwithstanding long-term consequences of such a policy. Financing of infrastructure for biofuel is problematic, because financial institutions do not want to take risks without a definite rate of return model to guarantee the payback. Thus, industries have been greatly dependent on governmental guarantee or assurance of price support through of mandates, etc. It is hoped that with time biofuel competitiveness with petroleum fuel will improve as a result of greater efficiency in feedstock production, higher yield, engineering advancement in post-harvest processing, and development of biofuels with drop-in ability.

Biofuel Sustainability—USA as Case Study

Sustained Rural Economy Impetus for Biofuel

Plant-based fuel was envisioned to be an important source of energy to drive automobiles when they were invented. Henry Ford's first automobile prototype in 1896 ran on ethanol. Ford Motor Company in 1908 released Model T which could run on gasoline, ethanol, or a mixture of the two. Rudolf Diesel ran his diesel engine on seed oil. The agrarian segment thus came to assume that while their participation in the old transportation industry ceased when horse drawn carriages were replaced with automobiles, they would participate in the new transportation industry by supplying fuel to run automobiles. In the United States, the alcohol tax levied on ethanol during the civil war period was lifted in 1906 with the help of President Theodore Roosevelt to make it affordable for use as fuel. In the beginning, gas stations were rare in rural America and many farmers made their own low-cost ethanol to run vehicles. World War I (1914–1918) saw a surge in the production of industrial alcohol reaching 50–60 million gallons yr^{-1}. The infamy of ethanol as a vice crippled its use for any other purpose when the eighteenth amendment to the US Constitution prohibited the manufacture, sale, or transport of alcohol beginning in 1920. The use of alcohol for fuel was still legal but had to be mixed with gasoline to prevent its use for drinking. Since the use of ethanol was more prevalent in rural areas and distillation was carried out on individual farms, it was very difficult to establish lawful purpose; thus, prohibition essentially ended the first era of ethanol-driven cars in the United States. Cheap and plentiful petroleum dominated the fuel for transportation vehicles through the twentieth century. However, rural Americans had not given up on their re-entry into transportation fuel market and farm state legislators continued in their attempts to pass legislation to help bring back biofuel. Their quest became urgent as the need to find alternative income sources in rural communities became essential to arrest the tide of emptying rural communities and constant addition to ghost townships starting in 1960s. Farmers lost farms at alarming rate due to depressed commodity prices. Globalization had lowered their ability to pass on added expenses while at the same time prices of external inputs like fuel, fertilizer, and labor., kept escalating.

Opportunity for the re-entry of biofuel into transportation arena arrived with the 1973 Arab Oil embargo. The petroleum crisis created by the embargo drew the world's attention to finding alternatives to petroleum. Brazil took the lead in deploying ethanol for running vehicles and remains the only country that produces cars that run on ethanol and filling stations with ethanol since the late 1970s.

Jimmy Carter was the President of the United States at the time of oil embargo. He came from a rural state and a rural community, with the first-hand knowledge of the suffering on the farms and surrounding communities. He was receptive to the proposition of easing the demand for import by boosting the oil inventory by adding ethanol. With the help of farm-state Senators, Congress passed legislation providing subsidy to ethanol produced for this purpose. Blenders mixed 10% ethanol with the gasoline and received a 50 cent tax break for every gallon of ethanol. Farmers also benefitted from government subsidy. The ethanol industry was shielded against competition from countries like Brazil through high tariffs. However, as energy crisis of 1970s eased, incentive to invest in alternative energy sources ebbed but ethanol remained the only holdover from the 1970s energy crunch, saved by the legislation to oxygenate gasoline to reduce pollution.

The Clean Air Act Amendments (CAA) of 1990 was the first environmental policy benefitting renewable fuels. It established the Oxygenated Fuels Program and the Reformulated

Gasoline Program in order to control levels of carbon monoxide and ozone in the air. Fuels under both programs were required to contain 2% oxygen. Since ethanol is 35% oxygen, blending ethanol to gasoline became popular to meet the mandated oxygen requirement under CAA. The Energy Policy act of 2005 eliminated the 2% oxygen requirement for reformulated gasoline and replaced it with a credit trading system. The new provision gave suppliers flexibility to add less renewable fuel to gasoline than required by the CAA by purchasing credits from suppliers who supplemented gasoline with more biofuel than required.

The new push for ethanol came under the administration of President George W. Bush (from Texas, a rural state). Congress passed the Energy Policy Act in 2005, which set a minimum requirement for ethanol use for automotive fuel, followed by another piece of legislation in 2007 that raised the threshold ethanol volume requirement sharply. The 2007 legislation mandated the use of 1.3 million barrels a day of ethanol by 2015 and 2.4 million barrels a day by 2022. To put this in perspective, the United States consumed 8.9 million barrels/day of gasoline for transportation in 2009.

United States Environmental Protection Agency (EPA) promulgated low sulfur diesel fuel standards for highway vehicles beginning July 2006 and nonroad vehicles beginning June 2010. Lowering sulfur in diesel also lowers its lubricity, the restoration of which is essential for full fuel efficiency. Biodiesel is an excellent lubricant and adding only 1–2% to the ultra-low-sulfur biodiesel is needed to restore the full lubricity. Lubricity additive could turn out to be a substantial market for biodiesel.

In summary, the credit for existence of current US biofuel industry belongs to legislations that provided incentives, mandates, and compliance requirements to promote biofuel production and use. The Energy Tax Act of 1978, the Energy Policy Act of 2005, and the Energy Independence and Security Act of 2007 (EISA) are landmark legislations affecting biofuels. Two US Presidents, Jimmy Carter and George W. Bush piloted landmark biofuel bills and signed them into law. Biofuels legislation passed by the US Congress is no different from any other legislation introduced and passed, in that those too had the impetus of lobbying, which in this case was from farming groups and driven by the desire of legislators from the farming states to their constituents' welfare. However, frequent oil crises and a desire for secure domestic supply of this essential commodity, together with the perception of environmental benefit from adding biofuel to gasoline or shifting to biofuel in a major way, have been motivation behind broader support for these energy bills.

Legislative Support and Guided Movement toward Sustainability

In early attempts to jump-start the biofuel industry, obviously, the known and proven methods of its production were adopted. We had the knowledge in producing grain crops and possessed experience of converting grain to alcohol. Similarly, oil from seed crops had been in use for centuries for burning to produce light. The food crops had been improved genetically through history for higher yield and better resistance to pests, and their method of cultivation have been continually advanced to optimize growth. Thus, when occasion arose for an expeditious increase in biofuel production, food crops were the only feedstock resource that was amply available and had a well-established supply chain. Maize was the crop of choice in the United States, because it has proven to yield highest among grain crops and already been used extensively for industrial purposes. Moreover, maize desperately needed another outlet to bring it out of the price doldrums. On the other hand, Brazil opted for sugarcane, because it has long been mainstay of its economy.

The recent bioenergy-related legislations in the United States and elsewhere show continued legislative support for biofuel, but they now come with stipulations to prevent its excessive intrusion into food crops and require measurable proof of benefits to the environment. The US Energy Policy Act of 2005, for the first time, set a renewable fuel standard (RFS) for automotive fuels. Under its provisions, fuel suppliers must blend 4 billion gallons of renewable fuel into gasoline in 2006, increasing annually to reach 7.5 billion gallons in 2012. The EPA was required to set minimum ratio for renewable content after that period not to be less than the ratio in 2012. The Act also contained a provision to encourage the production of cellulosic ethanol, which stated that every gallon of ethanol produced from biomass equaled to 2.5 gal toward satisfying the RFS. The Energy Independence and Security Act of 2007 further expanded the RFS previously set by the Energy Policy Act. The new RFS started at 9 billion gallons in 2008 rising to 36 billion gallons in 2022. Beginning 2016, all the new increases in RFS target will have to be met with advanced biofuels derived from sources other than maize starch, with explicit allocations for cellulosic biofuels and biomass-based diesel. Under this mandate of the total 36 billion gallons, 21 billion gallons will have to be obtained from cellulosic ethanol and advanced biofuels. Although the two legislations did not provide direct loans or grants for the construction of biofuel plants, provisions for market guarantee ensured growth of the biofuel industry. The laws clearly were intended to serve as a bridge to transit maize-based ethanol into the next-generation technologies.

The Energy Independence and Security Act of 2007 directed EPA to develop time-based mandate for different categories of biofuel under the law (RFS2). It applied to all refiners, blenders, and importers of transportation fuel. Both domestically produced and imported biofuel were covered under the regulation. It also provided legal definition for renewable fuels and renewable biomass. Biofuels were separated into four categories on the demonstrated minimum GHG reduction standards in comparison to the petroleum fuel they replaced and verified by life-cycle assessment. The EISA emission computation is very stringent and it must include direct emissions during the entire fuel cycle from feedstock production, distribution, and final use by consumer as well as emissions from land-use change.

EPA used models to assign feedstocks and processes of production into different GHG emission threshold categories (Table 1.1).

The definition of renewable biomass includes planted crops and crop residue, planted trees and tree residue, animal wastes, algae, and yard and food wastes. To ensure that no virgin land is brought under cultivation for biofuel, planted crops and crop residue are permitted only from agricultural land that was cleared or cultivated prior to December 19, 2007, and actively managed or fallow, and nonforested. The cropland, pastureland, and USDA Conservation Reserve Program (CRP) land qualify under the definition but not rangeland, federal land, or other rural land. Similarly, only tree plantations existing on nonfederal or tribal land prior to December 19, 2007 qualified for biofuel certification. Biomass from slash and pre-commercial thinnings

Table 1.1. Classification of biofuels into different categories based on GHG emission.

Feedstock and process	GHG emission category (%)
Ethanol from maize starch at new natural gas fired facility	20
Butanol from maize starch	20
Ethanol from sugarcane	50
Diesel from soy oil, waste oils, fats, and greases	50
Diesel from algal oil	50
Cellulose ethanol	60

are only permitted from nonecologically sensitive nonfederal and nontribal forestland. The law also puts requirement of record keeping to the parties covered under the mandate. As Raghu *et al.* (2011) put it, "the emerging biofuel economy is likely to result in the single largest reconfiguration of the agricultural landscape since the advent of industrial agriculture," EISA wants to make sure that during such reconfiguration due diligence is exercised to preserve the environment and conserve forests and other natural ecosystems that provide critical services to different sections of population.

According to EPA (2011), the steps taken under EISA to produce 36 billion gallons of biofuel by 2022 will replace 13.6 billion gallons of fossil fuel, approximately 7% of the expected transport fuel consumption at that time. Biofuel replacement will save US$41.5 billion in oil import and result in reduced gasoline and diesel cost by 2.4 and 12.1 cents gal^{-1}, respectively. In addition, it is expected to reduce GHG emission by 138 MT, equivalent to taking about 27 million cars off the road. The farm income is projected to increase by US$13 billion in 2022. However, the negative effect of biofuel will show up in annual increased food cost by US$10 per person.

From the above, it appears clear that the governmental regulations support the rural economy through the biofuel industry, but also wants to make sure that biofuels are produced in sufficient volume to make a noticeable reduction in fossil fuel use, deliver GHG emission reduction, do not encroach upon natural ecosystem and are not in conflict with societal need for food. Another aspect related to biofuel is also apparent by now—that the push for biofuel thus far has needed governmental subsidy or mandate to support all levels of supply chain. From all indications, biofuel is not yet ready for open market competition.

Certain projections and recommendations have been made to heighten the commercial competitiveness of biofuel. The fossil fuel price is projected to increase in the future, putting biofuel in a better position to compete. It is to be remembered that one of the important reasons behind the economic pricing of fossil fuel is that fracturing allowed multiple product application of the feedstock and generation of valuable by-products. With the continuing improvement in biofuel technology, the possibility of valuable by-products to support lower biofuel price may be feasible. The search continues for microorganisms and enzymes that break down biomass into sugars as platform chemicals for a host of processes, biofuel being one of them. Laser *et al.* (2009) have provided a model-based analysis of the cost of production of biorefineries when they will be fully developed similar to today's petroleum refineries. They considered the two emerging processing approaches—biological and thermochemical—for the production of fuels, power, and/or animal protein. According to their calculation, the cost of production amounted to US$0.36–US$0.57 l^{-1} (US$1.37–US$2.16 gal^{-1}) petrol equivalent for a plant with the capacity of 4535 dry tonne feedstock/day, considering 12% internal rate of return, 35% debt fraction, and 7% loan rate. However, when the biological production of ethanol was combined with thermochemical production of fuels and/or power at the same scale and financial structure, cost came down to US$0.25–US$0.33 l^{-1} (US$0.96–US$1.24 gal^{-1}) petrol equivalent.

Biofuel Sustainability Outlook

Economic Value

Singh (2010) has discussed the importance of industrial outlets for farm produce in order to maintain long-term viable farm economy. Rural communities throughout the world heavily reliant on income generated from farming have been suffering from poor economies. Setting aside recent increases in grain price, the farm commodity prices have been stagnant for a long time. For example, the price of soybean has hardly changed between 1981 and 2005 from US\$208 tonne^{-1} in 1981 to US\$223 tonne^{-1} in 2005. As a consequence of price stagnation, the total cost of producing a hectare of soybean for American farmers in 2005 was US\$663 against the gross value of US\$654 of the harvest (SoyStats, 2007). Creating multiple outlets for currently produced crops or finding alternate crops with market demand will generate new income for farmers. Diversification of enterprises will also give farmers the ability to switch between markets that bring the best prices for the produce and select enterprises requiring fewer inputs in years of adverse climatic conditions. Some of the cellulosic biofuel crops can be grown on marginal land that is generating little or no income at present, thus opening new avenue for increasing farm output.

The effect of market squeeze has been devastating for developing countries. While agriculture provides less than 2% of income and employment in developed countries, it happens to be the source of 35% of the gross domestic product of the less developed economies (Watkins and von Braun, 2003). Poor countries are not able to protect the livelihood of their farmers in globalized markets because they are bound by international laws or prevented through treaties. A good example of this is NAFTA (North American Free Trade Agreement), a trilateral treaty between the United States, Canada, and Mexico. It lowered maize price in Mexico to the level that subsistence farmers were forced out of the only livelihood they had known for generations and many were compelled to migrate to urban centers and into the United States in search of livelihood. Greater industrial demand for biofuel crops will increase their production in the poor countries, benefitting subsistence farmers and improving their standard of living.

The biofuel industry as a whole should be good for rural employment. This transportation fuel source is the only one among energy providers that permit rural population to participate in the generation end of the enterprise and not only as consumers. Biomass needed for biofuel is produced by agriculture and forestry located in rural areas and because of their bulkiness, it is economical for biofuel refineries to locate in proximity of feedstock source in rural communities rather than close to cities. Thus, rural job growth because of biofuel industry should spread across the spectrum from growing, harvesting, storage, and transportation of feedstock to employment in the biorefinery. Shumaker *et al.* (2006) found that construction of a 375 million liter maize-based ethanol plant in Georgia, the United States created a one-time economic output impact of US\$130 million to the state economy. Economic activity related to construction generated US\$51.7 million in labor income for 1203 jobs. Production of ethanol created annual economic output impacts of US\$335.8 million. Plant operations accounted for 50.2% of the total output impact, while 49.8% was attributed to maize produced in Georgia. Ethanol production generated US\$37.6 million in labor income from 1030 jobs to the Georgia economy. It also contributed US\$3.8 million and US\$3.1 million in taxes to the state and local governments, respectively. The following statistics compiled by Worldwatch Institute (2007) shows the value of biofuels to job creation: (i) biofuel industries require about 100 times more workers per joule of energy produced than petroleum industries; (ii) in Germany, the biodiesel industry generates roughly 50 times more jobs per tonne of raw oil than similar amount of diesel; (iii) job creation cost in Brazil is 25 times less for the ethanol industry than in the petroleum industry; (iv) in sub-Saharan Africa, a region-wide blend of biofuels to the tune of 10% for gasoline and 5% for diesel would provide job to between 700 000 and 1.1 million people; and (v) Jatropha farms in India could yield 313 person-days per hectare in the first year of the plantation and 50 person-days per hectare over the next 30–40 years.

Crop residues are projected to be an important component of cellulosic feedstock mix. At present, they are mostly treated as crop waste. Thus, this new use of farm produce heretofore of no economic value will add to the income earned by farmers from raising crops. However, it may not have much scope for creating new employment as existing farm personnel can accommodate this task by rearranging the scheduling of different operations on the farm. On the other hand, farmers may find it difficult to manage both harvesting of grains and residues within the weather window, and may prefer to give this task to a dedicated outfit in residue collection and delivery, thereby providing business opportunities to a new group of entrepreneurs.

Several tropical countries may turn out to be good source for feedstock and refined biofuel because of their climatic advantage in growing a number of high-biomass-accumulating crops. As most countries in this region fall under developing category based on their economic conditions, feedstock and refined biofuel could become important commodities for export enabling them to earn valuable foreign currency for essential imports. This would also help them in strengthening infrastructure needed for economic development and providing job opportunities to their rural population. Furthermore, it would cut down the costly fossil fuel imports as they substitute it in part by biofuel. Several biofuel refinery processes such as ethanol from sweet sorghum and pre-refinery feedstock preparation will provide opportunities to small entrepreneurs and refinery constructions will bring new investments. In the final analysis, biofuel export–import trade will stimulate economic activity in both developing and developed countries.

The biofuel feedstock market is mandate- and incentive–driven, and is expected to show stable growth because of the commitment by different governments. OECD/FAO (2011) predicts continued increase in the ethanol and biodiesel demand and reach 155 billion liters and 42 billion liters, respectively, by 2020. Also, biofuel prices to major biofuel feedstock prices should remain stable and biofuel prices on average 80% higher from the previous decade in the case of ethanol and 45% in the case of biodiesel.

Energy

According to the US Energy Information Administration statistics (EIA, 2007) demand for liquid fuels will increase to 118 million barrels/day by 2030. Most of this oil will come from the Middle East-controlled OPEC cartel. This monopolistic situation permits price manipulation outside of the free market, thereby creating risk to energy security of the importing countries. The sustained high price trend of recent years has been especially hard to the poor countries and has severely depleted their already scarce foreign exchange reserves. The situation clearly calls for alternative energy options and remedial strategies. Biofuels appear to be the best present alternative before other transportation sources such as hydrogen and electricity come on board.

There has been constant increase in global biofuel production, growing from 16 billion liters in 2000 to more than 100 billion liters (volumetric) in 2010 (IEA, 2011). Biofuels currently account for approximately 3% of total road transport fuel (on an energy basis). It is projected to increase to 9% by 2030 and to 27% of world transport fuel energy by 2050 (Figure 1.3). Most of the growth in biofuel use will occur in the United States, Europe, China, and Brazil (IEA, 2010).

Environment

Environmental Protection Energy (2011) estimates that emission from a gallon of gasoline and diesel amounts to 8887 g CO_2 gal^{-1} (gasoline factor is based on a recent regulation establishing GHG standards for model year 2012–2016 vehicles (75 FR 25324, May 7, 2010)

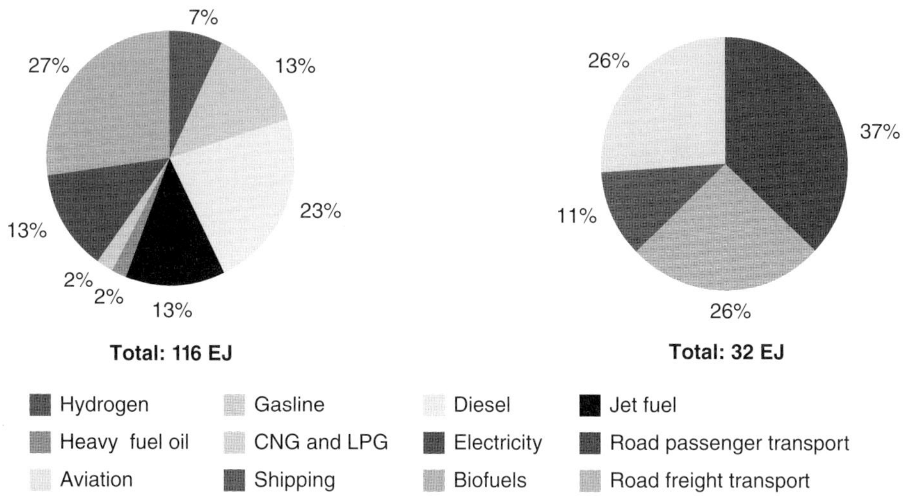

Figure 1.3. Global energy use in the transport sector (left) and potential of the use of biofuels in different transport modes (right) in 2050 (BLUE Map Scenario: it models future energy demand until 2050 based on the criteria of global long-term CO_2 concentration in the atmosphere of 450 parts per million). (Note: CNG, compressed natural gas; LPG, liquefied petroleum gas). *Source*: IEA (2010).

and 10 180 g CO_2 gal^{-1} (diesel factor is from the calculations that vehicle manufacturers use to measure fuel economy (40 CFR 600.113), respectively. Fossil-fuel-driven transport vehicles contribute approximately 25% of global CO_2 emissions. In general, plant-based feedstock should have an advantage over petroleum feedstock. Plants can fix the carbon, which is later released through the tail pipe, making them carbon neutral, while the release from fossil-based fuel is a one-directional load on the environment. According to IEA (2008), the replacement of fossil fuels by biofuel has the potential to mitigate GHG by 60–120% with the 70–110% mitigation level already observed for sugarcane ethanol. Sustainably produced biofuels can help reduce CO_2 emission by 2.1 gigatonnes (Gt) per year if their potential to provide 27% of total world transport fuel by 2050 is materialized (IEA, 2011).

The "well-to-wheel" comparison of biofuels produced from different biomass sources and technologies with the fossil fuels (IEA, 2011) is presented in Figure 1.4. The wide range of variation in GHG emission presented below results from the difference in crop production methods, amount of fertilizer used, and the details of the conversion technology applied. However, the superiority of cellulosic biomass and sugarcane for GHG mitigation is obvious. An improvement higher than 100% shown in the figure below is possible because of the benefits of co-products (notably power and heat).

Figure 1.5 clearly conveys the message that greatest reduction in GHG emission is expected from the incremental improvement in vehicular fuel efficiency. But at the same time, as biofuel feedstocks and technologies with low emission come on line, they will also contribute significantly to the lowering of GHGs.

Perennial feedstock, in general, is more environment friendly than annuals. It provides year-round cover to the soil, protecting it from the direct exposure to rain and wind that cause soil erosion while no crops are on the ground. This benefit could be substantial for erosion-prone soils. The ground cover also slows down the runoff after rain, thus facilitating

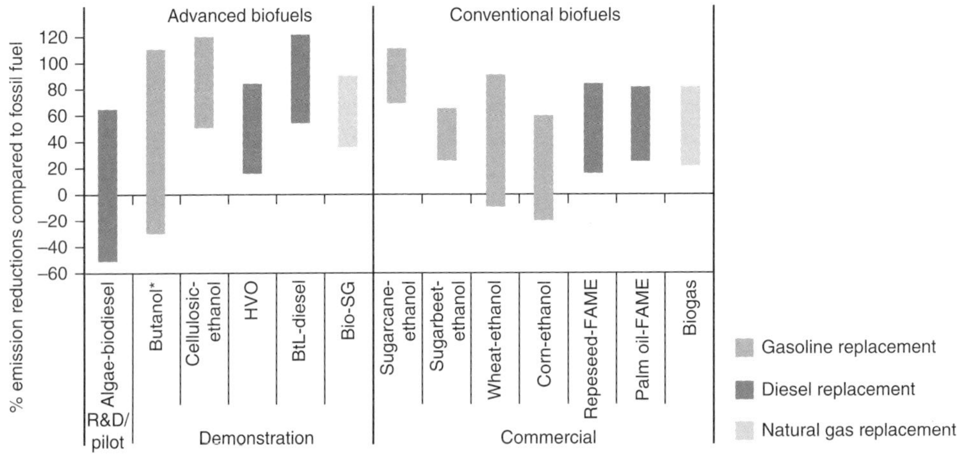

Figure 1.4. Life-cycle GHG balance of different conventional and advanced biofuels, and current state of technology. (Note: The assessments exclude emissions from indirect land-use change. Emission savings of more than 100% are possible through use of co-products. Bio-SG, bio-synthetic gas; BtL, biomass-to-liquid; FAME, fatty acid methyl esters; HVO, hydrotreated vegetable oil). *Source*: IEA (2011). For color detail, please see color plate section.

retention of greater portion of rainfall by the field soil. Perennial crops shed a portion of leaves during the annual growth process adding leaf litter carbon to the soil. Also, the root systems of perennials do not die over the years, allowing for sequential increase in carbon sequestration by the living plant tissues. Tilman *et al.* (2006) reported that low-input native grass mixtures sequestered 4.4 tonne ha^{-1} yr^{-1} of carbon dioxide in soil and roots, far exceeding the 0.32 tonne ha^{-1} yr^{-1} fossil carbon dioxide released during biofuel production.

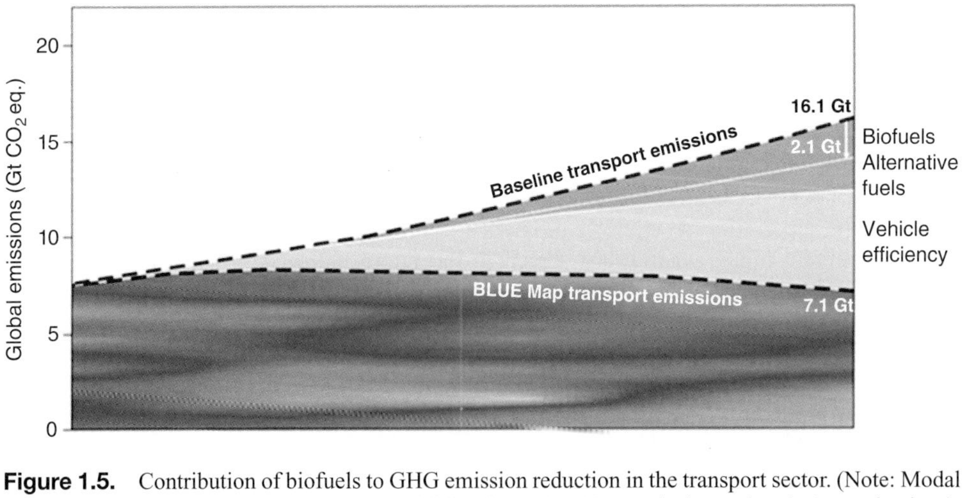

Figure 1.5. Contribution of biofuels to GHG emission reduction in the transport sector. (Note: Modal shifts (not included) could contribute an additional 1.8 Gt CO_2 equivalent of emission reductions). *Source*: IEA (2010).

de Vries *et al.* (2010) compared major current biofuel feedstock in different parts of the world for resource use efficiency (land, nitrogen, water, energy resources), soil quality, net energy production, and GHG emissions. The findings, based on nine production–ecological indicators, revealed that biofuel produced from oil palm (Southeast Asia), sugarcane (Brazil), and sweet sorghum (China) appeared most sustainable. The land-use change was not included among the indicators, probably with the assumption of no new land clearing to produce crops for biofuel. All the tree crops effectively reduced GHG emissions compared to fossil fuels. Maize (USA) and wheat (Northwest Europe) scored lowest for all indicators among the feedstocks evaluated. Sugar beet (Northwest Europe), cassava (Thailand), rapeseed (Northwest Europe), and soybean (USA) were in the middle.

Biofuel Sustainability Concerns

Food versus Fuel

According to OECD/FAO (2011), the use of cereal, sugar crops, and vegetable oil for biofuel production on global basis will continue to increase through 2020. Considering 2008–2010 as base, the increase will amount to from 11% to 12% for coarse grains, from 11% to 16% for vegetable oil, and from 21% to 33% for sugar crops. After being under pressure for decades, global food prices have shown significant upward trends in recent years. Of course, biofuels share a portion of the blame, such as in the case of maize. However, over all, a number of complex factors far beyond biofuels are behind recent grain price increases. The improving income level in the countries of Asia and sub-Saharan Africa has resulted in dramatic uptick in the demand of grains for animal feed, as it has become affordable for people to incorporate increased amounts of meat in their diets. Agriculture is greatly dependent on weather and 1 year of inclement weather in key production countries can turn world surplus into world deficit. Recent drought in major grain-producing countries has exacerbated the grain price situation. The cost of farm inputs, especially those derived from fossil fuel, as well as of labor has skyrocketed. The added production costs obviously ultimately show up in the price of the produce. As rural areas became depopulated as a result of the lack of livelihood, infrastructure investments for roads, irrigation, etc. have stalled, and funding for agricultural research has dwindled; as a net result, crop production has reached plateau unable to meet higher demand.

The feedstock choice for biofuel will gradually shift from grain to cellulosic crops by legislative mandates in several countries and societal pressure at large; as a result, the food versus fuel debate in time will become irrelevant. Biofuels, then, will become valuable to agricultural food production systems by providing a market stabilization function, which will add to farm economic sustainability. A glimpse of this important contribution of biofuel to agriculture is already visible in Brazil. The ethanol production model adopted by Brazil has been instrumental in finding an economically viable price for sugarcane by creating a balance between meeting the demand of sugar while diverting the surplus to ethanol conversion. Under this flexible policy, the market determines whether sugar is sold on the sugar market or converted to ethanol.

Reliability of Feedstock Supply

The reluctance of farmers to grow a crop, most of all, can be attributed to the absence of guaranteed sale outlets and a profitable price—both criteria mostly unmet for biofuel crops

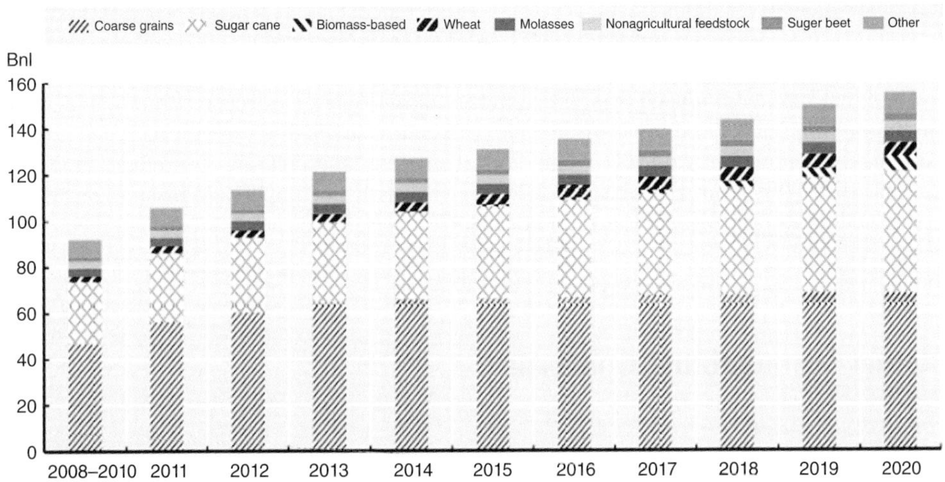

Figure 1.6. Ethanol feedstock source to the year 2020. *Source*: OECD/FAO (2011).

at the current juncture. Crop yields year to year are greatly influenced on the timeliness and adequacy of rainfall. Therefore, there is always the hanging question of how the biofuel feedstock demand will be satisfied in dry years.

Current biofuel feedstocks are mostly starch, oil, and sugar based, and drawn from the same crops as are used for food (Figure 1.6 and 1.7). It can be gathered from the two figures that until 2020, the heavy dependence on food crops will continue, although a shift will be evident toward the latter part of this period. Coarse grain use for biofuel, however, will start to stabilize beginning in 2012, the void being picked up by sugarcane. Much narrower shift is predicted for biodiesel feedstocks in the near term. Till cellulosic feedstocks become the

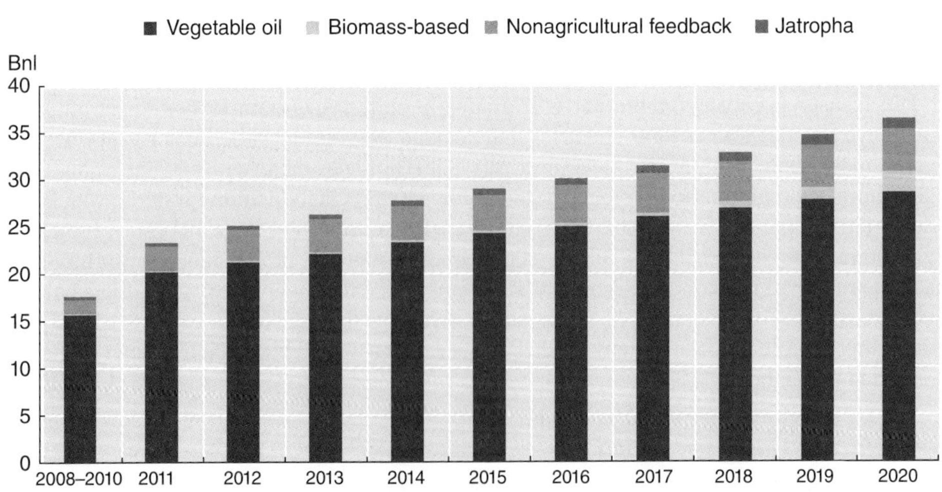

Figure 1.7. Biodiesel feedstock source to the year 2020. *Source*: OECD/FAO (2011).

mainstay for biofuels, two near-term approaches can be envisioned: shifting from major grain crop such as maize to minor grain crops like sorghum, cassava, etc. and from edible oil crops to inedible oilseed source such as Jatropha. Crops like sorghum are more drought hardy than maize and possess the ability to grow and produce optimum yields on marginal soils. Similarly, Jatropha, another hardy crop, provides an alternative for biodiesel production, although its full potential is not yet known. Comprehensive integration of used cooking oil and animal fats for biodiesel would also help relieve pressure on oilseeds. All indications suggest that the Brazilian sugarcane crop is adequate to support both sugar and ethanol needs.

The urgency of optimization of cellulosic conversion technology cannot be over emphasized in order to diminish dependence on grain and oilseed feedstocks. Plants produce about 180 BT of cellulose per year globally, making this polysaccharide the largest organic carbon reservoir on Earth. Cellulosic crops yield higher amounts of biofuel per unit land area because all biomass is converted to biofuel, which in turn, will also have favorable impact on feedstock availability. In addition, this feedstock option makes available a large amount of unutilized biomass in the form of bagasse and crop and forest residues. Utilization of forest understory biomass for biofuel feedstock will reduce the cost of its removal for forest health initiatives.

Shared Economic Prosperity

Among Nations

Many of the poorer tropical countries identified as potential targets for future investments currently lack the transportation and agricultural infrastructure to fully realize the potential of biofuels. Furthermore, trade barriers continue to block the development of a global biofuel market. But IEA (2008) predicts that by 2050 developing countries will produce more than half of the second-generation biofuel, 19% of which will come from China and India.

The ambitious biofuel goals set by the United States of 60 billion liters of second-generation biofuel per year and EU to replace 10% of transport fuel by renewable energy by 2020 may turn out to be a driver for second-generation biofuels in developing countries (IEA, 2010). Both entities may find substantial gap between domestic production and need, and have to resort to imports to make up the difference. The cost of second-generation biofuel plants fall in the range of US$125–US$250 million and the developing countries would require investment by foreign companies, foreign banks, or international assistance banks to supply the capital. The large biomass needs of commercial second-generation plants will mean that a new supply chain model will also have to be developed as land holdings in these countries are small and especially in some African countries, the land property rights are intricate. An alternate model probably can be envisioned for the developing countries. Under it, first-level biomass processing is carried out at the community level with a collection scheme for final refining at a central location. Every level of the supply chain accompanies employment opportunity.

Within Society

It will fall upon individual nations, nongovernmental organizations (NGOs), and other concerned groups to make sure that the benefit of the new industry flows to all segments of the society. Biofuel crops, such as sweet sorghum and Jatropha, which can be processed with ease at the village level by farm cooperatives, provide better participation opportunities to farmers at all levels of biofuel production. Forced evacuation of farmers from the productive farmland without providing suitable livelihood alternatives or adequate compensation to construct industrial complexes has been the cause of unrest in several developing countries. Biofuel-importing countries and other world bodies need to make sure that the same situation does not develop during biofuel industry establishment.

Environment

Water Availability

The water use by biofuel crops is no different from the water requirement of other plants. Thus, there is concern that biofuel crop production will divert irrigation water from food crops. Hoogeveen *et al.* (2009) have provided relevant statistics on this topic. Rapeseed, the dominant biofuel crop in Europe, is mostly grown on dryland while maize for biofuel, the main ethanol feedstock in the United States, accounts for about 3% of national irrigation water withdrawal. On the global basis, biofuel crops utilize about 1% of irrigation water. Even this miniscule share will see further decline as the feedstock source shifts to cellulosic crops in future. Furthermore, the perennial cellulosic crops will have three-to-seven times lower water requirements than annual crops like rapeseed and maize.

Nutrient Run Off

Singh (2010) drew attention to potential nutrient runoff problem from industrial crop fields. Maize cultivation is input intensive and heavily fertilized. Concern has been raised regarding water bodies' pollution from bringing additional acreage under cultivation for maize. Alexander *et al.* (2008) linked seasonal hypoxia in the Northern Gulf of Mexico mainly to N and to a lesser degree to P fluxes. Maize and soybean cultivation in Midwest USA was considered responsible for approximately 52% of N and 25% of P load of the Mississippi river. Oxygen-deficient water threatened the survival of marine life and productivity in the Gulf of Mexico, which is the source of a major percentage of US harvested shrimp and oyster and a significant percentage of commercial fish. These concerns are valid, and there has been substantial increase in maize acreage due to biofuels. But justifiably, Mississippi River nutrient load should be attributed in major part to grain production for feed. As long as the demand for meat remains high, the problem will persist and a comprehensive nutrient management program, including precision fertilizer application and nutrient trapping and reuse technologies, will be needed. As the emphasis of biofuel feedstock shifts from grain to perennial cellulosic crops, the nutrient runoff attributed to biofuels should substantially decline.

Land-use Change

There are three broad-based concerns regarding the impact of biofuels on land use. Gibbs *et al.* (2010) expressed the anxiety that, like in the 1980s and 1990s when large-scale deforestation in the tropics took place to make land available for agriculture, forests will be cleared to access land for biofuel crop production. Schoneveld (2010) is doubtful of the availability of sufficient land in the degraded or marginal category, as proposed for allocation toward biofuel cultivation, and the ability of these soils to provide economic biomass yields. There is also apprehension as to the willingness and ability of land planners and authorities to prevent cultivated lands from being taken away under acquisition schemes to meet huge feedstock requirements of biorefineries. Many farmers with smallholdings may lose land during consolidation and become unemployed.

Although in recent years countries in general have become aware of the need to preserve forests and develop land-use change guidelines, programs and policies are in the developmental stage. Careful mapping and planning will be needed to identify suitable lands for biofuel crops so that high yields and profitability can be obtained while other interests are not infringed upon. The Brazilian example set by *ZAE Cana* (Sugarcane Agroecological Zoning for Ethanol Production) provides a good example. It has three components: (i) agricultural policy, (ii) environmental sustainability, and (iii) food security. The scheme provides definitive guidelines for the incorporation of new land into sugarcane cultivation for ethanol production. Agricultural policy assures that the soil and environmental conditions are suitable for nonirrigated sugarcane production, and that the soil slope is less than 12%, so that mechanical harvesting is possible and the threat from soil erosion is not excessive. Environmental protection is ensured by preventing expansion into native vegetation, and areas excluded for tillage in the Amazon, Pantanal, and upper Paraguay River Basin. The sugarcane for ethanol is barred from direct competition with areas designated for food production. Surveys carried out by ZAE under these guidelines identified 34 Mha of land currently occupied by cattle breeding and degraded pasture fit for sugarcane production. Land allocation in several Asian and African countries will be particularly difficult because of the already existing pressure due to high population density. Collective and complex land ownership in several African countries may also make it difficult to set aside specific land areas for biofuel crops.

There is optimism regarding the decrease in the demand for new lands for biofuel production, as the productivity of crops for feedstock increase. The prospect of gaining significant gains in yields of biomass crops is especially promising considering their unimproved status.

Residue Diversion

The utilization of crop residues decreases the need of land for biofuel production and provides a new source of income to farmers. However, in many developing countries, crop residues are important animal fodder or sources of household fuel. These uses must first be considered during developing plans for using crop residues for biofuel production. Concern has also been voiced that using crop residues for biofuel will result in lowering of soil quality and increase erosion from vulnerable soils. Based on the currently available literature, there is scope of utilizing part of crop residue for biofuel without harming the soil. Confining the use to secondary residues should minimize the negative impact as these are seldom returned back to the soil. An estimation of plant nutrients that leave the field with the residues is essential, as their supplementation as fertilization may be essential to maintain soil productivity. Forest residues are projected as another major biofuel source. While logging residues may not pose soil quality problems, excessive removal of forest understory may have adverse consequences.

IEA (2008) has made projection of the amount of residues that will be available for biofuel use in 2030. Residues (crop and forest combined) could provide approximately 4.1% and 10.3–14.8% of the projected transport fuel demand in 2030 under the 10% and 25% collection scenarios.

Introduction of Invasive Species

Based on experiences in introduction of aggressive plant species, the danger of newly introduced feedstock species becoming invasive should be taken seriously. An introduced crop may grow more vigorously in a new environment than in the natural habitat because of the absence of natural enemies and encroach upon land reserved for the production of other crops. Elephant grass, an important biofuel feedstock candidate, provides a good example in this regard. It has been recognized as a weed in 25 countries, infesting 19 crops (Holm *et al.*, 1977). Infestation by elephant grass of sugarcane in Hawaii, fruit orchards in Israel, maize and rice in Nigeria, and coffee in Kenya have been reported (Farrell *et al.*, 2002). The threat posed to natural vegetation by elephant grass in the Amazon Basin has been observed by D'Antonio and Vitousek (1992).

Validity of GHG

Critics charge that in most of the GHG estimates, the impacts of land-use change have not been included, as a result of which reported values are distorted. For example, Searchinger *et al.* (2008) estimated that land converted from pasture to sugarcane can pay back the carbon emission caused by the disturbance of the soil for sugarcane planting in 4 years, but if as a result of this conversion, the cattle rancher has to resort to forest clearing to establish new pasture, the carbon payback period lengthens to 45 years. Fargione *et al.* (2008) calculated 17 years' payback duration for carbon debt when Brazilian savanna (the Cerrado) is converted to sugarcane. There is much concern regarding forest clearing for oil palm planting for biofuel use. It is projected that converting a peat swamp forest into an oil palm plantation will require between 420 and 900 years to pay back the carbon debt (Gibbs *et al.*, 2008; Fargione *et al.*, 2008). On the other hand, planting oil palm on grass land or degraded soil has a very short carbon payback time (Danielsen *et al.*, 2009) and a positive carbon sequestration benefit due to higher biomass of oil palm compared to the ecosystems replaced.

Berndes *et al.* (2010) pointed out that biofuel crop production could be held accountable for only about 1% of the global land-use-change-related emission, the rest is the result of the conversion of virgin lands for food, fodder, timber, road network, etc. Regulations put in place by the European Union (EU), United States, and other countries, certification schemes sponsored by NGOs, and environmental groups make it difficult for feedstock produced on converted land to find place in the biofuel supply chain.

Competitiveness

A major concern regarding biofuel sustainability is its competitiveness with fossil fuel. In most countries including the United States, the biofuel industry relies on out-of-market mechanisms to be viable. Brazil is the only country that appears to provide sugarcane-based sustainable ethanol model. The uniqueness of Brazil in this respect has been attributed to the abundance feedstock and a stable government policy to guide the industry toward sustainability. Figure 1.8 shows the projected biofuel mix to the year 2050. The main future biofuel types are expected to comprise ethanol sugarcane, ethanol cellulosic, and advanced biodiesel (BtL). Figure 1.9 indicates the price projection for different biofuel types under low-cost and high-cost feedstock scenarios. It is clear from the figure that biodiesel from oilseed crops like soybean and rapeseed have no possibility of becoming competitive with petroleum fuel under any circumstance, the main reason being low yields and high prices. But other biofuel types will become competitive to gasoline around 2030 at low production cost. Under a high-cost scenario, only sugarcane ethanol will become fully competitive around 2020 and biosynthetic gas somewhat competitive by 2040; the other two projected biofuel types, cellulosic ethanol and BtL, will lose out to gasoline. However, as sugarcane is adapted only to tropical climate, biofuel from

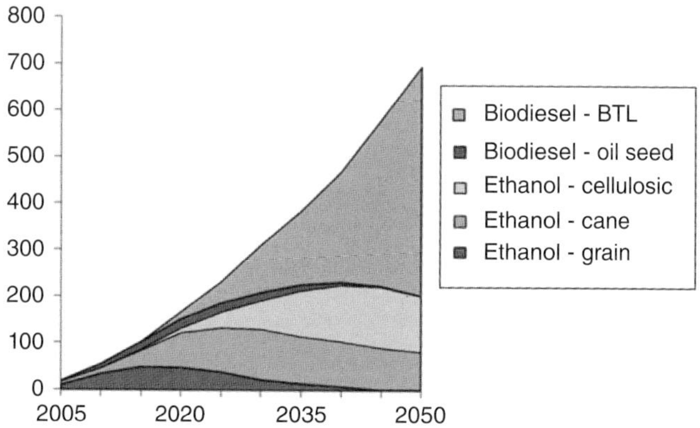

Figure 1.8. Trends in the production of different biofuels. *Source*: IEA (2008). For color detail, please see color plate section.

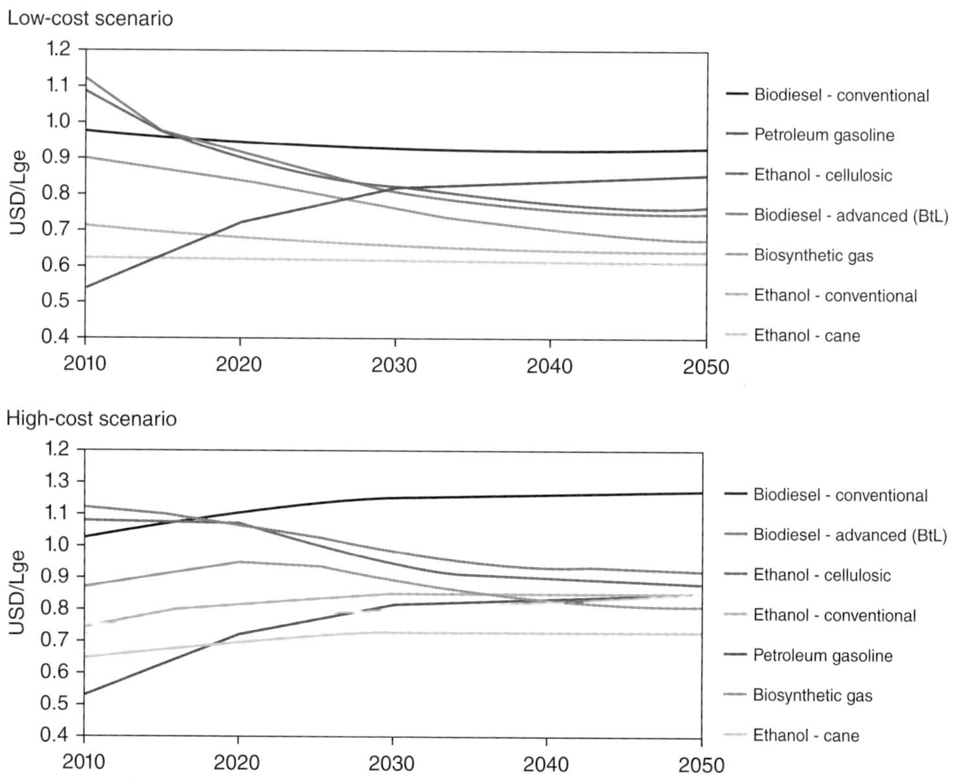

Figure 1.9. Costs of different biofuels compared to gasoline (BLUE Map Scenario: it models future energy demand until 2050 based on the criteria of global long-term CO_2 concentration in the atmosphere of 450 parts per million). (Note: costs reflect global average retail price without taxation. Regional differences can occur depending on feedstock prices and other cost factors). *Source*: IEA (2010).

cellulosic crops at competitive prices can only assure achieving target set for the replacement of biofuel in a timely manner.

Summary and Conclusions

It is expected that with the exception of sugarcane, the biofuel industry will incrementally transition into waste residues and perennial cellulosic crops. It is advisable that these perennial crop cultivation practices be based on the best management practices utilizing low-input technologies to lower production costs. As most of the perennial grasses that comprise the potential feedstocks for biofuel have not been developed for this specific purpose, research should focus on increasing their biomass production ability, which would reduce the land requirement for the feedstock production without affecting the output. Research should also focus in developing efficient harvesting and post-harvest handling procedures because these processes account for a considerable portion of production cost. Every effort should be made to preserve the existing multifaceted use such as wildlife habitat, recreation, soil conservation, and bioremediation or add these components to land planted into biofuel feedstock. The use of maize, soybean, palm oil, or rapeseed for biofuel should neither totally cease nor should such intention be part of public policy. The alternative option for these crops provides a relief outlet to farmers when the food crop prices drop below economical production levels. Of course, the primary purpose of producing food crops, as always, should be for food and biofuel should necessarily take secondary position. The land-use policy should preclude use of prime land for crops solely grown as biofuel feedstock. But the inclusion of biofuel crops in crop rotation that increases farm returns without affecting food crop production should not be barred; rather it must be encouraged. The adoption of biofuel production systems should be based on net life-cycle GHG emission reduction estimations. Proper nutrient management is important for the prevention of nutrient runoff from biofuel crop fields. The production system should be based on most product utilization to create least amount of waste. As marginal lands are targeted for the production of biofuel crops, attention would be required to prevent excessive soil erosion, as these soils are particularly prone to it.

It is a well-recognized fact that one of the important reasons for promoting biofuel is rural economic development. The input from all segments of stakeholders, therefore, is crucial during developing new biofuel projects. Sensitivity to the social and cultural uniqueness of native populations, especially in tribal dominated regions, is essential to preserving these traditions for not only those who practice them but also all others who are enriched by these experiences. During the industry development process, a special effort must be made that limited resource farmers are full participants in the resultant economic benefits. It is especially important for developing countries to ensure not only social justice but also perception of exclusion can lead to turmoil and violence. It is also imperative that policy and procedures should be developed to make production of biofuel crops a long-term enterprise not subject to boom and bust cycles depending on world fossil fuel market. In summary, biofuel industry faces many hurdles before success but hold the potential to compete against fossil fuel in time and provide substantial environmental and economic benefits.

References

Alexander, R.B., Smith, R.A., Schwarz, G.E., Boyer, E.W., Nolan, J.V., and Brakebill, J.W. (2008) Phosphorus and nitrogen delivery to The Gulf of Mexico from differences in the Mississippi River Basin. *Environmental Science and Technology* 42: 822–830.

Allen, P., Van Dusen, D., Lundy, L., and Gliessman, S. (1991) Integrating social, environmental and economic issues in sustainable agriculture. *American Journal of Alternative Agriculture* 6: 34–39.

Berndes, G., Bird, N., and Cowie, A. (2010) Bioenergy, Land-use change and climate change mitigation. IEA Bioenergy: ExCo:2010:03. Available at: www.ieabioenergy.com (accessed 13 September 2012).

Boyd, J. (2007) Nonmarket benefits of nature: what should be counted in green GDP. *Ecological Economics* 61: 716–723.

Buesseler, K.O., Doney, S.C., Karl, D.M., *et al.* (2008) Ocean iron fertilization: moving forward in a sea of uncertainty. *Science* 319: 136–144.

Carson, R. (1962) *Silent Spring*. Houghton Mifflin, Boston.

D'Antonio, C.M. and Vitousek, P.M. (1992) Biological invasions by exotic grasses, the grass/fire cycle, and global change. *Annual Review of Ecology and Systematics* 23: 63–87.

Danielsen, F., Beukema, H., Burgess, N.D., *et al.* (2009) Biofuel plantations on forested lands: double jeopardy for biodiversity and climate. *Conservation Biology* 23: 348–358.

de Vries, S.C., van de Ven, G.W.J., van Ittersum, M.K., and Giller, K.E. (2010) Resource use efficiency and environmental performance of nine major biofuel crops, processed by first-generation conversion techniques. *Biomass and Bioenergy* 34: 588–601.

Ehrlich, P. (1968) *The Population Bomb*. Ballantine, New York, 201p.

Environmental Protection Agency (EPA). (2011) Greenhouse gas emissions from a typical passenger vehicle. Available at: www.epa.gov/oms/climate/documents/420f11041.pdf (accessed 13 September 2012).

Fargione, J., Hill, J., Tilman, D., Polansky, S., and Hawthorne, P. (2008) Land clearing and the biofuel carbon debt. *Science* 319: 1157.

Farrell, G., Simons, S.A., and Hillocks, R.J. (2002) Pests, diseases and weeds of Napier grass, Pennisetum purpureum: a review. *International Journal of Pest Management* 48: 39–48.

Gari, I. (2002) Arabic treatise on environmental pollution up to the end of the thirteenth century. *Environment and History* 8: 475–488.

Gibbs, H.K., Ruesch, A.S., Achard, F., *et al.* (2010) Tropical forests were the primary sources of new agricultural land in the 1980s and 1990s. *Proceedings of the National Academy of Sciences* 107: 16732–16737.

Gibbs, H.K., Johnston, M., Foley, J.A., *et al.* (2008) Carbon payback times for crop-based biofuel expansion in the tropics: the effects of changing yield and technology. *Environment Research Letters* 3: 1–10.

Holm, L.G., Plucknett, D.L., Pancho, J.V., and Herberger, J.P. (1977) *The World's Worst Weeds*. East-West Center, Honolulu, Hawaii, USA, 609p.

Hoogeveen, J., Faurès, J-M., and Van de Giessen, N. (2009) Increased biofuel production in the coming decade: to what extent will it affect global freshwater resources? *Irrigation and Drainage* 58: S148–S160.

International Energy Agency (IEA). (2008) Energy technology perspective. Organization of Economic Cooperation and Development/International Energy Agency, Paris.

International Energy Agency (IEA). (2010) Sustainable production of second-generation biofuels. Organization of Economic Cooperation and Development/International Energy Agency, Paris.

International Energy Agency (IEA). (2011) Technology road map: biofuels for transport. OECD (Organisation for Economic Cooperation and Development)/IEA, Paris, 56p.

Laser, M., Larson, E., Dale, B., Wang, M., Greene, N., and Lynd, L.R. (2009) Comparative analysis of efficiency, environmental impact, and process economics for mature biomass refiing scenarios. *Biofuels, Bioproducts, and Biorefining* 3: 247–270.

Lehman, H., Clark, E.A., and Weise, S.F. (1993) Clarifying the definition of sustainable agriculture. *Journal of Environmental Ethics* 6: 127–143.

Morse, S. (2010) *Sustainability: A Biological Perspective*. Cambridge University Press, Cambridge, UK, 261p.

Odum, E.P. (1971) *Fundamentals of Ecology*, 3rd edn. Saunders, Philadelphia, USA, 546p.

OECD/FAO (2011) Biofuels. Agricultural Outlook 2011–2020. OECD Publishing and FAO, pp. 77–93.

Raghu, S., Spencer, J.L., Davis, A.S., and Wiedenmann, R.N. (2011) Ecological considerations in the sustainable development of terrestrial biofuel crops. *Current Opinion in Environmental Sustainability* 3:15–23.

Raup, D. and Sepkoski, J. (1982) Mass extinction in the marine fossil record. *Science* 215: 1501–1503.

Rohde, R.A. and Muller, R.A. (2005) Cycles in fossil diversity. *Nature* 434: 209–210.

Schoneveld, G.C. (2010) Potential land use competition from first-generation biofuel expansion in developing countries. Occasional paper no. 58. CIFOR, Bogor, Indonesia.

Searchinger, T., Heimlich, R., Houghton, R.A., *et al.* (2008) Use of US Croplands for biofuels increases greenhouse gases through emissions from land use change. *Science* 319: 1238.

Sepkoski, J.J. (2002) A compendium of fossil marine animal genera. *Bulletin of American Paleontology* 363: 1–560.

Shumaker, G.A., Morgan, A.L., and Mckissick, J.C. (2006) The economics of ethanol production in Georgia. Center for Agribusiness and Economic Development, The University of Georgia, Athens, Georgia, USA.

Singh, B.P., Granberry, D.M., Kelley, T., *et al.* (2005) Sustainable vegetable production, in *Vegetables: Growing Environment and Mineral Nutrition* (ed. R.D. Dris), WFL Publisher, Helsinki, Finland, pp. 1–38.

Singh, B.P. (2010) Overview of industrial crops, in *Industrial Crops and Uses* (ed. B.P. Singh), CABI, Wallingford, UK, pp. 1–20.

SoyStats. (2007) *Soy Stats Guide. American Soybean Association*, St. Louis, Missouri, USA.

Tilman, D., Hill, J., and Lehman, C. (2006) Carbon-negative biofuels from low-input high-diversity grassland biomass. *Science* 314: 1598–600.

United States Department of Agriculture (USDA). (1937) Abandoned farmstead in the Dust Bowl region of Oklahoma, showing the effects of wind erosion, 1937. Available at: http://www.britannica.com/EBchecked/media/96105/Abandoned-farmstead-in-the-Dust-Bowl-region-of-Oklahoma-showing (accessed 14 September 2012).

U.S. Energy Information Administration (EIA). (2007) International Energy Outlook, 2007. Available at: http://www.eia.doe.gov/oiaf/ieo/oil.html (accessed 20 August 2012).

United Nations Environment Program (UNEP). (1972) Declaration of the United Nations conference on the human environment. UNEP, New York. Available at: www.unep.org/documents/default.asp?documentid = 97&articleid (accessed 17 September 2012).

United Nations General Assembly. (1987) A/RES/42/187 Report of the World Commission on Environment and Development. Available at: www.un.org/documents/ga/res/42/ares42-187.htm (accessed 13 March 2012).

van Dam, J., Junginger, M., and Faaij, A.P.C. (2010) From the global efforts on certification of bioenergy towards an integrated approach based on sustainable land use planning. *Renewable and Sustainable Energy Reviews* 24: 2445–2472.

Watkins, K. and von Braun, J. (2003) Time to stop dumping on the world poor, 2002–2003 IFPRI Annual Report. International Food Policy Research Institute, Washington, DC, USA.
Watson, A.M. (1974) Arab agricultural revolution and its diffusion, 700–1100. *Journal of Economic History* 34: 8–35.
Watson, A.M. (1983) *Agricultural Innovations in the Early Islamic World.* Cambridge University Press, Cambridge, UK, 284p.
Worldwatch Institute. (2007) Implications for agriculture and rural development. in *Biofuels for Transport*. Earthscan, London, pp. 120–137.
Yunlong, C. and Smit, B. (1994) Sustainability in agriculture: a general review. *Agriculture, Ecosystem and Environment* 49: 299–307.

Chapter 2
Sustainable Production of Grain Crops for Biofuels

A.A. Jaradat

USDA-ARS Research Lab and Department of Agronomy and Plant Genetics, Morris, MN, USA

Introduction

Grain crops of the grass family (Gramineae) are grown for their edible starchy seeds; they include barley, maize, millets, oats, rice, rye, sorghum, and wheat. Their grain is used directly for human food, livestock feed, and as raw material for many industries. The rising demand for bioenergy has already caused a surge in the use of these crops for the production of biofuels. A fast growth in bioenergy demand and supply bears some risks for food security and for the environment. The greatest concern is land-use change (LUC) and the risk that large areas of natural habitats are converted to grow crops for biofuel production (Azar, 2011).

A roadmap for biofuels developed by the International Energy Agency (IEA) suggests that the contribution of biofuels can potentially increase from the current 2% to about 27% of total transportation liquid fuels by 2050 while meeting sustainability guidelines, only if appropriate technologies are deployed and policy incentives are implemented. However, the pursuit of increasing the share of biofuels in the global energy supply is occurring within the broad context of complex interlinkages between energy, food, land, water, and the environment, as well as their economic, social, and ecological implications (Tirado *et al.*, 2010).

The use of farmland and food grains for biofuel production is already sending warning signals, especially in the developing world. The price surge in 2008 is already the fifth such event witnessed in grain markets since the oil crises of 1973, despite the long-term declining trend of agricultural prices (Ajanovic, 2011). However, by 2020, taking current biofuel investment plans into account, international prices could increase by 26% for maize, 18% for oilseeds, 115% for sugar, and 8% for wheat. Comparing the expected grain price increase between 2000 and 2007, with or without the biofuel growth that occurred at the time, it was estimated that biofuels were responsible for 30% of the increase in weighted average grain prices.

The majority of current liquid biofuel production is based on grain crops that are usually used for food; therefore, their raw materials compete with food for fertile land and external inputs.

Biofuel Crop Sustainability, First Edition. Edited by Bharat P. Singh.

Currently, a small number of food crop species such as maize and wheat (i.e., first-generation energy crops) are used to produce biofuels (Lobell *et al.*, 2008). Potential availability of these and other grain crops is limited by land availability, soil fertility, productivity, and future climate change. The projected 2 °C average increase in mean global temperature may reduce output from the main grain-producing areas of the world by about 25% (Karp and Richter, 2011). However, proper crop residue retention (1 Mg ha^{-1} yr^{-1}) can, potentially, raise the annual world food grain production by 24–40 Mt (Lal, 2010).

Biofuels derived from grain crops rely on the fermentation of sugars to produce ethanol. It is generally well understood that grain crops are limited in their ability to achieve targets for oil-product substitution, climate change mitigation, and economic growth. For most grain crops, the annual change in above-ground carbon is equal to zero if the whole biomass is harvested for energy production. The cost and sustainability of grain crops have been criticized as expensive ways to meet environmental goals and provide energy alternatives (Wang and Yan, 2008). These limitations can be partly overcome by the utilization of lignocellulosic materials from their residues (Eisenbies *et al.*, 2009) and from dedicated bioenergy crops (Jaradat, 2011).

Global Demand for Food and Energy

By 2050, the predicted 9 billion people in the world would require 70% more food than today's population, and a large portion of the increased demand will come from developing countries (Yan *et al.*, 2011). It is estimated that more than half of the increased demand for cereals as a whole will come from maize. The allocation of grain crops to non-food uses, including animal feed, seed, bioenergy, and other industrial products, affects the amount of food available to the world. Since 1980, the annual rate of increase for grain crops is only ~1%, which is lower than the rate of world population growth of 1.1% (Pimentel *et al.*, 2009).

Globally, only 62% of grain crop production is allocated to human food (ranges from 25% in the Midwest United States to 90% in South Asia); 35% to animal feed; and 3% for biofuel, seed, and other industrial products. Nevertheless, bioenergy is seen as a substantial part of future energy systems, especially if climate policy aims at stabilizing CO_2 concentration at low levels to avoid climate change (Lotze-Campen *et al.*, 2010). Biofuel (mostly ethanol) production and demand have grown rapidly in recent years, largely driven by policies rather than by genuine market forces. In 2009, governments provided US$43 billion as subsidies to the renewable energy and biofuel industries combined; however, the IEA estimated the global fossil fuel subsidies for 2008 at US$557 billion (Robbins, 2011).

Worldwide, ethanol is currently equivalent to about 2% of the gasoline consumption and its share could reach 10–20% by 2030 (Walter *et al.*, 2008). Global biofuel production in 2008 included 66 BL of ethanol (a 34% increase from 2007) and 15 BL of biodiesel (a 48% increase from 2007) (Fargione *et al.*, 2010). The IEA predicted that, compared to 2008, biofuel consumption will increase by 17–220% in 2020 and by 250–620% in 2030. The higher end of these ranges is predicted if global atmospheric CO_2 target of 450 ppm is set, whereas it is estimated that 50% of the current global energy use of 450 EJ is potentially available in 2050 from bioenergy sources (Erisman *et al.*, 2010).

The rush to develop biofuels has led to tensions over land use (LU) and LUC to grow more grain crops for biofuels. In 2007, a record 375 000 km^2 of maize were planted in the United States, but only one-third of the crop was used to produce ethanol (Graham-Rowe, 2011). Even if all maize grain grown in the United States were converted to ethanol, it would have satisfied about 15% of the transportation needs (Bessou *et al.*, 2011).

Although the best lands have soils at low or moderate risk of degradation under annual grain production, these make up only 12.6% of global land area. Global food security depends on annual grains planted on almost 70% of all croplands; when combined, they supply a similar portion of the human calorie intake (Glover *et al.*, 2010).

Renewable fuel standards in the United States set clear, long-term targets for producing 136 BL of biofuels by 2022 while also reducing greenhouse gas (GHG) emissions. The EU is on target for biofuels to make up 10% of transport fuel by 2020, but progress varies widely between member states (Martin, 2011; Robbins, 2011). Therefore, it is expected that biofuels from grain crops will dominate and remain significant until 2030, after which biofuels from the "most sustainable" crops will remain important, then will be limited to specific crops with "high sustainability" factors between 2040 and 2050 (Murphy *et al.*, 2011).

Grain Crops: Food or Biofuel

The use of farmland and grains that could be consumed by humans for biofuel production is already sending warning signals in developed and developing countries around the world (Nigam and Singh, 2011) and the heated debate around whether "peaceful co-existence or permanent strife will emerge between biofuels and food sectors" will continue for the foreseeable future (Murphy *et al.*, 2011). The dominance of the current agro-industrial bioeconomy and its particular scientific base were questioned (Horling and Marsden, 2011) as to "*why more and more people are likely to go hungry while at the same time resource depletion and climate change threaten the Earth's growing population.*"

The growing emphasis on biofuels in large parts of the world is expected to create additional demand for grain crop production; this would lead to increased farm income and may enhance rural development (Hubert *et al.*, 2010). However, growing grain crops for bioenergy or biofuels is increasingly viewed as conflicting with food production (Armah *et al.*, 2009; Karp and Richter, 2011). Additionally, there is the bioethical issue of "*burning grain for fuel to run luxury automobiles*" when people go hungry or are undernourished (Gressel, 2008). Moreover, the actual or perceived negative impact of using grain crops for biofuel production on food prices may have tempered the enthusiasm about their role in reducing GHG emissions. Available data suggest that record grain prices in 2008 were not directly and exclusively caused by increased biofuel production, but were actually the result of a speculative bubble related to high oil prices, a weak US dollar, and increased volatility due to commodity index fund investment (Mueller *et al.*, 2011). Additionally, more than 80% of the US agricultural production goes into animal feed; a global increase in consumption of animal products is one of the main driving forces behind high food prices (Gopinathan and Sudhakaran, 2009). Biofuel production had, however, a modest (3–30%) contribution to the increase in commodity food prices observed up to mid-2008 (Hubert *et al.*, 2010).

Cereal grains make up an alarming 80% of the world's food supply (Pimentel *et al.*, 2009). Global food security depends on annual grain crops (Glover *et al.*, 2010), and the impact of bioenergy on food insecurity depends on the crop grown, the land used, the technology employed, and how the biofuel supply chain is integrated into agricultural, social, and economic systems. Currently, maize, wheat, and rice, as grain crops, make up ~75% of the ethanol potential (Johnston *et al.*, 2011). In less than 40 years, the demand for maize in the developing countries will double, and the nature of this demand is changing (Shiferaw *et al.*, 2011). Because of their direct use for human consumption, a price differential will likely continue to exist for wheat and rice grain in comparison to maize and sorghum mostly used for animal feed. Wheat and rice grain will probably be used for biofuel production only when prices are depressed by crop surpluses (Walter *et al.*, 2008).

The United States and the EU are major exporters and donors of relatively affordable food for developing countries. However, given the strong demand for grain crops for biofuel in the short- to medium-term, net cereal exports from these countries are projected to decline over the next decade. A major cause of soaring food prices in 2008 was attributed, in part, to the rapid growth in demand for biofuels, which has (temporarily) diverted land from food production (Hubert *et al.*, 2010). The prospects of continuing high grain market volatility is particularly threatening in developing countries, especially the Middle East and North Africa, given their high level of dependence on imports (Wright and Cafiero, 2011). The diversion of more maize (and other grain crops) to biofuel will further increase under current policies using subsidies and mandates, as well as protection from competition from more efficient biofuels (e.g., Brazilian sugar-derived ethanol).

The estimated, but far too optimistic, potential global food production at 32–47 Gt of grain equivalents, would provide an adequate supply for 16–24 billion people, let alone the 9 billion projected by 2050, with an affluent diet (Spiertz and Ewert, 2009). In reality, however, the shortage in market supply and concerns over the negative impact on food security has led some countries (e.g., China) to pose a ban on expanding maize- and other grain-based biofuel production (Yang *et al.*, 2009). Although grain crop production is now 2.7 times the amount that was being produced 50 years ago, a large proportion is used for livestock feed and for a growing biofuel industry (Horling and Marsden, 2011).

Globally, the area used for crops, prior to the large-scale production of biofuels, had been expanding at a rate of about 3.4 Mha yr^{-1}. The use of grain crops (mainly maize) and crop residues for biofuels has developed rapidly, particularly in the United States, since federal energy bills emphasizing biomass were passed by Congress in 2005. Global land demand for crop-based biofuels must be viewed as requiring additional cropland beyond the expansion required to meet future food demand (Fargione *et al.*, 2010).

Large-scale crop plantations for biofuels pose both opportunities and challenges, and will inevitably compete with food production for land, water, nutrient resources, and other inputs. In order for crops to be grown for biofuel production within the context of sustainable agro-ecosystems, the impact of biofuels on food prices as well as their potential to contribute to energy security, climate change mitigation through reduced GHG emissions, and rural development will remain the subject of considerable debate (Karp and Shield, 2008). Therefore, rather than focusing on the question of "food or biofuel" (Johnston *et al.*, 2011), a more challenging question is "how can increased demand for food and energy be met in the future?" particularly when water and land resources will be limited.

There are minimal food–biofuel trade-offs in a few coarse grain crops (e.g., sweet sorghum) (Kumar *et al.*, 2011) where the whole crop can be used in starch-to-ethanol, sugar-to-ethanol, and cellulose- or lignocellulose-to-biogas biofuel production technologies (Dahlberg *et al.*, 2011). Sorghum is one of the most important cereal crops widely grown for food, feed, forage, and fuel in the semiarid tropics of Asia, Africa, the Americas, and Australia (Kumar *et al.*, 2011). Although drought-tolerant sweet sorghum is a much better source for ethanol production than maize or wheat, it may compete with food crops for land and irrigation water resources, especially in parts of Africa, and in regions where soils are subject to erosion, degradation, and are of low productive capacity. Other coarse grain crops (e.g., barley, energy maize, oats, rye, millet, grain sorghum, and triticale) have the potential to play a similar crucial role in the global energy mix, especially under policies to reduce GHG emissions (Gillingham *et al.*, 2007).

Agricultural intensification is widely assumed necessary to meet projected food needs, given current economic and dietary trends. On the other hand, higher grain prices may intensify the use of both irrigation and chemical inputs that enhance crop yields. Nevertheless, yield gap still exists for most grain crops, especially for rice and maize (Walter *et al.*, 2008). Almost 60% of this increased production to fill the yield gap should come from conventional grain crops, and mainly from developing countries. However, additional crop production from agricultural intensification will clearly never be achieved for all crops and all countries due to differences in infrastructure and investment in research around the world (Johnston *et al.*, 2011).

Crop production systems differ as to their overall energy efficiencies expressed as output-to-input ratios; they ranged from 10 to 16 for conventional and no-till food production and from 7 to 11 for conventional and no-till fuel production, respectively. This analysis suggested that the use of cropland for food is more energetically efficient than for fuel production and large differences in efficiencies are attributed to management, which suggests multiple opportunities for improvement. Energy efficiencies for food-production systems follow the order, no-till > organic > conventional system, and the use of grain for biofuel production resulted in 30–40% lower net energy gain than when the crops were used for food production (Gelfand *et al.*, 2010).

Ecosystem Services

Biofuel targets (e.g., in the United States, EU, Brazil, and China) have been rushed through without full understanding of the wider impacts on LU, food prices (Robbins, 2011), and ecosystem services (Gressel, 2008). The production of grain crops often compromises essential ecosystem services, pushing some beyond sustainable boundaries (Glover *et al.*, 2010). There exists a close link between grain production, food insecurity, and climate change through the degradation of soils and desertification of ecosystems (Lal, 2010). Sustainable production of grain crops is dependent on the functioning and integrity of agro-ecosystems and, particularly, on ecosystem services related to soil, air, water, nutrients, and biodiversity; therefore, their use for biofuel production will affect many ecosystem services. To ensure future food production and to sustainably maintain other ecosystem services, farmers will need more options to produce grain crops. These options may be different from, and generally will be subjected to less favorable circumstances than those under which increases in grain production were achieved in the twentieth century (Glover *et al.*, 2010).

Positive effects on ecosystem services become more important when grain crops are deployed at a large scale in the landscape for biofuel production. To maximize GHG reduction, as a major ecosystem service and sustainability indicator of grain crops for biofuel production, the biofuel industry needs to target crops that have minimal processing requirements, minimize external inputs, utilize efficient energy conversion technologies, and increase use efficiency of grain and residue through multiple by-products and coproducts.

The demand for maize ethanol has changed production economics of grain crops dramatically; it is already contributing to loss of lands removed from crop production (e.g., Conservation Reserve Program (CRP) land in the United States) as contracts expire and lands are returned to agricultural (mostly maize) production (Wiens *et al.*, 2011). One strategy to reduce the pressure on this and similarly sensitive agro-ecosystems around the world is to develop more efficient conversion methods of grain into ethanol or into animal products (Azar, 2011). A long-term strategy is to develop and deploy perennial grain crops for biofuel production. Such crops, generally, will have advantages over annuals in maintaining and producing important ecosystem functions and services, particularly on marginal landscapes or where resources are limited (Glover *et al.*, 2010). To date, there are no perennial species that produce adequate grain harvests; however, breeding programs have been initiated to develop perennial wheat and sorghum using direct domestication and wide hybridization as two complementary approaches (Cox *et al.*, 2006).

The impact of the grain-derived ethanol industry on cropping systems, and the associated water and nutrient demand and the runoff of chemicals into surface and underground water bodies are important, but very complex issues (De La Torre Ugarte *et al.*, 2010). Some of these ecosystem functions are mediated by crop residue management. Crop residues are widely considered a source of lignocellulosic biomass, with 1 Mg being equivalent to 280 l of ethanol or 15–18 GJ (Lal, 2010); they have been identified as a near-term source of biomass for biofuels (Ghatak, 2011). However, seasonal variation in yield, bulkiness, and transport logistics are some of the drawbacks (Gressel, 2008). Nevertheless, under favorable socioeconomic and environmental conditions (e.g., in India), every tonne of crop residues utilized in biofuel production can save the emission of GHG to the extent of 0.25–0.35 tonne CO_2 equivalent, while providing renewable energy output 4–5 times the non-renewable energy utilized in the process (Ghatak, 2011).

Although crop residues could provide an additional source of revenue for grain producers, they play an important role in sustaining agro-ecosystems through managing soil erosion, nutrient loss, soil carbon, and soil moisture. The financial incentives to harvest crop residues (e.g., provided by the 2008 Farm Bill in the United States) may stimulate maize production due to demand for maize grain for feed and ethanol and maize residues for advanced biofuels (Nigam and Singh, 2011). If a reasonable portion of crop residues is removed for biofuel production, the extent of the damage to the soil will be negligible, at least in the temperate, but not the tropical or subtropical zones of the world (Gressel, 2008). On the other hand, removing maize cobs only would reduce the negative impact of stover removal on C-sequestration rate and reduce N-related emissions from the soil. If combined with no-till practices, removing maize cobs only may result in higher accumulation rates of organic soil carbon, lower fuel consumption, and lower N emissions (Kim *et al.*, 2009). Due to their chemical composition (low N, high C) and physical (e.g., high density) characteristics, maize cobs have an advantage for the production of biofuel depending on progress in harvest, transport, storage, and conversion technology (Jansen and Lübberstedt, 2011).

Approximately 0.7–0.9 billion Mg of crop residues are harvested every year, with maize, wheat, and rice contributing 75% of these residues; half of the residue can be used sustainably for biofuel production and, in addition to residues from sorghum, about 100 million Mg of ethanol yr^{-1} can be produced from residues of all four crops (Xie and Peng, 2011). In practice, however, it is not easy to assess the amount of residue that can be removed as it depends on grain yield, which depends on local conditions, among other factors. Data on long-term impact of crop residue removal on crops, soils, and climate are few. Recent studies show that indiscriminate crop residue removal harvesting has adverse impacts on soil properties, water quality, soil organic carbon (SOC) sequestration, and crop production, particularly in erodible and sloping soils (Blanco-Canqui, 2010).

The sustainable amount of crop residue that can be removed for biofuel production is environmentally dependent and is influenced by factors such as land slope, climate, precipitation, soil texture, organic matter, and wind conditions in production regions (Sarath *et al.*, 2008). However, most studies examining global biomass supply assumes that about 25% of the total available residue can be sustainably removed. Considering the significant role crop residue is expected to play in meeting the biofuel production goal, it would be necessary to implement best management practices for soil and water conservation to offset potential negative impacts associated with agricultural residue removal (Cibin *et al.*, 2011).

Biochar, a by-product of the maize- and wheat-based systems, can improve physical and chemical soil properties, avoid methane emissions, and sequester carbon in soils. Similarly, the application of untreated or carbonized rice husks in the wheat- and rice-based systems can increase total SOC, N, C/N ratio, and available P and K. The high C/N ratio of carbonized rice husks may have limited N availability, thereby slightly reducing grain yields in the first few years after application (Haefele *et al.*, 2011).

The growth of biofuel industry has important implications for biodiversity, the effects of which depends largely on which grain crops are being grown and the spatial extent and landscape pattern of land requirements for growing these crops (Wiens *et al.*, 2011). Significant land area expansion under grain crops will have detrimental effects on biodiversity and possibly on water resources (Woods *et al.*, 2010); on average, 140 l of water are embodied in each liter of ethanol; however, the range (5–2100 l) is enormous. On average, biodiversity is reduced by 60% in US maize and soybean fields compared to unconverted habitats (Fargione *et al.*, 2010).

Grain crops differ in their ability to support native biota or invasive species, and differences in harvesting practices may also affect a crop's conservation value. Crop residues left in the field after harvesting grains provide resources for migrating birds that other crops may not provide (Wiens *et al.*, 2011), whereas birds of greater conservation concern are more threatened by conversion to row crops (Fargione *et al.*, 2010).

Environmental Impact

Knowledge about the potential environmental impact of grain-based biofuel production systems is critical in evaluating their sustainability in the general context of GHG emissions and climate change (Persson *et al.*, 2010). The public is interested in knowing what the environmental impacts are on water and air quality, C-sequestration, and native plants and animals and their habitats; what the appropriate spatio-temporal scales are at which to examine environmental effects; and how potential trade-offs in environmental cost and benefits can be considered. These issues are hotly debated because their overall impacts with respect to wider ecological and socioeconomic issues are uncertain and difficult to assess (Havlík *et al.*, 2011).

Changing climate conditions will likely influence possible crop choice and energy requirements for biofuel production (Dale *et al.*, 2011). Biofuel production from grain crops has a variety of positive and negative effects on local and regional environments; its energy output and GHG balances differ depending on the crop, conversion and end use technologies, and system boundaries. Therefore, assessing the environmental performances of grain crops and their biofuels is a complex task (Boehmel *et al.*, 2008).

The uncertainty surrounding the relative environmental benefits of biofuels and their competition for land resources is a public concern (Lobell *et al.*, 2008; Jessup, 2009). It was estimated that utilizing the entire world's maize crop for bioenergy would replace only 6% of global fossil fuel demand, and offset less than 1% of global fossil fuel carbon emissions.

Agricultural cultivation of grain crops accounts for most of their environmental impact; therefore, a trade-off is needed between minimizing GHG emissions and reducing ecological impact, knowing that most biofuels that may reduce GHG by >30% have a greater environmental impact than fossil fuels. The overall environmental impact of biofuel production is largely determined by the scale of direct LUC and indirect land-use change (iLUC) (*see the next topic*), whether for total GHG balance or the conservation of natural resources and biodiversity (Searchinger *et al.*, 2008).

The extent to which grain-based biofuels can mitigate climate change depends on their GHG intensity relative to the liquid fossil fuels they displace; however, the largest environmental impact of biofuel production may well come from market-mediated LUC (Fargione *et al.*, 2010). For example, within 1 year, the area planted with maize in the United States increased by 6.2 Mha at the expense of soybean-planted area, which decreased by 4.4 Mha. Maize production causes more soil erosion, needs more N fertilizer, insecticides, herbicides, and water than any other crop (Pimentel *et al.*, 2009). This (and similar) LUCs would lead to increased GHG emissions compared with fossil fuels (Ravindranath *et al.*, 2011).

Although the amount of iLUC due to maize ethanol is not precisely known (*see section "Direct and Indirect Land Use Change"*), no doubt, it is large enough to be a significant factor in the overall sustainability assessment (Farber, 2011). The projected increased demand for maize may lead to the planting of maize continuously rather than in complex rotations; this may lead to increased N applications for several seasons and, consequently, to the increased export of dissolved N to surface and underground water of 10–34% due to ethanol production that will be required to meet current US renewable fuel standards (Fargione *et al.*, 2010). The additional grain yields will come at a high environmental cost due to overapplication of synthetic fertilizers, which might not be sustainable (Yan *et al.*, 2011). Therefore, nutrients will need to be recycled back to the soil to the extent possible.

Also high fertilizer application increases the decomposability of maize residue and soil C-sequestration becomes less efficient (Gallagher *et al.*, 2011). Benefits to maize grain yield from N fertilizer decline rapidly after the ecosystem's N demands are met; whereas maize residue shows only modest increases in carbohydrates yield. Heavier N applications will yield minimal grain benefits and almost no benefits in crop residue, while degrading the cellulosic ethanol feedstock quality (due to increased lignin) and soil C-sequestration capacity. Therefore, optimizing agricultural practices for carbohydrate yields may reduce the environmental impact of cellulosic ethanol production (Gallagher *et al.*, 2011).

Global production of crop residues is estimated at ~4 Pg (~0.5 Pg in the United States). The total removal of crop residues is not an option because of the negative environmental impact (Lal, 2010). The use of crop residues, as a coproduct, or instead of utilizing the grain, provides improvements over using the grain alone, and may provide some energy savings and GHG reductions (Karp and Richter, 2011). Environmental performances of the grain and its associated crop residues vary with the farm location due to crop management, soil properties, and climate conditions (Kim *et al.*, 2009).

Maize cobs and maize stover, for example, have better environmental performance than maize grain due to lower use of agrochemicals and fuel, and lower N-related emissions from the soil (N_2O, NO_x) and NO_3 leaching (Kim *et al.*, 2009; Jansen and Lübberstedt, 2011). Recent studies suggest that removing maize cobs only would reduce the negative impact of stover removal on SOC-sequestration rate and reduce N-related emissions from the soil (Kim *et al.*, 2009). Similarly, additional environmental benefits can be gained from using the residue left over after pressing sweet sorghum stems for juice; these residues can be used as a source of lignocelluloses for fermentation, or alternatively burned as a direct source of energy (Sarath *et al.*, 2008).

In the United States, about 12 l of wastewater must be removed per liter of ethanol produced, and this relatively large amount of sewage effluent has to be disposed of at energy, economic, and environmental costs (Pimentel *et al.*, 2009). These figures may have water availability and quality implications of biofuel production and could trigger local water shortages, along with soil erosion and nutrient runoff. However, we should note that many of the existing programs to reduce agricultural water pollution were developed in an era of low crop prices, before the expansion of maize ethanol, and rely on incentive payments for voluntary conservation measures (Martin, 2011).

Although GHG emissions during the production of grain crops could be large enough to offset the reduction of global warming achieved by using biofuels (Philippot and Hallin, 2011), the use of crop residues and other biomass for biofuels raises many environmental and ethical issues. The GHG emissions associated with the planting and processing of plant materials for biofuels are reported to reduce the nutritional quality of major world food crops (e.g., wheat, rice, barley, and soybean) (Pimentel *et al.*, 2009). The total GHG impact of growing 1 kg of harvestable above-ground plant dry matter, not including soil C changes, is 260 and 130 g CO_2 equivalent for maize and a perennial crop (switchgrass), respectively. If only maize grain is harvested, the impact is 390 g CO_2 equivalent kg^{-1} dry matter.

The environmental value of the use of a grain crop for biofuel production should be evaluated carefully, as some crops are environmentally preferable for qualities such as C-sequestration or efficient processing of nutrients and water. Grain and sweet sorghum, for example, are among the most climate-resilient crops that can better adapt to climate change conditions (Kumar *et al.*, 2011). Currently, these and other coarse grain crops contribute a small share to global biofuel production. The development of sweet sorghum as an energy crop lags far behind maize; the need is urgent to develop management practices and to study the complex heritability of sugar content in its stems (Zegada-Lizarazu and Monti, 2012).

Direct and Indirect Land-use Change

Biofuels currently made from grain crops (mainly in the United States) not only have high GHG emissions associated with field operations and the conversion of the crop into ethanol, but they also require large land areas. The conversion of land for more grain-based biofuel production would only increase the demand for crop production in other parts of the world (Azar, 2011). This iLUC is not an obvious direct effect of biofuels, and until recently, it was not obvious that its magnitude would be significant (Farber, 2011). Economic model estimates for iLUC associated with food-based biofuels range from 67 to 365 $\times 10^{-6}$ ha l^{-1}, leading to increased GHG emissions for decades compared to business as usual (Fargione *et al.*, 2010).

Decisions about LU and iLUC at the local level are affected by global factors; these include a growing human population, increasing wealth in developing countries, increased demand for grain as feed, and greater emphasis on biofuels as alternative energy sources (Wiens *et al.*, 2011). The land used to grow crops for biofuels increased from 13.8 Mha in 2004 ($\sim$1% of global cropland) to 26.6 Mha in 2007. For example, maize area increased 19% between 2006 and 2007 in the United States and resulted in reduced crop diversity in many parts of the Midwest (Searchinger *et al.*, 2008).

iLUC is a particular concern with maize-based ethanol, since maize is a major human food and livestock feed crop, and international food commodity markets are relatively inelastic (Farber, 2011). For example, almost 1 Mha left the CRP in 1 year, and most likely used for maize production in response to higher commodity prices ((Fargione *et al.*, 2010). If the current trend of increased demand for grain crops in the United States via biofuel requirements continues, it will reduce coarse grain and soybean exports, which may lead to increased iLUC in order to accommodate more grain crops, mostly in developing countries.

Biofuel production opportunities in developing countries are being encouraged by the relative availability of land to grow bioenergy crops; however, this raises concerns about potential social and environmental pressures, including environmental consequences resulting from iLUC. The results of an integrated assessment model that combines energy supply and demand and LU to examine the interaction of the energy system and LU (Gillingham *et al.*, 2007), suggested that the ultimate amount of bioenergy crops produced is sensitive to the amount of land needed for food production. It must be ensured, however, that any further land expansion for grain-based biofuel production will provide a positive contribution to climate change mitigation, knowing that iLUC is a source of the most significant GHG emissions.

In order to be a viable biofuel source, grain crops will have to environmentally and economically compete successfully with other LUs for a share of the finite land resources around the world. The extent to which grain-based biofuels displace other crops will influence global LU and the global agricultural system. Consequently, the continued development of grain crops for biofuel production could lead to land degradation as more land in natural ecosystems is hastily brought under cultivation, and the use of crop lands for grain-based biofuels may lead to increased GHG emissions from iLUC.

Genetic Improvement

The improvement of plants for both food and biofuel will require new breeding and selection emphasis that may differ from those targeted for grain yield during the Green Revolution (Jahn *et al.*, 2011). Classical breeding and genetic modification techniques are already available to develop crops with desired morphological, phenological, and biochemical traits and to maximize grain- and residue-based biofuel production (Baenziger *et al.*, 2006). These include large energy density, large C:N ratio, modified lignin biosynthesis, preprocessing *in planta* via expression of cellulases and cellulosomes, and cell wall traits that make the crop residue more amenable to processing by one or a combination of biological, physical, and chemical pretreatments. Several technologies, besides classical breeding, are also available to improve these traits, including genomic approaches to screen natural variation and the use of genetic modification to produce transgenic plants (Gressel, 2008).

Breeding crops with reduced carbon footprint holds great promise for sustainable food, feed, and biofuel production. Selecting or breeding a new generation of crops that reduce GHG emissions is a paradigm shift in agriculture with the possibility of securing productivity while mitigating the impact of agriculture on the climate (Philippot and Hallin, 2011). Targeted genetic improvement of grain crops for biofuel depends on identifying genetic variation in critical morphological, structural, and physiological traits (Jahn *et al.*, 2011). For example, the renewed interest in breeding sweet sorghum and energy maize as specialty crops for ethanol production was triggered by the discovery of the brown mid-rib (*bmr*) mutation in both crops. The mutants have lower lignin content (and a modified lignin subunit composition) and, therefore, much higher digestibility. Incorporation of *bmr* genes into sweet sorghum varieties could provide high-quality lignocellulose residues in addition to extracted juice (Sarath *et al.*, 2008). The selected sweet sorghum mutants displayed similar agronomic traits and grain yield to the wild type, but showed a remarkable alteration of cell wall composition, resulting in an increased rate of biomass degradation as compared to the wild type (Xie and Peng, 2011). This mutation was not found or identified in small grains, probably because the genes for lignin biosynthesis in small grains are in multigenic families, which are not amenable to single mutations (Gressel, 2008). Comparative analysis of sweet sorghum and field sorghum genomes is expected to identify genes or regulatory elements responsible for increased sugar production.

Researchers speculate that there are multiple targets for biomass improvement in rice; diverse rice varieties were found to achieve high biomass (i.e., high residue) production via independent developmental and physiological pathways (Jahn *et al.*, 2011). Genetic and genomic studies are underway to identify loci and networks underlying this functional variation; this will facilitate the improvement of biomass traits in other grasses being developed as biofuel crops. Elucidating the genetic basis of stem sugar and stem juice accumulation, modifying cell wall composition so that sweet sorghum biomass can be processed more efficiently, maximizing biomass yield for a given geographic area and production system, and understanding the different mechanisms underlying drought tolerance are the main focus areas among sorghum researchers that target bioenergy traits (Vermerris, 2011).

Ongoing genomic research has made significant advances in characterization of genomes of several plant crops including maize, rice, and sorghum that could be used to study and improve important agricultural traits in several biofuel crops (Li *et al.*, 2011). Maize is the most productive grain crop and occupies the largest land area in the United States; it continues to be an important target for genetic and genomic research and improvement by the private sector where most commercial breeding and seed production occurs (Brummer *et al.*, 2011). The rapidly expanding information from genomics and genetics, combined with improved genetic engineering technologies offer a wide range of possibilities for enhanced ethanol production from grain cobs, and stover of maize (Torney *et al.*, 2007). Several genes and their mutants have been identified for maize cob and ear characteristics, which could serve as candidate genes to manipulate cob chemical and physical characteristics (Jansen and Lübberstedt, 2011). The breeding of dual-purpose maize cultivars with combined high cob and grain yields requires a thorough understanding of factors influencing cob, ear, and grain development.

Maize breeders and geneticists demonstrated how breeding and genetic improvement can overcome biotic and abiotic stresses and play a key role in meeting future maize demand (Shiferaw *et al.*, 2011). For example, in response to the increasing cost of water to maize farmers, the private industry developed drought-tolerant maize cultivar(s); the value of which is being captured by the private seed sector, farmers, and society (Brummer *et al.*, 2011). As grain will likely remain the major product of maize, genetic improvement of maize for biofuel will need to focus on enhancing biomass quantity and quality without impacting grain yield and quality (Vermerris, 2011). Transgenic maize lines with modified expression of specific starch synthases, starch (de)-branching enzymes are being generated in attempts to produce starch granules with increased or decreased crystallinity, and thus altered susceptibility to enzymatic digestion (Torney *et al.*, 2007). Until recently, little has been done to exploit genetic variation in wheat starch properties for energy (Sarath *et al.*, 2008). Such variation exists and information from wheat genotypes with different starch properties could be used to develop grains that could be more efficiently converted to biofuel.

Based on current genetics, breeding, and physiological research and development efforts to understand the genome and genes of candidate crops, commercially viable perennial grain crops could be available within 20 years (Glover *et al.*, 2010). The super-domestication approach has been advocated (Vaughan *et al.*, 2007) to domesticate crops with dramatically increased yield that could not be selected in natural environments from naturally occurring variation without recourse to new technologies. Perennial grain crops can provide better opportunities than annual grain crops for grain- and biomass-based biofuel production. These crops can make the most efficient use of land, water, nitrogen, and energy resources, while they require minimum inputs in relation to the potential net energy produced. Breeding strategies of perennial grains for biofuel production need to combine multiple desirable traits in reliable regrowth and high grain yield and quality over multiple years and adaptation to multiple biotic and abiotic stresses (Glover *et al.*, 2010).

The ideotype concept was fundamental to understanding the physiological reasons behind the breeding success of grain crops during the last part of the twentieth century. A model plant was designed to yield a greater quantity of grain when developed as a crop cultivar. The phenotypic characteristics of a grain crop ideotype include a short stem, small, erect leaves, a low number of tillers, and a large and awned ear. This ideotype functioned as a weak competitor, with reduced intra-crop interference, and produced maximum yield per unit area (Chardon *et al.*, 2011).

Advancing appropriate genetic model systems for grain-based biofuel crop is essential in the development of systems' approaches to improve cell wall architecture and plant anatomies for the end use of biofuel production and for GCC adaptation and mitigation. Therefore, additional traits are suggested to develop future grain-based biofuel crops. These include traits that can maximize radiation interception and use efficiency, efficient nutrient recycling, high nitrogen-use efficiency (NUE), and water-use efficiency (WUE); architectural traits to allow for one-pass harvesting; and suitability of grains and crop residues for thermal, chemical, or biological conversion technologies.

Maize and sorghum are ideal genetic models for the improvement of C_4 grain-based biofuel crops, including perennial grain crops; both have close evolutionary relationship with future bioenergy perennial grasses, C_4 photosynthetic pathway, historical depth of genetic information, and a rapidly growing resource of genetic tools (Sarath *et al.*, 2008). Its small genome (730 Mb) makes sorghum an attractive model for functional genomics of C_4 grasses and its drought tolerance makes it especially important in dry regions, such as Northeast Africa, India, and the Southern Great Plains of the United States (Kumar *et al.*, 2011).

Rice is a candidate model plant for grass cell biology. The use of such model crops will elucidate, for example, how C_4 metabolism arose, how C_4 grasses partition C into sugar stores versus cell wall mass, and what are the genetic basis of several physiological and architectural plant traits (Murray *et al.*, 2009).The use of wheat starch as a biofuel is unlikely, but its value as a model to study starch properties on biofuel yields holds significant promise (Sarath *et al.*, 2008).

Life-cycle Analysis

Life-cycle analysis (LCA) is an important tool to identify the most appropriate technology for future utilization of feedstocks (grain or crop residue) for biofuel production and is mainly used to reveal the validity of biofuels as a means to reduce GHG emissions (Berndes *et al.*, 2003). Grain-based biofuel cropping systems vary as to the length of crop life cycle, grain and residue yield and conversion efficiencies, nutrient demand, soil carbon budget, and N losses. These factors affect the magnitude of the components contributing to net GHG emissions and nutrient losses (Heggenstaller *et al.*, 2008). Obviously, a thorough assessment of the GHG implications of LU and LUC to grain-based biofuel crops and cropping systems is a very complex and contentious issue.

Grain-based biofuel sustainability indicators need to focus on resource use efficiency, soil quality, net energy production, and GHG emissions, in addition to socioeconomic and biodiversity aspects, and LUC. In addition, grain-based biofuel assessment needs a relatively high spatial and technical disaggregation in order to adequately account for heterogeneous land qualities, technological differences, and possible adaptations (Havlík *et al.*, 2011).

Most LCA studies reported significant net reductions in GHG emissions and fossil energy consumption when grain-based ethanol was used to replace oil (Davis *et al.*, 2009). Yet, in some cases, grain-based biofuels performed worse than biomass-based biofuels in terms of deforestation, GHG emissions from land and irrigation water use, and relative price increase of crops (Havlík *et al.*, 2011). However, coproducts arising from crop (grain and residue) processing to biofuels are highly significant differentiating factors for all grain crops, when evaluating their GHG emissions, energy yield, and LU requirements within an LUC framework (Murphy *et al.*, 2011).

The wide range of the combined direct and iLUC estimates (e.g., 72–130 and 17–34 g CO_2 equivalent MJ^{-1} in the United States and Brazil, respectively) demonstrate the large variability within and among production systems converting maize (in the United States) and sugarcane (in Brazil) to ethanol; with GHG emissions from iLUC being more important than emissions from direct LUC (GEMIS, 2009). In addition, the N_2O emissions, due to N fertilizer production and field application, represent a real uncertainty source in the LCA results of many biofuel pathways (Nass *et al.*, 2007). It is important, in this regard, to notice that the United States imposes US$ 0.14 per liter tariff on Brazilian ethanol, even though it is widely considered more sustainable than domestic maize-based ethanol (Robbins, 2011).

Conversion Technologies

Grain-based biofuel (i.e., ethanol) is currently produced cost-effectively by conventional processes that depend on the edible, easily accessible fraction of the crop, which provides the substrates required for the conversion processes (Karp and Richter, 2011). Given the current agricultural practices and biofuel industrial conversion standards, the production of maize-grain-based ethanol yields ~25% more energy than the energy used to produce it. Ethanol made from starch (e.g., maize) needs more energy input than ethanol made from sugar (e.g., sweet sorghum) (Hattori and Morita, 2010). However, fermentation efficiency and ethanol recovery from Indian broken rice and pearl millet were much higher using a single-step, no-cook process than conventional multi-step, high temperature process (Gohel and Duan, 2011).

Grain-based biofuels require a relatively simple process to produce the finished fuel product; their sustainability, however, remains questionable due to their relatively high production cost compared to biomass-based biofuels and perceived competition with food production (Nigam and Singh, 2011). Therefore, it is necessary to modify biomass properties to reduce processing cost or increase biomass yield, and reduce inputs (Torney *et al.*, 2007).

Considerable research has been devoted to improving quantitative traits of importance to biofuel production. Recent research reported a 22% variation in sorghum ethanol yields and 9% variation in conversion efficiencies. Researchers noted significant differences in rates of fermentation and hypothesized that this would play an important role in biofuel production capacity (Sarath *et al.*, 2008). Near-infrared technologies have been developed to quickly and cost-effectively screen large numbers of accessions for compositional traits such as ash, lignin, glucan, xylan, and arabinan, all of which have unique properties related to various bioconversion technologies (Dahlberg *et al.*, 2011).

The modification of several crops with *in planta* hydrolytic enzymes or to convert cellulose into glucose showed some potential. The reduction or modification of lignin structure may reduce the need for physical or biochemical pretreatments (O'Brien and Mullins, 2009). Multiple transgenic strategies need to be developed to tackle complex traits, to engineer metabolic pathways, and to combine the expression of different genes (Torney *et al.*, 2007).Transgenic maize lines with modified expression of specific starch synthases and starch (de)-branching enzymes are being generated in an attempt to produce starch granules with increased or decreased crystallinity, and thus altered susceptibility to enzymatic digestion (Torney *et al.*, 2007). A modified starch with decreased gelatinization temperature might require less energy for the conversion process.

Until recently, little has been done to exploit genetic variation in starch properties for energy content. Such variation exists in small grains (e.g., wheat) and the differences in starch properties could be used to develop grains that could be more efficiently converted to biofuel (Sarath *et al.*, 2008). However, changes in starch composition to accelerate or optimize enzymatic conversion to sugars could be performed more quickly and easily by genetic engineering than by classical breeding (Gressel, 2008).

Innovations to Enhance Sustainability

Grain-based biofuels are expected to dominate and remain significant until 2030; then, those from the "most sustainable" crops will remain important. Finally, between 2040 and 2050 production will be limited to specific crops with "high sustainability" factors (Murphy *et al.*, 2011). To ensure the sustainable production of food, biofuel, and other ecosystem services, farmers need more options and innovations to produce grain crops under different, generally less favorable circumstances than those that have prevailed during the past 50 years (Glover *et al.*, 2010). Narrowing the large yield gap between developed and developing countries for most grain crops will contribute to achieving higher sustainability of food and biofuel production (Murphy *et al.*, 2011). On average, the US maize, sorghum, rice, and wheat production per hectare is 4.4-, 2.7-, 1.7-, and 1.2-fold, respectively, as compared to average yield in the top producing developed countries of these crops. For example, by closing the fiftieth percentile production gap (i.e., improving global yields to median level), the top 20 grain and oilseed crops could provide approximately 112.5 and 8.5 BL of ethanol and biodiesel production, respectively (Johnston *et al.*, 2011).

With increased competition for food and biofuels, non-food-grain crops (e.g., sorghum) become more important; however, their current supplies are not adequate to satisfy the growing, industrial-scale ethanol production market. Development of appropriate harvest, storage, and fermentation technologies could potentially make some of the non-food-grain crops (e.g., sweet sorghum) more sustainable biofuel crops (Sarath *et al.*, 2008). Therefore maize and, to some extent, wheat may have some advantage in the short term as grains for biofuel production, especially when value-added by-products are considered.

Grain yields could increase in response to biofuel production if demand leads to higher crop prices, and farmers were able to invest more in inputs; eventually, however, higher prices may lead to increased technological improvements (Fargione *et al.*, 2010). Although crop yields can be increased by means of supplementary chemical inputs, in the long term, it may not be economically and environmentally sustainable (Gressel, 2008). Also, expansion of land for both food and biofuel production will become limited by land scarcity unless agricultural intensification and advanced cropping systems and practices are developed and implemented at a large scale (Graham-Rowe, 2011). With biofuels coming under increasing pressure to slow or eliminate iLUC, the use of targeted intensification via established agricultural practices might offer an alternative for their continued growth (Johnston *et al.*, 2011).

The environmental performance of grain-based biofuels can be enhanced by utilizing more sustainable production practices that increase resource-use efficiency and integrate progressive management practices, reduced- or no-till, more efficient fertilization, and the use of cover crops. Future testing opportunities of soils and plants have been suggested to ensure sustainability of the biofuel industry (Karlen and Kerr, 2012). These testing opportunities will help monitor and guide production of grain-based biofuel production (e.g., from maize and sweet sorghum) and address other ecosystem services such as C-sequestration, environmental quality, wildlife habitat, animal production, and rural development.

Intensification of grain-based biofuel production can have substantial positive benefits on GHG emissions (Murphy *et al.*, 2011). Grain crops can benefit when included in well-planned crop rotations. Such rotations will help reduce external inputs, promote nutrient recycling and the effective use of natural resources, maintain long-term land productivity, mitigate biotic stresses, and consequently, increase crop yields and sustainability (Zegada-Lizarazu and Monti, 2011).

Crop productivity increases, larger than those achieved during the past 50 years, are needed to help avoid increasing GHG emissions; however, they need to be brought about by cost-effective interventions for mitigation, despite the use of inputs that may lead to increased emissions. Also, future intensification via improved practices needs to be compatible with protection of other sustainability and biodiversity benefits (Murphy *et al.*, 2011).

Increases in biomass in response to fertilization are not uniform across biochemical classes (i.e., carbohydrates, protein lipids, lignin), or tissues (i.e., leaf and stem, grain, reproductive support) (Gallagher *et al.*, 2011). The quality of crop residue declines (more lignin and less carbohydrates), and soil carbon sequestration becomes less efficient with increased N fertilization. Therefore, there is a need to manage fertilizer use more judiciously (Gaju *et al.*, 2011) and improve residue quality for biofuel production through genetic modification of cell wall structure (Xie and Peng, 2011).

A number of (eco)physiological traits, if manipulated, can contribute to higher grain yield and better sustainability of producing grain-based biofuels. The "stay-green" trait has been linked to changes in the balance between nitrogen demand and supply during the grain filling. Slower nitrogen translocation from the leaves to the grain should provide more carbohydrate for above-ground growth by prolonging photosynthesis and grain yield (Kant *et al.*, 2011). Genetic variation in leaf senescence is associated with post-anthesis N immobilization (in maize and sorghum) and that reducing N translocation efficiency might be a mechanism of delaying the onset of post-anthesis senescence; whereas raising N translocation efficiency associated with lower grain protein content is feasible in end-use markets for which a high grain starch-to-protein ratio is desirable (e.g., grain-based biofuel market) (Gaju *et al.*, 2011).

Net energy balance needs to be improved by using whole plant for biofuels and coproducts. Coproducts derived from crop processing to biofuels are highly significant differentiating factors for all grain-based biofuels when evaluating their GHG, energy density, and LU requirements (Murphy *et al.*, 2011). The net energy balance (NEB) of rice, for example, can be increased from 1.1 to 3.5 by utilizing different parts of the plant as different energy sources (Hattori and Morita, 2010). Similar improvements can be achieved by using residues from flour production for ethanol (wheat bran = ~23% of wheat production); using wheat straw for butanol production (Ferreira-Leitão *et al.*, 2010); or using the oleaginous fungus *Mortierella isabellina* (transforms sugar to storage lipid) in a semi-solid fermentation process for the production of biodiesel from sweet sorghum. The net energy ratio for sweet sorghum (7.06) is the highest among all other biofuel crops due to the high conversion efficiency of its juice to ethanol (Economou *et al.*, 2010).

Research Needs and the Future of Grain-based Biofuels

In spite of recent calls for the grain-based biofuels to be phased out and replaced by dedicated second-generation biofuel crops with multiple environmental benefits (Karp and Richter, 2011) and whether the nature of the whole industry is "*fragile*" (Selfa *et al.*, 2010), the key instruments widely adopted to advance production and increase consumption of grain-based biofuels are still in place. These include mandatory blending targets, tax exemption, and subsidies. Nonetheless, "*sustainable*" grain-based biofuels are expected to contribute to the overall energy portfolio up until mid-century (Murphy *et al.*, 2011); therefore, the debate over the domestic and international iLUC implications will continue (Oladosu *et al.*, 2011), and predictive models are needed to estimate future indirect effects of complex interactions among LU, crop production, crop exports, and biofuel policies.

Investments in research and development need to continue to close the yield gaps, develop widely adapted and stress-tolerant crop genotypes with high energy density, and optimize conversion technologies. At the same time, incentives should be provided for precision application of inputs to minimize negative environmental impact of these biofuels (Johnston *et al.*, 2011). This investment is supported by an emerging high value or multi-product strategy (Murphy *et al.*, 2011), whereby food and biofuel pathways are integrated. In the medium term, lignocellulosic biomass from small (e.g., barley, oats, rice, rye, triticale, and wheat straw) and coarse grain crops (e.g., maize and sorghum stover) could provide alternative sources of bioenergy by their conversion to cellulosic ethanol or use as a solid biofuel. However, inexpensive conversion technologies are yet to be developed and employed at a large scale. On the other hand, the crop residue properties need to be modified to reduce processing cost, increase yield, and reduce inputs (O'Brien and Mullins, 2009).

For the foreseeable future, however, the fate of the grain-based biofuel industry depends on crops with reduced carbon footprints (Philippot and Hallin, 2011), modified biochemical profiles (Gallagher *et al.*, 2011), and high energy ratio, and on being highly intertwined with the chemical and pharmaceutical industries (Hu and Phillips, 2011), even if it continues to be application oriented.

The need to optimize agricultural practices will continue in order to maximize carbohydrate yield and reduce the environmental impact of grain- and residue-based biofuel production (Gallagher *et al.*, 2011). Due to the wide variation in plant biochemical yield, research is needed to develop practices to manage crop biochemical profiles, especially in perennial grain crops, whether by altering N rates and timing, or by improving N translocation efficiency as the most important component of NUE (Gaju *et al.*, 2011).

A number of complex processes need to be better understood and quantified as a basis for improvements in crop management of unique cropping systems (e.g., for the rice–wheat cropping systems in SE Asia) that increase yields as well as NUE and WUE (Jing *et al.*, 2010); and develop a new generation of crop varieties for the same cropping systems that reduce CH_4 and N_2O emissions based on the discovery of large genotypic variation in nitrification rates in the rhizosphere of some rice genotypes (Philippot and Hallin, 2011).

In view of global climate change and higher temperatures, thermo- and photo-insensitive, non-lodging crop genotypes that are resistant to multiple biotic and abiotic stresses need to be developed, especially for new crops (e.g., sweet sorghum and perennial grain crops) and for certain geographical regions (e.g., the tropics and sub-tropics). Evaluation of GHG emissions and biofuel production under different climate change scenarios would increase our understanding of the role of grain-based biofuels in future energy systems.

Research on cell wall traits that make the crop residue more amenable to processing by pre-treatments is necessary to optimize energy production. There are >1000 genes that are related to plant cell wall biosynthesis, degradation, and regulation; therefore, diverse germplasm collections and cell wall mutants are needed for genetic research and manipulation (Xie and Peng, 2011).

Interspecific and intergeneric hybrids are being used to develop perennial grain crops and it is expected that commercially viable perennial grain crops could be available within 20 years and could help meet a wide range of domestic and international challenges, including food security, biofuel supplies, climate change, and environmental protection (Sarath *et al.*, 2008). Additionally, impressive progress in transforming plant breeding into crop engineering to increase crop yields with minimal environmental impact is being achieved through super-domestication (Vaughan *et al.*, 2007). Had large programs to breed perennial grains been started along with those of annual grain crops in the 1960s, farmers might well have had the seed of perennial grain crops today (Cox *et al.*, 2006). Research is needed to identify germplasm, select parents, develop perennial grain crops, and elucidate their agronomic characteristics and environmental impact to replace current annual grain crops (Glover *et al.*, 2010). Selection of genes within candidate species is needed for traits of domestication such as synchronous flowering and maturity, large non-shattering seed, and high seed yield per unit area (Cox *et al.*, 2006).

Conclusions

It will be a challenge to societies around the world to continue expanding the grain-based biofuel industry while abiding to environmentally sustainable production requirements. Recently, we witnessed that the grain-based (and grain-dependent) biofuel industry links events in world energy markets to food market behavior in a qualitatively new manner. For most grain-based biofuels, the net GHG emission reduction is marginal or unfavorable. The establishment of a maize-cob-based biofuel industry will not compete with food production and will increase sustainable biofuel production from biorenewable sources. However, the net impact is determined by many factors, including the grain crop, environmental and edaphic factors, management practices, and conversion technology. There are yet no large-scale production facilities of dedicated energy crops. Most countries continue to produce (if not expand) biofuels based on grains, sugar, or oilseed. The rapidly evolving grain-based (and increasingly biomass-based) biofuel industry will provide many opportunities for integrated system analysis to help ensure production, processing, and use of products and coproducts are sustainable. Therefore, the future of this industry will depend on a "landscape" vision that simultaneously addresses multiple challenges including biofuel production, C-sequestration, quality of natural resources, environmental health, animal production, and food availability and security.

An integrated agro-biotechnology system for food, feed, and biofuel production is likely to be a challenge from the regulatory, if not the technical, point of view, but will most certainly be the future for grain crops if they are to be bred and improved for biofuel production. Future intensification of grain-based biofuel production will require more attention to the efficiency of inputs and their environmental cost. More efficient use of chemical inputs, more sustainable alternatives, and breeding for efficiency will be required to reduce the C intensity (i.e., emissions per unit yield) of products, as well as reduce LU, iLUC, and inputs that damage environmental health. Continued expansion of grain-based biofuel production could exacerbate an already stressed water supply situation, particularly in irrigated areas. Current and future annual and perennial grain crops with high WUE, NUE, tolerances to biotic and abiotic stresses, and adaptation to future climate change, will have great potential for biofuel production.

References

Ajanovic, A. (2011) Biofuels versus food production: does biofuels production increase food prices? *Energy* 36: 2070–2076.

Armah, P., Archer, A., and Phillips, G.C. (2009) Drivers leading to higher food prices: biofuels are not the main factor. *In Vitro Cell Development and Biology – Plant* 45: 330–341.

Azar, C. (2011) Biomass for energy: a dream come true or a nightmare? *Climate Change* 2: 309–323.

Baenziger, P.S., Russel, W.K., Graef, G.L., and Campbell, B.T. (2006) Improving lives: 50 years of crop breeding, genetics and cytology. *Crop Science* 46: 2230–2244.

Berndes, G., Hoogwijk, M., and Broek van den, R. (2003) The contribution of biomass in the future global energy supply: a review of 17 studies. *Biomass and Bioenergy* 25: 1–28.

Bessou, C., Ferchaud, F., Gabrielle, B., and Mary, B. (2011) Biofuels, greenhouse gases and climate change: a review. *Agronomy for Sustainable Development* 31: 1–79.

Blanco-Canqui, H. (2010) Energy crops and their implications on soil and environment. *Agronomy Journal* 102: 403–410.

Boehmel, C., Lewandowski, I., and Claupein, W. (2008) Comparing annual and perennial energy cropping systems with different management intensities. *Agricultural Systems* 96: 224–236.

Brummer, E.C., Barber, W.T., Collier, S.M., *et al.* (2011) Plant breeding for harmony between agriculture and the environment. *Frontiers in Ecology and the Environment* 9: 561–568. doi:10.1890/100225 (10 pages).

Chardon, F., Noël, V., and Masclaux-Daubresse, C. (2011) Exploring NUE in crops and in Arabidopsis ideotypes to improve yield and seed quality. *Journal of Experimental Botany* 63: 3321–3323. doi:10.1093.jxb/err353 (12 pages).

Cibin, R., Chaubey, I., and Engel, B. (2011) Simulated watershed scale impacts of maize stover removal for biofuel on hydrology and water quality. *Hydrological Progress* 26: 1629–1641. doi:10.1002/hyp.8280 (13 pages).

Cox, T., Glover, J.D., Van Tassel, D.L., Cox, C.M., and DeHaan, L.R. (2006) Prospects for developing perennial grain crops. *BioScience* 56: 649–659.

Dahlberg, J., Berenji, J., Sikora, V., and Latkovic, D. (2011) Assessing sorghum [*Sorghum bicolor* (L) Moench] germplasm for new traits: food, fuels & unique uses. *Maydica* 56–1750: 85–92.

Dale, V.H., Kline, K.L., Wright, L.L., Perlack, R.D., Downing, M., and Graham, R.L. (2011) Interactions among bioenergy feedstock choices, landscape dynamics, and land use. *Ecological Applications* 21: 1039–1054.

Davis, S.C., Anderson-Teixeira, K.J., and DeLucia, E.H. (2009) Life cycle analysis and the ecology of biofuels. *Trends in Plant Science* 14: 140–146.

De La Torre Ugarte, D.G., He, L., Jensen, K.L., and English, B.C. (2010) Expanded ethanol production: implications for agriculture, water demand, and water quality. *Biomass and Bioenergy* 34: 1586–1596.

Economou, ChN., Makri, A., Aggelis, G., Pavlou, S., and Vayenas, D.V. (2010) Semi-solid state fermentation of sweet sorghum for the biotechnological production of single cell oil. *Bioresource Technology* 101: 1385–1388.

Eisenbies, M.H., Vance, E.D., Aust, W.M., and Seiler, J.R. (2009) Intensive utilization of harvest residues in Southern Pine Plantations: quantities available and implications for nutrient budgets and sustainable site productivity. *Bioenergy Research* 2: 90–98.

Erisman, J.W., van Grinsven, H., Leip, A., Mosier, A., and Bleeker, A. (2010) Nitrogen and biofuels; an overview of the current state of knowledge. *Nutrient Cycling in Agroecosystems* 86: 211–223.

Farber, D.A. (2011) Indirect land use change, uncertainty, and biofuel policy. *University of Illinois Law Review* 2011: 381–412.

Fargione, J.E., Plevin, R.J., and Hill, J.D. (2010) The ecological impact of biofuels. *Annual Review of Ecological and Evolutionary Systems* 41: 351–377.

Ferreira-Leitão, V., Gottschalk, L.M.F., Ferrara, M.A., Nepomuceno, A.L., Molinari, H.B.C., and Bon, E.P.S. (2010) Biomass residues in Brazil: availability and potential uses. *Waste Biomass Valor* 1: 65–76.

Gaju, O., Allard, V., Martre, P., *et al.* (2011) Identification of traits to improve nitrogen-use efficiency of wheat genotypes. *Field Crops Research* 123: 139–152.

Gallagher, M.E., Hockaday, W.C., Masiello, C.A., Snapp, S., McSwiney, C.P., and Baldock, J.A. (2011) Biochemical suitability of crop residues for cellulosic ethanol: disincentives to nitrogen fertilization in corn agriculture. *Environmental Science and Technology* 45: 2013–2020.

Gelfand, I., Snapp, S.S., and Robertson, G.P. (2010) Energy efficiency of conventional, organic, and alternative cropping systems for food, and fuel at a site in the U.S. Midwest. *Environmental Science and Technology* 44: 4006–4011.

GEMIS. (2009) Global Emission Model for Integrated Systems Version 4.5. Available at: www.iinas.org/gemis.html (accessed 25 January 2012).

Ghatak, H.R. (2011) Biorefineries from the perspective of sustainability: feedstocks, products, and processes. *Renewable and Sustainable Energy Reviews* 15: 4042–4052.

Gillingham, K.T., Smith, S.J., and Sands, R.D. (2007) Impact of bioenergy crops in a carbon dioxide constrained world: an application of the MiniCAM energy-agriculture and land use model. *Mitigation and Adaptation Strategies for Global Change* 13: 675–701. doi: 10.1007/s11027-007-9122-5 (29 pages).

Glover, J.D., Reganold, J.P., Bell, L.W., *et al.* (2010) Increased food and ecosystem security via perennial grains. *Science* 328: 1638–1639.

Gohel, V. and Duan, G. (2012) No-cook process for ethanol production using Indian broken rice and pearl millet. *International Journal of Microbiology* 2012. doi:10.1155/2012/680232 (9 pages).

Gopinathan, M.C. and Sudhakaran, R. (2009) Biofuels: opportunities and challenges in India. *In Vitro Cell Development and Biology – Plant* 45: 350–371.

Graham-Rowe, D. (2011) Beyond food versus fuel. *Nature* 474: S6–S8.

Gressel, J. (2008) Transgenics are imperative for biofuel crops. *Plant Science* 174: 246–263.

Haefele, S.M., Konboon, Y., Wongboon, W., *et al.* (2011) Effects and fate of biochar from rice residues in rice-based systems. *Field Crops Research* 121: 430–440.

Hattori, T. and Morita, S. (2010) Energy crops for sustainable bioethanol production; which, where and how? *Plant Production Science* 13: 221–234.

Havlík, P., Schneider, U.A., Schmid, E., *et al.* (2011) Global land-use implications of first and second generation biofuel targets. *Energy Policy* 39: 5690–5702.

Heggenstaller, A.H., Anex, R.P., Liebman, M., Sundberg, D.N., and Gibson, L.R. (2008) Productivity and nutrient dynamics in bioenergy double-cropping systems. *Agronomy Journal* 100: 1740–1748.

Horling, L.G. and Marsden, T.K. (2011) Towards the real green revolution? Exploring the conceptual dimensions of a new ecological modernization of agriculture that could 'feed the world.' *Global Environmental Change* 21: 441–452.

Hubert, B., Rosegrant, M., van Boekel, M.A.J.S., and Ortiz, R. (2010) The future of food: scenarios for 2050. *Crop Science* 50: S-33–S-50.

Hu, M-C. and Phillips, F. (2011) Technological evolution and interdependence in China's emerging biofuel industry. *Technological Forecasting and Social Change* 78: 1130–1146.

Jahn, C.E., Mckay, J.K., Mauleon, R., *et al.* (2011) Genetic variation in biomass traits among 20 diverse rice varieties. *Plant Physiology* 155: 157–168.

Jansen, C. and Lübberstedt, T. (2011) Turning maize cobs into a valuable feedstock. *Bioenergy Research* 5: 20–31.doi:10.1007/s12155-011-9158-y (12 pages).

Jaradat, A.A. (2011) Genetic resources of energy crops: biological systems to combat climate change. *Australian Journal of Crop Science* 5: 1072–1086.

Jessup, R.W. (2009) Development and status of dedicated energy crops in the United States. *In Vitro Cell Development Biology – Plant* 45: 282–290.

Jing, Q., Keulen van, H., and Hengsdijk, H. (2010) Modeling biomass, nitrogen and water dynamics in rice-wheat rotations. *Agricultural Systems* 103: 433–443.

Johnston, M., Licker, R., Foley, J., *et al.* (2011) Closing the gap: global potential for increasing biofuel production through agricultural intensification. *Environmental Research Letters* 6: 034028. doi:10.1088/1748-9326/6/3/034028 (11 pages).

Kant, S., Bi, Y-M., and Rothstein, S.J. (2011) Understanding plant response to nitrogen limitation for the improvement of crop nitrogen use efficiency. *Journal of Experimental Botany* 62: 1499–1509.

Karlen, D.L. and Kerr, B.J. (2012) Future testing opportunities to ensure sustainability of the biofuel industry. *Communications in Soil Science and Plant Analysis* 43: 36–46.

Karp, A. and Richter, G.M. (2011) Meeting the challenge of food and energy security. *Journal of Experimental Botany* 62: 3263–3271.

Karp, A. and Shield, I. (2008) Bioenergy from plants and the sustainable yield challenge. *New Phytologist* 179: 15–32.

Kim, S., Dale, B.E., and Jenkins, R. (2009) Life cycle assessment of corn grain and corn stover in the United States. *International Journal of Life Cycle Assessment* 14: 160–174.

Kumar, A.A., Reddy, B.V.S., Sharma, H.C., *et al.* (2011) Recent advances in sorghum genetic enhancement research at ICRISAT. *American Journal of Plant Science* 2: 589–600.

Lal, R. (2010) Managing soils for a warming earth in a food-insecure and energy-starved world. *Journal of Plant Nutrition and Soil Science* 173: 4–15.

Li, X., Hou, S., Yang, M., *et al.* (2011) Major energy plants and their potential for bioenergy development in China. *Environmental Management* 46: 579–589.

Lobell, D.B., Burke, M.B., Tebaldi, C., Mastrandera, M.D., Falcon, W.P., and Naylor, R.L. (2008) Prioritizing climate change adaptation needs for food security in 2030. *Science* 319: 607–610.

Lotze-Campen, H., Popp, A., Beringer, T., *et al.* (2010) Scenarios of global bioenergy production: the trade-offs between agricultural expansion, intensification and trade. *Ecological Modeling* 221: 2188–2196.

Martin, J. (2011) Don't foul the water. *Nature* 474: S17.

Mueller, S.A., Anderson, J.E., and Wallington, T.J. (2011) Impact of biofuel production and other supply and demand factors on food price increases in 2008. *Biomass and Bioenergy* 35: 1623–1632.

Murphy, R., Woods, J., Black, M., and McManus, M. (2011) Global developments in the competition for land from biofuels. *Food Policy* 36: 552–561.

Murray, S.C., Roony, W.L., Hamlin, M.T., Mitchell, S.E., and Kresovich, S. (2009) Sweet sorghum genetic diversity and association mapping for brix and height. *Plant Genome* 2: 48–62.

Nass, L.I., Pereira, P.A.A., and Ellis, D. (2007) Biofuels in Brazil: an overview. *Crop Science* 47: 2228–2237.

Nigam, P.S. and Singh, A. (2011) Production of liquid biofuels from renewable resources. *Progress in Energy and Combustion Science* 37: 52–68.

O'Brien, M. and Mullins, E. (2009) Relevance of genetically modified crops in light of future environmental and legislative challenges to the agri-environment. *Annals of Applied Biology* 154: 323–340.

Oladosu, G., Kline, K., Uria-Martinez, R., and Eaton, L. (2011) Sources of corn for ethanol production in the United States: a decomposition analysis of the empirical data. *Biofuels, Bioproducts and Biorefining* 5: 640–653.

Persson, T., Garcia, A., Paz, J.O., Fraisse, C.W., and Hoogenboom, G. (2010) Reduction in greenhouse gas emissions due to the use of bio-ethanol from wheat grain and straw produced in the south-eastern USA. *Journal of Agricultural Sciences* 148: 511–527.

Philippot, L. and Hallin, S. (2011) Towards food, feed and energy crops mitigating climate change. *Trends in Plant Science* 16: 476–480.

Pimentel, D., Marklein, A., Toth, M.A., *et al.* (2009) Food versus biofuel: environmental and economic costs. *Human Ecology* 37: 1–12.

Ravindranath, N.H., Lakshmi, C.S., Manuvie, R., and Balachandra, P. (2011) Biofuel production and implications for land use, food production and environment in India. *Energy Policy* 39: 5737–5745.

Robbins, M. (2011) Fuelling politics. *Nature* 474: S22–S24.

Sarath, G., Mitchell, R.B., Sattler, S.E., *et al.* (2008) Opportunities and roadblocks in utilizing forages and small grains for liquid biofuels. *Journal of Industrial Microbiology and Biotechnology* 35: 343–354.

Searchinger, T., Heimlich, R., Houghton, R.A., *et al.* (2008) Use of U.S. croplands for biofuel increases greenhouse gases through emissions from land-use change. *Science* 319: 1238–1240.

Selfa, T., Kulcsar, L., Bain, C., Goe, R., and Middendorf, G. (2010) Biofuels bonanza? Exploring community perceptions of the promises and perils of biofuels production. *Biomass and Bioenergy* 35: 1379–1389. doi:10.1016/j.biombioe.2010.09.008.

Shiferaw, B., Prasanna, B.M., Hellin, J., and Bänziger, M. (2011) Crops that feed the world 6. Past successes and future challenges to the role played by maize in global food security. *Food Science* 3: 307–327.

Spiertz, J.H.J. and Ewert, F. (2009) Crop production and resource use to meet the growing demand for food, feed and fuel: opportunities and constraints. *Netherlands Journal of Agricultural Sciences* 56: 281–300.

Tirado, M.C., Cohen, M.J., Aberman, N., Meerman, J., and Thompson, B. (2010) Addressing the challenges of climate change and biofuel production for food and nutrition security. *Food Research International* 43: 1729–1744.

Torney, F., Moeller, L., Scarpa, A., and Wang, K. (2007) Genetic engineering approaches to improve bioethanol production from maize. *Current Opinion in Biotechnology* 18: 193–199.

Vaughan, D.A., Balázs, E., and Heslop-Harrison, J.S. (2007) From crop domestication to super-domestication. *Annals of Botany* 100: 893–901.

Vermerris, W. (2011) Survey of genomics approaches to improve bioenergy traits in maize, sorghum and sugarcane. *Journal of Integrative Plant Biology* 53: 105–119.

Walter, A., Rosillo-Calle, F., Dolzan, P., Piacente, E., and da Cunha, K.B. (2008) Perspectives on fuel ethanol consumption and trade. *Biomass and Bioenergy* 32: 730–748.

Wang, Y. and Yan, L. (2008) CFD studies on biomass thermochemical conversion. *International Journal of Molecular Science* 9: 1108–1130.

Wiens, J., Fargione, J., and Hill, J. (2011) Biofuels and biodiversity. *Ecological Applications* 21: 1085–1095.

Woods, J., Williams, A., Hughes, J.K., Black, M., and Murphy, R. (2010) Energy and the food system. *Philosophical Transactions of the Royal Society B*. 365: 2992–3006.

Wright, B. and Cafiero, C. (2011) Grain reserves and food security in the Middle East and North Africa. *Food Science* 3: S61–S76.

Xie, G. and Peng, L. (2011) Genetic engineering of energy crops: a strategy for biofuel production in China. *Journal of Integrative Plant Biology* 53: 143–150.

Yan, J., Warburton, M., and Crouch, J. (2011) Association mapping for enhanced maize (*Zea mays* L.) genetic improvement. *Crop Science* 51: 433–449.

Yang, H., Zhou, Y., and Liu, J. (2009) Land and water requirements of biofuel and implications for food supply and their environment in China. *Energy Policy* 37: 1876–1885.

Zegada-Lizarazu, W. and Monti, A. (2011) Energy crops in rotation: a review. *Biomass and Bioenergy* 35: 12–25.

Zegada-Lizarazu, W. and Monti, A. (2012) Are we ready to cultivate sweet sorghum as a bioenergy feedstock? A review on field management practices. *Biomass and Bioenergy* 40: 1–12. doi:10.1016/j. biombioe.2012.01.048 (12 pages).

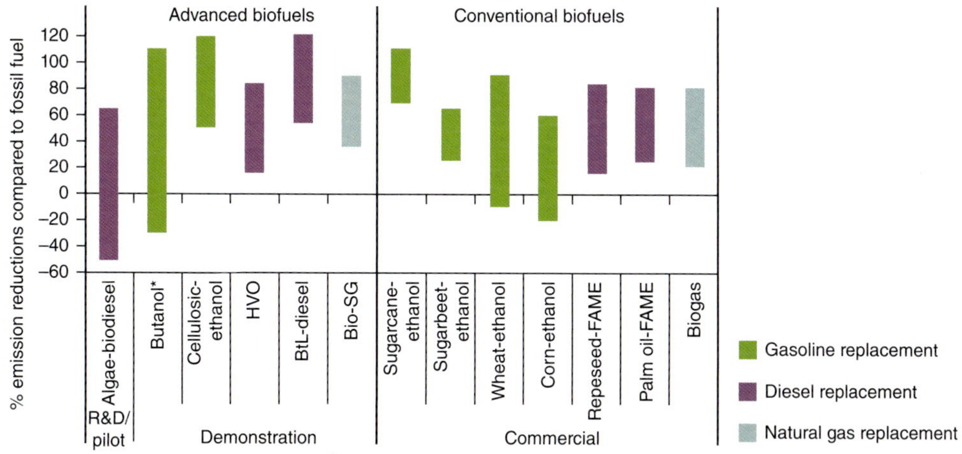

Figure 1.4. Life-cycle GHG balance of different conventional and advanced biofuels, and current state of technology. (Note: The assessments exclude emissions from indirect land-use change. Emission savings of more than 100% are possible through use of co-products. Bio-SG, bio-synthetic gas; BtL, biomass-to-liquid; FAME, fatty acid methyl esters; HVO, hydrotreated vegetable oil). *Source*: IEA (2011).

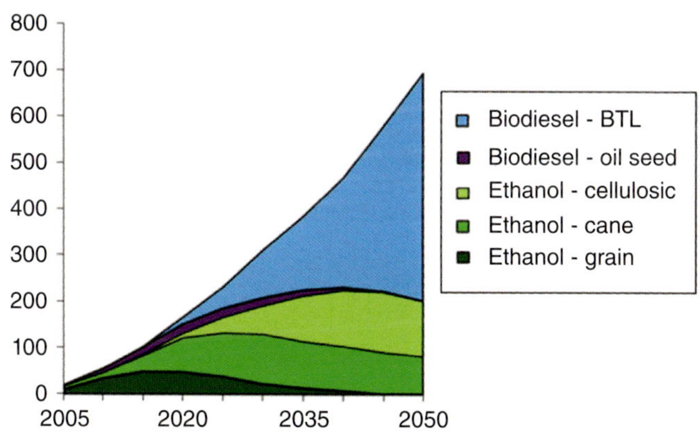

Figure 1.8. Trends in the production of different biofuels. *Source*: IEA (2008).

Biofuel Crop Sustainability, First Edition. Edited by Bharat P. Singh.

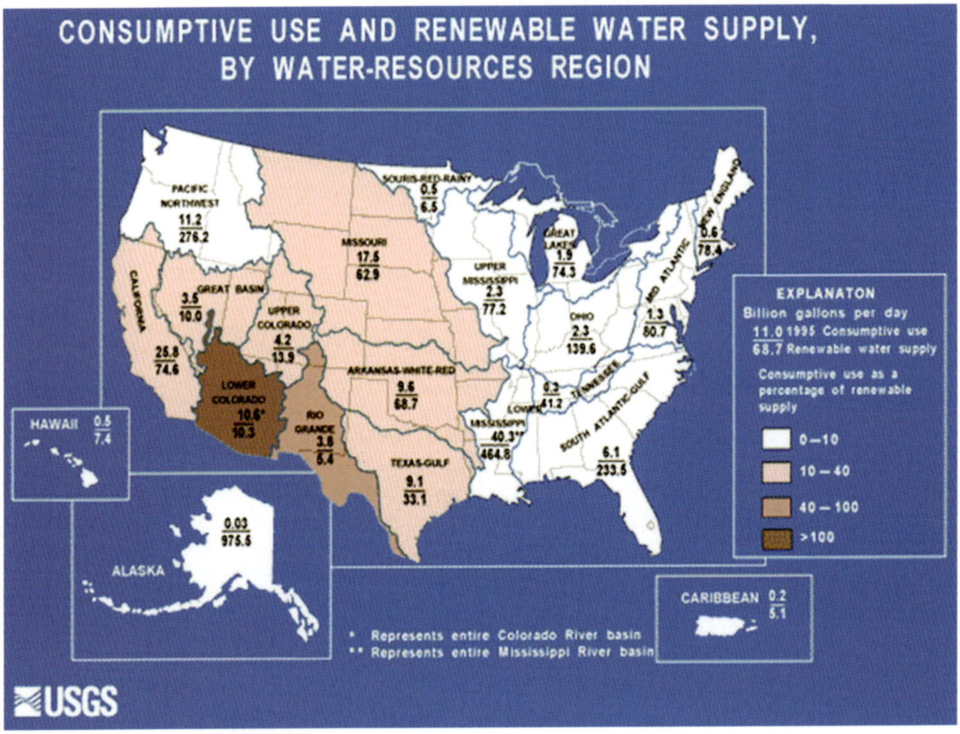

Figure 7.1. Comparisons of average consumptive use and renewable water supply for the 21water-resources regions of the United States, Puerto Rico, and US Virgin Islands. Adapted from USGS 1995 (http://water.usgs.gov/watuse/misc/consuse-renewable.html).

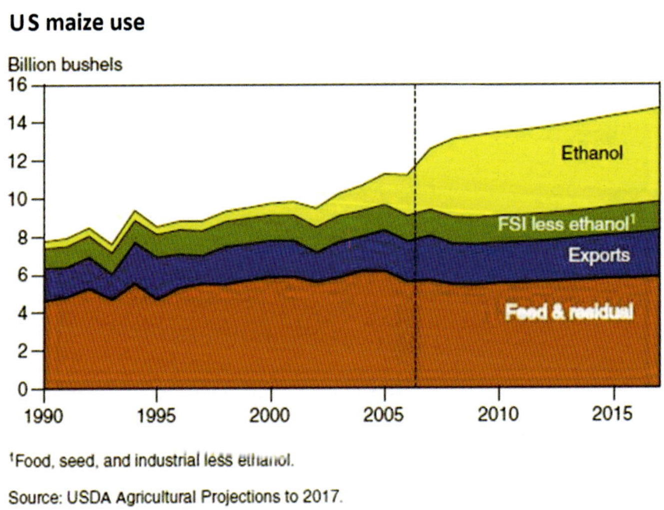

Figure 7.2. Projected U.S. Maize Use from USDA-Economic Research Service (ERS), 2008.

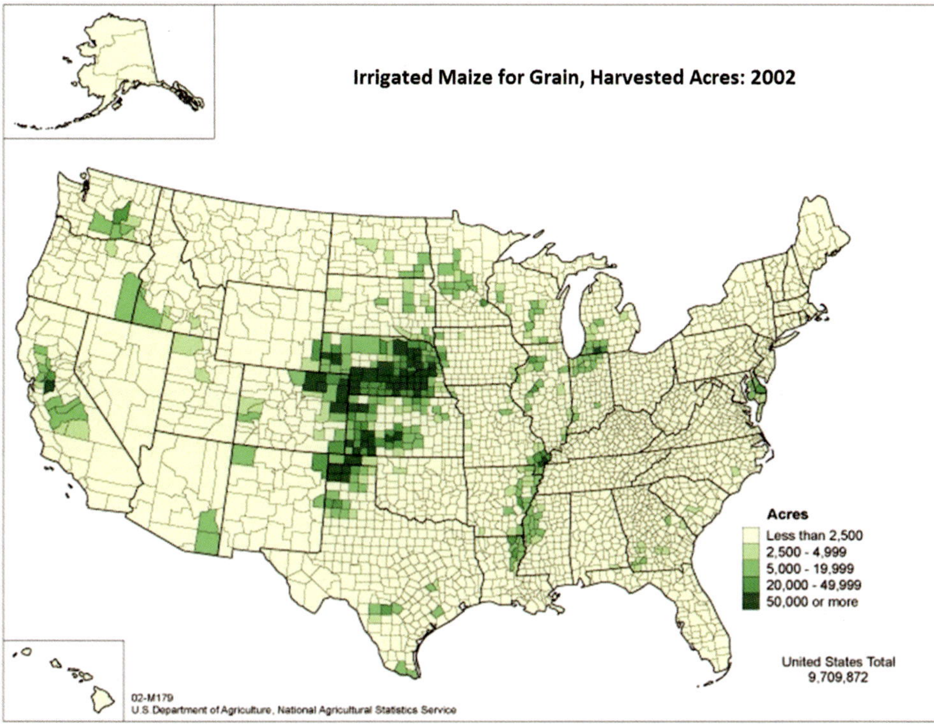

Figure 7.3. US Irrigation maize for grain (http://www.agcensus.usda.gov/Publications/2002/Ag_Atlas_Maps/Crops_and_Plants/index.asp).

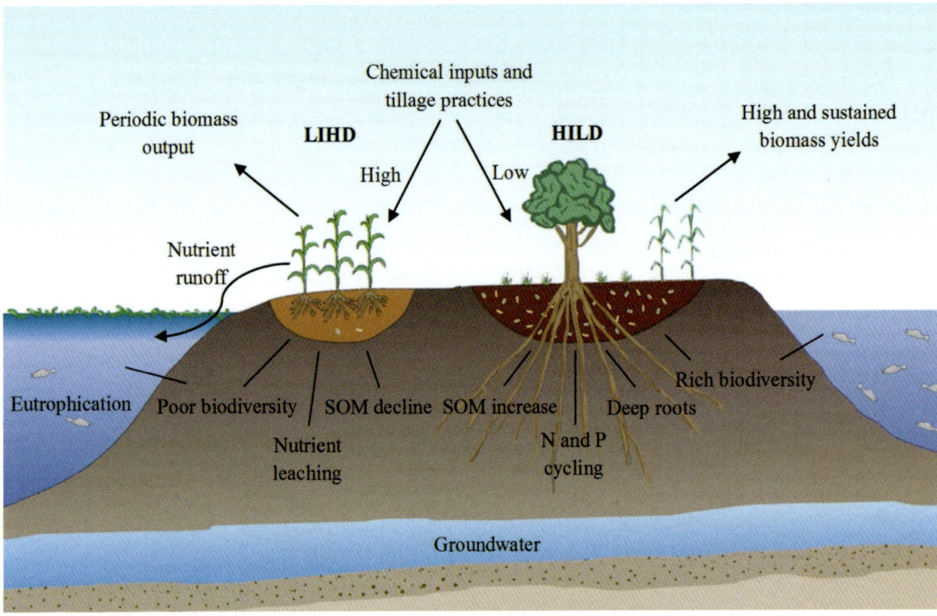

Figure 8.1. The impacts of HILD and LIHD crops on soil quality.

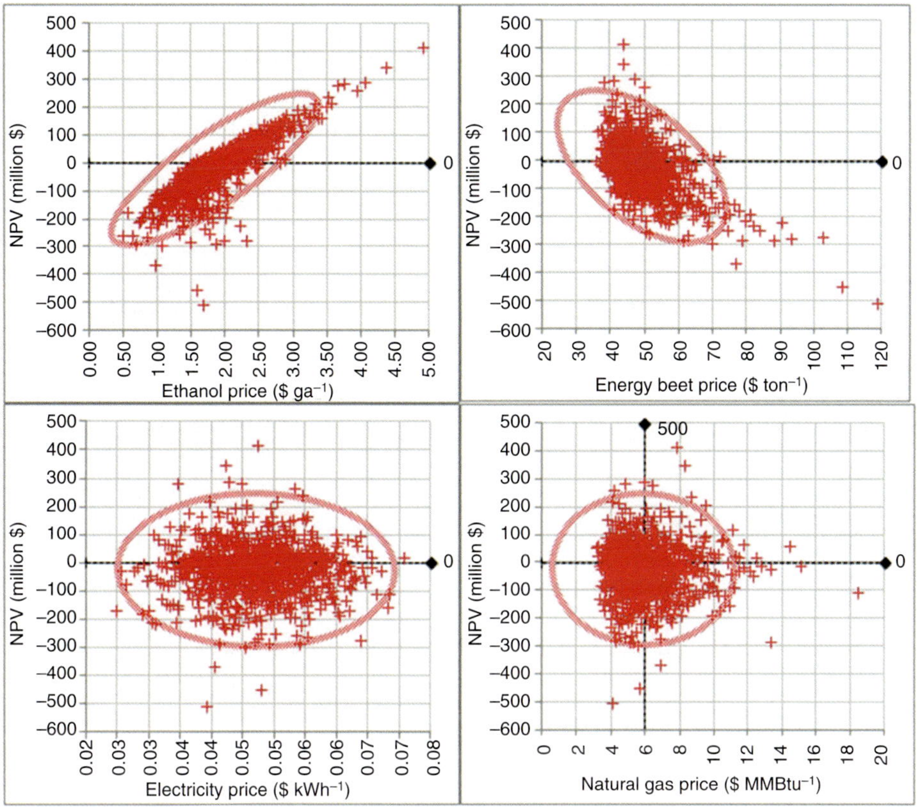

Figure 13.3. Net present value (NPV) versus ethanol, energy beet, electricity, and natural gas prices.

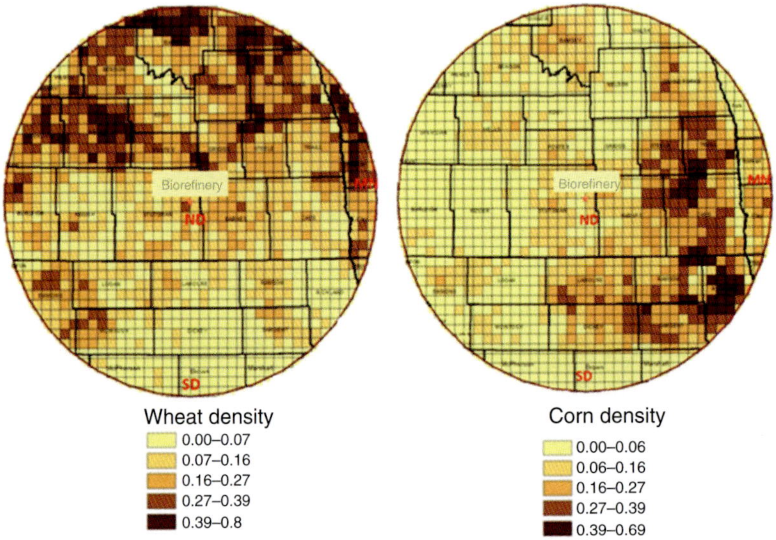

Figure 13.4. Wheat and maize concentration within 100-mile radius of biorefinery in Southeast North Dakota.

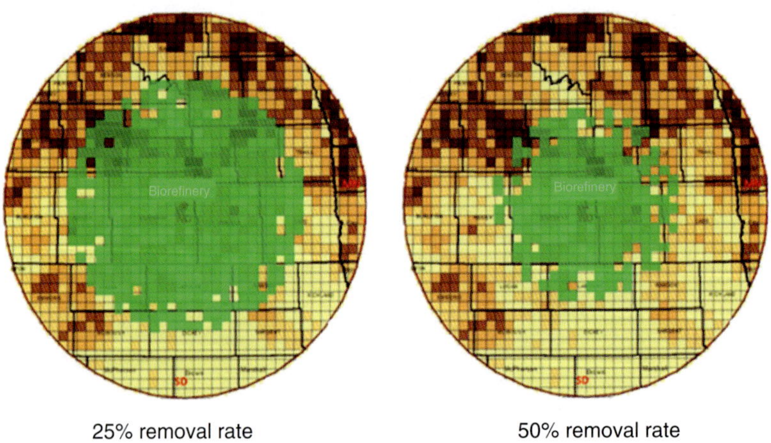

Figure 13.5. Estimated potential wheat straw supply sites.

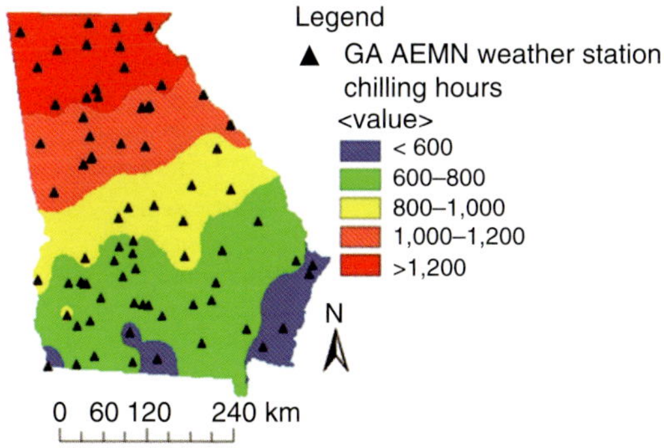

Figure 14.3. Chilling hour map for Georgia.

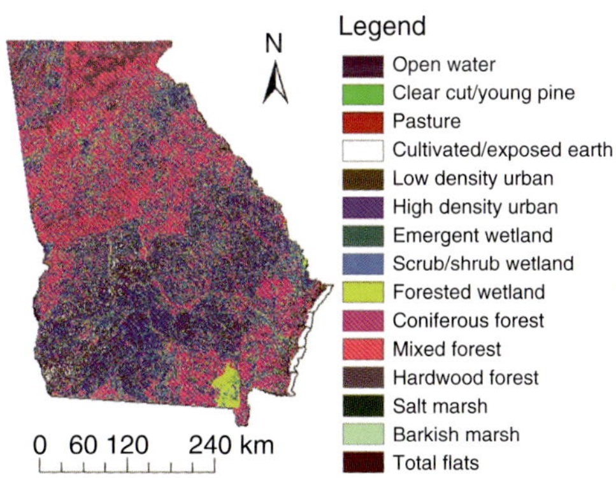

Figure 14.4. LULC analysis map of Georgia.

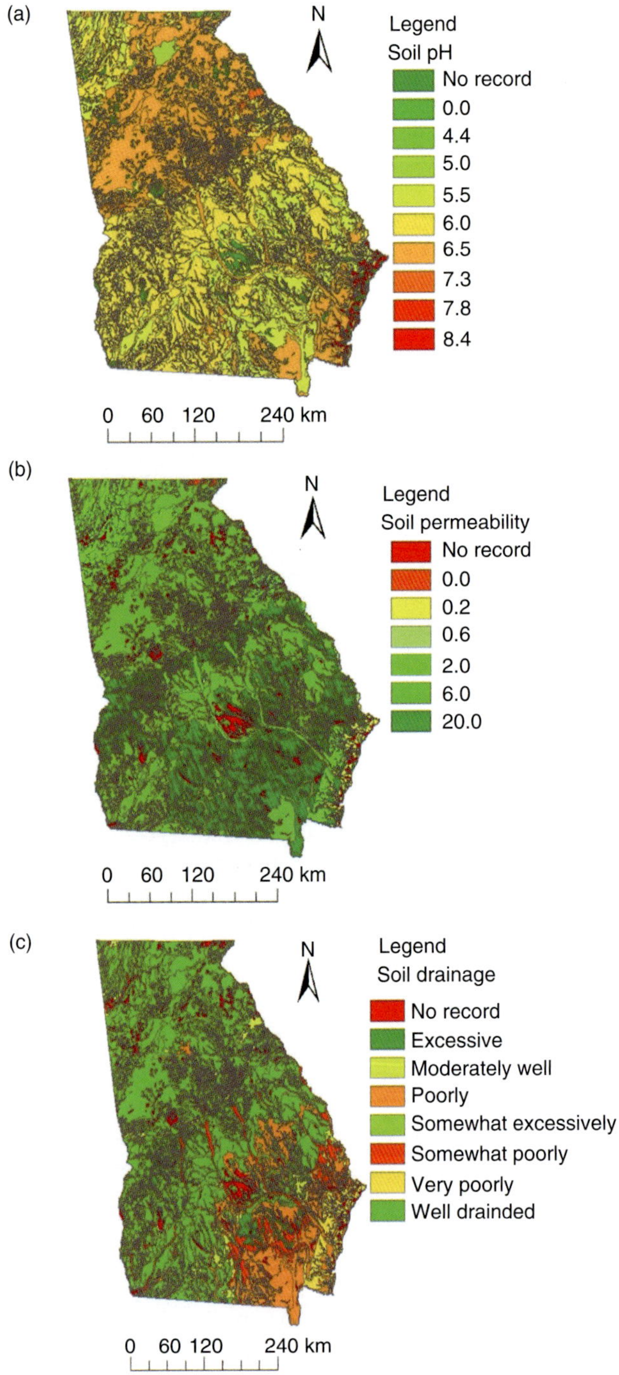

Figure 14.5. Maps showing the soil attributes that affect blueberry production, including (a) pH, (b) permeability, and (c) drainage.

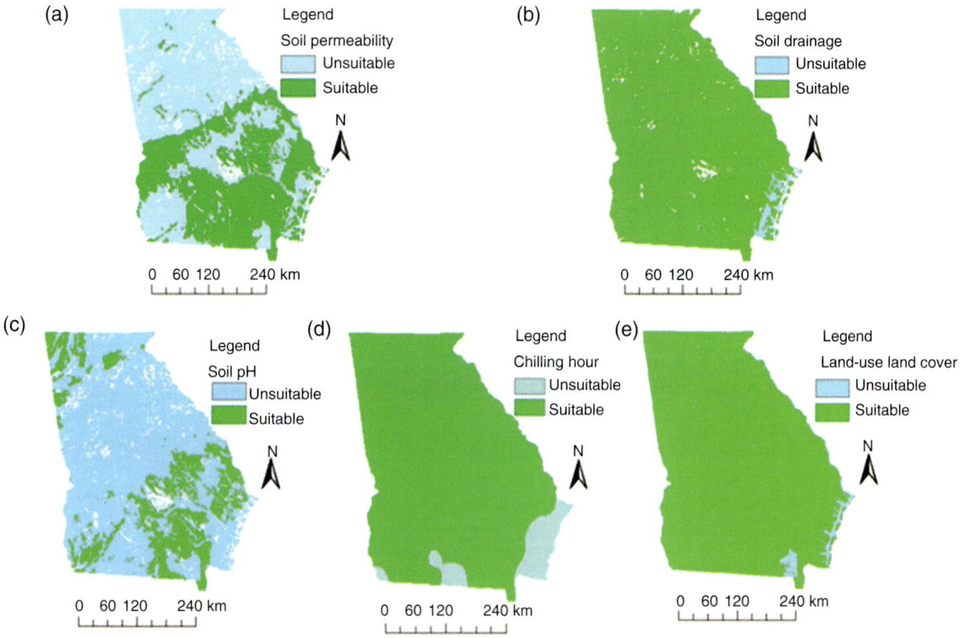

Figure 14.6. Maps showing the suitable areas in Georgia for various blueberry production parameters: (a) soil permeability, (b) soil drainage, (c) soil pH, (d) chilling hours, and (e) land use.

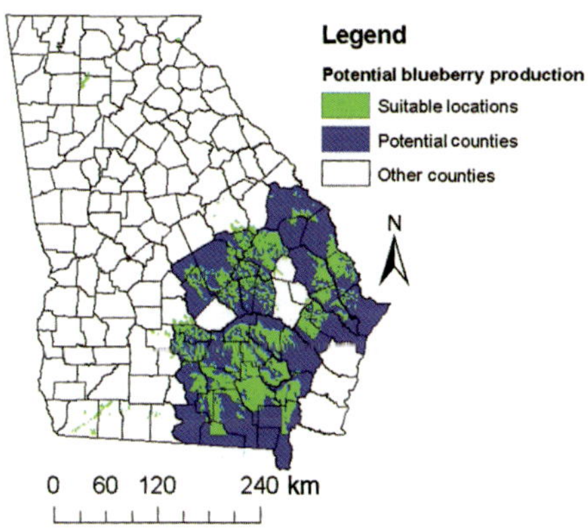

Figure 14.7. Blueberry production suitability map for Georgia.

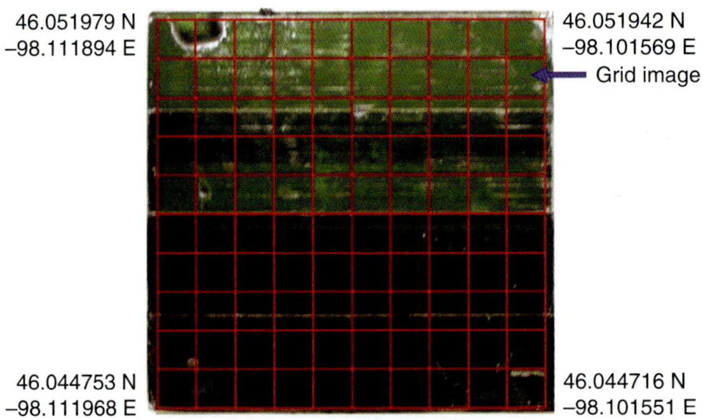

Figure 14.8. Aerial image of the OITA NW29 quarter in July 30, 1998, showing grid image pattern.

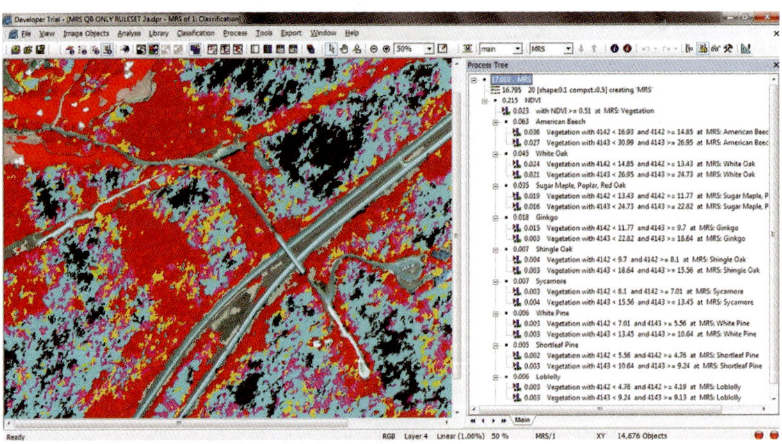

Figure 14.12. Advanced rule set (right) developed with OBIA software eCognition Developer 8 using MVSI imagery of the study area.

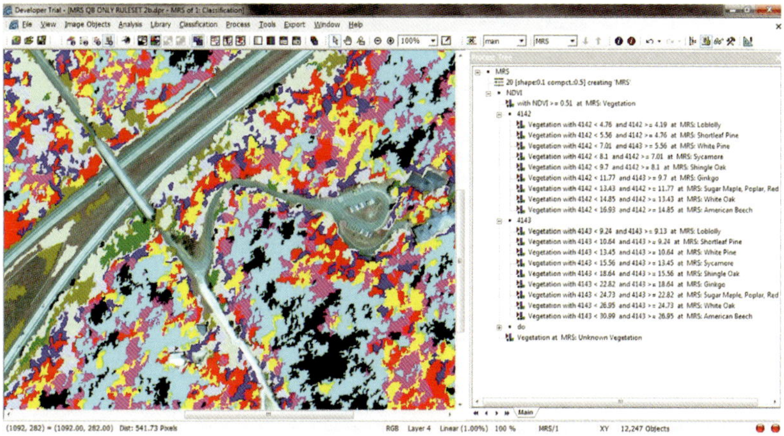

Figure 14.13. Final classified imagery of the study area to differentiate individual plant species.

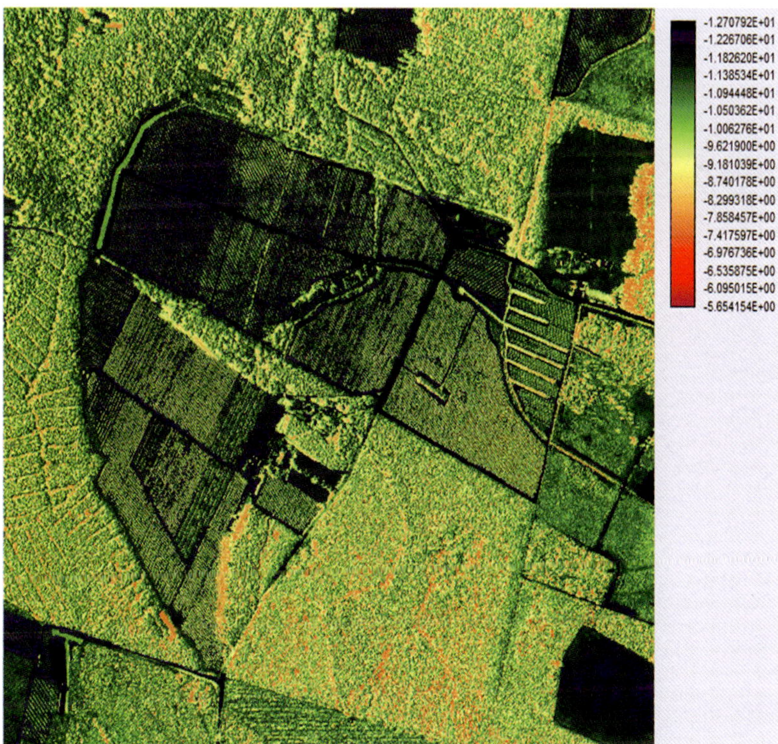

Figure 14.14. LAI raster spatial map of a blueberry orchard in Southeast Georgia developed using geospatial model.

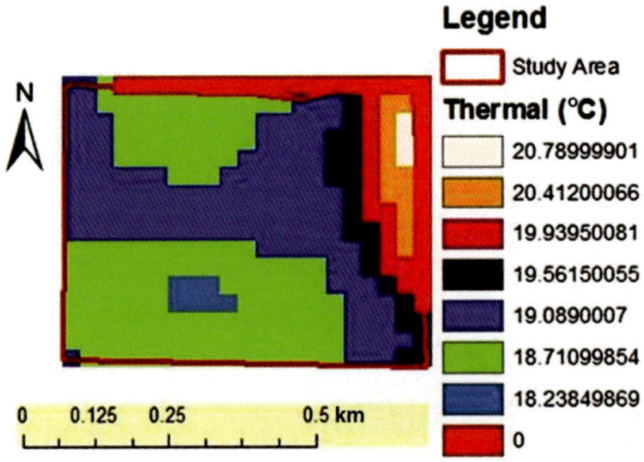

Figure 14.15. Thermal raster of showing the surface temperature variation of pine tree plantation in a study plot at Carteret, NC.

Chapter 3

Sugarcane as an Energy Crop: Its Role in Biomass Economy

R. van Antwerpen[1], S.D. Berry[2], T. van Antwerpen[3], J. Smithers[4], S. Joshi[5], and M. van der Laan[6]

[1] *University of Free State, Bloemfontein, South Africa*
[2] *Becker Underwood, Gillitts, South Africa*
[3] *South African Sugarcane Research Institute, Mount Edgecombe, South Africa*
[4] *University of KwaZulu-Natal, Pietermaritzburg, South Africa*
[5] *South African Sugarcane Research Institute, Mount Edgecombe, South Africa*
[6] *University of Pretoria, Hatfield, Pretoria, South Africa*

Introduction

Sugarcane genus, *Saccharum,* consists of 10 species of which *Saccharum officinarum*, *Saccharum robustum* and *Saccharum spontaneum* are the most well known. Sugarcane is a perennial tropical grass that tillers at the base to produce multiple stems that range from 2–3 m in height each approximately 3–5 cm in diameter. Its composition varies depending upon the climate, soil type, irrigation, fertilizers, pests, diseases, varieties, and harvesting period (Meade and Chen, 1977). The air-dry "millable" cane stalk makes up approximately 75% of the entire plant by weight and contains 11–16% fiber, 12–16% soluble sugars, 2–3% nonsugars, and 63–73% water (Van Dillewijn, 1952). Although the average yield of "millable" cane is 60 tonnes/ha/year (FAO, 1992), this figure can vary from 30 to 180 tonnes/ha/year (Perez, 1997).

In 2010, sugarcane was produced on about 23.8 Mha, in more than 90 countries, with a worldwide harvest of 1.69 billion tonnes (FAOSTAT, 2012). Brazil is the largest producer of cane sugar in the world and the next five major producers, in decreasing amounts of production, are India, China, Thailand, Pakistan, and Mexico.

Sugarcane is a perennial crop that ratoons (regrows) after each harvest: the whole aboveground biomass is harvested each year (or every second year in some climates) and harvests may continue for a number of years. Yields decline with ratooning and, after some years, the land is ploughed and new sugarcane is planted (Hartemink, 2008). Growing crops for biofuel require an abundant supply of nutrients and water (Tarimo and Takamura, 1998); any shortage will result in reduced biomass (biomass here is defined as the total aboveground dry matter produced by the crop). The management of sugarcane for maximum biomass is different from managing the crop for maximum sucrose. To maximize sugarcane for sucrose production the crop needs a ripening period, which is less important when growing the crop purely for biomass. In reality, sugarcane growers could probably manage the crop to optimize for both biomass and sucrose. These management decisions will affect water and nutrient

Biofuel Crop Sustainability, First Edition. Edited by Bharat P. Singh.

application rates. The main drivers of dryland yields are rainfall amount and distribution. Two other important drivers for potential yield and quality are solar radiation and temperature (van Den Berg *et al.*, 2009).

The rising global demands for energy, cost volatility, dwindling reserves of fossil fuels, and the impacts of the use fossil fuels on greenhouse gas (GHG) emissions have all resulted in biomass being considered as an important alternative source of energy (Botha, 2009; Duku *et al.*, 2011; Norris, 2011). The environmental impacts of the growing demand for petroleum-derived fuels can be mitigated by the use of lignocellulosic biofuels (Eranki and Dale, 2011). Not only is it being realized that biomass presents a potential renewable source of energy opportunity that could provide an alternative to the use of fossil resources and reduce human dependence on fossil fuels, but also that biofuels will have a positive impact on many environmental issues, including the minimization of the production of GHGs (Botha, 2009). The high production of biomass per unit of water by sugarcane, of up to 550 kg ha^{-1} d^{-1}, makes it an ideal energy crop (Botha, 2009).

The International Energy Agency (IEA) defines biomass as any plant matter used directly as fuel or converted into fuels (e.g., charcoal) or electricity and/or heat. Included here are wood, vegetal waste (including wood waste and crops used for energy production), ethanol, animal materials, and other organic wastes. Bioenergy from biomass is already making a substantial contribution to supplying global energy demand, and can make an even larger contribution, providing GHG savings and other environmental benefits as well as contributing to energy security, improving trade balances, providing opportunities for social and economic development in rural communities, and helping with the management of wastes, so improving resource management (www.ieabioenergy.com).

Sugarcane is a high biomass crop; the sugar stored in the stalk and the lignocellulosic residue remaining after sugar extraction called bagasse can both be used for the production of biofuel or other bioproducts. An average sugarcane annual yield of approximately 110 tonnes ha^{-1} (including stalks, tops, and leaves) produces more than 35 tonnes of dry solid biomass, corresponding to an output of 110 barrels of oil (Macedo *et al.*, 2004). Sugarcane is one of the most efficient C_4 grasses in the world, with an estimated energy in:energy out (I/O) ratio of 1:8 when grown for 12 months under tropical conditions and processed for ethanol instead of sugar (Heichel, 1974; Muchow *et al.*, 1994; Macedo *et al.*, 2004). Under more temperate environments, where temperature and sunlight are limited, I/O ratios of 1:3 are easily obtainable with current sugarcane cultivars if ethanol production from both sugar and cellulosic biomass is the goal (Tew and Cobill, 2008).

Compared to other crops with biofuel potential, sugarcane provides both sugar and cellulosic (bagasse) material for bioethanol production. Sugarcane bagasse consists of 420 g kg^{-1} cellulose, 250 g kg^{-1} hemicellulose, 200 g kg^{-1} lignin, and 50 g kg^{-1} ash on a dry mass basis. The balance is presumably soluble compounds, mostly sugars (Kim and Day, 2011; Lingle and Thomson, 2012). Brazil is the leading producer of biofuel with 24.5 billion liters of ethanol produced from sugarcane in 2008 (Renewable Fuels Association, 2010). Currently, most sugarcane bagasse in South Africa is burned at sugarcane factories to provide energy and steam to power the milling equipment. It is predicted that by 2020, Brazil will be planting around 14 Mha of sugarcane, producing more than 1 billion tonnes of cane, 45 MT of sugar, and 65 billion liters of ethanol. Additionally, the electricity produced by burning bagasse should equal or surpass the hydropower electricity produced in Brazil (Jank, 2008).

While all the energy needed for the sugarcane processing to ethanol comes from the bagasse, energy has to be traded in by the distilleries, most of which comes from fossil-derived fuels. This is the reason why the energy balance (i.e., the ratio of the energy contained in a liter of ethanol to the energy used in the process of preparation originating in fossil fuels) is 8:1 for sugarcane and 1.3:1 for corn. It means that ethanol from sugarcane is solar energy converted into a liquid, while ethanol from corn uses energy from fossil fuel (mainly coal in the United States) converted into a liquid. The consequence is that the GHG emissions resulting from the sugarcane route are much more favorable than from grains (Goldenberg, 2011).

Environmental Requirements

Climate

Sugarcane originates from Papua New Guinea (Irvine, 1981) and it is best suited for production in tropical regions. The distribution of sugarcane is, however, much wider than in the tropics and is mostly grown between latitudes 33°N and 33°S (Irvine, 1981) with yield decreasing rapidly as sugarcane is moved above latitudes of 30° (Muchow *et al.*, 1998). Sugarcane is cultivated as far north as 36.7° (Spain) and as far south as 31° (South Africa). In the United States, sugarcane is cultivated from Florida to Texas (Irvine, 1981) and is exposed to annual frost in Louisiana (Viator *et al.*, 2005). Ranging from warm temperate dry to moist, through tropical very dry to wet forest life zones, sugarcane is reported to tolerate annual precipitation of 470–4290 mm (mean of 58 cases = 1670 mm), annual temperature of 16°–29.9 °C (mean of 58 cases = 23.7 °C), and pH of 4.3–8.4 (mean of 49 cases = 6.3) (Duke, 1978, 1979).

Sugarcane is grown up to an altitude of 1600 m, close to the equator. At higher altitudes the growth cycle is longer: for example, in the Andean zone of Colombia, cane grown above 1500 m is harvested at the age of 18 months or older. As cane moves closer to its latitudinal limits the maximum altitude at which it is found decreases, and most sugarcane grown beyond the tropics of Capricorn and Cancer is at an altitude of less than 500 m.

In South Africa, sugarcane is grown from sea level on the east coast to altitudes of 1200 m, 70 km inland. Above altitudes of 300 m temperature becomes limiting and sugarcane is harvested at ages that gradually increase from 1 to 2 years at altitudes greater than 500 m. At altitudes above 500 m frost increases in severity and limit yields significantly.

Plants rely on the accumulation of heat (temperature) for their development and the rate of accumulation is season related. To calculate the accumulation of temperature (or thermal time) the mean daily temperature is determined (maximum plus minimum temperature per day divided by 2), from which a base temperature is subtracted. The calculation is started at a meaningful point (i.e., at plant) and accumulated. The base temperature is the temperature below which no growth and development will take place and is specific to plant species. Being a tropical crop, sugarcane has a relatively high base temperature. Base temperatures differ between various parts of the plant. For sugarcane, the following base temperatures have been published: germination—16 °C (Zhou *et al.*, 2000), root development—10 °C (van Antwerpen, 1998), stalk population and growth—16 °C (Inman-Bamber, 1994), leaf appearance or development—10 °C (Inman-Bamber, 1994), and canopy cover—16 °C (Singles and Donaldson, 2000). Base temperatures, however, may differ among cultivars. Sugarcane cultivars NCo376 and N25 have a base temperature value of 16 °C for canopy cover whereas the value for cultivar N26 is 17 °C (Singles and Donaldson, 2000).

Plants require specific amount of thermal time (TT) to reach a certain physiological stage. In Table 3.1 the thermal time to reach 50% shoot emergence was calculated for four cultivars from two trials and two base temperatures, that is, 10° and 16 °C. Cultivar N14 had the fastest

Table 3.1. Thermal time (°C day) to 50% shoot emergence of cultivars ZN6, ZN7, N14 and NCo376, using base temperatures of 10 °C and 16 °C for two experiments (Zhou *et al.*, 2000).

	Base temperature 10 °C		Base temperature 16 °C	
Cultivar	Experiment 3	Experiment 4	Experiment 3	Experiment 4
ZN6	654.1	378.7	262.4	228.7
ZN7	566.8	351	219.1	213
N14	531.3	317.2	199.6	191.2
NCo376	586.2	378.7	228.4	228.7

rate of emergence in both experiments, and cultivar ZN6 was the slowest. The difference in thermal time between the two experiments was least when using a base temperature of 16 °C, suggesting that this was a more likely base temperature for shoot emergence (Zhou *et al.*, 2000).

Soil

Sugarcane prefers a pH around 6.5 but the plant can be grown in soils with pH ranging from 5 to 8.5. It is relatively tolerant of soils with poor properties and has no requirement for a specific soil type (Tammisola, 2010). As a result of its versatility, sugarcane can be grown successfully under a wide range of soil conditions (Cornland *et al.*, 2001).

According to Tarimo and Takamura (1998), sugarcane is a heavy feeder crop and thus requires soils to have optimum properties for both physical and chemical characteristics. It was suggested that soil physical properties such as porosity, depth, bulk density, permeability, and water retention are more critical than the chemical characteristics to successful sugarcane production (Cornland *et al.*, 2001). This is due to the ease with which chemical properties can be modified by adding fertilizers or nutrients to the soil, whereas physical variables such as porosity are intrinsic and thus, cannot be modified (Cornland *et al.*, 2001).

For successful sugarcane cultivation, soils should be well aerated and granular in structure. This would ensure that the soil would have efficient water storage (approximately 150 mm m^{-1}) and drainage characteristics, thereby avoiding viruses, diseases, and bacterial infections associated with poorly drained and waterlogged soils (Schulze *et al.*, 1997).

Important soil physical properties include depth, bulk density, porosity, permeability, and water retention.

- *Depth:* Sugarcane grows well on deep, fertile soils, with the optimal depth for cane cultivation being in excess of 1 m (Schulze *et al.*, 1997). To ensure sufficient capacity to store water and nutrients and room for root development, soil depth needs to be at least 1 m for dryland production and 0.5 m when irrigated.
- *Bulk density:* Excessive dense soils (high bulk density) will restrict root development, water infiltration and redistribution, drainage of excess water and gas exchange (which is critical to ensure healthy life in soils) (Archer and Smith, 1972). The optimum soil bulk density value depends on a range of soil factors of which texture and soil organic matter content is probably the most important. Therefore any simple relationship with bulk density should be seen as merely a guideline (Table 3.2).

Table 3.2. Approximate soil bulk density for each soil texture class that favors plant growth and above which root growth is restricted (USDA, 2008).

Soil texture	Ideal bulk densities for plant growth (g cm^{-3})	Bulk densities that restrict root growth (g cm^{-3})
Sandy	<1.6	>1.8
Silty	<1.4	>1.65
Clayey	<1.1	>1.47

- *Porosity:* Porosity of soils is a function of texture and bulk density (Table 3.3). Ideally about 10% pores should be filled with air when soil water content is at field capacity in order not to restrict plant growth (Vomocil and Flocker, 1961).
- *Permeability:* Water tables above 0.8–1 m below the soil surface reduce production. Prolonged waterlogged conditions will also reduce yields significantly, and it is therefore essential to ensure that soils used for agricultural production is adequately drained. Soil water and air permeability is markedly reduced by high soil bulk densities and shallow impermeable layers.
- *Water retention:* Water storage capacity of soils is predominantly affected by soil texture, bulk density, and soil depth (Archer and Smith, 1972). Sandier soils have a lower capacity to store water per unit depth compared to clayey soils and this relationship is illustrated in Table 3.4.

Conditions for Optimal Growth

Nutrient

In environments that are conducive to the growth and development of sugarcane, large quantities of nutrients are taken from the soil to support maximum biomass growth (e.g., >200 tonnes sugarcane/ha; Tammisola, 2010). Meyer (2011) reported typical extraction rates of macro- and micronutrients for several countries for which the mean is given in Table 3.5. Numerous factors, such as genotype, environment, soil type, and water availability are responsible for the wide range of nutrients found in leaves. The nutrient content of millable stalks and leaves are comparable (Table 3.5) and are far more than that found in roots (Vered and Rao, 2012).

Table 3.3. Optimum bulk density at 10% air capacity, and maximum bulk density and total pore space at zero air capacity (Archer and Smith, 1972).

Soil	Clay (%)	Silt (%)	SOM (% mass)	Optimum[a] bulk density (g m^{-3})	Maximum[b] bulk density (g m^{-3})	Total[c] porosity (%)
Loamy sand	5.1	7.2	1.7	1.75	1.92	28
Sandy loam	15.2	14.7	2.4	1.50	1.70	36
Silt loam	18.8	17.7	2.7	1.40	1.56	41
Clay loam	35	22.1	5	1.20	1.32	50

[a]Optimum bulk density at 5 kPa with 10% of pores filled with air.
[b]Maximum bulk density at 5 kPa with zero air in pores.
[c]Total porosity at maximum bulk density.

Table 3.4. Available water capacity (AWC) ranges based on clay content (SASEX, 1999).

Clay content (%)	AWC range (mm m^{-1})
<7	<80
7–15	81–100
16–35	101–140
36–55	141–180
>55	<180

In order to grow optimally and produce a yield of 100 tonnes ha^{-1}, sugarcane needs to be fertilized with approximately 100–200 kg N/ha, 125–160 kg K/ha, and 20–90 kg P/ha (Tammisola, 2010). The variation may be attributed to differences in sampling techniques, sampling period, inherent soil conditions, fertilizer applications, and analytical methods (Hartemink, 2008).

Moberly (1971) evaluated four rates of N (N1: no nitrogen, N2: 112 kg, N3: 336 kg, and N4: 672 kg N/ha.) on irrigated sugarcane growing in a winter and summer cycle and harvested at the age of 12 and 18 months and concluded that

- Crops that had a winter start (May to May and May to November) had a greater need for higher nitrogen levels (>N3) in order to obtain maximum production.
- The 18-month-old crops benefit more from high levels of nitrogen (>N3) than the younger crops of 12 months of age.
- Water-use efficiency increased with increasing levels of applied nitrogen. The two highest levels of nitrogen (N3 and N4) caused soil water to be extracted at greater depths down the profile than the N1 and N2 levels.

Table 3.5. Nutrient removal rates by sugarcane (adapted from Meyer, 2011), optimum leaf tissue range (Anderson and Bowen, 1990) and quantity of macro- and micronutrients in the roots, stalks, and leaves of plant cane (Vered and Rao, 2012).

			Roots	Millable stalks	Leaves	Total
Macronutrients	Removal kg/100 tonnes	Optimum leaf range (%)	kg ha^{-1}			
N	116	1.80–2.40	8	83	77	168
P	19	0.19–0.25	1	15	8	24
K	200	1.05–1.45	4	109	105	218
Ca	31	0.15–0.25	2	30	45	77
Mg	36	0.08–0.18	1	29	18	48
S	31	0.12–0.20	2	25	22	49
Si	–	0.75–1.50	–	98	150	248
Micronutrients	kg/100 tonnes	mg kg^{-1}	g ha^{-1}			
Fe	5.450	75–200	4900	3800	7900	16,600
Mn	2.133	15–75	84	1170	1981	3235
Zn	0.398	13–20	72	437	336	845
Cu	0.108	3–7	13	201	105	711
B	0.160	2–7	34	214	144	392
Mo	0.001	0.05–0.15	–	4	10	14

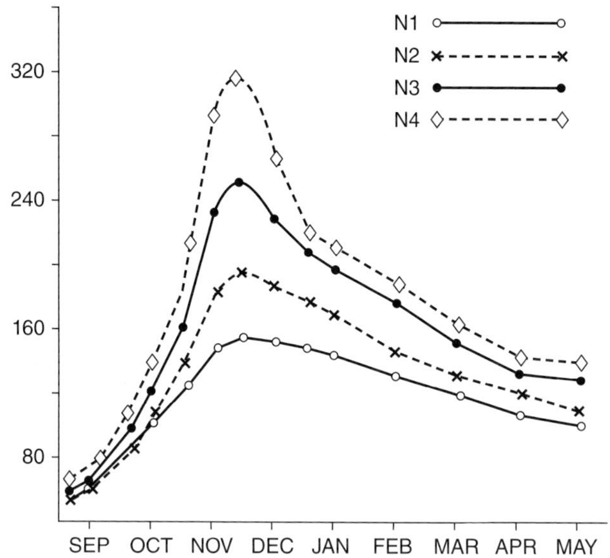

Figure 3.1. The effect of nitrogen levels on the stalk population of sugarcane harvested in winter at the age of 12 months (Moberly, 1971).

- Sugarcane yield increased with an increase in nitrogen application and this was also reflected in a marked increase in stalk population (Figure 3.1), an increase in stalk height, and consequently an increase in the individual stalk weight.

These results place a question mark as to the sensibility of applying the full N compliment to a crop that germinates over the winter period. Meyer (personal communication) recommended that a quarter to one-third of the N compliment be applied to these crops after harvest and the remainder in spring. This, of course, will result in reduced N volatilization losses and more efficient use of N which is probably more sustainable in the long run.

The uptake of nutrients is not equally spread over the cropping cycle of sugarcane. Vered and Rao (2012) published a generalized trend of N, P, and K uptake patterns during the various growth stages of a sugarcane plant. From this (Figure 3.2) it is evident that the highest rate of N and P uptake is during the tillering phase (which commences about 30–45 days after planting), while a substantial amount is also taken up during the grand growth period. Potassium continues to be taken up in larger quantities, which reaches a peak during the grand growth stage. However, because N uptake is improved in the presence of K, K needs to be applied at the same time as N.

Water

The sugarcane plant requires little water in the germination period (the soil does however need to be moist). Water requirement increases throughout the rapid growth phase until the crop reaches full ground cover and thereafter stays more or less constant. Lack of water especially during the early growth period will result in reduced cane tonnage. During the final 2 months before harvest the crop should not be irrigated in order to imply water stress (drying off) which

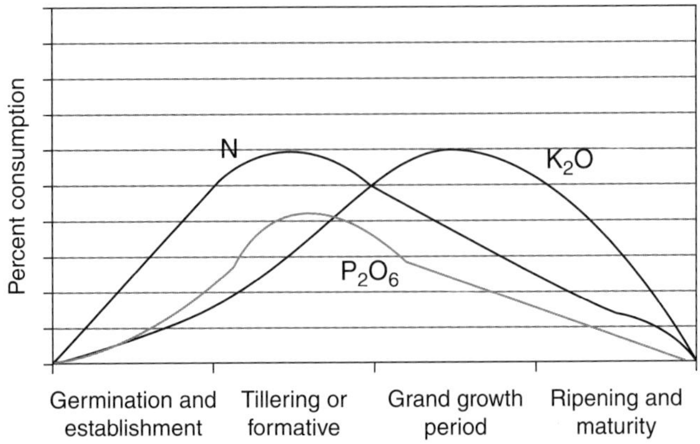

Figure 3.2. Variable nutrient uptake for different growth phases of sugarcane (Vered and Rao, 2012).

will cause crop growth to slow down, stalks to ripen, and thereby to accumulate maximum sucrose.

Sugarcane has the potential to produce large amounts of biomass (>200 tonnes ha^{-1}) (Tammisola, 2010) and to do this it requires 1500–2500 mm water (Inman-Bamber and Smith, 2005) spread throughout the growing period. If the mean annual rainfall is less than 800 mm, it is likely that the crop's water requirements will not be met and irrigation will become essential. Typically, sugarcane requires a maximum of 6 mm water per day in January (summer) to a minimum of 2.3 mm per day in June (winter) (Thompson *et al.*, 1963). In irrigated regions of Southern Africa, water is applied to supplement the annual rainfall, which could be as low as 500 mm/annum, to a total of approximately 1500 mm/annum.

Sugarcane is cultivated under both rainfed and irrigated conditions and yield is directly proportional to the amount of water used in the prevailing climatic conditions. About 37–330 kg of water is used to produce 1 kg of sugarcane and 1000–2000 kg water is required to produce 1 kg of sucrose (Inman-Bamber and Smith, 2005).

Some plant growth and physiological processes are affected by drought more than others. For example, plant extension rate is more sensitive to water stress than photosynthesis (Inman-Bamber, 1986). Plant extension ceases at leaf water potentials of between −0.4 to −0.9 MPa (indicating that water is becoming limited), while stomatal resistance increases rapidly only below water potentials of −1.2 MPa and peak resistance is encountered between −1.4 and −2.3 MPa. Inman-Bamber (1986) found that water stress promoted ripening when stress was sufficient to reduce plant extension but not to substantially decrease photosynthesis. Singels *et al.* (2000) observed that the fraction partitioned to sucrose (instead of biomass) increased when relative soil water content (RSWC = the fraction between field capacity and permanent wilting point) dropped below 55%. Biomass accumulation was reduced when RSWC dropped below 35% of available capacity. Partitioning between stalk and nonstalk components are not affected by water stress (Singels *et al.*, 2000).

Another form of water stress is that inflicted by salinity. Salinity is in general only found in irrigated areas and is characterized by an accumulation of cations (predominantly K, Ca, Mg, and Na). Increasing levels of salinity makes it more difficult for the crop to extract water from the soil leading to increasing levels of water stress and therefore reduced biomass production

Table 3.6. Effect of soil salinity on biomass production (SASEX, 1999).

Salinity class	EC_e (dS m^{-1})	Salinity effects on crops
Nonsaline	<2	Salinity effects are negligible
Slightly saline	2–4	Yields of very sensitive crops may be restricted
Moderately saline	4–8	Yields of many crops restricted
Very saline	8–16	Only tolerant crops' yield satisfactory
Extremely saline	>16	Only a few very tolerant crops' yield satisfactorily

(Table 3.6). Excessive levels of Na (sodicity) will lead to Na-toxicity, which might cause sugarcane stools to die in severe cases.

The water requirements of biomass from which energy is to be derived are about 70–400 times more than that of other energy carriers such as fossil fuels, wind, and solar energy (Gerbens-Leenes *et al.*, 2008). Gerbens-Leenes *et al.* (2008) estimated the average water requirements for fossil energy at about 1 cubic meter evapotranspiration (ET) per gigajoule of energy (m^3 ET GJ^{-1}) compared to 24–146 m^3 ET GJ^{-1} reported for bioenergy. Biodiesel and ethanol derived from conventional food crops typically require more water than biofuels based on lignocellulosic crops, and the range is wide for alternative crops (Table 3.7) (De Fraiture and Berndes, 2009).

Estimation of the land and water footprints of biofuel was conducted for a number of conventional feedstock crops in China (Table 3.8) by Yang *et al.* (2009). Among the bioethanol crops, sugarcane had a relatively low water and land footprints in comparison with other feedstocks. In contrast, maize and sugar beet had very high land footprints. Cassava had the second lowest land footprint among the crops considered and the highest water footprint amongst the bioethanol crops. For biodiesel, rapeseed, and soybean were the dominant feedstock. Overall, land and water footprints of biofuel made from rapeseed and soybean are significantly higher than those from feedstocks used for ethanol fuel (Yang *et al.*, 2009).

Table 3.7. Water requirements of different types of biofuel crops per unit of energy produced (De Fraiture and Berndes, 2009).

		m^3 ET GJ^{-1} Berndes (2002)[a]		m^3 ET GJ^{-1} WWF (2006)[b]	
Biofuel crop	Fuel	Low	High	Low	High
Rapeseed	Biodiesel	100	175	–	–
Oil palm	Biodiesel	–	–	46	250
Soybean	Biodiesel	–	–	143	500
Sugarcane	Ethanol	37	155	18	35
Sugar beat	Ethanol	71	18	48	76
Corn	Ethanol	73	346	100	323
Wheat	Ethanol	40	351	143	500
Sweet sorghum	Ethanol	–	–	56	233
Lignocellulosic crops	Ethanol	11	171	–	–
Lignocellulosic crops	Methanol	10	137	–	–
Lignocellulosic crops	Hydrogen	10	124	–	–
Lignocellulosic crops	Electricity	13	195	–	–

[a]Estimates by Berndes (2002) include liquid fuel, heat, and power; the lower range numbers include systems that deliver both biofuels for transport and heat/electricity.
[b]The WWF (2006) estimates the energy content per crop without specifying the energy carrier.

Table 3.8. Average water and land footprint of biofuel produced with different feedstocks (Yang *et al.*, 2009).

Biofuel type	Feedstock	Feedstock biofuel conversion ratio (kg/kg) (tonne/tonne)	Crop yield (kg ha^{-1})	CWR (m^3 kg^{-1})	Water footprint of biofuel (m^3 l^{-1})	Land footprint of biofuel (m^2 l^{-1})
Bioethanol	Maize	3	5001	0.84	2.01	4.75
Bioethanol	Cassava	6	16,226	0.55	2.64	2.93
Bioethanol	Sugarcane	15	62,563	0.12	1.47	1.9
Bioethanol	Sugar beets	14	20,196	0.2	2.24	5.49
Bioethanol	Sweet potato	10	20,968	0.23	1.83	3.78
Biodiesel	Rapeseeds	3.3	1836	2.02	5.82	15.67
Biodiesel	Soybean	5.6	1720	3.2	15.63	28.4

Sugarcane Disease and Pest Control

Sugarcane is increasingly becoming one of the world's most popular biofuels. With the expansion in the area grown with sugarcane, the growing threat of new pests and diseases can cause significant declines in sugarcane yield (sucrose as well as biomass). Sugarcane pests and diseases are already causing significant yield losses in many sugar industries worldwide. The diseases are caused by fungi, bacteria, viruses, and phytoplasmas. Common sugarcane pests in the South African sugar industry include thrips, white grubs, stalk borers, and nematodes.

Diseases and Pests

The most common fungal diseases that occur in sugarcane industries is smut (caused by *Ustilago scitaminea*), brown rust (caused by *Puccinia melanocephala*), and orange rust (caused by *Puccinia kuehnii*). Important viral diseases in sugarcane are sugarcane mosaic virus, Fiji disease virus, and sugarcane yellow leaf virus. Diseases caused by phytoplasmas include grassy shoot, green grassy shoot, white leaf, and Ramu stunt. Leaf scald (caused by *Xanthomonas albilineans*), ratoon stunt (caused by *Leifsonia xyli* subsp. *xyli*), and gumming (caused by *Xanthomonas axonopodis* pv. *vasculorum*) are some of the most common diseases caused by bacteria.

Sugarcane is prone to infection and damage from pathogenic organisms because of several characteristics of the crop which include the fact that it is vegetatively propagated. This means that stalks of seedcane are planted and not true seed as is the case in most other crops. Most of the important diseases such as smut, ratoon stunt, leaf scald, and sugarcane mosaic virus disease are systemic diseases, and the planting of infected seedcane can lead to infection in the plant crop. Sugarcane is grown as a perennial, so once it is planted a field is ratooned repeatedly. This provides an ideal opportunity for diseases to build up. Sugarcane is also grown as a monoculture in large areas of many sugarcane industries. This situation reduces the local biodiversity and natural pest and pathogen species, and substantially increases the potential of disease outbreaks. This situation favors the spread of diseases and the development of epidemics (Stewart and Cromey, 2011).

Worldwide, there are about 1500 different insect species which occur above ground as well as below ground in sugarcane fields (Goebel and Sallam, 2011). The vast majority of these pests have restricted geographical distributions (Goebel and Sallam, 2011). Important aboveground pest species are found in the orders and families of the Lepidoptera (Pyralidae, Noctuidae, and Castniidae), Homoptera (Aphididae, Cercopidae, Coccidae, Delphacidae, Diaspididae, and Pseudococcidae), and Orthoptera (Acridoidea) (Meagher, 2012). Belowground pests include Coleoptera (Curculionidae, Elateridae, and Scarabaeidae), Isoptera (Mastotermitidae, Rhinotermitidae, and Termitidae), Hymenoptera (Formicidae), Diptera (Stratiomyidae), Heteroptera (Cicadidae, Cydnidae, Margarodidae, and Pseudococcidae), and Orthoptera (Gryllidae and Gryllotalpidae) (Meagher, 2012). Stemborers (Pyralidae and Noctuidae) are pests in all sugarcane growing regions of the world (Meagher, 2012). *Eldana saccharina* Walker (Lepidoptera: Pyralidae) is a serious pest of sugarcane in various African countries.

Disease Control

An integrated system of disease control is being followed in the South African sugar industry because no single measure is the complete answer for the control of major sugarcane diseases. Disease-resistant varieties, disease-free seedcane, eradication of volunteers in a field, hot water treatment, and roguing are some of the most important parts of an integrated disease control system.

Varietal Resistance

The high cost of frequent applications of expensive pesticides generally precludes the use of chemicals for control of diseases in field crops. In the case of sugarcane, the sugarcane canopy would also hinder the effective spray cover of fungicides and bactericides. The most important way of controlling diseases in sugarcane in the long term is to reduce host susceptibility by planting resistant varieties. Each year, the South African Sugarcane Research Institute (SASRI) Plant Breeding programme produces one or two new sugarcane varieties. During the development of new sugarcane varieties, frequent inspections at the various selection stages, together with screening trials against major diseases such as leaf scald disease, smut, sugarcane mosaic virus, and rust, susceptible seedlings are eliminated. This ensures that varieties released to growers have an acceptable degree of resistance to most diseases. In addition, varieties are selected for the major agroclimatic regions in the sugarcane industry so that released varieties are directly suited to the range of conditions that are experienced in the South African sugar industry (Baker, 2012). A good disease control measure is to replace susceptible varieties with resistant or tolerant varieties. However, the displacement of susceptible varieties with resistant varieties is a slow process because only 10% of fields in South Africa are replanted annually.

Seedcane Quality

Because most of the important diseases such as ratoon stunt, smut, mosaic, and leaf scald are systemic (the pathogen is present in every part of the plant), they are present in the cane stalk and thus can be spread by planting infected seedcane. The planting of healthy seedcane is therefore essential for general disease control. As a first step toward achieving effective control of pests and diseases, nurseries of hot water treated, disease-free seedcane plays a vital role in the control of diseases and should be encouraged on each farm in the sugar industry (Stranack, 2012). Hot water treatment of sugarcane eradicates the bacterium *L. xyli* subsp. *xyli* from seedcane and increases the germination capacity of the seedcane. Reliable sources of clean seedcane can become a problem in future with expanded cultivation. The development of micropropagation, a methodology through which plants are multiplied rapidly by aseptic culturing meristem tip regions, can play an increasingly important role in disease control because it will provide healthy seed material for planting new varieties. Novacane® is a technique developed at SASRI for the rapid production of sugarcane plants via tissue culture. Certified disease-free stalks are used to produce thousands of plantlets from a single sugarcane meristem (Snyman *et al.*, 2008). The advantages of Novacane® micropropagated plantlets are that it can help in the rapid, disease-free, seed multiplication of newly released varieties.

Field Control Practices

Healthy seedcane must be planted into fields that are free from diseased volunteer regrowth. Seedcane fields must be inspected and diseased plants rogued to keep disease levels in a field low. The eradication of severely infected diseased fields can also contribute to reduce the amount of infective material. Sugarcane smut, as well as mosaic disease, is mostly controlled by destroying contaminated plants in the field (rouging).

Fungicides

Fungicides are not used readily in sugarcane fields. Pineapple sett rot is a disease in sugarcane and can be controlled by means of soaking sugarcane setts in fungicides such as Benlate, with the active ingredient Benomyl, and Eria (difenoconazole and carbendazim) in the South African sugarcane industry. According to Cheavegatti-Gianotto *et al.* (2011), the Brazilian sugarcane industry does not usually control sugarcane diseases in commercial fields, but recently, orange rust (caused by *P. kuehnii*) was detected in the country and as a result of this, the Brazilian Agriculture Department has registered products (azoxystrobin and ketoconazole) to control fungal diseases in sugarcane fields (Cheavegatti-Gianotto *et al.*, 2011). In South Africa, the fungicide Bayleton, with the active ingredient Triadimefon, is used to dip sugarcane setts in before planting, while in Brazil, the Triazole fungicides Triadimefon and Triadimenol are registered to treat sugarcane stalks before planting to prevent smut contamination caused by *U. scitaminea* (Cheavegatti-Gianotto *et al.*, 2011).

Pest Control

Global travel, the expansion of sugarcane regions and change in climatic conditions increase the risk of new pest outbreaks in many countries (Goebel and Sallam, 2011). Biosecurity and pest management plays an important role in the prevention of such outbreaks. In the South African sugarcane industry, the *Eldana* stalk borer is the most important pest and a strategy of integrated pest management has been adopted to combat its impact in the industry. The use of resistant varieties has been an important management tool in most sugarcane-growing regions around the world against stem boring pyralids (Mathes and Charpentier, 1969). In Louisiana, plant resistance has been a component of the sugarcane IPM program against sugarcane borer for many years and has been a successful management strategy when used alone or in combination with other strategies (Bessin *et al.*, 1990).

Nematode Control

In addition to herbivorous insects, parasitic nematodes may also grow into a severe problem in bioenergy crops as indicated by recent studies (Mekete *et al.*, 2009). In our research, current estimates of nematode yield loss to commercial sugarcane equates to ~7.6% of annual production and equals to ~1.6 MT cane per annum (Spaull and Cadet, 2003). Results from a variety of nematicide trial at Zinkwazi on the KwaZulu-Natal north coast showed that nematodes caused significantly much more loss than that what was currently estimated. Over four crops (plant crop + 3 ratoons), the average yield of the untreated N12 cultivar was 66 tc ha^{-1} compared to the recommended Temik® treatment (79 tc ha^{-1}, +20% increase) and the monthly Temik treatment (96 tc ha^{-1}, +45% increase) (Figure 3.3). For the more nematode-susceptible variety, N27, the yield losses were even greater. The average yield of the untreated N27 was 37 tc ha^{-1} compared to the recommended Temik treatment (50 tc ha^{-1}, +35% increase) and the monthly Temik treatment (86 tc ha^{-1}, +132% increase) (Figure 3.3).

As reported by Berry *et al.* (2004), monthly treatment with Temik kept the numbers of endoparasitic nematodes, particularly *Meloidogyne* and *Pratylenchus*, low throughout the plant and first ratoon crops. Six months after planting, the roots of N12 with multiple Temik application had only 13 endoparasitic nematodes per gram of dry roots and the roots of N27 with multiple Temik application had no endoparasitic nematodes,compared to the untreated plots of N12 and N27 which had 449 nematodes/gram dry weight and 2044 nematodes/gram dry weight, respectively. Removing most of the plant parasitic nematodes from the soil enabled us to investigate the full potential of these sugarcane varieties. The currently recommended rate of nematicide application (single application of Temik at 20 kg ha^{-1}) is based on economical response of sugarcane. At this rate, however, the nematicide does not completely eradicate the nematodes associated with sugarcane. Economically, it may be too expensive to treat every month for 12 months per cropping cycle. However, when growing cane for biomass, the economics may be very different and this type of monthly application may be possible. Work would be needed to investigate the most beneficial rates and times of application.

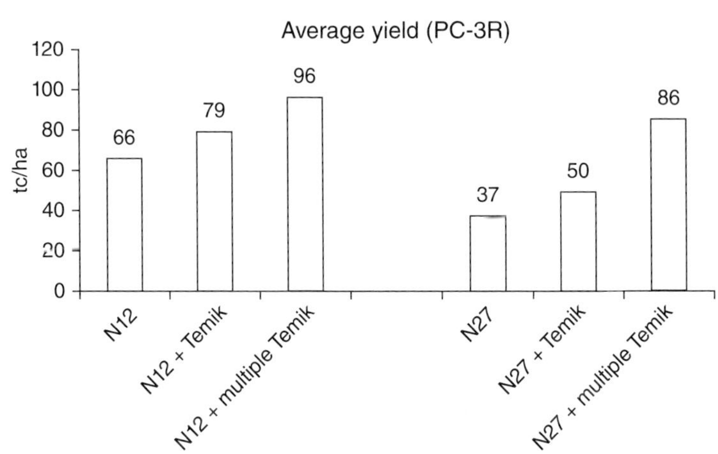

Figure 3.3. Average cane yield (tc ha^{-1}) for two varieties, N12 and N27, with single (recommended) application of Temik (at 20 kg ha^{-1}) and multiple (monthly) application of Temik (at 20 kg ha^{-1} mo^{-1}).

New Varieties, New Diseases and Pests, and their Management

New sugarcane varieties that will be propagated as biofuel crops will be exposed to all the current sugarcane disease threats that are found in all sugarcane industries worldwide. New diseases will also be a constant part of the sugarcane industries worldwide. According to Stewart and Cromey (2011), it is crucial that economic and life-cycle analyses of biofuel crops must include accurate assessments of likely disease risks. It is also important that priority diseases are identified and proactive management strategies are developed. According to Stewart and Cromey (2012), to be proactive is to be prepared, and to be prepared is to anticipate the certain disease issues that will almost certainly threaten biofuel crops. A study on viral pathogens that might become a problem in sugarcane in South Africa and the development of diagnostic tools for the detection of putative pathogens was done by van Antwerpen and Rutherford in 2008. Pathogen detection and identification are important, especially in quarantine procedures to prevent the introduction of exotic diseases into the South African sugar industry. The Potyviridae (transmitted by aphids and which include *sugarcane mosaic virus*, the causal agent of sugarcane mosaic) and the Geminiviridae (which include *sugarcane streak virus* (SSV) and *maize streak virus* (MSV)) are important pathogenic viral families. The Caulimoviridae, which include badnaviruses such as *sugarcane bacilliform virus* and *banana streak virus* (causing streak in bananas) can also become a threat to the sugar industry. The Closteroviridae cause diseases such as beet yellows (caused by *beet yellows virus*) and wheat yellow leaf (caused by *wheat yellow leaf virus*) that may also be a potential threat to sugarcane (van Antwerpen and Rutherford, 2008).

An unknown rust was observed in sugarcane in Swaziland and in the northern areas of South Africa (Mpumalanga, Umfolozi, Zululand as well as in the Midlands North area) in 2009 (Martin *et al.*, 2012). The rust has tentatively been identified as African sugarcane rust, which is caused by the fungus *Puccinia sparganioides* (Martin *et al.*, 2012). In South Africa, the new rust is sprayed with the fungicide Abacus® with the active ingredient pyraclostrobin, on 2–4-month-old sugarcane plantlets.

Orange rust (caused by the fungus *P. kuehnii*) has been found for the first time in Africa in Gabon and the Democratic Republic of the Congo in 2011 (McFarlane, 2011). The disease was considered to be of minor importance in the Australian sugar industry until variety Q124, a previously resistant variety, became severely infected in 2000 (McFarlane, 2011). Orange rust has not been reported in South Africa; however, the spores are spread mainly by the wind and can travel considerable distances in this way (McFarlane, 2011). Orange rust can cause serious disease losses in varieties in South Africa. Pathogens that cause diseases in crops like maize and sorghum might adapt to infect sugarcane, for example, MSV disease that was found in the South African sugar industry in 2007 (van Antwerpen *et al.*, 2008b).

Biosecurity strategies and disease prevention plans will have to be adopted. Biosecurity is commonly viewed as keeping new pests and diseases out of a country. However, this view takes the focus off the very serious problems that already exists in the South African sugar industry (Stranack, 2012). An example of such a problem is the rapid spread of the Eldana borer into the high altitude areas of the industry. Currently, more focus is put on the planting of new varieties that are resistant to the Eldana borer in these regions. However, most varieties that are resistant to Eldana tend to be susceptible to smut (Rutherford, personal communication). In the case of increasing temperatures due to global warming in the future, varieties that are resistant to smut will have to take priority over varieties that are resistant to Eldana. The reason for this is that when temperatures rise in the high-lying areas, the growing season will be shorter (from a 24-month cycle to an 18-month cycle). This will have an effect on the life cycle of Eldana. Smut resistant varieties need to be propagated in areas like this. Sugarcane mosaic virus resistance is often found in varieties that are smut resistant. In high-lying areas this will be the preferred choice of variety (Rutherford, personal communication). Ratoon-stunting disease as well as sugarcane mosaic virus disease is now also serious diseases in the irrigated northern areas of the South African sugar industry (Stranack, 2012). Cooperation across borders is vitally important to keep new threat at bay and to contain them at their source of origin (Stranack, 2012).

A new regional threat in the Southern parts of Africa is the spread of the spotted sugarcane stem borer *Chilo sacchariphagus* (Bojer) (Lepidoptera: Crambidae) (Rutherford and Conlong, 2010). These pests originated in Southeast Asia and are largely restricted to sugarcane as its host plant. *C. sacchariphagus* is currently present in Mozambique and is an eminent threat to the South African sugarcane industry (Rutherford and Conlong, 2010).

New threats of new diseases will emerge alongside existing diseases as biofuel crop monocultures become widespread (Stewart and Cromey, 2012). The disease management model currently employed for sugarcane production in countries like Australia and South Africa is to break the sugarcane monoculture in order to reduce populations of important pests and diseases of sugarcane. The use of biotechnology to improve disease resistance will be an important tool in the management and development of new disease-resistant sugarcane varieties. According to Rutherford and Conlong (2010), it is important to build resilience into the sugarcane agroecosystem and to refocus on control efforts into a biointensive area-wide integrated pest management approach. Such an approach will use conventional options with ecologically based new technologies such as delineation of within species populations, chemical ecology, push–pull strategies and sustainable IPM strategies (Rutherford and Conlong, 2010). According to Stewart and Cromey (2011), a proactive, multidisciplinary approach to manage diseases of biocrops are necessary. Traditional methods of control, as indicated above (resistant varieties, the breaking of the monoculture of sugarcane) will still form the backbone of disease management (Stewart and Cromey, 2011).

According to Stewart and Cromey (2011) there will be a significant shift in the use of old and new sugarcane varieties for bioenergy production over the next decade or two in most sugarcane industries in the world. With the increase of new varieties, the risk of sugarcane diseases will also increase substantially worldwide, which may cause substantial economic losses (Stewart and Cromey, 2011). One of the biggest challenges to be overcome in order to achieve sustainable production of bioenergy sugarcane varieties will be to convince growers to take a proactive approach to pest and disease management (Stewart and Cromey, 2011). It will be important to predict the priority disease threats in each part of the sugarcane-growing areas worldwide and to implement a range of crop management strategies to effectively manage these threats (Stewart and Cromey, 2011). The most important control strategy will be to use resistant varieties and thus the breeding for resistant varieties will be a high research priority. This can be supplemented by monitoring practices to detect diseases at an early stage and to have efficient biosafety incursion plans in place (Stewart and Cromey, 2011).

Weed Control

Weed interference (competition + allelopathy) with sugarcane growth may lead to yield losses of up to 50% (Pearson, 1961). McMahon *et al.* (2000) estimated the combined cost of control and production losses from weeds for the Australian sugar industry at US$70 million, far outweighing the impact of other pests and diseases. Weed control is particularly important during the early stages of crop growth before full canopy formation, as lack of radiation terminates weed growth below the canopy. Perennial and annual grasses are usually the most prolific weeds in sugarcane with the highest impact on yields. Sedges and broadleaf weeds are also important to control adequately.

A combination of mechanical and chemical control methods are commonly applied to control weeds in sugarcane cropping systems, with the latter method playing an ever-increasing role in integrated control programs as a result of shifts to reduced tillage systems. Deep soil cultivation is usually practiced before sugarcane re-planting to control perennial weeds. This is often followed by shallow soil cultivation to kill any aboveground vegetation, sometimes accompanied with the application of a broad-spectrum herbicide (Van Rijn, 2000). Several additional cultivations may then be required between planting/ratooning and cane re-growth.

Herbicides such as hexazinone, diuron, atrazine, alachlor, paraquat, and metribuzin are commonly used in sugarcane. Selection of herbicide(s) will depend on the weed spectrum, timing of application required (pre- or postemergence) and the length of control required. Glyphosate is commonly used to manage difficult to control weeds such as the grasses *Cynodon dactylon* and *Digitaria scalarum,* but evidence of lack of sufficient control has led to concerns of weed resistance developing as a result of the frequent and widespread use of this compound. The use of mixtures of herbicides is also becoming much more widespread in order to achieve better efficacy. Ideally, any small weed patches are spot-sprayed using a knapsack before they expand. As a result of inherent differences in sugarcane varieties their response to a specific herbicide is not alike; however, sugarcane is broadly categorized as tolerant or susceptible (McMahon *et al.*, 2000).

The risk of herbicide leaching from cultivated soils to water bodies can be reduced by carefully following product guidelines, not applying before heavy rainfall is expected, not over-irrigating and reducing erosion.

The amount of control required in trash systems is frequently reduced as the trash smothers weed growth. Since the adoption of green cane trash blanketing, vines have become more prolific, including cucumber (*Cucumis* spp.), morning glory (*Ipomoea* spp.), convolvulus (*Ipomoea* spp.), passion vine (*Passiflora* spp.), and star of Bethlehem (*Ipomoea quamoclit*) (McMahon *et al.*, 2000).

The use of intercropping is not uncommon in sugarcane to control weeds. Under this system, sugarcane is planted in wide rows so that it does not form a full canopy before 3 months thus allowing the early maturing companion (e.g., soybean, sweet potato, tomato) to achieve satisfactory yields (Van Rijn, 2000). Precision agriculture, allowing the use of a GPS system to map and target weed problems in the field using variable sprayer systems and species-specific herbicides, shows excellent potential in improving the effectiveness of chemical weed control.

Harvesting

The sugarcane industry has been moving from sugar production to sugar and energy production with the development of cogeneration and ethanol plants and this will result in the energy content of cane and energy yield per hectare and per tonne of cane to be important parameters which will be used for a variety of evaluations in the future and be part of future payment systems (Corcodel and Roussel, 2010). The agronomic benefits and the potential energy resource of trash have been recognized and the development of new harvesters has promoted the utilization of these benefits (Paes and de Oliveira, 2005). In a study to investigate diversification options for the sugar industry in Australia, whole crop harvesting to maximize electricity cogeneration was identified as the venture with the highest priority (Thorburn *et al.*, 2006a). The potential for the sugar industry in South Africa to mitigate against climate change by producing renewable cogenerated electricity and fuel ethanol was highlighted by Govender (2011).

Sugarcane bagasse is the most abundant crop residue produced globally and this resource can be increased by harvesting the sugarcane leaves and tops (Bocci *et al.*, 2009). With the improvements in sugarcane harvesting and cogeneration technology, bagasse and sugarcane agricultural residues (SCAR) have become important sources of bioenergy (Alonso-Pippo *et al.*, 2009b). The use of SCAR has the added benefit of not competing as a food source and has a similar energy content as bagasse per unit weight, but is frequently burnt off to facilitate harvesting of the stalks (Alonso-Pippo *et al.*, 2009b). SCAR or cane trash (leaf material separated from the cane stick at harvest), which is currently mainly burnt prior to harvest, is a significant source of energy (Schembri *et al.*, 2002; Seabra and Macedo, 2011) and sugarcane has the highest rate of energy per hectare (0.5–2 GJ ha^{-1}) (Bocci *et al.*, 2009). However, despite the potential to generate heat and power from bagasse, the potential for cogeneration of electricity remains largely unexploited (Bocci *et al.*, 2009).

Over 9 GW of worldwide generation capacity is currently provided by biomass-based electricity schemes. Sugarcane bagasse and sugarcane trash have the potential to provide a significant amount of biomass for electricity production, and the potential becomes much higher with more advanced cogeneration technologies (Mirza *et al.*, 2008). Sugarcane leaf biomass has a high calorific value which has a low production of micropollutants when processed with improved technology and thus represents a significant source of energy which can reduce environmental problems associated with energy production (Sartori and Florentino, 2007). It has been estimated in Brazil that energy from residual biomass can supply energy to 7 million people and that 5.5 million people can be supplied with energy from bagasse (Sartori and Florentino, 2007), and sugarcane biomass is seen as one of the most readily available energy sources in Brazil (Michelazzo and Braunbeck, 2008). However, high recovery costs will be incurred in gathering, baling, transportation, chopping, and technology to utilize the residues on a large scale (Michelazzo and Braunbeck, 2008). The current sugarcane transport infrastructure can be used to make the trash an economically viable energy source for boilers, irrigation power, and export to the local area (Norris, 2011).

Not burning sugarcane prior to harvest has environmental benefits and the residual biomass can help reduce the global energy deficit, but the feasibility of extracting the energy from the currently unutilized biomass has not been proven (Sartori and Florentino, 2007). Many countries are considering the use of sugarcane as a potential source of renewable energy. For example, in El Salvador, it is projected that 55 MW of power can be generated during the 4-month cane crushing season from 346 503 tonnes of sugarcane (WIIAD, 1994). In Mauritius, 10 out of 11 sugarcane processing factories export electricity to the grid during the milling season. This generation of electricity was promoted by government incentives which focused on the optimum use of bagasse (fibrous fraction of cane) to generate electricity and to investigate the use of other fractions of the sugarcane biomass (cane tops, leaves, and dry trash) for electricity generation (Deepchand, 2005). Efficient energy production from biomass has the potential to offer new development paths, sustain rural livelihoods, reduce dependence on imported energy sources, and reduce GHG emissions in the sub-Saharan Africa region, particularly in Malawi, Mozambique, Tanzania, and Zambia which have the greatest potential to achieve this after accounting for food production and resource constraints (Johnson *et al.*, 2007). An analysis by Johnson *et al.* (2007) showed that in South Africa, 1.2% of the energy provided by petrol could be supplied by ethanol in 2005 and this could rise to 7.4% by 2030.

In order for cogeneration to be successful, clear long-term pricing and contract terms for power must be established (WIIAD, 1994) and electricity pricing either provides incentives or disincentives (WADE, 2004). Government support and subsidies are generally necessary to recover energy from sugarcane biomass residues in developing countries when the price of crude oil is below US$50/barrel (Alonso-Pippo *et al.*, 2009a). Examples of these are the introduction of biomass feed-in tariffs in countries such as Brazil and parts of India (WADE, 2004). The escalating real cost of energy and advances in technology to produce bioenergy is starting to make the bioenergy cost competitive and this presents opportunity to the sugar industry in South Africa to generate income from cogeneration and fuel alcohol (Wienese and Purchase, 2004). Government commitment to develop renewable sources of energy, protection of jobs, reducing GHG emissions from fossil fuels, and the need for energy security will aid the sugar industry to produce bioenergy (Wienese and Purchase, 2004).

One of the barriers to energy production from sugarcane is the seasonal characteristics of the crop and an alternative biomass supply feedstock is necessary during the off-season (Alonso-Pippo *et al.*, 2009a, 2009b). One solution is to ensure that boilers can utilize other fuels, such as wood or coal (WADE, 2004). Year-round energy production is seen to be a requirement in the clean energy development mechanism to obtain carbon credits (Alonso-Pippo *et al.*, 2009b). The potential to store bagasse and SCAR for year-round energy production is an expensive, difficult, and risky operation due to their self-combustion properties (Alonso-Pippo *et al.*, 2009b), but stored sugarcane trash will be necessary for year-round power generation by sugarcane mills in order to supplement the power generated from bagasse (de Carvalho Macedo *et al.*, 2001).

With the exception of Brazil, sugarcane is produced primarily as a food crop although many countries are investigating other energy options as well (Leal, 2007). Thorburn *et al.* (2007) discuss the impact of changing the production of sugarcane from a purely sugar focus to a sugar and energy focus, which may require changes to farming systems and to the use of feedstock from other biomass to ensure year-round cogeneration potential. Integrated supply chain models are necessary to enable the assessment of the impacts and to benefit from participation by stakeholders (Thorburn *et al.*, 2007). Harvesting, loading, and transport are approximately one-third of the cost of cane at mills in Brazil (Hassuani, 2005).

Green Cane versus Burnt Cane Harvesting

Harvesting is the largest cost in cane production (Beattie and Crossley, 2007). It is estimated that more than 50% of the international cane crop is burnt prior to harvesting although green cane harvesting is being increasingly adopted internationally. Approximately 80% of the cane is burnt prior to harvest in South Africa (Meyer and Fenwick, 2003), while van Antwerpen *et al.* (2008a) estimate that only 10% of the cane in South Africa is trashed.

Green Cane Harvesting

The negative environmental effects of burning has resulted in many sugar industries drawing up industrial guidelines and codes of practice to address this issue (de Beer *et al.*, 1996). Economics, rising labor costs, low labor productivity, increase in labor accidents, agronomic, environmental pollution, community factors, legislation in certain countries (e.g., in Brazil and Colombia), and the potential energy in sugarcane trash are the major drivers for the adoption of green cane production and harvesting systems (Hassuani, 2001; Whiteing *et al.*, 2001; Hassuani, 2005; Kingston *et al.*, 2005; Meyer *et al.*, 2005; Johnson *et al.*, 2007).

An advantage of green cane harvesting is that less deterioration of cane prior to crushing can be expected (Rein, 2005). Mill trials have shown that green harvested cane has a higher raw sugar quality compared to burnt cane (Sandell *et al.*, 2002) and Mcguire *et al.* (2011) report increases in commercial cane sugar (CCS) from 0.4 for burnt cane to 1.4 CCS for green cane.

The use of green cane harvesting and the conservation of a trash blanket are required for sustainable sugarcane production systems (Norris *et al.*, 2007). Positive net benefits were predicted by Thorburn *et al.* (2006a) for full and partial trash blankets, but a negative net benefit was expected for complete removal of the trash. Norris *et al.* (2007) report that relative to burnt cane harvesting, green cane harvesting results in: increased yields, particularly under stress conditions; environmental improvements; increases in sugar recovery; increased organic matter accumulation; improved soil moisture retention; improved weed control; and reduced impacts from cane pests. The advantages of green cane harvesting in South Africa include reduced cane deterioration, increased yields under stress conditions from improved water infiltration and weed control, and environmental improvements (Meyer *et al.*, 2005). Manechini *et al.* (2005) found that trash quantities above 7.5–9 tonne ha^{-1} (dry basis) distributed uniformly on the soil resulted in effective controlling of annual weeds with efficiencies greater than 90%, while 3 tonne ha^{-1} are recommended in India (Prabhakar *et al.*, 2010).

The benefits of a trash blanket include the following: soil protection against erosion, reduced soil temperature variations, increased biological activity and water infiltration in the soil, more water available due to the reduction in water evaporation, and weed control (de Beer *et al.*, 1996; Hassuani *et al.*, 2005; Manechini *et al.*, 2005).

Núñez and Spaans (2007) report that the cost of weed control was reduced by 35% and irrigation by 10% (equivalent to 85 mm irrigation saving per season) when trash was left in the field. Despite the cost advantage of burnt cane harvesting, Núñez and Spaans (2007) suggest that if the agronomic advantages of green cane harvesting with trash left in the field are included, green cane harvesting would be the preferred harvesting system.

Reductions in sugarcane yield and pests in green harvested cane increase the motivation toward trash recovery which has additional economic benefits (Hassuani *et al.*, 2005). Manechini *et al.* (2005) noted a reduction in sugarcane yield as the amount of trash increases in the field. Other issues related to the maintenance of a trash blanket include fire hazard, difficulty in performing mechanical operations (e.g., cultivation, fertilization, weed control) through the blanket, delayed and poor growth after harvesting, and increased pests which thrive under the blanket (Manechini *et al.*, 2005). van Antwerpen *et al.* (2008a) estimated that trashing is not suitable for approximately 30% of the sugar industry in South Africa as a consequence of wet soils, soils with a high water table and frosting.

The disadvantages of green cane harvesting include delayed cane emergence, greater impact of frosts, moisture problems with certain soils, and the dry matter providing a high carbon to low nitrogen concentrations. The manual and mechanical harvesting rates and vehicle payloads are reduced as a consequence of lower bulk density and additional extraneous matter delivered to the mill (de Beer *et al.*, 1996). In addition, in wet and cool climates, residues can reduce yield by 4.5–13.5 tonnes ha^{-1} as a consequence of excessive soil moisture and lower soil temperatures, which are exacerbated by a trash blanket (Viator *et al.*, 2005).

The disadvantages of mechanical harvesting of green cane versus burnt cane harvesting include the following (Meyer *et al.*, 2005):

- The performance of a mechanical harvester in green cane is reduced with increased operating and repair and maintenance costs.

- The trash blanket limits the operator's ability to travel on the correct path and GPS guidance systems are recommended.
- Recumbent cane can impact harvester performance.
- Potential billeting losses, which are inversely related to the billet length.
- Dependent on harvesting conditions, the harvester extraction system typically removes 50–80% of the pre-harvest trash and leaves.

Burnt Cane Harvesting

Burning prior to harvesting improves manual cutter performance, chopper harvester pour rates, mechanical loading rates, and vehicle payloads due to the removal of unwanted vegetative material (Meyer *et al.*, 2005). The impacts of burnt cane harvesting include losses of N and C from leaves and tops, potential for increased soil erosion, more herbicide use, more rapid deterioration of cane after harvest, and impact on surrounding communities (Núñez and Spaans, 2007).

Impact of Change from Burnt Cane to Green Cane Harvesting on Harvesting Performance

Changing to green cane harvesting from burnt cane requires substantial investment and changes to supply chains (Thorburn *et al.*, 2006b). The operational issues to be considered include the efficiency and performance of both manual and mechanical harvesting, handling of sugarcane residues, and machine availability and suitability (Meyer *et al.*, 2005). Increases ranging from 11.5% to 17% in the weight of harvested cane billets and trash was estimated by Thorburn *et al.* (2006a).

The harvesting rate in green cane is typically 65–70% of that in burnt cane and 40–50 tonnes h^{-1} can be averaged under very good field conditions with a fuel consumption of approximately 0.5 l $tonne^{-1}$ (de Beer *et al.*, 1996). Meyer *et al.* (2005) report that in Sudan, chopper harvester output in green cane was reduced by between 38% and 50% and fuel consumption increased by between 75% and 128% compared with harvesting burnt cane. Average harvester speeds were reduced from 6.87 to 4.43 km h^{-1} when harvesting burnt and green cane, respectively (SRDC, 2011). The rate of harvesting green cane is reduced by 30–40% compared with burnt cane (Meyer, 1997), while results presented by Meyer *et al.* (2005) showed an average decrease in harvester performance of 25% when harvesting green versus burnt cane, with an increase of 13.5% in fuel consumption for the same mass of cane cut. Loss in sucrose was similar for green and burnt cane harvesting (Meyer *et al.*, 2005).

In Ecuador, six sites were mechanically green harvested and six sites were burnt and then mechanically harvested. The machine productivity was reduced by 43% for the green cane harvesting and the trash content in the green cane delivered was higher by 38% compared to the burnt cane (Núñez and Spaans, 2007). A reduction in productivity of 68% for field laborers with a consequent increase in cost by 213% when harvesting green cane resulted in manual harvesting of green cane not being feasible for a mill in Ecuador (Núñez and Spaans, 2007).

Economics of the Two Harvesting Methods

Wynne and van Antwerpen (2004) investigated the economics of trashing versus burning and recognized the benefits of trashing to include reduced sucrose loss and improved cane quality, weed inhibition with reduced herbicide costs, improved soil moisture retention, and improved soil health; however, they recommended that trashing should not be performed in wet and cooler areas as it results in increased stool rotting and inhibits ratooning. Trashing also results in a slower harvest.

The model developed by Wynne and van Antwerpen (2004) assumed that there is a faster linear reduction in the long-term cane yield in burnt cane compared to trashed cane, and a 30-year period with a discount rate of 5% was used to calculate a consolidated net present value (NPV) of cane yield. The thickness of the trash blanket, the total available moisture in the soil, the geographic region in which the cane is grown, and general growing conditions affected by weather and management are factors incorporated to estimate the impact on yield of trashing. The simple relationships used to simulate the impact on yield are largely based on empirical evidence and require further investigation (Wynne and van Antwerpen, 2004). van Antwerpen *et al.* (2008a) showed that the model is relatively insensitive to ambient temperature and years of trashing, but the results obtained could be logically explained.

Wynne and van Antwerpen (2004) used the model developed by Loubser (2002) to estimate the impact of delays on cane quality. The lower harvesting productivity in trashed cane and the additional cost of burning cane were included in the model. Generally, the delays associated with trashed cane are less than in burnt cane and, when coupled with a yield increase from trashing, results in trashed cane having a higher income per hectare. The impact on transport costs are included in the model by using the relative densities of burnt and trashed cane estimated using the fiber content of the delivered cane.

The application of the model to a study area showed that both millers and growers could benefit from trashing, but this conclusion is dependent on many assumptions (Wynne and van Antwerpen, 2004). The results of the study indicated that trashing should be practiced where appropriate and recommended that further research into the following aspects should be undertaken:

- yield benefits of trashing under different circumstances,
- cane composition of different varieties,
- rates of cane deterioration,
- the influence of density on transport costs,
- labor productivity under different trashing regimes,
- the costs associated with alleviating bottlenecks at the sugar mills should trashing be widely adopted,
- the temperature effect in higher altitude areas, and
- the economics of trashed cane under irrigation.

Harvesting Systems

Sugarcane harvesters have been traditionally viewed as retrievers of sugar bearing stalks and discarding as much of the extraneous material as possible, but are increasingly seen as sugarcane sorters as well with the ability to collect part or all the trash (de Beer *et al.*, 1996). Hassuani (2005) report that prior to 2003 mechanical harvesting of cane in Brazil accounted for less than 20% of the harvested cane, with manual harvesting and mechanical loading of the cane harvested onto trucks being the most common system.

Green cane harvesting can be done by two systems (Meyer *et al.*, 2005):

- manual harvesting of whole stalks followed by manual or mechanical loading, or
- a fully mechanized chopper harvesting system.

Manual sugarcane harvesting is not an attractive form of employment and labor is becoming a constraint for manual harvesting in South Africa (Meyer and Fenwick, 2003; Lyne *et al.*, 2007). This is a result of rising labor aspirations, growth in the industrial sector, and the impact of HIV/AIDS in South Africa. Meyer *et al.* (2005) report similar labor shortages in India and Reunion Island. Manual harvesting is often preferred in South Africa due to the cost of mechanical harvesters and the unsuitability of mechanical harvesters to operate on steep slopes (Lyne *et al.*, 2007).

Manual Harvesting

In South Africa two basic systems are used to harvest the crop manually (Meyer and Fenwick, 2003): cut and stack green or burnt cane, or cut and windrow green or burnt cane for subsequent mechanical loading.

When harvesting green cane manually, the trash is removed and the stalks are topped while for burnt cane only the tops are removed (Meyer, 1997). The cut cane is then bundled for transport (Meyer, 1997). Some success has been achieved to pre-trash upright cane (de Beer *et al.* 1996). The human energy expenditure for cutting green cane is 10% more than for cutting burnt cane (Smit *et al.*, 2001).

Semi-mechanized Harvesting

A number of semi-mechanized harvesting systems are available which include the following (Meyer, 1996): whole stalk transverse windrowing machines, whole stalk linear windrowing machines, whole stalk bundling machines, and chopped cane harvesters.

Cutting, topping, and trashing operations are performed manually and the cane is then either placed in 3–6 tonne stacks and the stack is crane winched onto tractor-drawn side or rear self-loading trailers by a winch mechanism. Alternatively, the cane from four to six rows is placed in windrows at right angles to the row direction and the windrowed cane is loaded mechanically using slewing or nonslewing loaders which can be self-propelled or mounted onto a tractor (Meyer, 1997).

Many sugarcane industries require a harvesting system which falls between manual and fully mechanized harvesting (Meyer *et al.*, 2005). Relatively simple cutting attachments mounted on standard tractors or self-propelled machines can be used to mechanize the cutting operation. The cane can be deposited in windrows at right angles to the row direction or left in linear windrows parallel to the row direction. In both cases, the cane can be collected manually or mechanically as described above (Meyer, 1997).

Fully Mechanized Harvesting

Fully mechanized harvesters include whole stalk and chopper harvesters, with chopper harvesters the most widely used.

Self-propelled whole stalk harvesters cut and top one or two rows of cane and deposit the whole sticks of cane from four to six rows into a single windrow using a flexible piling arm. After windrowing, the cane is burnt and mechanically loaded into road transport vehicles using high capacity push-pile loaders (Meyer, 1997).

Whole stalk harvesters have been used extensively in Louisiana and comparisons with chopper harvesters showed no significant differences in harvesting rates, although the losses from whole stalk harvesters were larger in wet weather and when the cane was lodged (Richard *et al.*, 1996). The use of in-field mechanical de-trashers resulted in an additional in-field operation (Richard *et al.*, 1996).

Modern chopper harvesters are capable of harvesting all but very heavy green cane crops (Meyer *et al.*, 2005). Combine chopper harvesters base cut and top one or two rows of cane and chop the cane stalks into billets of approximately 300–450 mm in length. Cane tops missed by the primary topper, trash, and other extraneous matter are removed by harvester extractor fans and the billets are delivered into trailers (haul-out vehicles) traveling alongside the harvester. Both burnt and green canes can be harvested by chopper harvesters which are capable of handling yields in excess of 150 tonnes ha^{-1}. Chopper harvesters handle lodged cane better than whole stalk harvesters, but harvesting rates of green cane are reduced by 30–40% compared with burnt cane (Meyer, 1997). The average fuel consumption of chopper harvesters have been reported to be 2.6 l $tonne^{-1}$ (Norris *et al.*, 2010).

Performance of Harvesting Systems

Manual Harvesting

Meyer and Fenwick (2003) report that improvements in productivity in other countries have been achieved by using cutting, trashing, topping and piling operations, and by introducing time incentive schemes, and by splitting the cut and stack activities into two independent operations, that is, cut only and stack only.

From a survey conducted by Meyer and Fenwick (2003) in South Africa, no distinct trends in terms of type of knife were detected for the different manual harvesting systems. As shown in Table 3.9 the performance of the cutters was more affected by the harvesting system than if the cane was burnt or not.

From time and motion studies, Meyer and Fenwick (2003) concluded that cutter performance would increase significantly if the trash and topping tasks were removed and they recommended that the use of mechanical cutting tools and pre-trashing and topping devices should be investigated.

De Beer *et al.* (1989) investigated the impacts of harvesting burnt and green cane with no trashing of the unburnt cane. They found that the performance of cutters when harvesting green cane was 90% of that when harvesting burnt cane. When manual topping was not performed, the cutter performance increases by 18% and 26% in burnt and unburnt cane, respectively. Meyer *et al.* (2005) report that manual harvesting rates were reduced by 28% when cutting and stacking, and by 15% when cutting and bundling green cane compared to rates for burnt cane.

Table 3.9. Average cane cutter performance for various harvesting systems (Meyer and Fenwick, 2003).

Harvesting system	No. of samples	Average cane yield (tonne ha^{-1})	Cutter output (tonnes d^{-1})	Cutters per 1000 tonnes
Cut and stack (green)	10	72.50	3.45	1.79
Cut and stack (burnt)	18	69.60	4.20	1.44
Cut and bundle (green)	6	73.94	5.58	1.07
Cut and bundle (burnt)	22	69.93	6.56	1.08
Cut and windrow (burnt)	30	92.87	8.01	0.99

Table 3.10. Manual burnt and green cane harvesting systems (Meyer *et al.*, 2005).

System	Output (tonne/man day)	EM (%)	Cost (% of burnt)
Burnt cane	6	2.53	100
Normal green cane	4	3.1	123
Clean green cane	2.5	0.78–1	130

De Beer *et al.* (1989) and Meyer *et al.* (2005) provide performance standards for various harvesting systems which are similar to cutter output rates in Brazil (Hassuani, 2005) of 7 tonnes d^{-1} in burnt cane and 3 tonnes d^{-1} for green cane and cutter output rates in Ecuador (Núñez and Spaans, 2007) of 7.5 and 2.4 tonnes d^{-1} in burnt and green canes, respectively (Tables 3.10 and 3.11).

Machine Harvesting

Harvester performance is influenced by the following (Meyer, 1999): terrain (steep versus flat), soil type and moisture level, cane variety, cane yield, cane condition (green or burnt and erect or recumbent), field condition (field layout, row spacing. row length, headland space, rocks), transport haul-out distance, number and capacity of infield transport, operator proficiency, and machine condition and setting.

Results presented by Meyer (1996) emphasized the impact on harvester performance of row length, infield transport, and maintenance and adjustment of machinery.

Meyer (1999) measured actual harvesting rates of burnt and unburnt cane ranging from 7.7 to 39.8 tonnes h^{-1}, with a potential rate between 11.7 and 40.4 tonnes h^{-1} for bundle/whole stalk harvesters, and from 25.1 to 51.8 tonnes h^{-1}, with a potential rate between 39.9 and 68.1 tonnes h^{-1} for chopper harvesters with average rates of 22 and 37 tonnes h^{-1} for the bundle and chopper harvesters, respectively.

The performance of harvesting machinery is reported in Table 3.12–Table 3.14. The productivity of a mechanical harvester was 60 tonnes h^{-1} in burnt cane and 34 tonnes h^{-1} in green cane, respectively (Núñez and Spaans, 2007).

Cane Losses during Harvesting

Post-harvest deterioration of cane occurs once it is cut or burnt. In cool, dry, conditions up to 20 h of cut-to-crush delay may not result in significant deterioration, but in hot, humid conditions significant deterioration can occur in 10 h (Sandell *et al.*, 2002).

The average infield losses for manually cut and mechanically loaded cane ranged from 1.53% to 3.89% (Meyer, 2001; Meyer *et al.*, 2002). Sugarcane losses in mechanical harvesting using base cutters, choppers, and extractors during harvesting are a major cost to the sugar industry. These losses have been difficult to quantify as they comprise shredded pulp or splattered juice which increase with increased fan speed.

The extraneous matter is related to the pour rate, which is related to forward speed, and cane loss is primarily a function of fan speed with increases in fan speed resulting in increased cane loss (Whiteing *et al.*, 2001; Sandell *et al.*, 2002). Sichter *et al.* (2005) and Mcguire *et al.* (2011) report that chopper harvesters result in sugar losses estimated to range from 5% up to 15% as a consequence of the chopper mechanism and the extractors. The losses are dependent on the length of the billet with average extractor losses of 6.47 tonnes cane/ha. Whiteing (2002)

Table 3.11. Manual cane cutter output (De Beer *et al.*, 1989).

Activity	Output			
	Total mass (tonne/man hours)		Excluding trash and tops	
	Trash and tops%	tonne/man hours		t/man day[a]
1. *In unburnt cane*				
1.1 Manual cutting				
Cut, top, trash, bundle	5.3	0.983	0.931	7.5
Cut, top, trash, stack	5.3	0.435	0.412	3.3
Cut, top, not trashed, stack	9.7	0.447	0.404	3.2
Cut, not topped, not trashed, stack	22.2	0.594	0.462	3.7
Cut, top, not trashed, bundle	9.7	1.030	0.930	7.4
Cut, not topped, not trashed, bundle	22.2	1.510	1.170	9.4
1.2 With mechanical cutting				
Not topped, not trashed, stack	22.2	1.27	0.990	7.9
Not topped, not trashed, bundle	22.2	2.12	1.650	13.2
Not topped, trashed, bundle	18.7	1.92[b]	1.617	12.9
Topped, trashed, bundle	5.3	1.36[b]	1.294	10.4
Mechanically topped, not trashed, bundle	9.7	2.09	1.890	15.1
Topped, trashed, stacked	5.3	0.82[b]	0.780	6.2
2. *In burnt cane*				
2.1 With manual cutting				
Cut, top, bundle	3.2	1.06	1.030	8.2
Cut, not topped, bundle	8.8	1.33	1.210	9.7
Cut, topped, stack	3.2	0.71	0.687	5.5
2.2 With mechanical cutting				
Not topped, bundle	8.8	3.13	2.85	22.8
Topped, bundle	3.2	1.90	1.84	14.7
Topped, stacked	3.2	1.25[b]	1.21	9.7
Topping bundled cane	3.2	6.32	6.12	49

[a]1 man day = 8 man hours.
[b]By deduction.

Table 3.12. Bell loader output (excluding tops and trash) (De Beer *et al.*, 1989).

Activity	Output (tonnes h^{-1})	
	Instantaneous	Overall
1. *Loading only*		
1.1 *In unburnt cane*		
Not topped, not trashed, from bundles	26.14	16.88
Topped, not trashed, from bundles	32.30	25.76
Topped, trashed, from bundles	51.55[a]	37.26[a]
Not topped, not trashed, from linear windrow	17	14.08
1.2 *In burnt cane*		
Topped from bundles	41.79	33.07
Not topped, from bundles	41.68	29.07
Topped (cut and topped mechanically) from linear windrow	17.94	15.24
2. *Loading and topping*		
2.1 *In unburnt trashed cane*		
Topping and loading from bundles	22.14[b]	1845[b]
2.2 *In burnt cane*		
Topping and loading from bundles	30.01	23.71

[a]Exceptionally good field conditions.
[b]By deduction.

Table 3.13. Payload (excluding tops and trash) (De Beer *et al.*, 1989).

Material	Payload (tonnes)	
	5 tonnes basket trailer	Standard hilo truck
1. *Unburnt cane*		
Not topped, not trashed, Bell loaded from linear windrow	2.57	12.62
Not topped, not trashed, Bell loaded from bundles	2.80	14.13
Topped, not trashed, Bell loaded from bundles	3.76	18.77
Topped, trashed, Bell loaded from bundles	4.73	23.63
Not topped, not trashed, hand stacked, self-loading trailer	3.27	16.38
2. *Burnt cane*		
Not topped, Bell loaded from bundles	4.55	22.75
Topped, Bell loaded from bundles	5	25.21
Topped, Bell loaded from linear windrow	4.54	22.70

Table 3.14. Output of front end cutter (excluding tops and trash) (De Beer *et al.*, 1989).

Field management	tonnes/field hour
Unburnt cane	19.8
Burnt cane	26.5

reports harvester losses of greater than 20 tonnes ha^{-1}. Hassuani *et al.* (2005) report a 5.4% reduction in cane loss when turning off the extractor fans, while Whiteing (2002) reports a reduction in cane losses of 15.6% through the extractor fans. The trash removal efficiency of chopper harvesters was found to be as low as 50% for high-yielding fields (>160 tonnes ha^{-1}) (Whiteing *et al.*, 2001). Akachi (2007) found that trash separation improved and harvesting losses decreased with increased engine power of harvesters.

Front-end pick-up losses may be up to 5 tonnes ha^{-1} in adverse conditions, but are generally very low (Sandell *et al.*, 2002). As billet lengths increase, cane and juice losses decrease, but juice losses of 2–5 tonnes ha^{-1} are possible and increase with pour rates (Sandell *et al.*, 2002).

Cane losses from chopper harvesters could be reduced by whole crop harvesting as there is no trash separation system in operation on the harvester, which could discard billets of cane with the separated trash. The gains would need to be offset against the additional processing costs associated with reduced sugar recovery and quality in the factory, or to compensate for the increased separation costs at the mill. Results of experiments conducted in the Condong and Broadwater districts in Australia showed that whole crop harvesting resulted in a 14% increase in sucrose yield, approximately 24% more biomass with 40% extra dry, ash-free fiber than conventional burnt cane harvesting (Mcguire *et al.*, 2011).

Meyer *et al.* (2002) conducted trials of semi-mechanized versus chopper harvesting in Swaziland over three seasons. Compared with cane harvested by a chopper harvester, infield cane losses were statistically significantly lower when the cane was cut manually and loaded mechanically, while there were no statistically significant differences in the quality of cane delivered to the mill by the two harvesting systems. The performance of the mechanical loader and chopper harvester varied considerably and was dependent on the type of machine, cane yield, crop and field conditions. The average infield losses for the chopper harvester ranged from 3.69% to 5.56% in burnt cane and 5.5% in green cane. Meyer *et al.* (2005) report that for chopper harvesters in Sudan green cane losses ranged from 2.9 to 7.3 tonnes ha^{-1} compared with 1.3 to 1.4 tonnes ha^{-1} in burnt cane. These results clearly indicate that infield cane losses markedly increased when using fully mechanized compared with semi-mechanized harvesting systems (Meyer, 2001; Meyer *et al.*, 2002).

Increased infield machinery travel associated with green cane harvesting has resulted in increased damage to cane stalks and cane stools and increased compaction (Meyer *et al.*, 2005) and uncontrolled traffic of mechanized and semi-mechanized harvesting equipment resulted in cane yield losses of 30 tonnes ha^{-1} in Colombia, mainly due to stool damage (Meyer *et al.*, 2005). The average loss of productivity of cane fields as a consequence of the effects of soil compaction and ratoon damage resulting from the operations to recover the trash by baling was found to be approximately 11% and 5% in areas of clay or sandy soils, respectively (Filho, 2005).

It is necessary to match infield machinery tracks to fit inter-row spacing and use controlled traffic paths in order to reduce direct cane stool damage and to limit compaction to the inter-row, although stool damage has a greater impact on yield than compaction (Torres *et al.*, 2010).

Torres *et al.* (2010) report that the use of 3.9 m wide rows followed by a set of three dual rows of cane at 0.8 m between rows and 2.1 m in between two adjacent pairs of dual rows resulted in a lower cane yield (10–20 tonnes ha^{-1}) in plant cane as compared to the single-row planting spacing at 1.75 m spacing, but resulted in a more stable production environment. The use of the 3.9 m rows facilitated the placement of trash and the separation of the trash from the cane windrows during manual cutting which resulted in a 1% better recovery of sugar. Sandell *et al.* (2002) report reduced yields of 10–25% as a result of soil compaction with 80% of the damage occurring in the first pass.

For chopper harvesters, Sandell *et al.* (2002) report that under dry conditions with erect cane, extraneous matter is usually less that 2% and can increase to 14% in wet conditions with lodged cane while for burnt cane typical levels of extraneous matter ranges from 3% to 7%.

Harvesting Costs

The cost of harvesting includes the cost of labor and machinery. Contractors are widely used to cut cane in South Africa and labor shortages to cut sugarcane have been experienced in the South African sugar industry (Meyer and Fenwick, 2003; Murray, 2008). Murray (2008) report average costs of R 0.40/m, R 0.49/m, and R 0.4/m, (R = South African Rand) for burnt, seed, and lodged canes, respectively. A survey conducted by Murray (2008) indicated that cost, reduction in administration, and removal of the stress of managing labor were the reasons growers used contractors for harvesting operations.

In Brazil, harvesting, loading, and transport represent approximately one-third of the cost of cane delivered to the mill (Hassuani, 2005) while in Columbia up to 82% of costs are related to trash chopping and harvesting (Amill *et al.*, 2005). Machinery and equipment costs can be as high as 40% of cane production costs (Meyer *et al.*, 2005).

Using the performance of labor from earlier tables, Meyer (1996) compared the labor costs of two harvesting systems and his results are contained in Table 3.15. The cost of three harvesting systems were compared which ranged from (i) manual cane cutting, topping, and windrowing, (ii) mechanically cut cane which is and placed in linear windrows and is manually topped and placed into small bundles, and (iii) cane which is fully mechanically harvested. The results obtained emphasized that the most economical system is dependent on labor costs and productivity, the performance of the machines used, and the total tonnes cut.

Trash Recovery Systems

Reductions in sugarcane yield and pests associated with green cane harvesting means that trash recovery has to have an economic benefit (Hassuani *et al.*, 2005). However, the equipment, labor, and other farmer priorities limit the recovery of crop residues by farmers, which also conflicts with efforts to maintain soil productivity (agronomic benefits) and to minimize soil erosion. Hence, trash recovery by farmers must be directly linked to economic benefits (Sarto and Hassuani, 2005).

Various efficiencies of trash recovery have been reported in the literature. Schembri *et al.* (2002) assumed that 50% of the trash could be recovered from the field, Filho and Badr (2004) estimate that at least 25% of the field residues can be recovered, Rein (2005) estimate the recovery efficiency to range from 56% and 84%, and Seabra and Macedo (2011) estimate the trash recovery efficiency to be 40%.

Trash recovery systems include baling, hay harvesters, whole stick harvesting, and trash collected by the sugarcane harvester (Hassuani *et al.*, 2005). Both de Carvalho Macedo *et al.*

Table 3.15. Labor costs for two cane harvesting systems (Meyer, 1996).

Burnt cane						Green cane					
System 1: Manual cut, top, and bundle						System 1: Manual cut, trash, and bundle					
		Percentage of typical labor costs						Percentage of typical labor costs			
		85	100	115	140			85	100	115	140
Labor performance (tonnes/man day)		Cost per tonne				Labor performance (tonnes/man day)		Cost per tonne			
Low	5.47	3.69	4.36	5.03	6.03	Low	5.25	4.03	4.76	5.49	6.59
Medium	8.20	2.58	3.05	3.52	4.22	Medium	7.50	2.82	3.33	3.85	4.62
High	10.66	1.98	2.35	2.71	3.25	High	9.75	2.17	2.56	2.96	3.55
System 2: Mechanical cut, manual top, and bundle						System 2: Mechanical cut, manual top, trash, and bundle					
Low	10.29	2.06	2.43	2.80	3.36	Low	7.28	2.91	3.43	3.96	4.75
Medium	14.70	1.44	1.70	1.96	2.35	Medium	10.40	2.03	2.40	2.77	3.33
High	19.11	1.11	1.31	1.51	1.81	High	13.52	1.56	1.85	2.13	2.56

(2001) and Hassuani (2005) report on the testing and analysis of four trash recovery systems, as shown in Figure 3.4.

Routes A and B involve whole cane harvesting with the cane and trash separated at the mill (Route A) or separated in the field with the trash collected with another infield operation (Route B). Routes C and D involve the use of chopper harvester, with the trash and cane either separated in the field by the harvester (Route C), with an additional infield operation to collect the trash, or the extractor fans on the harvester set such that the loss of cane and trash are both minimized and the combined trash and cane are transported to the mill and only separated at the mill (Route D). Routes A and D require equipment to separate cane from trash at the mill and also require equipment to process the trash at the mill, either in loose or baled form.

Yield

Sugarcane is one of the most efficient crops on earth to convert sunlight energy into biomass and is cultivated in all tropical and subtropical regions (Thompson, 1978; Perez, 1997; Waclawovsky *et al.*, 2010). In terms of biomass harvested (and transported), sugarcane is the world's largest crop with 1.69 billion tonnes of cane stalks (FAOSTAT, 2012). In 1974, the world produced 592 MT cane; 434 MT milk; 354 MT wheat, 324 MT maize, and 225 MT rice (Irvine, 1981). Sugarcane yields from the three countries with the greatest amount of sunlight (Waclawovsky *et al.*, 2010) namely Australia, Colombia, and South Africa, reveal extremely high cane fresh weight yields that averaged 84 tonnes/(ha yr) (Table 3.16). The commercial maximum cane yield for these three countries averaged 148 tonnes/(ha yr) and the experimental maximum averaged 212 tonnes/(ha yr). The theoretical maximum cane yield was 381 tonnes/(ha yr) (Waclawovsky *et al.*, 2010).

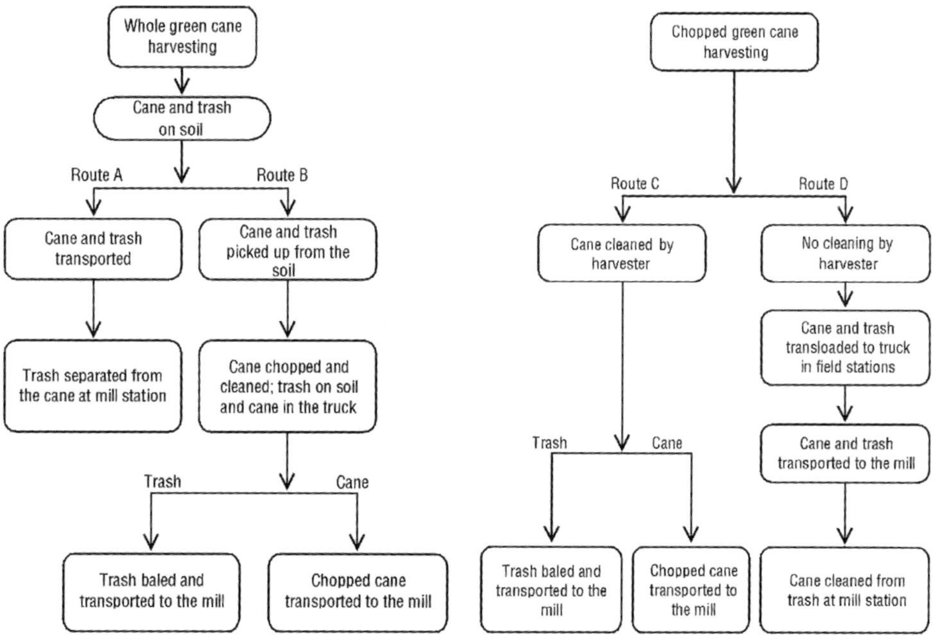

Figure 3.4. Selected routes for sugarcane harvesting with trash recovery (de Carvalho Macedo *et al.*, 2001).

Genetics and Breeding

Molecular Resources for Sugarcane Improvement, Genome Challenges, and Opportunities

Modern sugarcane varieties that are cultivated for sugar production are complex interspecific hybrids (*Saccharum* sp.) between the species *S. officinarum* and *S. spontaneum* with contributions from *S. robustum*, *Saccharum barberi*, *Saccharum sinense* and related grass genera such as *Miscanthus*, *Erianthus,* and *Narenga* (Daniels and Roach, 1987). *S. officinarum* is low in fiber and accumulates very high levels of sucrose in the stem, but has poor disease resistance

Table 3.16. Average, maximum and theoretical sugarcane yields (Australia, Colombia, and South Africa) and total dry matter production (Waclawovsky *et al.*, 2010).

	Cane yield	Biomass[a]	
Type of yield	tonne/(ha yr)	tonne/(ha yr)	g/(m^2 d)
Commercial average	84	39	10.7
Commercial maximum	148	69	18.8
Experimental maximum	212	98	27
Theoretical maximum	381	177	48.5

[a]Cane yield was converted to biomass dry matter first by calculating stalk dry weight (tonne cane ha^{-1} yr^{-1} × 0.30) then adding the proportion of trash dry weight [0.65 (stalk dry weight)] as calculated from Thompson (1978).

(Sreenivasan *et al.*, 1987). *S. spontaneum* is more genetically diverse than *S. officinarum* and has greater stress tolerance than *S. officinarum* (Sreenivasan *et al.*, 1987).

Although conventional breeding programs in sugarcane are relatively recent, compared to other major crops, interspecific hybridization within the genus *Saccharum* led to significant improvements in yield, ratooning ability, sugar content, and disease resistance; while maintaining acceptable fiber levels for milling (Jackson, 2005; Lakshmanan *et al.*, 2005; Ming *et al.*, 2006). Most modern sugarcane cultivars originate from crosses between a relatively small numbers of original progenitor clones compared with the large number of basic clones that exist in the *Saccharum* genus, resulting in a narrow gene pool (Jackson, 2005). Modern cultivars obtained by the process of nobilization have inherited most of their chromosomes from *S. officinarum*, and a minority from *S. spontaneum* or resulting from recombination between progenitor chromosomes.

Current sugarcane breeding programs depend on typically crossing elite parental cultivars followed by selection of superior progeny—a process that takes more than one decade until commercial release (Jackson, 2005; Matsuoka *et al.*, 2009). It remains challenging to exploit the large genetic variation existing among clones of different *Saccharum* species (Ming *et al.*, 2006). Molecular markers may assist breeders in incorporating useful genes from sexually compatible sources into the gene pool of the advanced cultivars. The highly polyploid, aneuploid, and heterozygous nature of sugarcane imposes severe complications on genetic mapping, QTL identification and, consequently, MAS. High polyploidy and heterozygosity allow small individual contribution of multiple alleles present at a single hom(e)ologous locus to affect phenotypic variation.

Also, in comparison to diploid species, determining the number and frequency of allele types at a single locus requires more sophisticated methods (Cordeiro *et al.*, 2006). In addition, most traits of interest seem to be of quantitative nature, and QTL identified so far make only small contributions to phenotypic variation (Hoarau *et al.*, 2002; McIntyre *et al.*, 2005; Reffay *et al.*, 2005; Aitken *et al.*, 2006, 2008). Loci with dominant alleles present in single copies are more easily mapped in sugarcane. Accordingly, so far only three genes have been mapped (Daugrois *et al.*, 1996; Raboin *et al.*, 2006). One of them is *Bru1*, a major gene controlling the important rust-resistance trait, identified as a dominant, Mendelian gene in the progeny of cultivar R570 (Asnaghi *et al.*, 2000; 2004, Costet *et al.*, 2012).

Despite the difficulties in mapping genes in sugarcane, it is possible to take advantage of comparative mapping with other Poaceae species for which high density maps and sequenced genomes are available, such as rice, corn, and especially sorghum, whose genome shows the most colinearity with that of sugarcane (Grivet *et al.*, 1994; Dufour *et al.*, 1997; Glaszmann *et al.*, 1997; Guimarães *et al.*, 1997; Jannoo *et al.*, 2007), to facilitate mapping in sugarcane. In this regard, mapping of *Bru1* with the help of sorghum-derived markers illustrates the importance of comparative genomics and the application of synteny for genetic mapping in species such as sugarcane bearing complex genomes.

Association mapping is a method to associate specific DNA polymorphisms with traits of interest based on linkage disequilibrium. This approach is powerful when there is no known population structure, or when individuals are distantly related (Rafalski, 2002; Yu and Buckler, 2006). Casa *et al.* (2008) described a sorghum resource suitable for association mapping based on a set of 377 diverse sorghums and Butterfield (2007) explained in sugarcane.

Initiatives to understand the complex sugarcane genome have been made in recent years; large breeding programs will be likely based on marker-assisted selection (MAS), since it can greatly accelerate breeding rates and high-throughput genotyping is becoming less expensive than phenotyping (Bernardo, 2008). MAS is used routinely in breeding programs at Monsanto

(Eathington *et al.*, 2007; Edgerton, 2009). Sugarcane breeding, which takes an exceedingly long time, could benefit greatly from MAS strategies.

Initial efforts have been made at marker-assisted quantitative trait locus (QTL) introgression in sugarcane (Aitken *et al.*, 2002). Due to the genetic complexity of sugarcane, deployment of MAS requires the discovery of a number of single nucleotide polymorphism (SNPs) that is not met by the current availability of public sugarcane expressed sequence tag (ESTs), which were produced largely by cDNA cloning and dideoxy sequencing. Recently, second-generation sequencing efforts for the identification of SNPs in sugarcane were reported (Bundock *et al.*, 2009). Second-generation sequencing, in conjunction with high-throughput technologies for SNP genotyping and array-based platforms (Wenzl *et al.*, 2004; Gupta *et al.*, 2008), will accelerate the rate of marker discovery and validation. Large amounts of marker and phenotypic data need to be combined for the identification of marker trait associations (Bernardo, 2008) that, ultimately, will allow the utilization of MAS to accelerate sugarcane breeding programs.

In the near future, next generation sequencing technologies are expected to allow the complete sequence of the sugarcane genome to become available through organized efforts involving groups in South Africa, Australia, Brazil, and the United States (Sugarcane Genome Sequencing Initiative, 2009; Souza *et al.,* 2011). Ever-increasing knowledge of the sugarcane genome and transcriptome will continue to boost breeding and biotechnology approaches aiming at increasing sucrose and biomass yields.

In recent times, whole-genome sequences generally represent a single genotype, so that genome-wide comparisons to dissect specific traits are not possible. Even if multiple genotypes were sequenced, complementary approaches such as expression profiling or the use of mapping populations would have to be included. The genome sequence data would be an ideal source to identify additional molecular markers. Better understanding of the genes involved in cell wall biogenesis, and the precise role of different members of multigene families will ultimately enable targeted modification of cell wall architecture in such a way that processing characteristics are optimized without compromising plant performance in the field. In addition to elucidating the function of species-specific genes, comparative genomics approaches in the grasses will be of help when addressing fundamental questions about genome evolution, morphological diversity, growth habit, and traits such as perenniality (Paterson *et al.*, 2009).

Relevant bioenergy traits in sugarcane are sugar yield, biomass yield, modified cell wall composition that renders biomass processing more efficient, and drought tolerance (Zhang *et al.*, 2006; Waclawovsky *et al.*, 2010). The explanation of the genetic basis of sugar accumulation will help sugarcane genetics, as this knowledge would enable much more efficient selection for high sugar yield and at the same time will help in understanding and breeding for high biomass genotypes.

Forward and Reverse Genetics Approaches to Sugarcane Improvement

Naturally occurring mutations have played a prominent role in genetics since the discipline was founded, and the ability to induce mutations with chemicals such as ethyl methanesulfonate (EMS), diethylsulfate (DES), physical via gamma rays, X-rays, and via biological route such as transposons, T-DNA agents, has propelled the field. Forward genetics is based on the identification of mutants with a phenotype of interest, with the ultimate goal being the identification of novel genes involved in the process or pathway under study. Reverse genetics relies on the identification of mutations in specific genes with known sequences, with the goal of better understanding the role of these genes in a process or pathway. Aside from helping to establish gene function in general, reverse genetics approaches can be particularly useful to study tissue-specific or developmental regulation of gene expression, especially when the gene of interest is part of a multigene family. Mutagenesis is typically not a viable option for polyploids because of the genetic redundancy. Sugarcane is therefore at a disadvantage when it comes to the implementation of forward and reverse genetics.

With the increasing availability of high-throughput expression profiling, also referred to as RNA-seq technology (Mortazavi *et al.*, 2008), additional reports on the elucidation of cell wall and sucrose metabolism are expected in the near future. The technology requires access to high-tech equipment and trained staff, but is becoming more routine, with cost dropping as output increases. One of the benefits of expression profiling is that it can be implemented for any species. A challenge is the vast amount of data that need to be analyzed, whereby experimental designs including replicates are critical. Even though expression profiling is helpful in narrowing down the list of genes responsible for major phenotypic differences, identification of the genes underlying sugar accumulation will require additional experiments that investigate the causal relationships between the presence of certain alleles and the accumulation of sugars or the effect on cell wall composition, for example, with the use of genetic mapping studies (see the previous section) when feasible, or the use of transgenic strategies (see later).

Transgenic Route for Enhancing Biomass Production

Transgenic sugarcane has been produced routinely in many laboratories since the early 1990s. Currently, many cultivars can be transformed, and several biolistic and agrobacterium-mediated protocols of relatively high efficiency are available (Bower and Birch, 1992; Manickavasagam *et al.*, 2004; Arencibia and Carmona, 2006). Engineered traits have included herbicide, virus, and insect resistance, enhanced sugar content and drought tolerance, flowering control, and molecular farming of biopolymers and pharmacological proteins (reviewed by Lakshmanan *et al.*, 2005; D'Hont *et al.*, 2008). Collectively, to date, the most commonly and successfully targeted traits are herbicide resistance (Gallo-Meagher and Irvine, 1996; Falco *et al.*, 2000; Leibbrandt and Snyman, 2003) and insect herbivory resistance via expression of *Bacillus thuringiensis* Cry proteins (Arencibia *et al.*, 1997; Braga *et al.*, 2001; Braga *et al.*, 2003; Weng *et al.*, 2006), proteinase inhibitors (Falco and Silva-Filho, 2003; Christy *et al.*, 2009) and lectins (Nutt *et al.*, 1999). Herbicide and insect resistance most likely will be the first traits to be commercialized and bring to sugarcane the benefits observed in several other major crops.

Frequent occurrence of transgene silencing and lack of a suite of robust promoters for controlling transgene expression have hindered sugarcane biotechnology and analysis of gene function. These constraints usually require screening of a prohibitively high number of events for the identification of those with appropriate expression levels. Reports indicate that promoters derived from banana streak virus (Schenk *et al.*, 2001), sugarcane bacilliform virus (Braithwaite *et al.*, 2004), and a tandem CaMV-35S: corn Ubi-1 promoter (Groenewald *et al.*, 2000; Groenewald and Botha, 2008) could provide relatively high levels of expression in mature plants. However, stable stalk-preferred transgene expression driven by endogenous promoters in sugarcane was reported recently (Damaj *et al.*, 2010). Adoption of agrobacterium-mediated methods for sugarcane transformation, in addition to the manipulation of silencing mechanisms combined with the identification of regulatory elements for gene expression, will likely be necessary to overcome current limitations.

Despite the biological and commercial importance of sucrose accumulation in sugarcane, few transgenic strategies targeting genes involved directly with carbohydrate metabolism that have successfully generated genotypes with higher sucrose content or yields have been reported. Nonetheless, recent studies indicate that sugarcane plants have enough photosynthetic capacity to support additional growth and sugar accumulation. A rare successful attempt to increase sugarcane sucrose content by altering the expression of an endogenous gene central to carbohydrate metabolism was targeted at pyrophosphate: fructose 6-phosphate 1-phosphotransferase (PFP)—an important glycolytic enzyme (Groenewald and Botha, 2008). In sugarcane internodes, PFP activity is inversely correlated with sucrose content and directly related to respiration activity (Whittaker and Botha, 1999). Accordingly, down-regulation of PFP activity via antisense or co-suppression strategies enhances sucrose accumulation in immature internodes of several transgenic lines (Groenewald and Botha, 2008). In addition, fiber content is increased in PFP-down-regulated lines, possibly caused by the diversion of carbon units from glycolysis and respiration toward storage and structural carbohydrates. These findings highlight the great potential of transgenic manipulation of key regulators of source–sink relationships and their underlying sensing mechanisms for enhancing sugarcane biomass and sucrose yields.

Breeding for Energy Cane

Energy cane refers to high biomass sugarcane (*Saccharum* spp.) hybrids with fiber content at least 20% higher and juice content at least 10% lower than that of conventional cane. Energy cane can potentially yield 10 tonnes to more than 30 tonnes dry matter per acre in diverse environments (St. John *et al.*, 2007). Typically, sugarcane contains 30–32% solids (fiber and Brix). Sugarcane breeders and geneticists aim to develop two types of energy canes, that is, Type I and Type II (Tew and Cobill, 2008). Type I would have slightly more fiber (14–18%) than conventional sugarcane varieties and approximately the same amount of Brix (14–18%). Type II would have 20–26% fiber on a fresh weight basis, with 6–12% Brix. Type II generally corresponds to the earlier generation hybrids (F_1 and BC_1 hybrids) that have higher biomass yields and more stress tolerance, and hence are better adapted to climates outside the traditional cane growing areas. In more temperate climates, it would be possible for growers to grow both Type I (BC_2 and BC_3 hybrids) and Type II varieties, with the Type I varieties being harvested early in the summer and the Type II varieties in late summer and early winter. Type II varieties would have the longer growing season; hence, they would be expected to have the greatest input to output energy ratios.

Breeding programs need to be directed to grow energy cane on marginal soils that may be prone to flooding, insect, nematode, and disease damage, and a further exploitation of some sugarcane varieties is required; for example, encouraging symbiotic relationships with nitrogen-fixing bacteria. Improvement of sugarcane for increased energy efficiency and adaptability to a wide range of environments is considered by many breeders and geneticists as similar to the concept of genetic base broadening where utilization of wild *Saccharum* germplasm, particularly *S. spontaneum* and close relatives, such as Erianthus spp., in sugarcane breeding programs is practiced (Ming *et al.*, 2006).

S. spontaneum, produces both viable seed and rhizomes and is adapted well to diverse climatic conditions. This makes it an excellent source of a number of valuable genes (Mukherjee, 1950; Panje and Babu, 1960; Panje, 1972; Roach, 1978). The lack of cold tolerance characteristic of sugarcane has begun to be addressed by introgression efforts involving *S. spontaneum* (Wang *et al.*, 2008a). Through the use of photoperiod facilities, *Saccharum* hybrids can also be crossed successfully with their close relatives Erianthus (Cai *et al.*, 2005) and Miscanthus (Lo *et al.*, 1986). Erianthus species, especially *Erianthus arundinaceus*, have been found to be vigorous, tolerant to adverse environmental conditions including water stress and waterlogging, and resistant to diseases and insects. Introgression breeding programs involving these species could offer value feedstock for future bioenergy production systems. However, despite this reputation, there have been few successful studies published to support this status. Further studies, particularly comparing Erianthus and Spontaneum accessions with other high profile candidates for biomass production such as Switchgrass and Miscanthus could help validate the value of Erianthus and Spontaneum for potential future bioenergy production systems. Success with these crosses could also potentially further expand the genetic diversity of this crop, and should open the door to further improvements in production efficiency.

Sorghum is also within the *Saccharum* complex, and demonstrated Saccharum–Sorghum hybrids (Gupta *et al.*, 1978) suggest great potential to introgress drought tolerance and large seed size from sorghum into high-biomass, perennial energy canes. The rich source of genetic diversity and the plasticity of autopolyploid genomes also offer a wealth of opportunities for the application of genomics and related technologies to increase biomass production of sugarcane (Lam *et al.*, 2009).

By enhancing the level of stress tolerance through conventional, molecular breeding, and genomics techniques, the geographic area of distribution of the crop can be expanded to more temperate regions. It will be conceivable that the area devoted to this crop could be tripled, thus making it a more attractive market for biotechnology companies with patented genes to further enhance the level of stress tolerance, or introducing genes for the production of profitable by-products without the category restrictions encountered in food crops.

Sugarcane Physiology

Photosynthetic efficiency is not just a matter of light and CO_2 availability, but is controlled by a whole plant source–sink balance (Paul and Foyer, 2001). The observation that *S. spontaneum*, which exhibits low sucrose content in the stalk, has a 30% higher photosynthetic rate than higher sucrose-accumulating *Saccharum* spp. hybrids, suggests that source–sink relationships are crucial in shaping sucrose accumulation in stalks (Irvine, 1975). The increased photosynthetic capacity in sugarcane as a consequence of the increase in sink strength was demonstrated by McCormick *et al.* (2006).

The differences in sucrose content can be the result of photoassimilate partitioning between the stalk and other organs in the plant (Inman-Bamber *et al.*, 2008). By limiting expansive growth and maintaining photosynthesis (no significant difference in net photosynthesis was observed between low and high clones), both low and high sucrose clones were able to increase sucrose accumulation, with low sucrose clones presenting some internodes with the same sucrose amount as high sucrose clones (Inman-Bamber *et al.*, 2009). The authors observed that low sucrose clones produced more stalks and allocated more dry matter to leaves than high sucrose clones. The additional demand for structural carbohydrates in low sucrose clones could be one of the reasons for the lower sucrose accumulation in the stalk. A proposed model to determine the contribution of each trait revealed that stalk number, plant extension rate, and photosynthesis were most responsible for the variation in sucrose accumulation rate and content between low and high sucrose clones (Inman-Bamber *et al.*, 2009).

The sugarcane stalk is made of internodes of different ages, and the crop undergoes several growth phases during the year (Lingle, 1999). Prior to harvest, vegetative growth slows and ripening process starts where the sugar content of the stalk increases. Recently, a study has been done on how cell wall composition changes during sugarcane internode development, or whether there are composition changes during different growth stages of the crop (Lingle and Thomson, 2012). The results suggest that as sugarcane internodes develop, they initially build structure through elongation and thickening of cell walls and thus increase cellulose and lignin content. Sugars, and especially sucrose, accumulate after elongation is complete. Lignin appears to continue to accumulate after elongation ceases. The ripening process has a great effect on sugar content and cell wall composition. The sucrose content of younger internodes increases during ripening, while the hemicellulose content decreases and cellulose and lignin contents increase. Delaying harvesting to maximize sucrose yield may reduce the quality of the bagasse by increasing lignin content. This information is important for bioenergy feedstock harvest if biomass quality constraints were in place (Lingle and Thomson, 2012).

In many crops biomass accumulation slows down well before harvest due to phenomena that have been defined as RGP (van Heerden *et al.*, 2010). Available evidence shows that RUE is not constant during the full cropping cycle in sugarcane. Park *et al.* (2005) and Donaldson *et al.* (2008a) estimated substantial losses in biomass production in sugarcane crops affected by RGP. Aboveground biomass production is especially important in the case of energy cane where achieving highest biomass yields is the main objective (Waclawovsky *et al.*, 2010). To overcome the RGP barrier, a better understanding of the factors involved is crucial. Further research in this area in the future will definitely provide directions for such physiological manipulations, which will be an important step to overcome sugarcane yield restrictions due to reduced growth phenomenon.

Life-cycle Assessment of Sugarcane Biofuel Production

The rapid growth rates and high potential biomass production that can be achieved by sugarcane most often requires substantial water and agrochemical input, and this has resulted in concerns on the impact of this crop on the environment. In a number of regions around the world, pressure to reduce water consumption and environmental pollution is mounting (Inman-Bamber *et al.*, 2000; Hurst *et al.*, 2004; Thorburn *et al.*, 2011). Life-cycle assessment (LCA) is a useful tool for establishing the environmental profile of complex sugarcane production systems, including alternative mill products and by-products. LCA is useful for determining the influence of different management practices while monitoring for any "problem shifting" (Finnveden *et al.*, 2009) or "pollution swapping" (Thorburn and Wilkinson, 2012). LCA studies with varying objectives have been completed for sugarcane products in Australia (Reouf and Wegener, 2007; Renouf *et al.*, 2008), Mauritius (Ramjeawon, 2004), Cuba (Contreras *et al.*, 2009), South Africa (Mashoko *et al.* 2010; Mashoko *et al.* van der Laan *et al.*, submitted), and Brazil (Seabra *et al.*, 2011). Popular impact categories considered are nonrenewable (fossil) energy consumption, GHG emissions (or global warming potential), eutrophication and acidification potential, and water consumption. Abiotic resource depletion, human and ecosystem toxicity, and land use are often also considered.

The actual cultivation of sugarcane (cradle to farm gate) is the largest contributor to the overall environmental impact of sugarcane-derived products. For example, in Thailand, Yuttitam *et al.* (2011) estimated GHG emissions of 490 kg CO_2 equivalent/tonne sugar for sugarcane cultivation compared to 60 kg CO_2 equivalent/tonne sugar for milling processes. On-farm GHG emission estimates are also where the highest uncertainty lies. Based on coefficients of variation, uncertainty ranges of 40% for fuel use, 7% for manufacture of synthetic N fertilizers, and 70% for soil N_2O emissions have been reported (Williams *et al.*, 2006; Tuomisto *et al.*, 2012). Seabra *et al.* (2012) estimated GHG emissions of 56.3 kg CO_2 equivalent/tonne cane for rainfed production and 86.2 kg CO_2 equivalent/tonne cane for irrigated production in South Africa. Despite higher yields achieved under irrigation, the larger C footprint is a result of electricity used to pump irrigation water generated using fossil fuels. Reouf and Wegener (2007) reported GHG emissions relating to raw cane sugar production ranging from 540 to 710 kg CO_2 equivalent/tonne sugar for Australia.

Energy required to manufacture, package, and transport agrochemical inputs (such as fertilizers and pesticides), to pump irrigation water, and diesel consumption during mechanical soil preparation, harvest and transport of biomass to the mill are the main consumers of nonrenewable energy in sugarcane production. Energy inputs of 112–235 MJ/tonne green cane for rainfed sugarcane production in Queensland, Australia were reported by Reouf and Wegener (2007). Higher values can be expected for sugarcane produced under irrigation, for example, Reouf and Wegener (van der Laan *et al.*) (submitted) reported energy consumption of 257 MJ/tonne green cane or 1685 MJ/tonne sugar for cane grown under pivot irrigation in Pongola, South Africa. For the predominantly rainfed Brazilian Centre-South Region, Seabra *et al.* (2011) reported energy input requirements of 1109 MJ/tonne sucrose for sugarcane farming activities and a further 508 MJ/tonne sucrose for agrichemical production. For the 1970s, Ricaud (1980) reported energy inputs of around 8600 MJ/tonne sugar and 645 MJ/tonne green cane, owing to lower yield and sucrose content. While the Ricaud (1980) study also accounted for energy embedded in farm machinery (about 13% of total), this large difference emphasizes progress in improving energy efficiency over the last three decades (van der Laan *et al.*, submitted).

Reported sugarcane water-use efficiencies can range widely from 7.4 to 16.9 tonnes cane per 100 mm of ET and 0.5 to 1.9 tonnes sucrose per 100 mm of ET (Kingston, 1994; Inman-Bamber *et al.*, 2000). As freshwater availability diminishes in many regions across the world, sugarcane water consumption will come under increasing scrutiny, beginning with water scarce catchments. Both blue water (surface water and groundwater that can be diverted to a range of human activities) and green water (water in the soil originating from rainfall) consumption are considered in LCA. For irrigated systems, savings on blue water consumption (irrigation) through strategies to use green water more effectively reduce the environmental impact of sugarcane production. Compared to water consumption during sugarcane ET, consumption of water in other sugarcane production and processing activities is extremely low. In addition to water consumption, sugarcane cropping systems can have a negative impact on water quality, mostly through nonpoint source nutrient, pesticide and sediment pollution, and salinization of water bodies. The release of nutrients and pesticides are accounted for in LCA by the eutrophication potential and eco- and human toxicity potential impact categories, but suitable methodology to quantify the impact of sediment losses and salinization effects is currently lacking. Concerns have arisen on the increasing demand for biomass-derived biofuels on water resources (quantity and quality) due to the high quantities of water required to grow biofuel crops and the impact of these cropping systems on water quality (Dominguez-Faus *et al.*, 2009)

In a case study for irrigated sugarcane produced in Pongola, South Africa, Dominguez-Faus *et al.*, van der Laan *et al.* (submitted) estimated that nonrenewable energy consumption, global warming potential and eutrophication potential can be reduced by 20%, 25%, and 45%, respectively, through intensified irrigation water and N fertilizer management by farmers. In an LCA of sugar beet production in Germany, Brentrup *et al.* (2001) demonstrated the environmental benefits of using calcium ammonium nitrate over urea as fertilizer N source, mostly due to decreased volatilization losses.

Comparing conventional and organic sugarcane farming in Queensland, Australia, Renouf *et al.* (2005) observed no significant differences for energy input and GHG emissions between the two systems. While organic farming was shown to provide water quality and ecotoxicity benefits through the elimination of fertilizer and pesticide use, the lower yields achieved while converting to organic systems result in a major influence on the overall environmental burden per tonne of cane produced. This burden will be reduced if yields can eventually match those of conventional systems.

The cogeneration of surplus electricity using bagasse can play a fundamental role in avoiding CO_2 emissions in countries where the energy mix is based primarily on fossil fuels (Mapako *et al.*, 2012). Results for an LCA of electricity generation from bagasse in Mauritius showed that bagasse-derived electricity performs well for the nonrenewable energy, GHG emission, and acidification potential impact categories, but performs poorly for water consumption and eutrophication potential (Ramjeawon, 2008). In addition to electricity cogeneration, Contreras *et al.* (2009) found that producing alcohol, biogas, animal food, and fertilizers from by-products can lead to significant benefits in the abiotic resource depletion impact category.

The environmental benefits of sugarcane-based ethanol over petroleum-based gasoline are broadly acknowledged (Seabra *et al.*, 2011). For example, GHG emission reductions of over 80% have been estimated in various reports (Wang *et al.*, 2008b). Comparing sugarcane-derived ethanol produced in Brazil with petroleum gasoline for light-duty vehicle use in the United States, Wang *et al.* (2008b) reported potential reductions of 78% for GHG emissions and 97% for fossil energy use. Seabra *et al.* (2011) conducted LCA studies for sugar and ethanol produced from sugarcane in Brazil's Center-South Region. Fossil energy use and GHG emissions were estimated at 721 kJ kg^{-1} and 234 g CO_2 equivalents kg^{-1}, respectively, for sugar production, and 80 kJ MJ^{-1} and 21.3 g CO_2 equivalents MJ^{-1}, respectively, for ethanol production. These values are considerably lower than for countries with mills where ethanol and surplus electricity are not produced. A study by Wang *et al.* (2008b) for sugarcane ethanol produced in Brazil using the GREET model showed that pre-harvest burning has the largest contribution to GHG emissions, followed by emissions from the ethanol manufacturing plant, fertilizer production, and N_2O emissions from sugarcane fields (Figure 3.5).

In a comparison between Australian sugarcane with United States maize and United Kingdom sugar beet as sugar sources for fermentation, Renouf *et al.* (2008) observed clear advantages for sugarcane for the energy input and GHG emissions impact categories due to high saccharide yield and the generation of renewable energy using sugarcane fibers. The authors raised concerns regarding potentially high N_2O emissions from Australian sugarcane for certain regions with conditions conducive to high emissions, however.

For the United States, Evans and Cohen (2009) used a comparative LCA to assess energy balances and water resource implications for four dedicated ethanol feedstocks—maize, sugarcane, sweet sorghum, and southern pine in Florida and Georgia—two Southeastern US states foreseen to play a key role in biofuel production into the future. Net energy benefit ratios of 1.26 for maize, 1.94 for sweet sorghum, 2.51 for sugarcane, and 2.97 for southern pine were reported. Ultimately, to meet the Energy Independence and Security Act (EISA) of 2007 renewable fuel standards production goals of 136 billion l of biofuel by 2022 in the United States (Sissine, 2007), all four feedstocks require substantial land (10.1, 3.1, 2.5, and 6.1 Mha for maize, sugarcane, sweet sorghum, and pine, respectively), annual N fertilization (3230, 574, 396, 109 million kg N, respectively), and annual total water (54 400, 20 840, 8840, and 14 970 million m^3, respectively) resources (Evans and Cohen, 2009). This production will offset only 17.5% of regional gross gasoline consumption, and the authors concluded that careful scrutiny of environmental trade-offs is required before aggressive production mandates are embraced and pursued.

LCA is expected to play a key role in quantifying the potential impacts of sugar production and benefits of sugarcane by-products. Differences in functional unit (FU) (e.g., tonne cane

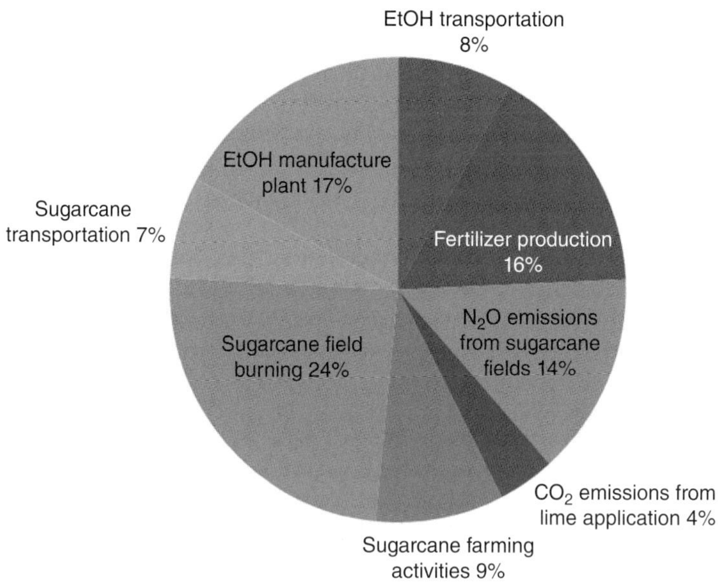

Figure 3.5. Relative GHG emissions for ethanol pathway activities from sugarcane produced in Brazil (Wang *et al.*, 2008).

vs. kg sugar) used and the processes accounted for in the LCA often makes comparisons between studies difficult. LCA methodology is applicable from smaller scales to assess on-farm mitigation measures up to catchment and basin scales to fully account for the environmental trade-offs of different policies.

Sustainability

Factors for the sustainable production of sugarcane and its by-products include soil quality, water quality and quantity, availability of fertilizers, successful control of pests and diseases, economical transport, job security, working conditions, and secured markets.

The sustainable use of different types of land (marginal and degraded, as well as good quality agricultural and pasture land) depends on the success of accelerating the improvements in current agricultural management practices, and integrating biomass production in a sustainable way into current land-use patterns. Understanding of how this can be achieved from region to region is often limited (Junginger *et al.*, 2006).

Sugarcane that is burnt at harvest is not able to maintain the high soil organic carbon (SOC) levels found in native land. Virgin land that originally contained 5.5% SOC and converted to sugarcane which was burnt at harvest lost 1.5% and 3.5% SOC within the first 10 years from a Rhodic Ferrasol (FAO) with a 62% clay content and an Ochric Cambisol (FAO) with a 18% clay content, respectively (Dominy *et al.*, 2001). Similarly, the conversion of uncultivated land to biofuel agriculture resulted in significant SOC losses—an effect that was most pronounced when native land was converted to sugarcane agriculture (Anderson-Teixeira *et al.*, 2009). By contrast, if appropriate bioenergy farming systems could be operated on degraded land, as in the case of sugarcane in Brazil, soil carbon and land restoration for sustainable use could increase (Fritsche *et al.*, 2006).

In South Africa, the concept of minimum tillage has been promoted since the 1970s as a practice to reduce land degradation due to erosion on steep fields (Iggo and Moberly, 1976). The protection of the soil's surface against crusting by a layer of sugarcane leaves left in the field after harvest has also been recognized as essential to promote water infiltration and to reduce runoff and erosion (Platford, 1982). Other benefits resulting from the use of trash blankets is an increase in water infiltration rates, increased soil water content, reduced periods of water stress (Thompson, 1966), increased soil SOC content, and improved soil microbial activity (Dominy *et al.*, 2001; Haynes and Graham, 2004; Wallis *et al.*, 2010). Furthermore, perennial bioenergy crops, such as sugarcane, could improve soils and reduce erosion on arable land currently in use by creating year-round soil coverage (Fritsche *et al.*, 2006).

Soil quality is threatened by post-harvest residue removal for energy production via cogeneration at the mill. Farmers must therefore be made aware that not all of this residue should be allowed to be removed from fields as maintenance of soil quality depends on a significant retention of residues in fields to boost soil organic matter levels. The green leafs are less desirable than brown leafs for the production of energy at the mill, but are more beneficial than the brown leafs to re-circulate nutrients as a source of soil organic matter.

Following green cane harvesting, a lot of residues (e.g., brown and green leaves) are left in the fields. The dry mass of post-harvest residues is roughly equal to between 20% and 30% of the fresh stalk mass (Donaldson *et al.*, 2008b). Dale (2007) estimated that up to 80% of the residues could be utilized for raw materials without compromising sustainable sugarcane production. In sugarcane-cultivating countries the quantity of biomass available from sugar production may vastly exceed the combined quantity of other potential biomass sources, including municipal wastes. For example, in Australia, four times more biomass is available from sugar industry wastes than all other sources combined (Dale, 2007).

Large-scale production of modern biofuels could provide a major opportunity for many rural regions around the world to generate income and employment. Given the size of the global market for transport fuels, the benefits could be vast, for example, by reducing oil imports and exporting biofuels. It is, however, far from certain that those benefits will accrue to the rural populations and smallholder farmers (Junginger *et al.*, 2006).

Nevertheless, biofuels and renewable energy in general, hold similar promise for advancing rural areas. In many cases, it is prohibitively expensive to build electricity lines or to transport fuel to remote areas. By producing their own fuel, rural residents will not have to dispose hard-earned income on oil products in order to get their crops to market. Thus, because it would provide them with fuel for transportation, producing biofuels could make taking goods to the marketplace more affordable (Laney, 2006). In addition, growing feedstock crops for biofuels could also help increase the incomes of farmers in developing economies, as there could be new cash crop opportunities. Since developed countries consume much more energy than developing ones, a biofuels export market is ripe for exploiting, and developing countries have the climates and growing seasons to take advantage of this opportunity. Several developing countries have climates suited to growing sugarcane or oil palm trees, currently the most efficient feedstocks for conversion to ethanol and biodiesel, respectively (Laney, 2006).

Pesticides are the primary cause of health risks for agricultural workers. Air pollutants caused by field burning can lead to adverse health effects, especially as a result of the cultivation of sugarcane and palm oil (Fritsche *et al.*, 2006). In São Paulo state, Brazil, it has been shown that biomass burning in sugarcane fields is detrimental to respiratory health of two vulnerable populations: children and the elderly (Uriarte *et al.*, 2009). Air quality data collected at selected monitoring stations suggest that burning of sugarcane fields leads to greater concentrations of pollutants associated with fires such as smoke, inhalable particulate matter, and total level of suspended particulates. In response to public pressure, municipalities have been pressuring sugarcane producers to stop burning. According to the law, mechanizable areas (<12% slope and >150 ha) must be converted to green cane harvesting by 2020. Nonmechanizable areas are expected to be phased out of production by 2030 (Uriarte *et al.*, 2009).

Future Initiatives

There will be a significant increase in the production of new bioenergy crops worldwide over the next 10–20 years and the versatility and productivity of sugarcane will continue to support agriculture through multiple energy products for developing, emerging, and developed economies. There are too many alternatives for the conversion of biomass to ethanol or other more directly used forms of energy. All conversion technologies are neither fully developed nor optimized. There is also the possibility of the integration of first and second generations of these technologies. Overall, despite these technologies not being developed enough, the need of development of double purpose or energy cane is a high demanding task, considering the long time frame needed (average of 12 years) to raise a new cultivar for each of these different purposes and the prevision that in 20 years we can reach the maximum efficiencies in energy conversion by enzymatic hydrolysis. This means crossing our modern hybrids with wild *Saccharum* species and related genera, such as *S. spontaneum*, *Erianthus,* and further process of nobilization will be crucial (Loureiro *et al.*, 2011). Because phenotypic characterization of cell wall properties becomes crucial in these breeding efforts, we need to accomplish a complex analysis of plant cell wall structure in parallel to improvement efforts. Lignin and phenolics biosynthesis or degradation, and even structure (lignin) is far from well known, particularly in grasses. The risks from diseases and pests will also increase substantially with major economic losses predicted.

The development of successful breeding programs for biotic resistance will be crucial for sustainable production of bioenergy crops. One of the greatest risks is that breeding programs will rely on a narrow genetic base of resistance. Stacked genetic resistance within clonal crops or strategic management of multiple strains of clonal crops containing single dominant genes must be available to ensure that pathogen threats do not overcome the resistance strategies. Potential improvements include higher sucrose content and longer harvest seasons along with enhanced sustainability. The latter would include varieties that use water and nitrogen more efficiently and by decreasing the dependence on applied chemicals to control pests and diseases (Rohwer *et al.*, 2007).

The industry also needs precision agriculture to link management to the spatial distribution of pest and disease outbreaks at large scales and allow early warning predictions. Ecologically based solutions combined with creative science-based landscape and area-wide management strategies will be required to reduce biotic pressure regardless the production focus. These industry changes are generating key challenges for pest and disease management that will require scientists, agronomists, and producers to work together (Goebel and Sallam, 2011).

References

Aitken, K.S., Hermann, S., Karno, K., Bonnett, G.D., McIntyre, L.C., and Jackson, P.A. (2008) Genetic control of yield related stalk traits in sugarcane. *Theoretical and Applied Genetics* 117: 1191–1203.

Aitken, K.S., Jackson, P.A., and McIntyre, C.L. (2006) Quantitative trait loci identified for sugar related traits in a sugarcane (Saccharum spp.) cultivar *Saccharum officinarum* population. *Theoretical and Applied Genetics* 112: 1306–1317.

Aitken, K.S., Jackson, P.A., McIntyre, C.L., and Piperidis, G. (2002) Marker assisted introgressing of high sucrose genes in sugarcane. Proceedings of the 12th Australasian Plant Breeding Conference, Perth, Australia, p. 120.

Akachi, T. (2007) Performance and characteristics of the harvester under green cane harvesting in Okinawa. *Proceedings of the International Society of Sugar Cane Technologists* 26: 181–184.

Alonso-Pippo, W., Luengo, C.A., Fonseca, F.F., Garzone, P., and Conacchia, G. (2009a) Cogeneration and bio-oil production starting from sugarcane biomass residues: barriers, challenges and opportunities. *The Open Fuels and Energy Science Journal* 2: 34–39.

Alonso-Pippo, W., Luengo, C.A., Fonseca, F.F., Garzone, P., and Conacchia, G. (2009b) Energy recovery from sugarcane biomass residues: challenges and opportunities of bio-oil production in the light of second generation biofuels. *Journal of Renewable and Sustainable Energy* 1: 063102 (15 pages); http://dx.doi.org/10.1063/1.3259170 (accessed 25 September 2012).

Amú, L.G., Briceño, C.O., Moreno, E., and Gomez, L. (2005) Estimating costs of handling sugarcane trash for use as boiler fuel. *Proceedings of the International Society of Sugar Cane Technologists* 25: 310–315.

Anderson, D.L. and Bowen, J.E. (1990) *Sugarcane Nutrition*, Potash and Phosphate Institute, Atlanta, GA. ISBN 0-9629598-0-4.

Anderson-Teixeira, K.J., Davis, S.C., Master, M.D., and Delucia, E.H. (2009) Changes in soil organic carbon under biofuel crops. *GCB Bioenergy* 1: 75–96.

Archer, J.R. and Smith, P.D. (1972) The relation between bulk density, available water capacity, and air capacity of soils. *Journal of Soil Science* 23: 475–480.

Arencibia, A.D. and Carmona, E.R. (2006) Sugarcane (*Saccharum* spp.). *Methods of Molecular Biology* 344: 227–235.

Arencibia, A., Vázquez, R.I., Prieto, D., *et al.* (1997) Transgenic sugarcane plants resistant to stem borer attack. *Molecular Breeding* 3: 247–255.

Asnaghi, C., Paulet, F., Kaye, C., *et al.* (2000) Application of synteny across Poaceae to determine the map location of a sugarcane rust resistance gene. *Theoretical and Applied Genetics* 101: 962–969.

Asnaghi, C., Roques, D., Ruffel, S., *et al.* (2004) Targeted mapping of a sugarcane rust resistance gene (Bru1) using bulked segregant analysis and AFLP markers. *Theoretical and Applied Genetics* 108: 759–764.

Baker, C. (2012) SASRI's sugarcane varieties. *South African Sugar Journal* 96: 72.

Beattie, R. and Crossley, R. (2007) Achieving world's best practice harvesting and transport costs for the NSW sugar industry. SRDC project number: NSC 006, Sugar Research and Development Corporation, Brisbane, Australia, 26p.

Bernardo, R. (2008) Molecular markers and selection for complex traits in plants: learning from the last 20 years. *Crop Science* 48: 1649–1664.

Berndes, G. (2002) Bioenergy and water—the implications of large-scale bioenergy production for water use and supply. *Global Environmental Change* 12: 253–271.

Berry, S., Cadet, P., and Spaull, V.W. (2004) Estimation of nematode damage to sugarcane in sandy soil in KwaZulu Natal. *African Plant Protection* 10: 125

Bessin, R.T., Moser, E.B., and Reagan, T.E. (1990) Integration of control tactics for management of the sugarcane borer (Lepidoptera: Pyralidae) in Louisiana sugarcane. *Journal of Economic Entomology* 83: 1563–1569.

Bocci, E., Di Carlo, A., and Marcelo, D. (2009) Power plant perspectives for sugar cane mills. *Energy* 34: 689–698.

Botha, F.C. (2009) Energy yield and cost in a sugarcane biomass system. *Proceedings of Australian Society of Sugarcane Technologists* 31: 1–9.

Bower, R. and Birch, R.G. (1992) Transgenic sugarcane plants via microprojectile bombardment. *The Plant Journal* 2: 409–416.

Braga, D.P.V., Arrigoni, E.D.B, Burnquist, W.L., Silva-Filho, M.C., and Ulian, E.C. (2001) A new approach for control of Diatraea saccharalis (Lepidoptera: Crambidae) through the expression of an insecticidal Cryla(b) protein in transgenic sugarcane. *Proceedings of the International Society of Sugar Cane Technologists* 24: 331–336.

Braga, D.P.V., Arrigoni, E.D.B, Silva-Filho, M.C., and Ulian, E.C. (2003) Expression of the Cry1Ab protein in genetically modified sugarcane for the control of Diatraea saccharalis (Lepidoptera: Crambidae). *Journal of New Seeds* 5: 209–221.

Braithwaite, K.S., Geijskes, R.J., and Smith, G.R. (2004) A variable region of the sugarcane bacilliform virus (SCBV) genome can be used to generate promoters for transgene expression in sugarcane. *Plant Cell Reports* 23: 319–326.

Brentrup, F., Küsters, J., Kuhlmann, H., and Lammel, J. (2001) Application of the Life Cycle Assessment methodology to agricultural production: an example of sugar beet production with different forms of nitrogen fertilizers. *European Journal of Agronomy* 14: 221–233.

Bundock, P.C., Eliott, F.G., Ablett, G., *et al.* (2009) Targeted single nucleotide polymorphism (SNP) discovery in a highly polyploid plant species using 454 sequencing. *Plant Biotechnology Journal* 7: 347–354.

Butterfield, M. (2007) Marker assisted breeding in sugarcane: a complex polyploid. PhD thesis, University of Stellenbosch.

Cai, Q., Aitken, K.S., Deng, H.H., *et al.* (2005) Verification of the introgression of Erianthus arundinaceus into sugarcane using molecular markers. *Plant Breeding* 124: 322–328.

Casa, A.M., Pressoir, G., Brown, P.J., *et al.* (2008) Community resources and strategies for association mapping in sorghum. *Crop Science* 48: 30–40.

Cheavegatti-Gianotto, A., Couto de Abreu, H.A., Arruda, P., *et al.* (2011) Sugarcane (Saccharum X officinarum): a reference study for the regulation of genetically modified cultivars in Brazil. *Tropical Plant Biology* 4: 62–89.

Christy, L., Arvinth, S., Saravanakumar, M., *et al.* (2009) Engineering sugarcane cultivars with bovine pancreatic trypsin inhibitor (aprotinin) gene for protection against top borer (*Scirpophaga excerptalis* Walker) *Plant Cell Reproduction* 28: 175–184.

Contreras, A.M, Rosa, E., Pérez M., Van Langenhove, H., and Dewulf, J. (2009) Comparative Life Cycle Assessment of four alternatives for using by-products of cane sugar production. *Journal of Cleaner Production* 17: 772–779.

Corcodel, L. and Roussel, C. (2010) Energy content: A new approach to cane evaluation. *Proceedings of the International Society of Sugar Cane Technologists* 27: 1–7.

Cordeiro, G., Eliott, F., McIntyre, C., Casu, R., and Henry, R. (2006) Characterisation of single nucleotide polymorphisms in sugarcane ESTs. *Theoretical and Applied Genetics* 113: 331–343.

Cornland, D., Johnson, F.X., Yamba, F., *et al.* (2001) *Sugarcane Resources for Sustainable Development: A Case Study in Luena, Zambia*, Stockholm Environment Institute, Stockholm, p. 94.

Costet, L., Le Cunff, L., Royaert, S., *et al.* (2012) Haplotype structure around Bru1 reveals a narrow genetic basis for brown rust resistance in modern sugarcane cultivars. *Theoretical and Applied Genetics* 125: 825–836. doi:10.1007/s00122–012-1875-x.

Dale, J. (2007) *Cellulosic Ethanol: Huge Potential but Challenging. Centre for Tropical Crops and Biocommodities*, Queensland University of Technology, Australia.

Damaj, M.B., Kumpatla, S.P., Emani, C., *et al.* (2010) Sugarcane DIRIGENT and O-METHYLTRANSFERASE promoters confer stem-regulated gene expression in diverse monocots. *Planta* 231:1439–1458.

Daniels, J. and Roach, B.T. (1987) Taxonomy and evolution, in *Sugarcane Improvement Through Breeding* (ed. D.J. Heinz), volume 11, Elsevier, Amsterdam, The Netherlands, pp. 7–84.

Daugrois, J.H., Grivet, L., Roques, D., *et al.* (1996) A putative major gene for rust resistance linked with a RFLP marker in sugarcane cultivar 'R570'. *Theoretical and Applied Genetics* 92: 1059–1064.

De Beer, A.G., Boast, M.M.W., and Worlock, B. (1989) The agricultural consequences of harvesting sugarcane containing various amounts of tops and trash. *Proceedings of the South African Sugar Technologists Association* 63: 107–110.

De Beer, A.G., Hudson, C., Meyer, E., and Seigmund, B. (1996) Green cane harvesting and trash management. *Proceedings of the International Society of Sugar Cane Technologists* 22: 133–141.

De Carvalho Macedo, I., Leal, M.R.L.V., and Hassuani, S.J. (2001) Sugar cane residues for power generation in the sugar/ethanol mills in Brazil. *Energy for Sustainable Development* 5: 77–82.

Deepchand, K. (2005) *Sugar Cane Bagasse Energy Cogeneration—Lessons from Mauritius, Parliamentarian Forum on Energy Legislation and Sustainable Development*, Cape Town, South Africa.

De Fraiture, C. and Berndes, G. (2009) Biofuels and water, in *Biofuels: Environmental Consequences and Interactions with Changing Land Use* (eds R.W. Howarth and S. Bringezu), Proceedings of the Scientific Committee on Problems of the Environment (SCOPE) International Biofuels Project Rapid Assessment, 22–25 September 2008, Gummersbach Germany. Cornell University, Ithaca, NY, pp. 139–153. http://cip.cornell.edu/biofuels/ (accessed 21 September 2012).

D'Hont, A., Souza, G.M., Menossi, M., *et al.* (2008) Sugarcane: a major source of sweetness, alcohol, and bio-energy, in *Genomics of Tropical Crop Plants* (eds P.H. Moore and R. Ming), Springer, New York, pp. 483–513.

Dominguez-Faus, R., Powers, S.E., Burken, J.G., and Alvarez, P.J. (2009) The water footprint of biofuels: a drink or drive issue? *Environmental Science and Technology* 43: 3005–3010.

Dominy, C.S., Haynes, R.J., and van Antwerpen, R. (2001) Long term effects of sugarcane production on soil quality in the South Coast and Midlands areas of KwaZulu-Natal. *Proceedings of the South African Sugar Technologists Association* 75: 222–227.

Donaldson, R.A. (2009) Season effects on the potential biomass and sucrose accumulation of some commercial cultivars of sugarcane, PhD thesis, University of KwaZulu-Natal, Pietermaritzburg, South Africa, p. 173.

Donaldson, R.A., Redshaw, K.A., Rhodes, R., and van Antwerpen, R. (2008a) Season effects on productivity of some commercial South African sugarcane cultivars, I: biomass and radiation use efficiency. *Proceedings of the South African Sugar Technologists Association* 81: 517–527.

Donaldson, R.A., Redshaw, K.A., Rhodes, R., and van Antwerpen, R. (2008b) Season effects on productivity of some commercial South African sugarcane cultivars: II. Trash production. *Proceedings of the South African Sugar Technologists Association* 81: 528–538.

Dufour, P., Deu, M., Grivet, L., *et al.* (1997) Construction of a composite sorghum genome map and comparison with sugarcane, a related complex polyploid. *Theoretical and Applied Genetics* 94: 409–418.

Duke, J.A. (1978) The quest for tolerant germplasm, in *ASA Special Symposium 32, Crop Tolerance to Suboptimal Land Conditions*. American Society of Agronomy, Madison, WI, pp. 1–61.

Duke, J.A. (1979) Ecosystematic data on economic plants. *Quarterly Journal of Crude Drug Research* 17: 91–110.

Duku, M.H., Gu, S., and Hagan, E.B. (2011) A comprehensive review of biomass resources and biofuels potential in Ghana. *Renewable and Sustainable Energy Reviews* 15: 404–415.

Eathington, S.R., Crosbie, T.M., Edwards, M.D., Reiter, R.S., and Bull, J.K. (2007) Molecular markers in a commercial breeding program. *Crop Science* 47: 154–163.

Edgerton, M.D. (2009) Increasing crop productivity to meet global needs for feed, food, and fuel. *Plant Physiology* 149: 7–13.

Eranki, P. and Dale, B.E. (2011) Comparative life cycle assessment of centralized and distributed biomass processing systems combined with mixed feedstock landscapes. *GCB Bioenergy* 3: 427–438.

Evans, J.M. and Cohen, M.J. (2009) Regional water resource implications of bioethanol production in the Southeastern United States. *Global Change Biology* 15: 2261–2273.

FAO. (1992) *Production Yearbook*, FAO, Rome.

FAOSTAT. (2012) Available at: http://faostat.fao.org/site/567/DesktopDefault.aspx? PageID=567#ancor (accessed 5 May 2012).

Falco, M. and Silva-Filho, M. (2003) Expression of soybean proteinase inhibitors intransgenic sugarcane plants: effects on natural defense against Diatrea saccharalis. *Plant Physiology and Biochemistry* 41: 761–766.

Falco, M.C., Tulmann Neto, A., and Ulian, E.C. (2000) Transformation and expression of a gene for herbicide resistance in a Brazilian sugarcane. *Plant Cell Reports* 19: 1188–1194.

Filho, J.P.R. (2005) Trash recovery cost, in *Biomass Power Generation: Sugar Cane Bagasse and Trash. Programa das Nações Unidas para o Desenvolvimento* (eds S.J. Hassuani, M.R.L.V. Leal, and I. de Carvalho Macedo), Centro de Tecnologia Canavieira, Piracicaba, Brazil, pp. 75–85.

Filho, P.A. and Badr, O. (2004) Biomass resource for energy in North-Eastern Brazil. *Applied Energy* 77: 51–67.

Finnveden, G., Hauschild, M.Z., Ekvall, T., *et al.* (2009) Recent developments in life cycle assessment. *Journal of Environmental Management* 91: 1–21.

Fritsche, U.R., Hünecke, K., Herman, A., Schulze, F., and Wiegman, K. (2006) Sustainability Standards for bioenergy. WWF, Germany, Frankfurt am Main, 80p. Available at: www.wwf.de/fileadmin/fm-wwf/Publikationen-PDF/Sustainability_Standards_for_Bioenergy.pdf (accessed 5 May 2012)

Gallo-Meagher, M. and Irvine, J.E. (1996) Herbicide resistant transgenic sugarcane plants containing the bar gene. *Crop Science* 36: 1367–1374.

Gerbens-Leenes, P.W., Hoekstra, A., and Van der Meer, T.H. (2008) Water footprint of bioenergy and other primary energy carriers. Value of Water Research Report Series No. 29. UNESCO-IHE, Delft, The Netherlands.

Glaszmann, J.C., Dufour, P., Grivet, L., *et al.* (1997) Comparative genome analysis between several tropical grasses. *Euphytica* 96: 13–21.

Goebel, F.R. and Sallam, N. (2011) New pest threats for sugarcane in the new bioeconomy and how to manage them. *Current Opinion in Environmental Sustainability* 3: 81–89.

Goldenberg, J. (2011) The role of biomass in the world energy system, in *Routes to Cellulosic Ethanol* (eds M.S. Buckeridge and G.H. Goldman), Springer Science+Business Media, London, pp. 3–14.

Govender, M. (2011) COP17 and renewable energy. *South African Sugar Journal* 95: 202–203.

Grivet, L., D'Hont, A., Dufour, P., Hamon, P., Roques, D., and Glaszmann, J.C. (1994) Comparative genome mapping of sugar cane with other species within the Andropogoneae tribe. *Heredity* 73: 500–508.

Groenewald, J.H. and Botha, F.C. (2008) Down-regulation of pyrophosphate: fructose 6-phosphate 1-phosphotransferase (PFP) activity in sugarcane enhances sucrose accumulation in immature internodes. *Transgenic Research* 17: 85–92.

Groenewald, J.H., Hiten, N.F., and Botha, F.C. (2000) The introduction of an inverted repeat to the 5′ untranslated leader sequence of a transgene strongly inhibits gene expression. *Plant Cell Reports* 19: 1098–1101.

Guimarães C.T., Sills, G.R., and Sobral, B.W.S. (1997) Comparative mapping of Andropogoneae: Saccharum L. (sugarcane) and its relation to sorghum and maize. *Proceedings of National Academy of Sciences of the United States of America*, 94: 14261–14266.

Gupta, S.C., de Wet, J.M.J., and Harlan, J.R. (1978) Morphology of Saccharum–Sorghum hybrid derivatives. *American Journal of Botany* 65: 936–942.

Gupta, P., Rustgi, S., and Mir, R. (2008) Array-based high-throughput DNA markers for crop improvement. *Heredity* 101: 5–18.

Hartemink, A.E. (2008) Sugarcane for bioethanol: soil and environmental issues. *Advances in Agronomy* 99: 125–182.

Hassuani, S.J. (2001) Sugarcane trash recovery for use in power generation. *Proceedings of the International Society of Sugar Cane Technologists* 24: 192–196.

Hassuani, S.J. (2005) Evaluation of agronomic routes to unburned cane harvesting with trash recovery, in *Biomass Power Generation: Sugar Cane Bagasse and Trash. Programa das Nações Unidas para o Desenvolvimento* (eds S.J. Hassuani, M.R.L.V. Leal, and I. de Carvalho Macedo), Centro de Tecnologia Canavieira, Piracicaba, Brazil, pp. 44–45.

Hassuani, S.J., Silva, J.E.A.R., and Neves, J.L.M. (2005) Sugarcane trash recovery alternatives for power generation. *Proceedings of the International Society of Sugar Cane Technologists* 22: 394–402.

Haynes, R.J. and Graham, M. (2004) Composition of the soil microbial community under sugarcane production as indicated by phospholipid fatty acid analysis. *Proceedings of the South African Sugar Technologists Association* 78: 331–342.

Heichel, G.H. (1974) Comparative efficiency of energy use in crop production. Connecticut Agricultural Experiment Station Bulletin no.739, 26p.

Hoarau, J-Y., Grivet, L., Offmann, B., *et al.* (2002) Genetic dissection of a modern sugarcane cultivar (*Saccharum* spp.). II. Detection of QTLs for yield components. *Theoretical and Applied Genetics* 105: 1027–1037.

Hurst, C.A., Thorburn, P.J., Lockington, D., and Bristow, K.L. (2004) Sugarcane water use from shallow water tables: implications for improving irrigation water use efficiency. *Agricultural Water Management* 65: 1–19.

Iggo, G.A. and Moberly, P.K. (1976) The concept of minimum tillage in sugarcane. *Proceedings of the South African Sugar Technologists Association* 50: 141–143.

Inman-Bamber, N.G. (1986) The reaction of sugarcane to water stress. Unpublished PhD thesis, University of the Orange Free State, Bloemfontein, South Africa.

Inman-Bamber, N.G. (1994) Temperature and seasonal effects on canopy development and light interception of sugarcane. *Field Crops Research* 36: 41–51.

Inman-Bamber, N.G., Bonnett, G.D., Spillman, M.F., Hewitt, M.L., and Jackson, J. (2008) Increasing sucrose accumulation in sugarcane by manipulating leaf extension and photosynthesis with irrigation. *Australian Journal of Agricultural Research* 59: 13–26.

Inman-Bamber, N.G., Bonnett, G.D., Spillman, M.F., Hewitt, M.L., and Xu, J. (2009) Source-sink differences in genotypes and water regimes influencing sucrose accumulation in sugarcane stalks. *Crop Pasture Science* 60: 316–327.

Inman-Bamber, N.G. and Smith, D.M. (2005) Water relations in sugarcane and response to water deficits. *Field Crops Research* 92: 185–202.

Inman-Bamber, N.G., Zund, P.R., and Muchow, R.C. (2000) Water use efficiency and soil water availability for sugarcane. *Proceedings of the Australian Society of Sugar Cane Technologists* 22: 264–269.

Irvine, J. (1975) Relations of photosynthetic rates and leaf canopy characters to sugarcane yield. *Crop Science* 15: 671–676.

Irvine, J.E. (1981) Sugarcane. Saccharum hybrids, in *CRC Handbook of Biosolar Resources. Resource Materials* (eds T.A. McClure and E.S. Lipinsky), CRC Press, Boca Raton, FL, pp. 211–229.

Jackson, P.A. (2005) Breeding for improved sugar content in sugarcane. *Field Crops Research* 92: 277–290.

Jank, M.S. (2008) Cane for sugar, ethanol and bioelectricity: a global economy. UNICA, the Brazilian Sugarcane Industry. Available at: http://www.unica.com.br (accessed May 2012).

Jannoo, N., Grivet, L., Chantret, N., *et al.* (2007) Orthologous comparison in a gene-rich region among grasses reveals stability in the sugarcane polyploid genome. *Plant Journal* 50: 574–585.

Johnson, F.X., Woods, J., Rosillo-Calle, F., *et al.* (2007) Cane energy for sustainable development and economic competitiveness in Southern Africa. *Proceedings of the International Society of Sugar Cane Technologists* 26: 1246–1256.

Junginger, M., Faaij, A., Rosillo-Calle, F., and Wood, J. (2006) The growing role of biofuels—opportunities, challenges and pitfalls. *International Sugar Journal* 108: 618–629.

Kim, M. and Day, D.F. (2011) Composition of sugar cane, energy cane, and sweet sorghum suitable for ethanol production at Louisiana sugar mills. *Journal of Industrial Microbiology and Biotechnology* 38: 803–807.

Kingston, G. (1994) Benchmarking yield of sugarcane from estimates of crop water use. *Proceedings of the Australian Society of Sugar Cane Technologists* 16: 201–209.

Kingston, G., Donzelli, J.L., Meyer, J.H., *et al.* (2005) Impact of the green cane harvest and production system on the agronomy of sugarcane. *Proceedings of the International Society of Sugar Cane Technologists* 25: 521–531.

Lakshmanan, P., Geijskes, R.J., Aitken, K.S., Grof, C.P.L., Bonnett, G.D., and Smith, G.R. (2005) Sugarcane biotechnology: the challenges and opportunities. *In Vitro Cellular and Developmental Biology—Plant* 41: 345–363.

Lam, E., Shine, J. Jr., daSilva, J., *et al.* (2009) Improving sugarcane for biofuel: engineering for an even better feedstock. *GCB Bioenergy* 1: 251–255.

Laney, K. (2006) Biofuels: Promises and constraints. International Food and Agricultural Trade Policy Council. Available at: http://www.agritrade.org/Publications/DiscussionPapers/IPC_Biofuels_Promises%20and%20Constraints.pdf (accessed 5 May 2012).

Leal, M.R.L.V. (2007) The potential of sugarcane as an energy source. *Proceedings of the International Society of Sugar Cane Technologists* 26: 23–34.

Leibbrandt, N.B. and Snyman, S.J. (2003) Stability of gene expression and agronomic performance of a transgenic herbicide-resistant sugarcane line in South Africa. *Crop Science* 43: 671–677.

Lingle, S.E. (1999) Sugarcane, in *Crop Yield, Physiology and Processes* (ed. D.L. Smith), Springer, Berlin, pp. 287–310.

Lingle, S.E. and Thomson, J.L. (2012) Sugarcane internode composition during crop development. *Bioenergy Research* 5: 168–178.

Lo, C.C., Chen, Y.H., Huang, Y.J., and Shih, S.C. (1986) Recent progress in Miscanthus nobilization program. *Proceedings of the International Society of Sugar Cane Technologists* 19: 514–521.

Loubser, R.C. (2002) Model for estimating effects of harvesting practices on factory output. *Proceedings of the South African Sugar Technologists Association* 76: 42–50.

Loureiro, M.E., Barbosa, M.H.P., Lopes, F.J.F., and Silverio, F. (2011) Sugarcane breeding and selection for more efficient biomass conversion in cellulosic ethanol, in *Routes to Cellulosic Ethanol* (eds M.S. Buckeridge and Goldman), Springer, London, pp. 199–239.

Lyne, P.W.L., Langton, M.I., Bezuidenhout, C.N., and Smithers, J.C. (2007) Performance and economics of the illovo mechanical cane cutter. *Proceedings of the International Society of Sugar Cane Technologists* 26: 168–172.

Macedo, I.C., Leal, M.R.L.V., and da Silva, J.E.A.R. (2004) Assessment of greenhouse gas emissions in the production and use of fuel ethanol in Brazil. Sao Paulo. Available at: http://www.unica.com.br/i_pages/files/pdf_ingles.pdf (accessed 11 September 2012).

Manechini, C., Júnior, A.R., and Donzelli, J.L. (2005) Benefits and problems of trash left in the field, in. *Biomass Power Generation: Sugar Cane Bagasse and Trash. Programa das Nações Unidas para o Desenvolvimento* (eds S.J. Hassuani, M.R.L.V. Leal, and I. de Carvalho Macedo), Centro de Tecnologia Canavieira, Piracicaba, Brazil, pp. 27–35.

Manickavasagam, M., Ganapathi, A., Anbazhagan, V.R., *et al.* (2004) Agrobacterium-mediated genetic transformation and development of herbicide-resistant sugarcane (*Saccharum* species hybrids) using axillary buds. *Plant Cell Reports* 23: 134–143.

Mapako, M., Farioli, F., and Diaz-Chavez, R.A. (2012) Sustainability assessment of energy production from sugar cane resources, in *Bioenergy for Sustainable Development and International Competitiveness—the Role of Sugar Cane in Africa* (eds F.X. Johnson and V. Seebaluck), Routledge, New York, pp. 255–283. ISBN: 978-1-84971-103-6

Martin, L., McFarlane, S.A., and Rutherford, R.S. (2012) Update on a new rust infecting sugarcane. *SASRI Link* 21: 12–13 (January edition).

Mashoko, L., Mbohwa, C., and Thomas, V.M. (2010) LCA of the South African sugar industry. *Journal of Environmental Planning and Management* 5: 793–807.

Mathes, R. and Charpentier, L.J. (1969) Varietal resistance in sugar cane to stalk moth borers, in *Pests of Sugar Cane* (eds J.R. Williams, J.R. Metcalfe, R.W. Mungomery, and R. Mathes), Elsevier Publishing Co, Amsterdam, pp. 175–188.

Matsuoka, S., Ferro, J., and Arruda, P. (2009) The Brazilian experience of sugarcane ethanol industry. *In Vitro Cellular and Developmental Biology—Plant* 45: 372–381.

McCormick, A.J., Cramer, M.D., and Watt, D.A. (2006) Sink strength regulates photosynthesis in sugarcane. *New Phytologist* 171: 759–770.

McFarlane, S.A. (2011) Orange Rust. *South African Sugar Journal*, 95: 27.

Mcguire, P.J., Inderbitzin, M., Rich, B., and Kent, G. (2011) The effect of whole crop harvesting on crop yield. *Proceedings of the Australian Society of Sugar Cane Technologists* 33: 1–12.

McIntyre, C.L., Whan, V.A., Croft, B., Magarey, R., and Smith, G.R. (2005) Identification and validation of molecular markers associated with Pachymetra Root Rot and Brown Rust resistance in sugarcane using map- and association-based approaches. *Molecular Breeding* 16: 151–161.

McMahon, G., Lawrence, P., and O'Grady, T. (2000) Weed control in sugarcane, in *Manual of Cane Growing* (eds D.M.H. Hogarth and P. Allsopp), Bureau of Sugar Experiment Stations, Indooroopilly, pp. 241–261.

Meade, G.P. and Chen, J.C. (1977) *Cane Sugar Hand Book*, John Williamson Limited, New York/London, p. 925.

Meagher, R.L. (2012) Sugarcane IPM, in *Radcliffe's IPM World Textbook* (eds E.B. Radcliffe, W.D. Hutchison, and R.E. Cancelado), URL: http://ipmworld.umn.edu, University of Minnesota, St. Paul, MN. Available at: http://ipmworld.umn.edu/chapters/meagher.htm (accessed 11 September 2012).

Mekete, T., Gray, M.E., and Niblack, T.L. (2009) Distribution, morphological description and molecular characterization of Xiphinema and Longidorus spp. associated with plants (Miscanthus spp. and Panicum virgatum) used for biofuels. *GCB Bioenergy* 1: 257–266.

Meyer, E. (1996) A review of mechanised cane harvesting systems in Southern Africa. *Proceedings of the South African Sugar Technologists Association* 70: 77–84.

Meyer, E. (1997) Factors to consider when implementing a fully mechanised sugarcane harvesting system. *Proceedings of the South African Sugar Technologists Association* 71: 79–81.

Meyer, E. (1999) Some measurements of mechanical sugarcane harvester performance. *Proceedings of the South African Sugar Technologists Association* 73: 16–19.

Meyer, E. (2001) The performance of machinery for mechanical harvesting and loading of sugarcane. *Proceedings of the South African Sugar Technologists Association* 75: 43–45.

Meyer, J.H. (2011) The soil and its environment, in *Good Management Practices Manual for the Cane Sugar Industry* (eds J.H. Meyer, P. Rein, P. Turner, K. Mathias and C. McGregor), International Finance Corporation, Sao Paulo, Brazil.

Meyer, E. and Fenwick, E.J. (2003) Manual sugarcane cutter performances in the southern African region. *Proceedings of the South African Sugar Technologists Association* 77: 150–159.

Meyer, E., Govender, N., and Clowes, M.St.J. (2002) Trials comparing semi-mechanised and chopper harvesting methods over three seasons in Swaziland. *Proceedings of the South African Sugar Technologists Association* 76: 120–134.

Meyer, E., Norris, C.P., Jacquin, E., Richard, C., and Scandaliaris, J. (2005) The impact of green cane production systems on manual and mechanical farming operations. *Proceedings of the International Society of Sugar Cane Technologists* 25: 294–303.
Michelazzo, M.B. and Braunbeck, O.A. (2008) Analysis of six systems of trash recovery in mechanical harvesting of sugarcane. *Revista Brasileira de Engenharia Agrícola e Ambiental* 12: 546–552.
Ming, R., Moore, P.H., Wu, K.K., *et al.* (2006) Sugarcane improvement through breeding and biotechnology. *Plant Breeding Reviews* 27: 17–118.
Mirza, U.K., Ahmad, N., and Majeed, T. (2008) An overview of biomass energy utilization in Pakistan. *Renewable and Sustainable Energy Reviews* 12: 1988–1996.
Moberly, P.K. (1971) The nitrogen requirements of ratoon crops of sugarcane in relation to age and time of harvest. *Proceedings of the South African Sugar Technologists Association* 45: 123–130.
Mortazavi, A., Williams, B.A., McCue, K., Schaeffer, L., and Wold, B. (2008) Mapping and quantifying mammalian transcriptomes by RNA-Seq. *Nature Methods* 5: 621–628.
Muchow, R.C., Spilman, M.F., Wood, W.W., and Thomas, M.R. (1994) Radiation interception and biomass accumulation in a sugarcane crop under irrigated tropical conditions. *Australian Journal of Agricultural Research* 45: 3–49.
Muchow, R.C., Wood, A.W., Higgins, A.J., and McDonald, L.M. (1998) Developing cane supply and harvesting schedules that enhance whole industry profitability. *Proceedings of the South African Sugar Technologists Association* 72: 3–9.
Mukherjee, S.K. (1950) Search for wild relatives of sugarcane in India. *International Sugar Journal* 52: 261–262.
Murray, J.J. (2008) Environmental challenges facing the sugar manufacturing industry. *Proceedings of South African Sugar Technologists Association* 81: 422–429.
Neves, J.L.M. (2005) Development and test of "Copersucar Two Rows Whole Stalk Cane Harvester", in *Biomass Power Generation: Sugar Cane Bagasse and Trash. Programa das Nações Unidas para o Desenvolvimento* (eds S.J. Hassuani, M.R.L.V. Leal, and I. de Carvalho Macedo), Centro de Tecnologia Canavieira, Piracicaba, Brazil, pp. 46–48.
Norris, C. (2011) *Trash Separation*, Norris ECT, Brisbane, Australia.
Norris, C., Lyne, P., and Choonhawong, K. (2010) Sugarcane mechanisation for profitability and sustainability under environmental conservation. *Proceedings of the International Society of Sugar Cane Technologists* 27: 1–6.
Norris, C., Mirande, J.C., Guillen, R., *et al.* (2007) Productivity improvement under green cane harvesting conditions. *Proceedings of the International Society of Sugar Cane Technologists* 26: 82–88.
Núñez, O. and Spaans, E. (2007) Evaluation of green-cane harvesting and crop management with a trash-blanket. *Proceedings of the International Society of Sugar Cane Technologists* 26: 131–142.
Nutt, K.A., Allsopp, P.G., McGhie, T.K., *et al.* (1999) Transgenic sugarcane with increased resistance to canegrubs. *Proceedings of the Australian Society of Sugar Cane Technologists* 21: 171–176.
Paes, L.A.D. and de Oliveira, M.A. (2005) Potential trash biomass of the sugar cane plant, in *Biomass Power Generation: Sugar Cane Bagasse and Trash. Programa das Nações Unidas para o Desenvolvimento* (eds S.J. Hassuani, M.R.L.V. Leal, and I. de Carvalho Macedo), Centro de Tecnologia Canavieira, Piracicaba, Brazil, pp. 19–23.

Panje, R.R. (1972) The role of Saccharum spontaneum in sugarcane breeding. *Proceedings of the International Society of Sugar Cane Technologists* 4: 217– 223.

Panje, R.R. and Babu, C.N. (1960) Studies in Saccharum spontaneum. Distribution and geographical association of chromosome numbers. *Cytologia* 25: 152–172.

Park, S.E., Robertson, M., and Inman-Bamber, N.G. (2005) Decline in the growth of a sugarcane crop with age under high input conditions. *Field Crops Research* 92: 305–320.

Paterson, A.H., Bowers, J.E., Bruggmann, R., *et al.* (2009) The Sorghum bicolor genome and the diversification of grasses. *Nature* 457: 551–556.

Paul, M.J. and Foyer, C.H. (2001) Sink regulation of photosynthesis. *Journal of Experimental Botany* 52: 1383–1400.

Pearson, C.H.O. (1961) The effects of timing of weeding operations on the yield of sugarcane and their economic significance. *Proceedings of the South African Sugar Technologists Association* 35: 141–144.

Perez, R. (1997) Feeding pigs in the tropics. Food and Agriculture Organization of the United Nations. Available at: http://www.fao.org/docrep/003/w3647e/W3647E00.htm (accessed 5 May 2012).

Platford, G.G. (1982) The determination of some soil erodibility factors using a rainfall simulator. *Proceedings of the South African Sugar Technologists Association* 56: 130–133.

Prabhakar, N., Raju, D.V.L.N., and Vidya Sagar, R. (2010) Cane trash as fuel. *Proceedings of the International Society of Sugar Cane Technologists* 27: 11.

Raboin, L., Oliveira, K., Lecunff, L., *et al.* (2006) Genetic mapping in sugarcane, a high polyploid, using bi-parental progeny: identification of a gene controlling stalk colour and a new rust resistance gene. *Theoretical and Applied Genetics* 112: 1382–1391.

Rafalski, J.A. (2002) Novel genetic mapping tools in plants: SNPs and LD-based approaches. *Plant Science* 162: 329–333.

Ramjeawon, T. (2004) Life Cycle Assessment of cane-sugar on the Island of Mauritius. *International Journal of Life Cycle Assessment* 9: 254–260.

Ramjeawon, T. (2008) Life cycle assessment of electricity generation from bagasse in Mauritius. *Journal of Cleaner Production* 16: 1727–1734.

Reffay, N., Jackson, P.A., Aitken, K.S., *et al.* (2005) Characterisation of genome regions incorporated from an important wild relative into Australian Sugarcane. *Molecular Breeding* 15: 367–381.

Rein, P.W. (2005) The effect of green cane harvesting on a sugar mill. *Proceedings of the International Society of Sugar Cane Technologists* 25: 513–520.

Renewable Fuels Association. (2010) Statistics. Available at: http://www.ethanolrfa.org/pages/statistics#E (accessed 11 September 2012).

Renouf, M.A., Antony, G., and Wegener, M. (2005) Comparative environmental Life Cycle Assessment (LCA) of organic and conventional sugarcane growing in Queensland. *Proceedings of the Australian Society of Sugar Cane Technologists* 27: 312–323.

Reouf, M.A. and Wegener, M.K. (2007) Environmental life cycle assessment (LCA) of sugarcane production and processing in Australia. *Proceedings of the Australian Society of Sugar Cane Technologists* 29: 385–400.

Renouf, M.A., Wegener, M.K., and Nielsen, L.K. (2008) An environmental life cycle assessment comparing Australian sugarcane with US corn and UK sugar beet as producers of sugars for fermentation. *Biomass and Bioenergy* 32: 1144–1155.

Richard, C., Jackson, W., and Waguespack, H. (1996) The Louisiana whole-stalk, soldier harvesting system and innovations/modifications to improve its efficiency. *Proceedings of the International Society of Sugar Cane Techologists* 22: 196–200.

Roach, B.T. (1978) Utilization of *Saccharum spontaneum* in sugarcane breeding. *Proceedings of the International Society of Sugar Cane Technologists* 16: 43–58.

Rohwer, J.M., Uys, L., Meyer, K.L., Botha, F.C., and Hofmeyer, J.H. (2007) Plant-scale kinetic and structural modeling of sucrose accumulation in sugarcane. *Proceedings of the International Conference on Systems Biology* 8: 84.

Rutherford, R.S. and Conlong, D.E. (2010) Combating sugarcane pests in South Africa: from researching biotic interactions to bio-intensive integrated pest management in the field. *Proceedings of the International Sugar Cane Technologists* 27: 1–17.

Sandell, G.R., Agnew, J.R., Stainlay, G.T., *et al.* (2002) *The Harvesting Best Practice Manual for Chopper Extractor Harvesters*, Bureau of Sugar Experiment Stations, Indooroopilly, Australia, 97p.

Sarto, C.A. and Hassuani, S.J. (2005) Trash recovery: Baling machines, in *Biomass Power Generation: Sugar Cane Bagasse and Trash. Programa das Nações Unidas para o Desenvolvimento* (eds S.J. Hassuani, M.R.L.V. Leal, and I. de Carvalho Macedo), Centro de Tecnologia Canavieira, Piracicaba, Brazil, pp. 57–60.

Sartori, M.M.P. and Florentino, H.O. (2007) Energy balance optimization of sugarcane crop residual biomass. *Energy* 32: 1745–1748.

SASEX Soil Identification and Management Working Group. (1999) Identification and management of the soils of the South African sugar industry. South African Sugar Association Experiment Station (3rd edn), ISBN 1-874903-17-4.

Schembri, M.G., Hobson, P.A., and Paddock, R. (2002) *The Development of a Prototype Factory-Based Trash Separation Plant*. PK Editorial Services, Brisbane, Australia, pp. 12–18.

Schenk, P.M., Remans, T., Sági, L., *et al.* (2001) Promoters for pregenomic RNA of banana streak badnavirus are active for transgene expression in monocot and dicot plants. *Plant Molecular Biology* 47: 399–412.

Schulze, R.E., Maharaj, M., Lynch, S.D., Howe, B., and Melvil-Thomson, B. (1997) South African Atlas of Agrohydrology and Climatology. Water Research Commission Report, TT82/96, p. 276.

Seabra, J.E.A. and Macedo, I.C. (2011) Comparative analysis for power generation and ethanol production from sugarcane residual biomass in Brazil. *Energy Policy* 39: 421–428.

Seabra, J.E.A., Macedo, I.C., Chum, H.L., Faroni, C.E., and Sarto, C.A. (2011) Life cycle assessment of Brazilian sugarcane products: GHG emissions and energy use. *Biofuels Bioproducts and Biorefining* 5: 519–532.

Seabra, J., Macedo, I., Eustice, T., and van Antwerpen, R. (2012) Sugar cane energy production and climate change mitigation: a case study for South Africa, in *Bioenergy for Sustainable Development and International Competitiveness—the Role of Sugar Cane in Africa* (eds F.X. Johnson and V. Seebaluck), Routledge, New York, pp. 304–328. ISBN: 978-1-84971-103-6

Sichter, N.J., Whiteing, C., and Bonaventura, P. (2005) Estimation of harvester losses by determination of sugar in harvest residue. *Proceedings of the Australian Society of Sugar Cane Technologists* 27: 8.

Singles, A. and Donaldson, R.A. (2000) A simple model of unstressed sugarcane canopy development. *Proceedings of the South African Sugar Technologists Association* 74: 151–154.

Singels, A., Kennedy, A.J., and Bezuidenhout, C.N. (2000) The effect of water stress on sugarcane biomass accumulation and partitioning. *Proceedings of the South African Sugar Technologists Association* 74: 169–172.

Sissine, F. (2007) *Energy Independence and Security Act of 2007: A Summary of Major Provisions*, Congressional Research Service, Washington, DC.

Smit, M.D.L., Coetsee, M.F., and Davies, S.E.H. (2001) Energy expenditure and economy of movement of sugarcane cutters in burnt and green cane. *Proceedings of the South African Sugar Technologists Association* 75: 46–50.

Snyman, S.J., Baker, C., Hucket, B.I., *et al.* (2008) South African Sugarcane Research Institute: embracing biotechnology for crop improvement research. *Sugar Technology* 10: 1–13.

Souza, G.M., Berges, H., Bocs, S., *et al.* (2011) The sugarcane genome challenge: strategies for sequencing a highly complex genome. *Tropical Plant Biology* 4: 145–156.

Spaull, V.W. and Cadet, P. (2003) Impact of nematodes on sugarcane and the benefit of tolerant varieties. *Proceedings of the South African Sugar Technologists Association* 7: 230–238.

SRDC. (2011) *Improving the Harvesting and Transport of Biomass for Sugar and Power Production in NSW*, Sugar Research and Development Corporation, Australia, 37p.

Sreenivasan, T.V., Ahloowalia, B.S., and Heinz, D.J. (1987) Cytogenetics, in *Sugarcane Improvement Through Breeding* (ed. D.J. Heinz), Elsevier, Amsterdam, pp. 211–253.

Stewart, A. and Cromey, M. (2011) Identifying disease threats and management practices for bio-energy crops. *Current Opinion in Environmental Sustainability* 3: 75–80.

Stewart, A. and Cromey, M. (2012) Biofuels: the need for disease risk management. *Biofuels* 3: 1–3.

St. John, J.B., Boethel, D., and Breaux, J. (2007) *Notice of Release of High Fiber Sugarcane Variety Ho 00–961*, Agricultural Center, Louisiana State University, Baton Rouge, Louisiana.

Stranack, R. (2012) Biosecurity. *South African Sugar Journal* 96: 86–87.

Sugarcane Genome Sequencing Initiative. (2009) Proceedings of Plant and Animal Genome XVIII Conference, San Diego, CA.

Tammisola, J. (2010) Towards much more efficient biofuel crops—can sugarcane pave the way? *GM Crops* 1: 181–198.

Tarimo, A.J.P. and Takamura, Y.T. (1998) Sugarcane production, processing and marketing in Tanzania. *African Study Monographs* 19: 1–11.

Tew, T.L. and Cobill, R.M. (2008) Genetic improvement of sugarcane (Saccharum spp.) as an energy crop, in *Genetic Improvement of Bioenergy Crops* (ed. W. Vermerris), Springer, New York, pp. 249–272.

Thompson, G.D. (1966) The production of trash and its affects as a mulch on the soil and on sugarcane nutrition. *Proceedings of the South African Sugar Technologists Association* 40: 333–342.

Thompson, G.D. (1978) The production of biomass by sugarcane. *Proceedings of the South African Sugar Technologists Association* 52: 180–187.

Thompson, G.D., Pearson, C.H.O., and Cleasby, T.G. (1963) The estimation of the water requirements of sugarcane in Natal. *Proceedings of the South African Sugar Technologists Association* 37: 134–142.

Thorburn, P.J., Archer, A.A., Hobson, P.A., *et al.* (2006a) Integrated value chain scenarios for enhanced mill region profitability. SRDC Project CSE010, Sugar Research and Development Corporation, Brisbane, Australia, 87p.

Thorburn, P.J., Archer, A.A., Hobson, P.A., *et al.* (2006b) Value chain analyses of whole crop harvesting to maximise co-generation. *Proceedings of the Australian Society of Sugar Cane Technologists* 28: 12.

Thorburn, P.J., Archer, A.A., Hobson, P.A., *et al.* (2007) Evaluating regional net benefits of whole crop harvesting to maximise co-generation. *Proceedings of the International Society of Sugar Cane Technologists* 26: 1815–1824.

Thorburn, P.J., Biggs, J.S., Webster, A.J., and Biggs, I.M. (2011) An improved way to determine nitrogen fertilizer requirements of sugarcane crops to meet global environmental challenges. *Plant Soil* 339: 51–67.

Thorburn, P.J. and Wilkinson, S.N. (2012) Conceptual frameworks for estimating the water quality benefits of improved agricultural management practices in large catchments. *Agricultural Ecosystems and Environment.* doi:10.1016/j.agee.2011.12.021.

Torres, J.S., Villegas, F., and Durán, A. (2010) Modified dual row planting system for green cane management in the tropics. *Proceedings of the International Society of Sugar Cane Technologists* 27: 1–7.

Tuomisto, H.L., Hodge, I.D., Riordan, R., and Macdonald, D.W. (2012) Comparing energy balances, greenhouse gas balances and biodiversity impacts of contrasting farming systems with alternative land uses. *Agricultural Systems* 108: 42–49.

Uriarte, M., Yackulic, C.B., Cooper, T., *et al.* (2009) Expansion of sugarcane production in São Paulo, Brazil: implications for fire occurrence and respiratory health. *Agriculture, Ecosystems and Environment* 132: 48–56.

USDA (United States Department of Agriculture) (2008) Soil quality indicators: Bulk density. Available at: http://soils.usda.gov/sqi/assessment/files/bulk_density_sq_physical_indicator_sheet.pdf (accessed 5 May 2012).

van Antwerpen, R. (1998) Modelling root growth and water uptake of sugarcane cultivar NCo 376. Unpublished PhD thesis, University of the Orange Free State, Bloemfontein, South Africa.

van Antwerpen, T., McFarlane, S.A., Buchanan, G.F., *et al.* (2008b) Disease notes: First Report of Maize streak virus field infection of sugarcane in South Africa. *Plant Disease* 92: 982.

van Antwerpen, R., Rhodes, R., and Wynne, A.T. (2008a) Economics of trashing: improvements and sensitivity analysis of this decision support program. *Proceedings of the International Society Sugar Cane Technologists* 81: 486–681.

van Antwerpen, T. and Rutherford, R.S. (2008) Increased risk on new virus infections in the South African sugarcane industry: preparing for the future. *Proceedings of the International Society Sugar Cane Technologists* 81: 365–380.

van Den Berg, M., Singels, A., Armitage, R.M., Gillitt, C.G., Way, M.J., and Mcfarlane, S.A. (2009) South African sugarcane production and quality in the 2008–2009 milling season. *Proceedings of the International Society Sugar Cane Technologists* 82: 30–49.

van der Laan, M., Jumman, A., and Perret, S.R. Quantifying the environmental benefits of improved water and nitrogen management in irrigated sugarcane: a combined crop modelling and life cycle assessment approach in Pongola, South Africa. *Agricultural Systems* (accepted for publication).

Van Dillewijn, C. (1952) *Botany of Sugarcane*, Veenman and Zonen, Wageningen, The Netherlands, p. 371.

Van Heerden, P.D.R., Donaldson, R.A., Watt, D.A., and Singles, A. (2010) Biomass accumulation in sugarcane: unravelling the factors underpinning reduced growth phenomena. *Journal of Experimental Botany* 61: 2877–2887.

Van Rijn, P.J. (2000) *Weed Management in the Humid and Sub-humid Tropics*. Royal Tropical Institute, Amsterdam.

Vered, E. and Rao, V.P. (2012) Sugarcane (Agronomic practices, Fertigation). Available at: http://www.sugarcanecrops.com/ (accessed 5 May 2012).

Viator, R.P., Johnson, R.M., and Richard, E.P. (2005) Challenges of post-harvest residue management in the Louisiana sugarcane industry. *Proceedings of the International Society of Sugarcane Technologists* 25: 238–245.

Vomocil, J.A. and Flocker, W.J. (1961) Effect of soil compaction on storage and movement of soil air and water. *Transactions of American Society of Agricultural Engineers* 4: 242–246.

Waclawovsky, A.J., Sato, P.M., Lembke, C.G., Moore, P.H., and Souza, G.M. (2010) Sugarcane for bioenergy production: an assessment of yield and regulation of sucrose content. *Plant Biotechnology Journal* 8: 263–276.

WADE. (2004) Bagasse Cogeneration—Global Review and Potential, World Alliance for Decentralized Energy, Edinburgh, Scotland, 72p.

Wallis, P.D., Haynes, R.J., Hunter, C.H., and Morris, C.D. (2010) Effect of land use and management on soil bacterial biodiversity as measured by PCR-DGGE. *Applied Soil Ecology.* doi:10.1016/j.apsoil.2010.06.006.

Wang, L.P., Jackson, PA., Lu, X., *et al.* (2008a) Evaluation of sugarcane x Saccharum spontaneum progeny for biomass composition and yield components. *Crop Science* 48: 951–961.

Wang, M., Wu, M., Huo, H., and Liu, J. (2008b) Life-cycle energy use and greenhouse gas emission implications of Brazilian sugarcane ethanol simulated with the GREET model. *International Sugar Journal* 110: 527–545.

Weng, L.X., Deng, H., Xu, J.L., *et al.* (2006) Regeneration of sugarcane elite breeding lines and engineering of stem borer resistance. *Pest Management Science* 62: 178–187.

Wenzl, P., Carling, J., Kudrna, D., *et al.* (2004) Diversity arrays technology (DArT) for whole genome profiling of barley. *Proceedings of the National Academy of Science USA* 101: 9915–9920.

Whiteing, C. (2002) Facilitation of best practice to reduce extraneous matter and cane loss. Final Report—SRDC Project BSS189, Sugar Research and Development Corporation, Brisbane, Australia, 33p.

Whiteing, C., Norris, C.P., and Paton, D.C. (2001) Extraneous matter versus cane loss: finding a balance in chopper harvested green cane. *Proceedings of the International Society of Sugar Cane Technologists* 24: 276–282.

Whittaker, A. and Botha, F.C. (1999) Pyrophosphate: D-fructose-6-phosphate 1-phosphotransferase activity patterns in relation to sucrose storage across sugarcane varieties. *Physiologum Plantarum* 107: 379–386.

Wienese, A. and Purchase, B.S. (2004) Renewable energy: An opportunity for the South African sugar industry? *Proceedings of the South African Sugar Technologists Association* 78: 39–54.

WIIAD. (1994) Energy from sugarcane cogeneration in El Salvador. Report No. 94–03, Winrock International Institute for Agricultural Development, 86p.

Williams, A.G., Audsley, E., and Sandars, D.L. (2006) Determining the Environmental Burdens and Resource Use in the Production of Agricultural and Horticultural Commodities, Main Report. Defra Research Project IS0205, Bedford, United Kingdom, 84p.

WWF (World Wildlife Fund). (2006) A first estimate of the global supply potential for bioenergy. A briefing study commissioned by WWF, July 2006.

Wynne, A.T. and van Antwerpen, R. (2004) Factors affecting the economics of trashing. *Proceedings of the South African Sugar Technologists Association* 78: 207–214.

Yang, H., Zhou, Y., and Liu, J. (2009) Land and water requirements of biofuel and implications for food supply and the environment in China. *Energy Policy* 37: 1876–1885.

Yu, J. and Buckler, E.S. (2006) Genetic association mapping and genome organization of maize. *Current Opinion in Biotechnology* 17: 155–160.

Yuttitam, M., Gheewala, S.H., and Chidthaisong, A. (2011) Carbon footprint of sugar produced from sugarcane in eastern Thailand. *Journal of Cleaner Production* 19: 2119–2127.

Zhang, S.Z., Yang, B.P., Feng, C.L., *et al.* (2006) Expression of the Grifola frondosa trehalose synthase gene and improvement of drought-tolerance in sugarcane (*Saccharum officinarum* L.). *Journal of Integrative Plant Biology* 48: 453–459.

Zhou, M.M., Singels, A., and Savage, M.J. (2000) Physiological parameters for modelling differences in canopy development between sugarcane cultivars. *Proceedings of the South African Sugar Technologists Association* 74: 610–621.

Chapter 4
Sustainable Cellulosic Grass Crop Production

J.H. Fike, D.J. Parrish, and W.B. Fike
Virginia Polytechnic Institute and State University, Blacksburg, VA, USA

Introduction

Transportation and industry in the past were mainly based on renewable, sustainable energy supplies—powered heavily by herbaceous and woody biomass. Well into the early portions of the twentieth century, forage grasses and grain "concentrates" provided the energy draft animals needed to power most conveyances and many machines (Vogel, 1996), and woody biomass was used to power steam engines for transportation and industry and to produce heat. Some nations continue to rely on biomass for a significant portion of their energy mix; however, internal combustion engines and various other fossil-fueled power units have generally made traditional, sustainable, biomass-based systems "obsolete" and ushered in an age of petroleum and coal dependence.

A growing political and public awareness of dwindling fossil fuel supplies—brought into sharper focus by periodic geopolitical "energy crises"—along with concerns about unintended consequences of fossil fuel extraction and use are spurring interest in developing truly sustainable energy-producing systems. Biomass-based systems are, of course, only one approach to regaining sustainability in our energy supply (IPCC, 2011). Many others are under consideration or in various stages of commercialization; but we will focus here perhaps on the ironic return to an "obsolete" energy source—forages and related grasses. As Vogel (1996) so aptly puts it, "back to the future..."

For argument's sake, we shall contend that to be "truly sustainable" energy-producing systems must be totally renewable, that is, must not deplete nonrenewable resources nor overtax slowly renewed resources. (While we have used the phrase, we note that "truly sustainable" is not unlike "very unique" or "somewhat pregnant"—systems either are or are not sustainable, and no modifier belongs in front of "sustainable".) Along with obviously finite resources such as mineral nutrients, water, soil, air, and biodiversity also can be threatened by agricultural production systems. Energy cropping systems that cause net losses of resources or disrupt vital ecosystem services are, by definition, not sustainable—even when implemented in the name of

Biofuel Crop Sustainability, First Edition. Edited by Bharat P. Singh.

sustainability. That sets a very high, perhaps impossible, "deep ecology" standard by which to compare species; but, it is an essentially Hippocratic yardstick—first, do no harm. Our intention is to examine grass-based bioenergy systems for their potential to be environmentally—not just economically—sustainable. Traditional slash-and-burn, or swidden, cropping systems that use only *in situ* resources and that allow for sufficient fallow periods can hypothetically be sustainable, that is, perpetual *ad infinitum* (Dove, 1983); however, even they can be very damaging to the environment when overpopulation forces expansion and shortening of fallow periods (Raintree and Warner, 1986). Most agriculture inherently depends on resources—nutrients, fuels, implements, etc.—that ultimately must be factored into the analysis of whether a cropping system is one that could be carried on "forever," that is, is indeed sustainable.

A Sustainable Energy Crop Ideotype

We shall begin by listing some properties that might characterize an "ideal," or "truly sustainable," energy crop. Possessing all or most of these traits would theoretically augur well for a candidate species—suggesting that its culture and subsequent utilization might constitute a sustainable energy-producing system. We will move from that to a consideration of several grasses that have been studied for their potential as cellulosic energy crops, and we shall try to look at each of those candidates in the light of our ideal type.

Defining a Sustainable Energy Crop Ideotype

The notion of ideotype is familiar to plant breeders (Donald, 1968). It suggests a hypothetical idealized plant—one possessing a combination of characteristics that should allow a cultivar possessing them to fit optimally into its agroenvironmental niche. Such ideotypes serve as goals or targets for breeding efforts. We will use the concept in a similar fashion, while focusing on characteristics that would impact the sustainability of an energy cropping system (Table 4.1). We are using ideotype in a rather naïve way. Many of the traits that we cite are not easily addressed in breeding, that is, they are often controlled by undefined and/or multiple, interacting genes, and some of them are not genetic at all (e.g., farmer acceptance). However, to the degree that these characteristics are already present in a species or might be improved by breeding efforts, they will make that species or cultivar a stronger candidate as an energy crop. We have previously offered such a list (Parrish and Fike, 2009), but we shall expand on it here.

For our characterization of an ideotype, we are assuming biofuels will be produced at distributed processing points scattered across agricultural landscapes—a frequently used model (Cundiff *et al.*, 2009). In this scenario, biomass produced by contracting farmers is hauled to a biorefinery for processing. All such energy cropping systems put a premium on crop characteristics like yield and adaptation to a biorefinery's region, so we can be rather specific about what would make a crop "ideal" for those traits. Because the chemical processes proposed for yet-to-be-built biorefineries are quite diverse, we can only generalize about biochemical components of biomass that might make for optimal feedstock quality. (Furthermore, we shall consider as beyond the scope of this chapter how by-products of biorefineries might well affect the economic aspect of sustainability.) Thus, our overall goal is to examine several key parameters of a hypothetical ideal energy crop species, that is, an ideotype, and to see how they would contribute to sustainable production of second-generation biofuels.

The first nine traits in Table 4.1 speak directly or indirectly to the biological limits of a plant. The next three characteristics are agronomic in their focus, but biology partially underlies two

Table 4.1. A list of biological, agronomic, and environmental factors that will affect a crop's potential for sustainable biomass production. For each factor, there is a continuum of potential values. The farther the value is to the right on the continuum, the more favorable, or positive, that value would be, that is, the more it might contribute to overall sustainability. A sustainable energy crop ideotype would carry traits expressing all of the values in the right column.

	Effect of Trait on Potential for Sustainable Production	
Trait, Characteristic, or Factor	← Limiting	Favoring →
Establishment	Slow and/or inconsistent	Rapid and reliable
Adaptation to target site	Poor or low	Very good
Geographic/stress range	Narrow or limited	Broad
Pest susceptibility	Many and/or serious pests	Few and minor pests
Weediness	Highly invasive	Not invasive
Biomass yield	Low	High
Feedstock quality	Low or recalcitrant	Many high-quality factors
Seed/propagule production	Cannot be commercialized	Already or readily developed
Breeding potential	Limited or recalcitrant	High or malleable
Grower acceptance/perception	Resistant or negative	Willing or positive
Input requirements	High and costly or frequent	Low and infrequent
Management requirements	Novel and/or inflexible	Familiar and flexible
Other environmental impacts	Many or major negatives	Many or major positives

of these as well. The three deal with factors that can affect farmer acceptance or adoption of an energy crop as part of their enterprise (as well as having other impacts). The final factor overlaps with several of those above it but looks more specifically at how production of an energy crop could affect the environment and ecosystem services (either positively or negatively). For each factor, or characteristic, in the left column of Table 4.1, a plant (either a species or a cultivar under consideration as an energy crop) will lie somewhere along a continuum of the values enumerated in the other two columns. Some plants may rank highly for one characteristic (e.g., able to tolerate diverse environments) and poorly on another (e.g., not readily adapted to standard agricultural practices). The hypothetical ideotype would rank very highly for all 13 of the traits or factors. We will begin by discussing each of the 13, to show how it might affect overall sustainability. Then we will make reference to some of these as we later discuss the specifics of various energy crop candidates from the grass family.

The ideal energy crop would be *easily and reliably established.* Also impinging positively on this factor would be establishment using familiar, readily available machinery. A perennial crop might have some advantage for this characteristic, since it would not have to be established yearly, but some perennial crops are slow to come to their full production potential. The ideotype would produce good stands (using modest/economical seeding rates) under a reasonable range of environmental and agronomic conditions, and those stands would produce to their full potential ideally in the year of planting or by no later than the second year.

An energy crop must be *well adapted* to the environment (soils, day-length, climate, etc.) under which it is to be grown. This is an obvious prerequisite for any crop and for producing the yields for which a species is biologically capable. The ideal energy crop will perform maximally (or without significant yield reductions) in the area surrounding the biorefinery that it will supply. This is, without doubt, a very complex capability—not a single trait—one that would have to be addressed point by point in breeding or management programs. As an example, switchgrass (*Panicum virgatum* L.) accessions or cultivars from lower latitudes are sometimes planted at higher latitudes because of a positive effect that longer days can have on their vegetative (biomass) growth (Parrish and Fike, 2005). However, the limited cold hardiness of southern-derived lines limits how far north they can survive. Examples of limited adaptation abound in all crops. We will address some as we discuss various species below.

Besides developing site-specific adaptation for an energy crop ideotype, there could be merit in developing an energy crop that is *well adapted to a range of environmental conditions*: that could tolerate stresses associated with variations in growing conditions. If a species is widely adapted, the value of any improvements made to that species through breeding or other approaches would be multiplied. Plasticity or breadth of adaptation might also be highly desirable in an era of global climate change. With anticipated changes in temperature and rainfall patterns, plants with broader tolerances to heat and water stress could provide a hedge against yield reductions. This perhaps partially underlay the decision by the US Department of Energy (DOE) in 1991 to focus herbaceous biomass research efforts solely on widely adapted switchgrass (Lewandowski *et al.*, 2003; McLaughlin and Kszos, 2005). It was to serve as a "model species" but it had already been established by DOE-funded work that it could be grown with good biomass yields throughout much of the United States (Parrish *et al.*, 2012).

The next two factors in Table 4.1 address more key biological and ecological characteristics affecting the sustainability of an energy crop—its potential to be harmed by pests, such as insects and disease, and its potential to be a pest as a weed. The ideal energy crop would *not be subject to pests that reduce yield or processing value*. Many scenarios envision that energy crops will eventually be grown on extensive acreages—rivaling the land base devoted to maize and wheat (*Triticum aestivum* L.) (Perlack *et al.*, 2005). Unfortunately, widespread plantings increase the likelihood for outbreaks of insects or diseases; organisms that may have been minor problems in small or isolated plantings could produce epidemics or scourges when their hosts proliferate. Plant pathologists and entomologists have the unenviable task of trying to predict which "minor" or occasional pests of species X might become "major." Of course, there is potential for "minor" disease-causing bacteria or viruses to increase in their virulence almost overnight. In an ideotype-development or crop-selection stage, we can only look at currently described pests. Species with clear susceptibilities to substantial attack from foliage feeders, stem borers, smuts, rusts, viruses, etc. would rank poorly for this sustainability factor.

To assure its sustainability, an energy crop must *not be weedy or invasive*. The more easily or aggressively a species might spread, the greater could be its threat to farming systems and to natural ecosystems. Some species that have been or are being considered for biofuel production can spread easily and inadvertently—for example, giant reed (*Arundo donax* L.) (Lewandowski *et al.*, 2003) and Johnson grass (*Sorghum halepense* L.) (DiTomaso *et al.*, 2007). This does not automatically exclude them from consideration, but it should be weighed carefully in any overall decision. More will be said later in comments on general ecosystem worthiness of a biomass species. Choosing a candidate species that is truly native to an area might serve as a logical way to "screen" for potential weediness. The logic is that a native plant that exhibits some ecological balance—showing no tendency to increase in frequency in its natural habitat—would not likely be weedy if cultivated in a similar environment.

The logic and logistics of biorefineries require that sufficient feedstock (*high yield*) be produced within a reasonable hauling distance (Cundiff *et al.*, 2009). Lower-yielding crops or residues implicitly expand the radius required to obtain enough biomass to keep a biorefinery operating, and that radius becomes unviable at some point. (Frequent assumptions are that biorefineries must be of some minimum capacity—several million liters per year (Perlack *et al.*, 2005)—and must operate more or less continuously (Cundiff *et al.*, 2009). A system relying on a too-low-yielding crop could expend more energy in hauling the required biomass than would be netted in its utilization as a biofuel, or shutdowns due to feedstock shortages could make the biorefinery unsustainable economically.

The yield threshold for an energetically sustainable energy crop, that is, more energy produced than consumed, will depend on multiple factors—most especially the amount of land available for dedicated energy crop production. To be a net energy producer, the system must offer enough suitable and "contractable" land within a radius that represents the maximum feasible hauling distance (Cundiff *et al.*, 2009). Maximizing yield will minimize the land requirement. Furthermore, too low yields logically will not provide sufficient biomass within a feasible circumference even if all of the land within that radius might be dedicated to energy crop production. That minimum threshold will depend partially on the efficiency with which the biorefinery can convert biomass in biofuel—something that is beyond the scope of this chapter.

Just as with adaptation, "high yielding" is an oversimplified ideotype characteristic. Yield is complex and difficult to address directly; it is a product of many different biochemical and developmental processes. Assigning an ideotype characteristic that is more specific than "high yielding"—but that would help drive higher yields—would be desirable. For example, some efforts at breeding for increased biomass yield in switchgrass have focused on yield as a product, where the biomass is dissected morphologically. Different ideotypes have been examined: in one case, plants with increased tiller numbers (Das *et al.*, 2004); in another, plants with increased tiller size (Smart *et al.*, 2004). Each approach showed merit. Not unsurprisingly, it has not been possible to achieve both in the same plant. Efforts to increase biomass production in sorghum (*Sorghum bicolor* L.) have similarly focused on more-specific ideotype targets (Rooney *et al.*, 2007). Again, manipulating one trait can result in compensation that results in less than the hoped for overall benefit.

While we know intuitively that higher biomass yields redound to greater potential *within* species, our focus below will be more on differences in biomass production and sustainability attributes *among* species. With the efforts of breeders, those interspecies differences may be narrowed (or widened), but we will choose to look at where we seem to be today (to the degree that we can).

We shall give relatively little coverage to ideal *feedstock qualities* of a sustainable energy crop; but it could someday be one of the most crucial factors in determining whether a crop can be grown sustainably from an economic perspective. Processors will want the biomass used in their biorefineries to have a maximum of quality factors (properties that improve biofuel production) and a minimum of antiquality factors (components that reduce the efficiency of fuel-yielding processes); and that may well translate to higher prices for the biomass. Ideotype developers may be somewhat frustrated when they address this criterion, however. Their frustration will come from the multiplicity of platforms and processes that might be used in biorefineries (Cherubini *et al.*, 2009). Some biorefineries might primarily use biochemical or enzymatic processes to turn biomass into fuels, using low temperatures and pressures and relying on enzymes found in various organisms to "digest" complex biomass and then further process components to produce ethanol or other fuels. Others will take a more conventional chemical ("wet chemistry") approach, using higher temperatures and/or pressures to saccharify biomass and separate out resulting components prior to fermentation. Still others may rely heavily on thermochemical or pyrolytic processes, using high temperatures and pressures and/or limited O_2 to disaggregate the biomass and then condense and process the resultant gases into hydrocarbons. Into this process stew, expect to add highly proprietary variations between seemingly similar biorefineries. Because of this factorial explosion of what might constitute "ideal" feedstock qualities at the refinery, an ideotype might best be developed based on biorefinery types, keying on the specific requirements for each operation.

In order for an energy crop to succeed, *a seed or propagule source should be readily available* and/or readily ramped up for large-scale production. This is both a biological and an economic issue. Biologically, the species must lend itself to producing sufficient seeds to allow planting on large acreages; ample seed increase must be able to follow quickly after suitable cultivars are developed. From an economic perspective (which clearly has sustainability implications), developing seed markets is a chicken-and-egg conundrum. Without sufficient demand for seeds of species X, there is no incentive to produce them. If seeds are not available in sufficient supply to keep their price from escalating, the crop may not as readily be adopted. Thus, for example, the price of switchgrass seeds more than tripled in the mid-1980s, when a few million hectares of Conservation Reserve Program (CRP) land "suddenly" needed to be planted with suitable perennial species. And prices escalated again in the early 2000s when a state supported the planting of thousands of hectares.

Promising energy crop lines that might not breed true or that might be sterile, that is, do not produce viable seeds, would presumably require cloning or some form of asexual propagation. Assuming that the species' biology would allow for commercial-scale production of vegetative propagules, establishing markets for the propagules could be equally or more problematic than for seeds. When Jerusalem artichoke (*Helianthus tuberosus* L.) was being examined as an energy crop in the 1980s (Parrish *et al.*, 1985), some interesting (and seemingly Ponzi-like) schemes were developed to ramp up tuber seed stock production; but the crop never achieved "escape velocity" (developing enough interest to break out of an experimental mode to become a viable, industry-supported enterprise).

Breeding potential determines how easy or difficult it would be to achieve desired character modifications in the species under consideration. Inherent in the notion of ideotypes is the presumption that it could be altered genetically to produce the ideal phenotype. Such alterations could perhaps be done by recombination of genes/alleles within the species; that is, "traditional" plant breeding that might take advantage of the gene pool available via intraspecific crosses. If the species might hybridize with another (by cross-pollination, plus perhaps "embryo rescue," or by protoplast fusion), a new gene pool might thereby be introduced. The broader and deeper the gene pool, the greater the hypothetical potential for introducing ideotypic traits.

We shall group under this same heading of genetic malleability the overall biotechnical manipulability of a species. For example, the ability to develop or maintain some traits might require micropropagation at key points. Not all species are equal in this regard. Some are much more amenable to regeneration of whole plants from tissue culture, while some are deemed recalcitrant (Ochatt *et al.*, 2010).

This leads to the bigger—and more contentious—consideration of the potential for an ideotypic trait to be manipulated by transgenic means. We shall save the contention for a section below. The ability to use transgenic methods broadens and deepens the pool of genes available almost infinitely. Traits that were heretofore found only in bacteria, fungi, other plants, or even animals might be introduced into a bioenergy species. Holding in abeyance for the moment the potential danger of using genetically modified organisms (GMOs), genetic engineering might hold tremendous potential for modifying traits related to yield, feedstock quality, weediness, and pest potential. Considering that GMOs may well be employed in bioenergy systems, it is worth noting that some species are more readily engineered than others. For example, the grasses have only slowly yielded to transformation via *Agrobacterium* plasmids. Some species are not good candidates for particle bombardment or electroporation for introduction of transgenes, because, as noted above, regeneration of whole plants from their cells (transgenic or not) is difficult (Ochatt *et al.*, 2010). Unless such biotechnological barriers can be overcome, the affected species would be rated as lower in potential for development (if the GMO avenue is open for other species).

Perhaps as important as environmental adaptation is cultural adaptation. From a pragmatic standpoint, a promising energy crop should be *readily adopted or adapted by the agricultural community*. Adaptation relates to biological limits or potential, while ready adoption depends on some rather subjective sociological, or cultural, factors. Both adaptation and "adoptation" must be taken into consideration in assessing the sustainability of an energy cropping enterprise. In a survey of Virginia farmers, 57% of the respondents indicated they would not be interested in growing switchgrass for energy even if it would provide them a profit (Wen *et al.*, 2009). Of course, one might expect such a number to erode as early adopters prove an enterprise successful; but the chicken-and-egg nature of biorefineries-and-biomass requires that sufficient producers be fully on board at the outset.

Acceptance or adoption by growers might be more ready if familiar practices or technologies are used in producing the biomass and if perceptions of success are positive. Being able to use implements and practices that are common and readily available (e.g., haymaking equipment) might help overcome landowners' resistance to changes in their agricultural enterprises. Flexibility in required management practices could also increase likelihood of adoption. One aspect of energy crop management that has sustainability implications both for growers and for processors is the desirability of a crop that lends itself to *flexible management*. For growers who grow energy crops as one element of a diversified operation (and for whom the energy crops are likely a less valuable component), being able to schedule biomass harvests around other operations could be advantageous. For the processors at a biorefinery, being able to harvest some portion of a season's biomass on an as-needed basis (to limit the amount of storage required on site) could be advantageous (Cundiff *et al.*, 2009). A crop with characteristics that would allow it to remain harvestable into the winter could have an advantage when considering the sustainability of the system.

Some of the ideotype factors discussed previously could affect adoption of an energy crop because they deal with the perceived potential for success of the enterprise. Farmers want—and need—crops that can be reliably and readily established, to include coming into full production quickly in the case of perennials. Switchgrass has a reputation for being difficult or slow to establish, with failures frequently reported. That reputation is not totally deserved perhaps (Parrish and Fike, 2005), but perception may be more important than reality when planting of a novel crop is under consideration. These are pragmatic matters that could adversely affect the ability of a potential energy crop to "hit escape velocity."

For economic, agronomic, and ecological reasons, *input requirements* for a sustainable energy crop would ideally be low, although we are quick to remind the reader of TINSTAAFL—"there is no such thing as a free lunch." Lower labor and materials (seed, nutrients, fuel, etc.) costs per unit of biomass produced can only favor economic sustainability if the yields in such idealized, low-input systems are sufficient to sustain livelihoods. Mass and energy balances are the crux. Typically, nutrients (fertilizers), agrichemicals (pesticides, growth regulators, etc.), and energy (usually fossil fuels) are externally supplied to maintain production at a level that will be economically viable. To the degree that those external inputs can be minimized, the system will come closer to meeting the goal of being "sustainable," simultaneously protecting or conserving key nonrenewable resources.

For ecological or environmental reasons as well, we might wish to minimize some inputs. Nutrients and various agrichemicals applied to biomass-producing fields that inadvertently get into surface or groundwater could pose threats to water resources. Fuel not burned will avoid emissions of greenhouse gases (GHGs).

The final—and, in some other ways, the ultimate—arbiter in determining the sustainability of an energy cropping system is its *effect on the environment*. A sustainable crop must "tread lightly"; its culture should not incrementally impact soil, air, or water quality or biodiversity in adverse ways, because such increments will eventually cause the system to fail or otherwise reduce the environment's ability to support life. We shall discuss briefly and intuitively some of the more obvious concerns; but we shall also observe how energy crops could have "redeeming" values—providing positive impacts on agroecosystems and the overall environment.

With regard to the soil resource, grasses—especially grasses grown in swards—are generally considered soil conserving. Their dense foliage and mass of fibrous roots can protect against erosion. Perenniality could be a plus in this regard, reducing the time during which fields might be essentially bare during the nongrowing season. Protection of the soil resource could also be aided by growing crops using minimum- or no-till technology, thereby reducing the soil erosion potential. With suitable species and good management, soil erosion could likely be held to sustainable levels in many energy cropping systems; that is, soil losses would not outstrip soil-formation processes. Indeed, with perennial grasses, it might be possible for the soil to be improved through addition of organic matter (Sartori *et al.*, 2006) and facilitation of soil-forming processes. A very positive by-product of increases in organic matter content would be the building of C into some forms that might remain sequestered in the soil (and out of the atmosphere) for scores or even hundreds of years (Lal, 2004).

With regard to water resources, we will dismiss out of hand the notion that irrigating to produce biomass for energy production might be sustainable, except in perhaps a few very limited cases. Neither surface nor groundwater supplies are sufficient in many areas to sustain irrigation *ad infinitum*. In many agricultural settings, already groundwater removal exceeds recharge; and, even if the annual deficit is very small, it is only a matter of time until the water table will be too depressed to allow pumping—it is not sustainable. Likewise, surface water is siphoned from many streams and rivers to the extent that the streams no longer provide or sustain ecosystems with key services. Fisheries and navigation are but two of the ecosystem services "competing" with agriculture. It would seem difficult to justify abrogating those or other ecosystem services in the name of producing more biomass.

By the same token, energy cropping systems must not do damage to the quality of the water resource. Again, many grasses have an inherent advantage over other species in this regard. Minimizing soil runoff has clear downstream (pun fully intended) consequences. Likewise, reduced sediment runoff generally coincides with reduced nutrient runoff and reduced eutrophication of various bodies of water; but to the degree that energy cropping systems would impact adversely water quality (to include potential groundwater contamination), those systems' sustainability would have to be suspect.

The air resource must likewise be protected from potential negative impacts of energy cropping systems—or, better yet, improved wherever possible. The production of biofuels is inherently intended to be protective of the atmosphere—reducing emissions of fossil C. In case of bioenergy crops, CO_2 is not the only GHG in play; N_2O and CH_4 can be released during the production and conversion of biomass. The potential for release of N_2O from energy cropping systems is significant (Smith and Searchinger, 2012). Besides protecting or conserving soil, water, and atmospheric resources, a sustainable energy crop must have minimal—or preferably positive—impacts on biodiversity. Taking highly diverse ecosystems out of production in order to establish energy crop monocultures would very adversely impact biodiversity *ipso facto*. On the other hand, if previously cleared agricultural lands were employed, there should be no net loss of biodiversity; and, with good management, there might even be increased biodiversity. Energy cropping systems that provide wildlife habitat—and that can be managed to minimize adverse impacts on wildlife they attract—gain an advantage in the search for sustainable systems. Grassland species of fauna would be favored in a suitably managed energy cropping system (Roth *et al.*, 2005). Multiple-species plantings—mixed intercrops—might add niches for greater biodiversity as well as increasing yield potential in some cases (Tilman *et al.*, 2006).

We will speak only briefly about the biodiversity (and therefore sustainability) implications of GMOs as energy crops, but modifying energy crops with transgenes introduces major concerns (Carpenter, 2011). Some of those concerns are so great that transgenetic approaches should probably be rejected out of hand for at least some species. The very real and reasonable concerns generally center around "escape" of transgenes with unforeseeable and potentially disastrous consequences for native species and, indeed, for entire ecosystems. The risk is greater if GMOs are open pollinating and might cross with wild types—or cultivars—of their own species or with closely related species. Such unintended introgression of transgenes, sometimes called "genetic pollution," has been documented in numerous cases, but the ecological consequences are not always so readily deciphered. The contention is that unintended consequences are essentially inevitable, and some of those possible outcomes are so unacceptable that the risk should/must be avoided. Some worst-case scenarios envisioned include "super weeds" or other fitness-favoring traits that result in annidation and a swamping out of wild types. As a minimum, the introgression of transgenes into a natural population would, by definition, impinge on natural biodiversity and thereby affect sustainability.

Grass Feedstocks of Interest

Many species hold promise for biomass-to-bioenergy production systems. A simple, imprecise "survey" using the CAB Abstracts database was conducted with the key word phrase "bioenergy or biofuel" coupled to individual genus names. The searches turned up two dozen genera of perennial grasses that in some form have been considered (and perhaps rejected) for bioenergy production (Table 4.2). The list certainly is not comprehensive for perennial grasses, and the search terms were based on a minimalist "dragnet" approach to try to reduce

Table 4.2. Genera of perennial grasses that have been explored (and perhaps rejected) for potential bioenergy and biofuel production based on a search of the CAB Abstracts database. Numbers in parentheses represent hits associated with the genus name and the search terms "biofuel or bioenergy."

Andropogon (27)	*Erianthus (7)*	*Saccharum (727)*
Arundo (57)	*Festuca/Schedonorus (48)*	*Sorghastrum (5)*
Bothriochloa (3)	*Lolium*[a] *(56)*	*Sorghum*[b] *(430)*
Bromus (19)	*Miscanthus (405)*	*Spartina (39)*
Cynodon (23)	*Panicum (529)*	*Thinopyrum (2)*
Dactylis (27)	*Pennisetum*[a] *(58)*	*Tripsacum (8)*
Elymus (12)	*Phalaris (129)*	
Eragrostis (5)	*Phragmites (28)*	

[a]Excludes annual Lolium (*L. multiflorum*) or *Pennisetum* (*P. glaucum*).
[b]Includes *S. bicolor*, which is a perennial that does not perenniate in temperate climates and *S. halepense* (Johnson grass), which perenniates but is of less interest due to potential invasiveness.

"stray" hits. We have ignored a few "one hit wonders," but the list certainly could be expanded by including species outside the Poaceae that may prove important contributors in time—for example, *Medicago* had about 90 hits in our search, and Voigt *et al.* (2012) have recently reviewed a list of possible herbaceous feedstocks. The list would certainly be longer still, if the genera of interest were segregated by species or if it included annual grasses (to include cereal and grain crops). This is a chapter about sustainable cellulosic grass crop production, and we view that domain primarily in the realm of the perennials.

Of the species listed in Table 4.2, several may eventually be deployed for energy cropping, but it is likely that only a few will play a major role in the greater scheme of things. We reiterate that the list is not intended to be comprehensive or exacting, but it should provide some sense of the breadth of genera of interest, and of the interest within a given genus relative to the other feedstocks.

Our aim for the remainder of this section is to provide some detailed discussion of the three most prominent individual feedstock crops for warm temperate zones: *A. donax*, *Miscanthus* spp., and *P. virgatum*. We will provide some background on their history, origins and adaptations, and production and sustainability issues and attributes. Harvest and processing issues to a great degree have common characteristics across feedstocks, and this will be addressed in the next section.

While there appear to be challenges with each feedstock—and greater for some than for others—the reader should recognize that changes in technology throughout the production and supply pipeline have potential for significant impacts on the sustainability metrics of a given feedstock species. In an attempt to maintain feedstock neutrality, the descriptions that follow are presented roughly in alphabetical order by genus. We note, too, that a discussion of giant reed (*A. donax*) may be an appropriate place to begin our consideration of sustainable grass-based bioenergy crops given that the species embodies so much of the promise—and perhaps the perils—of energy cropping.

Arundo donax L.—Giant Reed

Description

Giant reed, a tall, rapidly growing, perennial species is commonly found in much of the warm temperate and subtropical regions of the world (Figure 4.1). The plant is one of the world's largest herbaceous species and can reach heights of more than 10 m with culms up to 4 cm in diameter (Perdue, 1958; Saltonstall *et al.*, 2010). The culms are hollow, with walls 2–7 mm thick; multiple nodes are found along these stems, which are glossy and quite woody in nature. Leaves are glabrous but with hairs at the base. Leaf lamina are about 5–8 cm wide at the base and 30–70 cm long and taper to a narrow point. Leaves along most of the stem have planophile leaf angle and are arranged in a noticeably distichous pattern. If culms are left standing (intact) over winter, new shoots will develop from buds on the stem nodes. If the culms are harvested during the dormant season, new shoots must regrow from meristematic buds on the plant's massive rhizomes. Culm production is indeterminate, but shoots that grow later in the season often die (Lewandowski *et al.*, 2003). Giant reed's large, plume like panicles may be 30–60 cm long (Perdue, 1958). Except for variegated forms, the leaves and stems of giant reed generally are glaucous while panicles are silver to purple in color.

Giant reed typically is considered sterile in the United States (Boland, 2006; Johnson *et al.*, 2006), Australia (Virtue *et al.*, 2010), and Europe (Lewandowski *et al.*, 2003). Although Perdue (1958) reported that seeds collected in the Near East have been viable and generated plants,

Figure 4.1. Giant reed (right) and miscanthus (left) on a slope position in Blacksburg, Virginia and grown with no management inputs. The youth in the photo is about 1.6 m tall.

no recent studies document the development of plants from fertile seed. Work by Johnson *et al.* (2006) suggests some ovaries may be viable, and seed production has been reported by Sharma *et al.* (1998), but the authors indicated that germination and development in the field was precluded by allelopathy. Although the plant has largely increased via asexual means, Saltonstall *et al.* (2010) noted that this does not preclude rare production of viable seed.

Giant reed has a base chromosome number of 12 (2n = 110); aside from the variegated forms that are considered subspecies (*A. donax* var. *variegata*; 2n = 40), the species has only limited genetic diversity (Lewandowski *et al.*, 2003), and this constrains opportunities to find new varieties. Some limited genetic variability is reported for populations in Southern Europe (Lewandowski *et al.*, 2003) and California (Khudamrongsawat *et al.*, 2004). Conflicting research by Ahmad *et al.* (2008) suggests that the giant reed populations in North America essentially are of uniform genetics, although this may simply reflect differences in analytical technique (Saltonstall *et al.*, 2010).

Among the "very large grasses" being considered for bioenergy cropping, giant reed is unique in deploying the C_3 photosynthetic pathway. Rossa *et al.* (1998) reported that the species displays very high photosynthetic and transpiration rates. Giant reed also can survive at low temperatures, although it is subject to die back if late-season frosts occur after emergence.

Origins, Distribution, and Adaptations

Among many uses, humans long have exploited giant reed as a resource for thatch and building structures, for fishing poles and stakes, baskets and mats, and for musical instruments and the sounding devices thereof (Perdue, 1958; Mariani *et al.*, 2010). Some of the earliest (albeit indirect) references to the species may come from biblical allusions to "reeds," which Perdue (1958) asserted would almost surely have included giant reed. The plant has many common names and most of those of English origin reference "reed" or "cane" in some form (Perdue, 1958), which may give some indication of its preference for wet habitats.

The plant's current broad distribution to all of Earth's subtropical and warm temperate regions (Perdue, 1958) is at least in part a testament to its usefulness to earlier human societies as our species scattered across the globe. Perhaps, because of its adaptation and history of use around the Mediterranean, giant reed was historically considered a native to that region (Perdue, 1958; Mariani *et al.*, 2010), although that view has changed in the past several years. Analyses of plant materials collected from 80 sites across the Mediterranean and Asia suggest that the giant reed more likely has its origins in East Asia (Mariani *et al.*, 2010). These conclusions were based on findings of lower genetic diversity and the incidence of rare polymorphisms in plants from around the Mediterranean region compared with greater diversity in plants originating in East Asia.

Giant reed is adaptable to a wide range of climatic and edaphic conditions in the warm temperate to subtropical zones. Soil type, per se, does not appear to be a major factor in giant reed production (Perdue, 1958), although survival rates for new plantings were higher in loam than other soil types (Ceotto and Candilo, 2010).

Lewandowski *et al.* (2003) stated that the plant, classified by suitable site conditions, is a "Mesophyte or almost a Hydrophyte or Xerophyte" (p. 351), because it is capable of surviving long periods of time under very wet and very dry conditions. The capacity for mature plants to survive extended drought is considered a function of giant reed's massive rhizomes (Figure 4.2) and root system. The plant's roots are reported capable of penetrating soils to a meter depth (Sharma *et al.*, 1998).

Giant reed's water demand is very high relative to other plants; water availability is a large factor in its productivity (Watts and Moore, 2011) and will be an important factor in accounting the plant's sustainability. Giant reed plants grown near streams typically are taller and have more leaves than those found at distance from stream edges at the same elevation (Spencer *et al.*, 2005; Watts and Moore, 2011). Although such findings may be confounded by higher soil nutrient status near the stream, the data are supported by the work of Williams *et al.* (2008) who reported that yields were increased by a factor of >3 for irrigated versus dryland reed (51 vs. 15.4 tonne ha^{-1}) harvested 10 months after planting. Such high water use is a key driver of the efforts to shrink *A. donax* populations in arid regions such as the Lower Rio Grande Valley of Texas (Seawright *et al.*, 2009). For example, Seawright *et al.* (2009) calculated that each public dollar invested to remove giant reed would return $4.38 in benefits and that removal would have net positive economic outcomes in terms of increased water availability for agricultural uses.

High water use may not be a problem in the right context, however. Giant reed's adaptation to, and high productivity in, wet sites has made it a candidate for use in various waste water reclamation schemes (Karpiscak *et al.*, 1996; Sudha and Vasudevan, 2009; Calheiros *et al.*, 2010; Abbassi *et al.*, 2011; Idris *et al.*, 2012a, 2012b). Interestingly, Tzanakakis *et al.* (2011) reported that NO_3-N levels in pore water under giant reed were lower than under other reclamation species. Their analysis indicated that this was not due to differences in N uptake

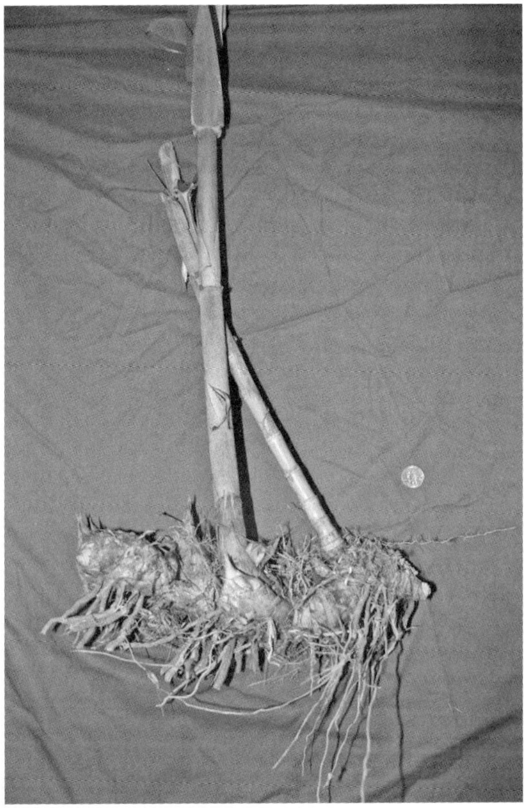

Figure 4.2. A giant reed rhizome of a 4–5 cm diameter size.

and removal, but could reflect differences in denitrification due to the unique environment of giant reed's rhizosphere. Giant reed also displays tolerance of heavy metals in contaminated soils (Papazoglou *et al.*, 2005; Papazoglou *et al.*, 2007; Papazoglou, 2009) and the species has been suggested as a biomonitor for contaminated water and sediments (Bonanno, 2012).

Concern about diseases in production systems seems minimal. Although giant reed has been distributed broadly across diverse environments, the species displays little sensitivity to pests and pathogens despite its limited genetic diversity. Giant reed is host to over 190 fungi (Farr and Rossman, 2012) but few are reported to cause serious damage. One, *Nigorsproa oryzae*, is associated with shoot tip death (Widmer and Kirk, 2005) and foliar and cane rot but likely requires insect transmission (Widmer and Kirk, 2005; Williams *et al.*, 2008). *N. oryzae* now is under investigation as a potential biological control agent.

Similar to fungal pathogens, few insect pests historically have been identified that have serious consequences for giant reed growth, despite its role as a host for several Lepidoptera species. Recent research efforts have, however, found promising insect species that may be useful for biological control. Predation by the diaspidid scale *Rhizasidiotus donacis* can reduce giant reed rhizome weights by about 50% and shoot growth rates by more than twofold (Cortés *et al.*, 2011a, 2011b). *R. donacis* has since become the first scale released as a giant reed control agent (Goolsby *et al.*, 2011), although it may be more effective in concert with the Arundo wasp (*Tetramesa romana* Walker) (Moore *et al.*, 2010), which also has been released in the region (Racelis *et al.*, 2010).

Crop Management and Production

Because giant reed does not produce viable seed, establishment requires some form of vegetative propagation, and this adds greatly to production costs relative to those of seeded crops. The primary component of this cost is the plant material and amount required, which is a function of planting density (Lewandowski *et al.*, 2003). Historically, bioenergy plantings have largely been established by rhizomes (Ceotto and Candilo, 2010), and although new plants can be produced from meristematic buds located at the nodes on the canes, this has often proved less successful (Lewandowski *et al.*, 2003).

More recently, Ceotto and Candilo (2010) reported that propagating new plantlets from stem nodes would be more successful with stems that were more than 1 year old. The lateral branches that form on older stems have greater propensity for root development and limited irrigation (245 mm over the season) applied on sandy soils was sufficient to support the developing crop. Despite this advance, future plant propagation schemes for large-scale establishment are more likely to be produced via tissue culture. Several recent reports have shown the feasibility of such techniques (Cavallaro *et al.*, 2011; Tzanakakis *et al.*, 2011).

Advances in propagule development are likely to significantly drop the cost of vegetative establishment, although stand density—plantlets established per land area—will still be a critical factor. High stand density may allow more rapid ground cover, reducing weed competition and evaporative losses, but in some cases this can reduce plant size due to increased competition among plants for light and nutrients (Lewandowski *et al.*, 2003). Angelini *et al.* (2005) tested productivity as a function of rhizome planting density (20 000 vs. 40 000 plants ha^{-1}) and reported that the greater stand density supported higher biomass dry matter yields in the first 2 years following planting. However, dry matter yields did not differ by stand density treatment from the third year of production. Rhizome pieces planted for this study weighed approximately 500 g and would have had substantial reserves for subsequent growth or competing with weeds, but the feasibility and cost of very large plantings of such large rhizomes remain in question. Resource inputs (e.g., weed control, irrigation) required to support establishment with vegetative or micropropagules across an array of potential environments bear further elucidation.

As noted above, giant reed is of interest because of its high potential yields. For example, Lewandowski *et al.* (2003) noted a Spanish study in which biomass production ranged from about 30 to 63 Mg DM ha^{-1} (mean = 45.6 Mg DM ha^{-1}). In Italy, in a study spanning more than a decade, yields peaked at over 50 Mg DM ha^{-1} in the year following establishment and approached 38 Mg DM ha^{-1} when averaged over 11 years (Angelini *et al.*, 2009). Seasonal fertility inputs also were high, however, with the crop receiving annual inputs of 100–100–100 kg ha^{-1} N-P_2O_5-K_2O. Average annual productivity over time was high despite the fact that the final 4 (of 11) measurement years had yields of less than 30 Mg ha^{-1}. These and other data that display declining yields over time (Angelini *et al.*, 2005; Knoll *et al.*, 2012) suggest that some form of disturbance could improve annual yields.

Although low-input biomass cropping systems are given greater consideration and deference for being sustainable, giant reed may be challenged to fit this description, particularly when it comes to nutrient inputs. The literature has little information regarding the role of soil type or inherent soil fertility on plant production, and we are left with indirect comparisons here. When grown with only residual fertility and no irrigation in the sandy soils of the Southeastern

US coastal plain, average giant reed yields were quite low (6.4 Mg DM ha^{-1} yr^{-1}) (Knoll *et al.*, 2012). Much greater yields (19.4 Mg DM ha^{-1}) with no added N were reported by (Kering *et al.*, 2012b) in the Southern Great Plains of the United States, but giant reed yields were over 33 Mg DM ha^{-1} when supplied with 168 kg N ha^{-1}. In both of these studies, nutrient removal in harvested feedstock was high; this suggests that sustainability of annual yields may be challenged without some means to recover and reutilize these inputs. In addition, N use efficiency is likely lower than for other energy feedstocks (see Mantineo *et al.*, 2009).

Of course, water is a key contributor to giant reed production, as noted previously. Although crop water use is high, in some environments water use efficiency (WUE), calculated as the biomass yield relative to water input, may be favorable relative to other feedstock crops. Data from Mantineo *et al.* (2009) indicated that giant reed had about 56% greater WUE than *Miscanthus* × *giganteus* in Italy, as measured over 5 years and two fertility and irrigation regimes. However, one should be cautious in using such simple measures of WUE, because total water input (irrigation plus precipitation) is obviously not the only factor that drives productivity. For instance, *P. virgatum* produced higher yields than giant reed in Georgia (Knoll *et al.*, 2012) but was not as productive as giant reed in Oklahoma (Kering *et al.*, 2012b).

Miscanthus spp.

Miscanthus has received quite a bit of attention for energy cropping systems over the past decade and a half. A book and several reviews on miscanthus species in an energy context have been published in the past decade (Jones and Walsh, 2007; Atkinson, 2009; Heaton *et al.*, 2010; Zub and Brancourt-Hulmel, 2010). As with giant reed, the largest body of research on miscanthus comes from Europe, where efforts have been made to test biomass feedstocks in environments across much of the continent, but efforts in the United States are rapidly increasing. Before we embark on a specific discussion of the species, we make one note about convention/usage here: if not italicized, the word "miscanthus" is meant to designate all of the potential species of interest.

Description

All members of the miscanthus genus use the C_4 photosynthetic pathway. The genus belongs to the Saccharinae subtribe, along with taxa of the *Saccharum*, *Erianthus*, *Sclerostachya*, and *Narenga* genera (Amalraj and Balasundaram, 2006; Kim *et al.*, 2012). This grouping of genera form the "*Saccharum* complex" and hybridization amongst genera members is common and useful for crop development (Sobral *et al.*, 1994; Kim *et al.*, 2012). Miscanthus is considered the most primitive genus among the member of the complex (Amalraj and Balasundaram, 2006). The genus has a base chromosome number of x = 19 and both diploid (2n = 2x = 38) and tetraploid (2n = 4x = 76) species and hybridization within miscanthus is common. One hybrid, especially, the sterile triploid (2n = 3x = 57) identified as *M.* × *giganteus* Greef & Deuter ex Hodkinson & Renvoize (Hodkinson and Renvoize, 2001) has been of particular interest given its high biomass yield potential.

The miscanthus genus is comprised of 14 or more species, several of which have been investigated directly for bioenergy production or as sources of genetic variability to exploit for this purpose (Kim *et al.*, 2012). Similarities in appearance among some species can at times make distinction based on morphology difficult, as can, ironically, the range of morphological variation within species (Scally *et al.*, 2001). Vegetative morphology can also

Figure 4.3. *Miscanthus × giganteus* plants at Orange, Virginia (foreground) in November 2010. Senescing plants have begun to dry down. Note the giant reed (background) is still green.

be affected by environmental conditions that influence development, adding further difficulty to species delineations based on morphological characterization. Species distinctions historically have often come down to the fine points of the differences in inflorescences. Miscanthus inflorescences have racemic panicles on a long axis, and small differences in reproductive structures—such as the length or presence of awns and hairs—have often been the determining factor for species designation (Scally *et al.*, 2001; Sun *et al.*, 2010). These low-resolution methods have given way to the application of molecular techniques that can help scientists read miscanthus' history in its genetic code—and find useful genes for further crop development (Głowacka, 2011).

Miscanthus rhizomes generally are 1–2 cm in diameter and scaly. The rhizomes of *Miscanthus sacchariflorus* are broader and longer, with much greater creeping habit than the shorter, thinner rhizomes of *Miscanthus sinensis* (Lewandowski *et al.*, 2003; Sun *et al.*, 2010). Emerging shoots of miscanthus plants are pilose, sometimes with slight purple coloration. Leaves are bright green, glabrous, and typically with a strong, white midrib. Colorations vary, however, particularly with *M. sinensis* cultivars such as "Zebrinus" and "Variegatus" and others that primarily are used for ornamental purposes. Miscanthus leaf blades often are 1–1.5 m long, and leaf widths generally range from 2 to 4 cm. Narrower leaves are more typical of *M. sinensis*, which also are sharply serrate. Culms are reed-like and yellow-green during the growing season; culms turn golden yellow as the plants senesce (Figure 4.3). Leaves *M.* × *giganteus* typically drop after senescence while those of *M. sinensis* often are more rigid and more likely to remain intact over winter. Leaf retention characteristics may be of import relative to yield and conversion and we consider these later.

The more typical morphological characteristics of interest in miscanthus (and other) species that may be of value for energy cropping include (taller) heights and (larger) stem diameters (Zub and Brancourt-Hulmel, 2010; Yan *et al.*, 2012). Those miscanthus of interest usually are quite tall (often 3–4 m at maturity), with stem diameters of about 10 mm or more.

M. sacchariflorus is generally taller and more robust than *M. sinensis*, but the latter and its derivative hybrids often have many more (albeit smaller) stems per plant; tiller number, however, may not be predictive of biomass yield (Yan *et al.*, 2012).

Much of the miscanthus bioenergy research has been conducted with *M.* × *giganteus*, a hybrid of *M. sacchariflorus* and *M. sinensis* that is somewhat intermediate to these two species. In Germany, end-of-season canopy heights of mature *M.* × *giganteus* plants were ∼2.7 m, compared with ∼3.3 m for *M. sacchariflorus* and ∼2.2 m for *M. sinensis* or *M. sinensis* hybrids (Gauder *et al.*, 2012). In contrast, shoot density for *M. sinensis* was ∼225 shoots m^{-2} versus 170 shoots m^{-2} for *M. sinensis* hybrids and ∼70 shoots m^{-2} for both *M. sacchariflorus* and *M.* × *giganteus*. Mean stem diameters were similar for *M.* × *giganteus* (∼9.3 mm) and *M. sacchariflorus* (∼8.8 mm) but typically twice the stem diameter of *M. sinensis* or its hybrids (3.8–4.7 mm).

The reader is cautioned of generalities about species, because the relative morphological relationships are not always consistent across studies and likely reflect the variability within the genera being tested (Clifton-Brown *et al.*, 2001b; Yan *et al.*, 2012). These morphological relationships between *M. sacchariflorus* and *M. sinensis* did not hold in a comparison across environments where heights and tiller diameter often were similar. Stands were not mature, however, and significant site X species interactions were observed for these traits (Yan *et al.*, 2012). In the same study, *Miscanthus lutarioriparius* proved most productive, and it may provide additional genetic resources for biomass development in future.

Origins, Distribution, and Adaptations

As with giant reed, species of the miscanthus genus have their origins in the Orient and have a long history of human use and intervention. *M. sinensis* grasslands in Japan are thought to have been maintained for thousands of years by annual burning; without this intervention, these stands likely would have given way to forests (Stewart *et al.*, 2009). Historically, miscanthus has been used for thatch and building material for windbreaks and brooms, for toys, bags, and clothing, as a source of yellow dye, and as forage for horses and livestock (Chou, 2009; Stewart *et al.*, 2009). The building material theme has recurred of late, as the species has been studied for its potential as a component of lightweight concretes (Murphy *et al.*, 1995; Pude *et al.*, 2004; Pude *et al.*, 2005).

Miscanthus species are common to the tropical and subtropical regions of East and Southeast Asia along the Pacific Rim (Hodkinson *et al.*, 2002). In addition, two species are found in the Himalayan region and four in Southeastern Africa (Hodkinson *et al.*, 2002), although these have not been considered for energy cropping. The centers of origin of the East Asian species are typified by high rainfall and warm climatic conditions, and the range of some species is quite large. For example, *M. sinensis*, the origin of which is in China (Kim *et al.*, 2012), has a range extending from the subarctic Kuril Islands all the way to Taiwan (Stewart *et al.*, 2009). The northern limit of *Miscanthus floridulus*' range begins around Southern Korea and Southern Honshu (Japan) but its area of habitation extends over greater latitude to the Pacific islands in Melanesia and Polynesia (Garnock Jones, 1978; Hodkinson *et al.*, 2002).

M. sinensis grows best on sandy or loamy well-drained soils (Stewart *et al.*, 2009) while *M. sacchariflorus* and *M. floridulus* are wetland adapted (Pan *et al.*, 2012a, 2012b). *M. sacchariflorus* is not tolerant of flooding, however, likely due to its lower levels of aerhenchyma relative to other wetland species (Qin *et al.*, 2010). *M. floridulus* is a target species for use in constructed wetlands based on its potential to support microbial communities that can support nutrient removal (Ge *et al.*, 2011).

M. sinensis tolerates a wide range of soil pH (3.5–7.5) (Stewart *et al.*, 2009), and the plant's ability to grow in low-pH conditions may reflect its high Al tolerance and resistance to oxidative stressors (Ezaki *et al.*, 2008). This may also help explain the species' ability to colonize disturbed and degraded sites and sites with acid sulfate soils or toxic gas emissions (Ezaki *et al.*, 2008; Stewart *et al.*, 2009). However, differences among miscanthus species and their interactions with soil conditions bear further investigation. In contrast to the data for *M. sinensis*, evidence suggests that a higher pH may be important for the function of *M.* × *giganteus*, particularly on degraded sites. For example, Wilkins (1997) reported poor growth of *M.* × *giganteus* on acidic, metal-contaminated mine spoils unless the soils were limed. Small declines in stem and leaf biomass yields were reported for *M.* × *giganteus* grown on fly ash (pH ~8) with high concentrations of Cu, Pb, Co, and Cr, but plants had similar root and rhizome productivity as those grown on reference soils. Soil pH and substrates also affect *M. sinensis* root colonization by and community structure of arbuscular mycorhizzae (Ma *et al.*, 2007; An *et al.*, 2008). These fungi mediate soil reactions and metals uptake and can increase biomass production of *M. sinensis*. Although little explored, it is likely that mycorhizzae play the same role with other miscanthus species in terms of benefiting water and nutrient uptake, particularly on marginal sites.

The high productivity of miscanthus is closely linked to resource use efficiency. The capacity for photosynthesis at cooler temperatures is a trait typical of the miscanthus genus (Beale and Long, 1995; Carver *et al.*, 2001). Light interception efficiency in *M.* × *giganteus* is higher than for many other species in part because the plant has the capacity to develop a canopy earlier and maintain that canopy later into the growing season (Dohleman and Long, 2009). Greater interception efficiency can have a stronger effect on plant productivity than conversion efficiency, which is higher in maize (Dohleman and Long, 2009). Under cold temperatures, *M* × *giganteus* also displays high expression and activity of phosphate dikinase (Wang *et al.*, 2008) and high photosystem II operating efficiency (Farage *et al.*, 2006). The bottom line is that *Miscanthus* as a genus is capable of substantial productivity in environments that might typically be limiting for warm-season crop production. Moreover, opportunities to push productivity higher appear to exist given that there is room to increase *Miscanthus* photosynthetic conversion efficiencies (Kromdijk *et al.*, 2008; Dohleman and Long, 2009).

Miscanthus displays both high water use (Clifton-Brown *et al.*, 2001a; Hickman *et al.*, 2010) and high WUE, although different strategies for facing drought are observed among species (Clifton-Brown and Lewandowski, 2000). Under water deficit, young *M. sacchariflorus* and *M.* × *giganteus* senesced leaves but not an *M. sinensis* × *sinensis* hybrid. The hybrid had the lowest leaf conductance of the three species, suggestive of stomatal control. WUE did not differ among species on a total plant basis but did on the basis of aboveground biomass production. WUE was greater for *M. sacchariflorus* at all moisture levels, with comparable efficiency displayed by *M. sinensis* subject to drought. *M. sinensis* also had nil photosynthesis when moisture stressed (Clifton-Brown *et al.*, 2002). As these data suggest, the framework for the efficiency estimate may have important consequences for the conclusions drawn. For example, evaluating a crop on the single metric of WUE (calculated on the basis of biomass per unit of water) may be insufficient for assessment if other ecosystem services such as carbon sequestration, especially across a biome, are desired outcomes (VanLoocke *et al.*, 2012).

Although miscanthus has been perceived as a pest-free crop, this is largely due to a lack of knowledge about adapted pests and their consequences to biomass production (Prasifka *et al.*, 2012). While reports of extensive yield losses in biomass plantings are limited to date, several insect species can damage the crop and may have the potential to become economically important pests (Prasifka *et al.*, 2012). In field studies with artificial infestation,

M. × *giganteus* experienced low levels of stem-boring activity by the Southwestern corn borer *Diatraea grandiosella* Dyar (Crambidae), resulting in yield reductions of 12–30% (Prasifka *et al.*, 2012). *M.* × *giganteus* is not a host for armyworm *Mythimna* (*Pseudaletia*) *unipuncta* (Prasifka *et al.*, 2011). However, tests with fall armyworm (*Spodoptera frugiperda*) indicate 50% survival from hatch to pupation on *M.* × *giganteus* and variance by insect strain with the pests showing preference for young tissue (Prasifka *et al.*, 2009). *M.* × *giganteus* also can host western corn rootworm, *Diabrotica virgifera virgifera* (Spencer and Raghu, 2009; Gloyna *et al.*, 2011). These and several other stem-boring insects that attack miscanthus' cousins in the Andropogoneae—for example, sorghum (*S. bicolor*), maize (*Zea mays*), and sugarcane (*Saccharum officianarum*)—may some day present a threat to miscanthus crop yields and plant breeding for resistance may be needed (Prasifka *et al.*, 2012).

Miscanthus also can serve as a host for several species of aphids. Widespread damage has not been observed, but large populations of the aphid *Sipha flava* (Forbes) were reported on young *M.* × *giganteus* stands (Bradshaw *et al.*, 2010). Thus, as with the Lepidopteran species mentioned above, the threat to productivity should be recognized. Bird cherry oat aphids (*Rhopalosiphum padi*) were observed on young plantlets in Poland (Hurej and Twardowski, 2009), although work in the United Kingdom suggests that *R. padi* cannot complete development on miscanthus (Huggett *et al.*, 1999). In contrast, *Rhopalosiphum maidis* can feed on mature *M.* × *giganteus* and does so with preference over immature *M. sinensis* plants. *R. maidis* may be of greater concern as vectors for barley yellow dwarf virus, which in studies by Huggett *et al.* (1999) reduced biomass yield about 20%. Expanded miscanthus plantings may present concerns for sustainability of other agronomic crops, particularly as an overwintering host to insect pests such as *Aphis sacchari (Melanaphis sacchari)* Zhnt. (Wang *et al.*, 1961).

Several species of plant-parasitic nematodes have been isolated from *M.* × *giganteus* plots sampled across six states in the United States (Mekete *et al.*, 2011a,b). Of 35 plots tested, more than 90% were infested with *Pratylenchus* species (Mekete *et al.*, 2011a), which are thought to be the most damaging. Population thresholds for significant damage to *M.* × *giganteus* are unknown, but based on responses of other crops, yield losses due to nematodes would be expected (Mekete *et al.*, 2011b). Lower (36%) infestation rates of *M.* × *giganteus* by *Xiphinema americanum* and *Xiphinema rivesi* and by *Longidorus breviannulatus* were found, but the presence of *L. breviannulatus* was associated with root stunting and necrosis (Mekete *et al.*, 2009). *M. sinensis* used for ornamentals is sensitive to *Meloidogyne hapla* (LaMondia, 1996) and nematode infestation has been considered the primary factor in overwintering losses in new stands of *M.* × *giganteus* that followed maize (Orth and Klug, 1995). These reports suggest that nematodes could be an important limit factor for *Miscanthus* production, particularly if there ever arises an incentive to convert land from maize.

To date, miscanthus has been deemed relatively disease free, but recent reports suggest the potential for significant pathogen pressures in future. A basal stem rot and foliar blight caused by *Pythium sylvaticum* (W.A. Campb. & J.W. Hendrix) killed small patches of *M. sinensis* seedlings in a field study in Illinois (Ahonsi *et al.*, 2011). Also, a severe case of leaf blight caused by *Pithomyces chartarum* (Berk. & Curt.) M.B. Ellis, occurred on *M.* × *giganteus* in Kentucky (Ahonsi *et al.*, 2010). Although the effects were greater on 4–5-month-old plantings than on 2-year-old plants, disease incidence was 100% for both. Severe symptoms of miscanthus blight, a disease caused by *Leptosphaeria* sp. and its asexual spore state (*Stagonospora* sp.) also has been identified on ornamental *M. sinensis* varieties in Maryland (O'Neill and Farr, 1996). A search of the USDA's fungal database indicates *M. sacchariflorus*, *M. sinensis*, and *M. floridulus* are hosts for 12, 58, and 122 fungal species respectively, but for these almost 200 fungus–host combinations, the database notes only nine corresponding reports in the research literature. The fungal database records no specific cases of *M.* × *giganteus* as a host for fungal disease. However, rhizome rot due to an attack by *Fusarium* and *Mucor* fungi caused *M.* × *giganteus* establishment failure in Italy (Beccari *et al.*, 2010), and the possibility of pathogens affecting the sustainability of energy cropping cannot be ignored.

Reports of viral diseases in miscanthus are more limited. The miscanthus streak virus has been identified in *M. sacchariflorus* (Yamashita *et al.*, 1985) and barley yellow dwarf virus on both *M. sacchariflorus* and *M. sinensis* (Christian *et al.*, 1994). Sorghum mosaic virus has also been identified in *M. sinensis*, but in no cases have significant yield declines been reported.

Crop Management and Production

Species selection and the attendant method of establishment is perhaps one of the more important issues surrounding the sustainability of miscanthus production. Large-scale crop establishment that requires extensive vegetative planting—as in the case of *M.* × *giganteus* or other sterile hybrids—can be an expensive proposition, whether this is done using rhizomes or plantlets. The slow adoption of *M.* × *giganteus*—and of energy cropping in general—provides some space for the development of a "seed" industry, if not the demand. As energy systems come online, however, escape velocity for *M.* × *giganteus* or other vegetatively established feedstocks potentially could be hindered by a lack of available propagules needed to establish these systems over large acreages. The size of this problem—and the cost of establishment—could be mitigated in part by successfully lowering planting densities. However, typical recommendations are for higher planting densities to maximize early yields (Atkinson, 2009)—even though yield differences by stand density often are negligible after two or three growing seasons (Bullard, 1996; Bullard *et al.*, 2001; Miguez *et al.*, 2008). For example, Bullard (1996) reported mature stand stem densities would likely approach 100 m^2 regardless of initial planting density with yields of about 20 t DM ha^{-1} by year 4.

Much work on *M.* × *giganteus* has focused on establishment and expansion constraints. Rhizome cold tolerance will be a key issue for expansion of the miscanthus to greater latitudes, independent of the establishment issues we describe below. Selection for survival at colder temperatures is possible, however (Clifton-Brown and Lewandowski, 2002), and advances on this front are expected as new hybrids and cultivars are developed.

Variability of rhizome survival has been a concern, and very large postestablishment losses have been reported in some cases (e.g., 75% for Maughan *et al.*, 2012). Factors such as rhizome parent age, rhizome size, time in storage, planting method and depth, and postplant treatments to improve field moisture have all been explored for relationships to stand survival and productivity (Nixon *et al.*, 2001; Pude, 2003; Pyter *et al.*, 2010; Davies *et al.*, 2011; Płazek *et al.*, 2011; Poenaru *et al.*, 2012), and gains are being made.

Still, the constraints with rhizome establishment have supported the drive to find micropropagation techniques that can increase stand uniformity, dramatically reduce costs, and allow for more rapid plant development (Atkinson, 2009; Kim *et al.*, 2010; Gubišová *et al.*, 2012). Although propagation techniques sometimes affect plant morphology—for example, micropropagules often have more tillers versus taller, thicker culms from rhizomes—yield differences typically are small to nonexistent (Clifton-Brown *et al.*, 2007; Christian *et al.*, 2008). However, as with rhizomes, winter survival can sometimes be a limitation for micropagule establishment (Kim *et al.*, 2010).

The cost of establishment for clonal propagules has been justification for some to consider seeded miscanthus species, despite the potential invasive risk. A proposed solution to mitigate invasive issues is to develop photosensitive plants—much like efforts with the biomass sorghums—that could only produce viable seeds at much lower latitudes. In theory these plants would not develop viable seeds at the latitudes where the crops were grown for energy production. Efforts would likely be needed to lower the base temperatures for Miscanthus seed germination, however, as base temperatures are higher than for other warm-season grasses. Currently, seeded establishment at higher latitudes is limited without tillage practices that increase soil temperature (Clifton-Brown *et al.*, 2011) and this introduces a set of sustainability issues in its own right.

Extensive tillage—for either establishment system—has potential negative implications for environmental and economic sustainability. Because existing pasture and grasslands are likely to be target sites for new energy crop plantings, every effort should be taken to lower the impacts and costs of establishment during this transition phase. Specially engineered equipment may be needed to successfully establish no-till planting systems for miscanthus rhizomes or plantlets. However, strip till techniques suitable for other agronomic crops (see Truman *et al.*, 2012) should be feasible and used to good effect. No-till establishment practices, which can greatly reduce run-off, nutrient losses, and soil organic matter oxidation, should also be pursued if in fact seeded miscanthus is to be adopted—although as we have noted, this may be site limited at present.

With either feedstock/establishment system, site preparation methods are expected to have important carryover effects in terms of energy requirements and carbon balance. The greater the tillage and disturbance required, the greater the CO_2 losses at establishment. One estimate suggests it may take as many as 30 years to recoup the carbon lost during the conversion of natural vegetation to *M.* × *giganteus* plantations (Hughes *et al.*, 2010). This is of particular note when one considers that many of the stated expectations of stand longevity suggest a crop will be kept in place around 15–20 years.

For established miscanthus stands, input requirements generally are low relative to other crops, but the literature is mixed on this point. Some researchers have reported some positive yield responses to inputs such as fertilizer nitrogen (N) or manures (Ercoli *et al.*, 1999; Smith and Slater, 2010), while others have found year-to-year variability (Clifton-Brown *et al.*, 2007) or no yield response to N fertility in either short- or long-term experiments with *M.* × *giganteus* (Christian *et al.*, 2008; Dohleman *et al.*, 2012; Maughan *et al.*, 2012). In the latter cases, leaching losses with high N applications is probable (Curley *et al.*, 2010).

The low-to-nil response to exogenous N that is sometimes observed likely reflects a complex set of interactions. These may be functions of soils, climate, and management (Heaton *et al.*, 2010), as well the plant's high nutrient absorption and use efficiencies, considerable internal cycling between rhizomes and aboveground biomass, and the recycling of nutrients to belowground storage organs in fall (Cadoux *et al.*, 2012). Here we have explicitly used the term exogenous in reference to synthetic fertilizers, as there is growing evidence that miscanthus, among several species, benefits from biological N fixation (Eckert *et al.*, 2001; Kirchhof *et al.*, 2001; Davis *et al.*, 2010; Monteiro *et al.*, 2012).

Limited response to applied N likely is a function of the soils on which the plants were grown; our unpublished data from low-fertility ultisols suggest N inputs will provide at least moderate increases in crop production to a point. Cadoux *et al.* (2012) noted that uncertainties about *M.* × *giganteus*, nutrient requirements arise because there is little information regarding N derived from soils or the contributions of rhizomes to aboveground biomass, and tools for plant N diagnosis are limited. The degree to which biological associations can meet N needs for miscanthus is important, particularly in light of "upper limit" yields that approach 55 Mg ha^{-1} (Heaton *et al.*, 2010).

In contrast to N, minerals cannot be biologically synthesized, and long-term harvest without mineral nutrient replacement may limit biomass yields (Clifton-Brown *et al.*, 2001a; Clifton-Brown *et al.*, 2007). Annual replacement rates of 7–8 kg P ha^{-1} and 100–122 kg K ha^{-1} would be needed to prevent soil depletion in systems averaging almost 18 Mg ha^{-1} yr^{-1} (Cadoux *et al.*, 2012). Still, these nutrients too are finite resources, and they will need to be replaced with limited economic or environmental sustainability costs.

Panicum virgatum—Switchgrass

While more than one member of the *Panicum* genus has been tested for energy cropping, far and away the species in the spotlight is *Panicum virgatum*—switchgrass. The United States is "ground zero" for testing efforts given US interests in renewable fuels and that North America is essentially the species' home court. However, the potential of the species has led to its being tested for energy cropping in countries around the globe. As with Miscanthus, at least one book and numerous reviews have been written about the species to date, and doubtless there will be more that follow.

Description

The *Panicum* genus in the broadest sense (*sensu lato*) is large and diverse. Among its roughly 450 members, C_3, C_4, and C_3/C_4 intermediate species can be found. More recently, however, *Panicum* "membership" in the strict sense (*sensu stricto*) has been limited to the former *Panicum* subgenus (Aliscioni *et al.*, 2003), all of which deploy the C_4 photosynthetic pathway. This "new" genus has global distribution and contains about 100 species that are separated into five sections, including section *Virgata*, of which switchgrass is a member (Triplett *et al.*, 2012). The species is further divided out into "upland" and "lowland" categories that usually are described as "ecotypes"—although some ambiguous types may represent hybrids between the two (Zhang *et al.*, 2011a). Ecotypic groupings also have been designated as upland or lowland "cytotypes" based on a DNA polymorphism carried in their plastids (Vogel, 2004), and various molecular markers also now can be used to make this distinction (Hultquist *et al.*, 1996; Missaoui *et al.*, 2006).

The upland and lowland ecotypes are phenotypically distinct, with upland ecotypes characteristically being shorter and having finer stems than their more robust, lowland kin. Leaves are green or glaucous, and blades are up to 60 cm long and up to 1.5 mm wide, glabrous ro pilose at the base (Hitchcock, 1935). Upland ecotypes contain both tetraploid ($2n = 4x = 36$) and octoploid ($2n = 8x = 72$) genotypes; lowlands heretofore have been considered exclusively tetraploid, but this view may change as possible octoploids have been discovered (Zhang *et al.*, 2011a,b)

As might be expected from genotypic and ecotypic descriptions, switchgrass is morphologically heterogeneous. Plant heights range from about 0.5 to >3 m tall, and rooting depths to 3 m are reported (Porter, 1966; Moser and Vogel, 1995). Along with being taller, lowland types generally have thicker stems, longer, wider leaves with longer ligules, and larger panicles. Leaves of both ecotypes tend to be erectophile, but lowland ecotypes have a more bluish-green color, in contrast with the greener tones of upland (Porter, 1966; Casler, 2005). Lowland ecotypes have short rhizomes that are tightly gathered and result in a bunch form (Vogel, 2000), whereas upland plants have more active rhizomes and often are sod forming (Beaty *et al.*, 1978). Although root diameters frequently are larger for lowlands, root internodes and lengths generally are longer for uplands. Shoots of lowland plants originate solely from rhizome buds, but those of upland plants originate both from rhizomes and from basal nodes of culms produced in the previous year (Porter, 1966).

Origins, Distribution, and Adaptation

Tracking switchgrass' evolutionary timeline has provided interesting insights into the plant's history. Molecular clock estimates suggest switchgrass diverged from a common ancestor with maize about 22 or 23 million years ago (MYA) (Huang *et al.*, 2003; Zhang *et al.*, 2011a). Switchgrass, a polyploid, diverged from two diploid ancestors around 5–10 MYA (Zhang *et al.*, 2011b) and by 2 MYA had assembled the gene pool we now recognize (Huang *et al.*, 2003). The upland–lowland divergence probably occurred between 0.5 and 1.3 MYA (Young *et al.*, 2011; Zhang *et al.*, 2011b).

In the course of switchgrass' evolution, Earth's cooling and thawing through ice ages would have contracted and expanded the range of suitable sites for the species through time. All current switchgrass populations appear to have been derived from three centers of origin in North America. These centers developed out of Eastern and Western Gulf Coast refugia that harbored switchgrass during the last ice age (McMillan, 1959; Zalapa *et al.*, 2011) Upland types, as the name suggests, more typically occupy the drier, cooler habitats of upslope topographic positions and hail geographically from the central and Northern Great Plains and the Eastern savannas. In contrast, the taller lowland ecotypes are adapted to warmer, moister habitats, and their evolutionary histories are rooted to the Southern Great Plains and eastern Gulf Coast regions of the United States (Cortese *et al.*, 2010; Zalapa *et al.*, 2011). As the glaciers receded after the last ice age 11 000 years ago, switchgrass marched out from these preserves, adapting and recolonizing two-thirds of North America over the course of these few millennia. At the time of Europeans' arrival, the species' native range was from Central America to Canada and from the East Coast to western reaches of Nevada and Arizona (Hitchcock, 1935). Along with these latitudinal adaptations, switchgrass displays longitudinal differences in productivity (Hopkins *et al.*, 1995a,b; Madakadze *et al.*, 1998; Casler and Boe, 2003). This likely reflects differences in adaptation to the moisture and disease pressures that would change from east to west and vice versa (Cornelius and Johnston, 1941; Vogel, 2000).

Along with wide geographic distribution, switchgrass is highly adaptable to diverse edaphic conditions, although site suitability typically breaks out along ecotypic lines as discussed above. Switchgrass production appears little affected by soil type. However, soil texture appears to be of some importance, as both excessive and poor drainage can reduce productivity (Fike *et al.*, 2006b; Thelemann *et al.*, 2010).

Despite the production guides that recommend maintaining soil pH at 6 or higher, there is little data in the literature tying poor productivity to low soil pH. Switchgrass can tolerate high levels of acidity, although this may reduce seedling germination (Hanson and Johnson, 2005). Switchgrass also can tolerate many contaminants; in some cases its ability to remediate or remove herbicides and heavy metals such as atrazine and cadmium makes it a likely candidate for phytoremediation schemes (Chen *et al.*, 2011; Murphy and Coats, 2011).

Switchgrass benefits from fungal mutualisms that mediate responses to stresses such as drought, contaminants, and limited water and nutrient availability. Greater yields are sometimes with mycorrhizal associations, although these benefits vary by strain and source (Koslowsky and Boerner, 1989; Clark 2002). While mycotrophy's benefits often are greater at lower pH (Clark *et al.*, 1999a; Clark *et al.*, 1999b; Clark, 2002) and under conditions of environmental stress, increased plant production has been reported even in nonstressed environments (Ghimire and Craven, 2011). Such results suggest opportunities to exploit plant–microbe interactions for greater nutrient use efficiency, or even to tap systems of biological nitrogen (N) fixation.

Pests and diseases generally have been considered limited issues for switchgrass. Parrish and Fike (2005) speculated that this could be an increasing problem as plantings for energy crops were expanded, and since that time there have been several new publications on the subject. The degree to which this is a growing issue or is it simply due to increased awareness, we cannot say. However, from 2004 to 2012 the number of cataloged fungal diseases for which switchgrass is a host increased from 65 to more than 150 (Farr *et al.*, 2004; Farr and Rossman, 2012).

Rusts (*Puccinia* spp.) are one of the more common pathogen pressures for switchgrass (Zale *et al.*, 2008; Hirsch *et al.*, 2010), with northern adapted cultivars more susceptible (Cassida *et al.*, 2005b). Differences in heritability will be helpful for breeders, but this needs to take into account gene × environment interactions. For example, "rust-resistant" southern lowlands have sometimes displayed greater rust symptoms during very dry periods in Virginia, while their "rust-susceptible" counterparts did not (B. Zhao, personal communication). Recent reports of spot blotch caused by *Bipolaris sorkiniana* have come from diverse regions of North America (Tomaso-Peterson and Balbalian, 2010; Vu *et al.*, 2011; Waxman and Bergstrom, 2011). Studies by Vu *et al.* (2011) suggest the disease is seed borne. Similarly, leaf spot attacks by *Bipolaris oryzae* have been reported in West Virginia and North Dakota (Belesky and Fedders, 1995; Krupinsky *et al.*, 2004). Smuts caused by *Tilletia macalagani* have been an issue in long-term plantings in Iowa (Gravert *et al.*, 2000). Follow-up reports indicated that yield losses reach 40% in some fields (Thomsen *et al.*, 2008). Outbreaks also have occurred in New York and Texas (Carris *et al.*, 2008; Layton and Bergstrom, 2011).

Viral infections appear more limited than fungal diseases. Panicum mosaic virus (PMV) was reported in 1957 (Sill and Pickett, 1957) but subsequent research primarily has focused on PMV as a disease agent in other species. Switchgrass is susceptible to some strains of barley yellow dwarf virus (Garrett *et al.*, 2004), which are transmitted by aphids, and (Schrotenboer *et al.*, 2011) suggested that selection for biofuel crops should account for increasing virus susceptibility to avoid potential disease vulnerability. In a similar vein, Thomsen *et al.* (2008) noted that research on management approaches to deal with such issues are of critical importance for the long-term success of switchgrass for biomass production systems.

Insect pests may be of less concern than diseases, although pressure may mount when switchgrass for bioenergy systems scale up (Parrish and Fike, 2005). Insect pressures during the seedling stage may present the greatest insect threat to the switchgrass stand productivity, but there is little published data on the subject. The corn flea beetle *Chaetocnema pulicaria* has been a common source of damage in Virginia plantings, however (Dale Wolf, personal communication). Switchgrass is not a preferred host for many insect species (Davis, 1914; Walkden, 1943) and an inferior host relative to other warm-season crops (Nabity *et al.*, 2011; Prasifka *et al.*, 2011). Fall armyworm (*Spodoptera frugiperda* (J.E. Smith) (Lepidoptera: Noctuidae) has reduced survivorship on switchgrass compared with other grasses (Nabity *et al.*, 2011), although resistance varies by switchgrass strain (Dowd and Johnson, 2009). Grasshoppers (*Saltatoria*) are known to feed on switchgrass, but the extent of the damage has not been quantified (Parrish and Fike, 2005). Schaeffer *et al.* (2011), in a baseline study of insects in Nebraska switchgrass stands, found that about 60% of arthropods collected were of the orders Thysanoptera and Hymenoptera; leafhoppers, grasshoppers, grass flies, and wire worms were noted as the most abundant of potential pest species.

Switchgrass is the only known host for a stem-boring caterpillar, *Blastobasis repartella* (Dietz.), and its life stages, geographic distribution, and the symptoms of predation recently have been described (Adamski *et al*., 2010; Prasifka *et al*., 2010). So far, *B. repartella* has been found both in cultivated and natural switchgrass stands in eight northern states in the United States (Prasifka *et al*., 2010), but damage to tillers has been <10%. In addition, a new gall midge species, *Chilophaga virgate* Gagne, was recently discovered in South Dakota (Boe and Gagné, 2011). Tiller infestation in 10 switchgrass genotypes ranged from 7% to 22%; tiller mass was reduced 35% and produced no appreciable seed.

The damage thresholds for plant-parasitic nematodes currently are unknown (Mekete *et al*., 2011b). Based on thresholds for other monocotyledon hosts, nematode populations in several states in the United States have potential to reduce switchgrass yields (Mekete *et al*., 2011b). Nematode damage is most likely with young or stressed plants, and in many cases, nematode damage may be mistaken for environmental stress symptoms. Soil fumigation in Texas improved switchgrass establishment and seedling vigor compared to a weed-free check, indicating soil-borne pathogens were responsible for poor establishment (Evers and Butler, 2000), and nematode numbers are likely to be greater in higher rainfall regions (Cassida *et al*., 2005a).

Crop Management and Production

The relationship between switchgrass and humans prior to European settlement is perhaps not as clear as that described for miscanthus and humans in Japan (Stewart *et al*., 2009). That is, the degree to which the fires that maintained the North American grasslands were anthropogenically (vs. naturally) ignited is in contention (Barrett *et al*., 2005). Clearly, fire is an essential component of grassland ecosystems, and humans living in grasslands populated with switchgrass and other prairie species would, at the very least, have relied on the grass as a source of feedstuffs for the game they ate. The degree to which Native Americans used this as an intentional management tool is less clear.

Switchgrass' use as a crop is a relatively recent phenomenon. Early (and ongoing) research with switchgrass explored its utility as forage, but it has received attention for much broader assignments from the latter half of the twentieth century. Beyond its use for pasture or conserved forage, switchgrass is now deployed for: erosion control around flowing surface waters, filter strips to retain soil and absorb nutrients, reclamation or stabilization of disturbed sites, growing medium for mushrooms, nesting material in poultry houses, and for wildlife habitat. As switchgrass assumes a more "industrial crop" profile, it is finding use as an energy and chemicals feedstock (across several conversion platforms), for fiber or pulp production, for phytoremediation of contaminated sites, and for pharmaceuticals and biomaterials.

From a bioenergy perspective, switchgrass has some advantage over several of the other potential perennial biomass crops in that it can be established from seed. Appropriate cultivar selection is critical to the long-term productivity of a stand. In addition to on-site environmental conditions, one must consider photoperiodicity in relationship to the cultivar's origins. Switchgrass acts as a determinate plant, producing a flush of tillers that become reproductive after it receives the appropriate environmental cues; the plants cease biomass accumulation with the advent of floral development. The primary signal appears to be day length (Esbroeck *et al*., 2003), and this photoperiodic flowering pattern provides an important opportunity to increase switchgrass yields. Timing of flowering is genetically programmed and tied to some minimum night length. Thus, moving switchgrass cultivars from their place of origin to higher latitudes can increase biomass yield by prolonging vegetative growth and delaying the onset of inflorescence development. There are limits to this strategy, however; even if they have sufficient cold tolerance for their new environs, southern adapted lines may not get the signal in time to harden off in preparation for cold weather. In contrast, moving a northern line south, however, will have the opposite effect on yield by abbreviating the normal amount of time required to get to the critical night length.

Successful switchgrass stand establishment often requires greater attention to detail than is needed for other common agronomic crops. Small missteps can lead to stand failure. It is not our intention to give detailed recommendations on establishment here—several planting guides adapted to local conditions are available—but we simply point out some of the typical causes of failure. These include high seed dormancy, especially when coupled with inappropriate time of planting, excess planting depth, poor seed–soil contact due to excess residue, excess weed competition, and more. While we again make a plug for the sustainability merits of no-till establishment techniques, such methods can accentuate several of the problems mentioned above. Stand frequencies for successful establishment should approach at least 40% (Schmer *et al.*, 2006), and this may be possible with a range of seeding rates (Parrish *et al.*, 2008). Obviously, the lower the seeding rate for successful stand establishment, the more economical the production of switchgrass for biomass energy purposes (Perrin *et al.*, 2008).

Switchgrass seedling development can appear slow because much of the photosynthate is directed toward belowground growth. Seedlings also display little response to N during establishment (Sanderson and Reed, 2000). Weed control can be a major cause of establishment failure and is facilitated in high fertility sites because readily available nutrients stimulate the top growth of competitive annual weeds. Thus, most N fertility recommendations suggest deferring N inputs until the second or even third growing season.

Data on switchgrass N requirements (and attendant recommendations) are varied, likely reflecting the inherent demands and capacity for recycling, soil type and N status, precipitation and atmospheric deposition, and harvest timing among other factors. High yields in response to applied N are more typical in sandy soils with little capacity to retain nutrients (Ma *et al.*, 2001; Muir *et al.*, 2001), whereas little response to N was observed for switchgrass grown on soils with high levels of soil organic N (Stout and Jung, 1995). Biological nitrogen fixation and plant growth stimulation may also occur in some systems (Tjepkema, 1975; Riggs *et al.*, 2002; Ker *et al.*, 2010, 2012).

The capacity to translocate nutrients from belowground storage structures to support aboveground growth reduces the utility of exogenous N applications in switchgrass plantings. As growth progresses during the season, switchgrass N concentrations decline (Waramit *et al.*, 2011). Switchgrass sinks N in roots and rhizomes at the end of the growing season (Beaty *et al.*, 1978; Lemus *et al.*, 2008; Ker *et al.*, 2012), and postsenescence N concentrations commonly fall within 5 and 8 g kg^{-1} (Madakadze *et al.*, 1999; Fike *et al.*, 2006a, 2006b; Guretzky *et al.*, 2011), greatly lowering N removal rates relative to that removed in “green” feedstock.

Determining a set rate of fertility inputs, particularly for N, is problematic given the range of switchgrass yields and yield responses to the nutrient. One of the highest reported single-season yields with a single harvest under dryland agriculture was 27.4 Mg ha^{-1}, but more typical yields under such conditions are in the 10–15 Mg ha^{-1} range (McLaughlin and Kszos, 2005). Assuming long-term biomass removal rates increase with advances in cultivar development, average annual yields of 15 Mg ha^{-1} should not be a stretch. Thus, with feedstock N concentrations of about 6 g kg^{-1}, N replacement rates probably will need to approach 90 kg ha^{-1} with such management.

Limited inputs of P and K also will be required for mature switchgrass production systems, and the guidelines will follow suit. Responses to P and K inputs often are limited (Kering *et al.*, 2012a) and replacement rates will vary with harvest timing and soil fertility among other factors. Replacement rates for our assumed 15 Mg ha^{-1} will likely need to be in the ranges of 10–20 and 50–100 kg ha^{-1} for P and K, respectively.

Long-term sustainability likely will be best served by management practices that both minimize and balance fertility imports and exports (Lemus *et al.*, 2008), and nutrient use efficiencies will be higher when those inputs are low (Fike *et al.*, 2006a). Thus, low-input management with end-of-season harvest typically is recommended. Systems that have the highest efficiency may not be sufficiently productive to meet yield and economic sustainability metrics, however, and in some cases there may be economic incentive to harvest during the growing season to reduce logistic challenges associated with annual, one-off harvests (Cundiff 1996; Fike *et al.*, 2007; Cundiff *et al.*, 2009).

In both low- and (not-so-)high-input scenarios, management strategies that can utilize alternative nutrient sources may improve the long-term production and economic sustainability of these systems. Applications of animal manures may help with this approach (Lee *et al.*, 2007), although this may have other undesirable consequences in terms of soil P accretion—and the sustainability of the systems that generate the manure may be in question. Adding legumes to these systems may also reduce N fertility costs, although varied success has been reported with this strategy (Springer *et al.*, 2001; Bow *et al.*, 2008).

Fertility practices may have other sustainability consequences beyond the carbon costs associated with fertilizer N synthesis via the Haber–Bosch process. With added N fertility switchgrass partitions a greater proportion of nutrients to shoots than roots (Heggenstaller *et al.*, 2009; Garten *et al.*, 2011), although the N-fertilized switchgrass may also have fewer tillers, particularly under one-cut management (Muir *et al.*, 2001; Fike *et al.*, 2006a). These morphological changes may not affect yields (Muir *et al.*, 2001), but they may affect carbon sequestration and GHG emissions (Garten *et al.*, 2010). Thus, the relationships of N fertility to overall system sustainability in terms of carbon budgets bear further investigation (Jung and Lal, 2011).

Returning to the low-input theme, some have proposed adopting a "prairie model" as the best way to achieve yield stability, high productivity, and positive energy balance and GHG mitigation in bioenergy systems (Tilman *et al.*, 2006). This idea, particularly related to the value of diversity in grassland production systems has since been the subject of rather strong debate between the ecological and agricultural communities and perhaps within the agricultural community itself (Schmer *et al.*, 2008; Adler *et al.*, 2009; Fargione *et al.*, 2009; DeHaan *et al.*, 2010; Picasso *et al.*, 2011). Even the lowest levels of diversity—binary mixes—that provide appropriate functionality (as with the addition of suitable legumes) may increase aboveground net primary productivity (DeHaan *et al.*, 2010). Whether low-input–high-diversity systems can meet productivity goals in agricultural settings will require more research. Even so, conversion systems that require feedstocks to have uniform chemical attributes can be challenged by the feedstock variability attendant to such mixtures (Adler *et al.*, 2009). Regardless of the final word on this issue, switchgrass, unlike many bioenergy crops, may have the greatest potential for this type of management given that it comes from the mixed prairie community.

Harvest, Storage, Logistics, and Process Considerations of Perennial Grasses

There are direct links and important interactions among each of the components of grass-based bioenergy production systems, and these are best not studied in isolation (Fike *et al.*, 2007).

Integrating the whole system—agronomics, logistics, and processing—will be critical for building a functional, profitable, and sustainable industry. In this section we briefly consider these issues in a general manner, addressing specific feedstocks as needed.

Nutrient Management and Harvest Timing

We include a discussion on nutrient management in a section on harvest considerations because of the important interactions between the two. Among nutrients used in bioenergy systems, N is the most costly both economically and environmentally. Also, because N typically enters the system as a product of fossil-fuel-driven processes, and because it can leave the system via leaching or as an N_2O emission, it has great potential to impact water quality and GHG sustainability metrics (Mitchell and Schmer, 2012). Fertility applications can have important negative downstream effects on feedstocks by changing cell constituents or supporting greater leaf production (Allison *et al.*, 2012).

With switchgrass, maximum seasonal yields possible with a single harvest can be attained by cutting and collecting the biomass sometime between full panicle emergence and post-anthesis (Vogel *et al.*, 2002; Mitchell *et al.*, 2008). However, management recommendations for feedstocks such as switchgrass or miscanthus generally promote single, end-of-season harvests. The goal of such management is to maximize or optimize yields and feedstock quality while minimizing inputs of fertilizer, fuel, equipment, and labor (Sanderson *et al.*, 1999; Vogel *et al.*, 2002).

Harvesting biomass within a few weeks after a killing frost generally will result in lower moisture and nutrient concentrations with only moderate yield losses, but the degree of benefit is feedstock dependent. For example, plant moisture concentrations in the standing biomass of giant reed probably will not decline below 40% (Angelini *et al.*, 2005) because the plant's large, woody stems resist moisture loss. Of course, latitude will also play a role in the feasibility of "postfrost" harvest recommendations, because feedstocks grown in regions that get little or no freezing weather often do not completely shut down. In either case, the difficulty or inability to harvest a feedstock in dry form will play a large role in determining harvest, storage, and process methodologies, which we will mention later.

Along with moisture reduction, delaying harvest provides plants time to translocate nutrients from aboveground biomass to belowground storage organs. In the case of giant reed, stem N concentration declined from 10.3 to 3.8 g kg^{-1} from October to December in Oklahoma (Kering *et al.*, 2012b). Delaying harvest until late winter may also increase survival of feedstocks when they are grown in colder environments where they are marginally adapted.

While delaying harvest until sometime after the growing season can reduce yields, this will to some degree be compensated by the fertility savings and possibly with premiums for better feedstock quality. Feedstock N concentrations often will drop from around 10 to 6 g kg^{-1} or less. Losses incurred by deferring harvests to late winter can be significant—for example, 30% was reported for miscanthus (Heaton *et al.*, 2010) and 40% for switchgrass (Adler *et al.*, 2006), but this is largely due to leaves and inflorescences lost during winds and precipitation events. Losing this fraction of the biomass may have positive outcomes for feedstock quality depending on the process system (Clifton-Brown and Lewandowski, 2002; Adler *et al.*, 2006). For example, the elevated alkali elements and silicates in biomass (especially leaves) can reduce conversion efficiency in thermochemical process systems and may increase waste streams.

Feedstock losses may be greater in regions with more challenging winter weather conditions, although with switchgrass there is some limited evidence that these can be minimized by cutting in fall and harvesting in spring (Anonymous, 2008). In this case, feedstock quality was increased by leaching nutrients from leaf and stem tissues and leaf shattering losses were reduced. Whether this could work more broadly is in question, in part because the degree to which snow cover is needed to protect the feedstock is unknown.

Despite the potential benefits of single, late-season (or "over winter") harvests, some data suggest that for upland switchgrass cultivars, two harvests per season could improve total yield 30–40% (Fike *et al.*, 2006a,b). The suitability of such management would likely be driven exclusively by improved logistics considered below and must be weighed against the costs of added harvest, nutrient removal, and process efficiency (Fike *et al.*, 2007; Cundiff *et al.*, 2009).

Harvest Methods and Storage Issues

Forage choppers or rotary mowers will be sufficient for most bioenergy cropping operations. However, much available mowing equipment has not been optimized for use with the large, stiff stalks common to many biomass crops. Recent research suggests there may be opportunities to greatly reduce harvest input requirements and costs by modifying existing harvest equipment (Lewandowski *et al.*, 2000; Johnson *et al.*, 2012). Methods for collecting and distributing biomass are essentially limited to baling or chopping systems. Each scheme has a number of variations on a theme, and each comes with different merits and challenges.

Baling systems provide particular advantage in terms of breaking the link between harvest and in-field hauling because bales can be left on the ground for later recovery. Not all feedstocks and balers are compatible, however. Following cutting, miscanthus can be baled, but bundling the material sometimes can be a challenge with round balers. Large, square balers may be more effective at producing bundles (bales) of greater density, but the balers require increased horsepower (i.e., larger tractors) and may not be as well suited to existing farm operations in regions with small, irregular-shaped fields of uneven terrain. Large square bales also require some form of storage in humid regions, because they do not shed moisture and storage is expected for miscanthus (Nolan *et al.*, 2009). In contrast, large round bales of switchgrass wrapped with net wrap can be stored outside with low (~5%) storage losses (Cundiff and Marsh, 1996) due to their propensity to shed moisture. This may vary by feedstock, as grasses with large stalks (Figure 4.4) may not repel water, particularly if bale densities are low. Along with added storage costs, low bale density will increase the energy demands and GHG emissions in our sustainability "ledger" by increasing the number of bales and adding to handling, storage, and transport needs of these bulky materials. A further challenge for feedstocks packaged as bales is the need for size reduction systems to process the material.

In contrast to baling, chopping systems perform in-field particle size reduction that creates a flowable product on the go. Large feedstocks such as giant reed or elephantgrass (*Pennisetum purpureum*) that dry slowly—if at all—will likely be better suited for harvesting with chopping systems. However, chopping systems are more labor and equipment intensive because two persons are needed at all times to keep the system moving. All operations must necessarily stop when load–haul equipment is not at hand in the field to collect the harvested biomass. Moreover, these systems may require more expensive storage infrastructure, especially for wet feedstocks that cannot be immediately delivered to a refinery and need to be put in anaerobic storage.

Logistics and Process Considerations

Bioenergy production research too often has been an "either-or" undertaking; most work has been dedicated to growing crops for, or converting them to energy, fuels, or products. While

Figure 4.4. Left to right: switchgrass, giant reed, and miscanthus bales. Large stems, especially with giant reed, are difficult to dry down—even when standing over winter—and the coarse form facilitates water absorption.

certainly critical, these links are interdependent and they are dependent on the development of well-conceived delivery systems that can move biomass from farm to refinery and product from refinery to market (Fike *et al.*, 2007; Cundiff *et al.*, 2009; Miao *et al.*, 2012). These issues are critical to sustainability, because transporting biomass is a major component of system financial and energy costs, and improvements may substantially improve industry competitiveness. Although a discussion of logistics issues could consume its own chapter, we will briefly consider the challenges attendant to moving biomass, especially in terms of agronomics and sustainability.

Scaling a biomass industry to a size that will make a significant contribution to US energy supplies promises to be a huge undertaking. Some anticipate that biorefineries will store no more than about 72 h worth of biomass at a given time (Resop *et al.*, 2011). To meet the demands of the refiner day in and day out, the other 362 days' worth of feedstock must be growing or safely in storage at field edges or in satellite facilities. These feedstock depots must be readily accessible in all manner of weather, must minimize storage losses (Sanderson *et al.*, 1997), and must maintain feedstock composition in a desirable state (Hess *et al.*, 2007). Of course, the value to the system as a whole must be weighed against the economic and environmental costs of building this infrastructure—and may lead to some not-so-obvious conclusions.

As conceived based on the generally accepted recommendation of a single, end-of-season harvest, biomass systems will need to gather a year's worth of feedstock in relatively small (fall–winter) window of time. Such an approach magnifies the infrastructural demands (in equipment, roads, and storage), particularly when one considers the restrictions to harvest and handling operations that will be caused by variable weather and field conditions—especially on marginal sites. While opening the harvest window may be anathema to many agronomists, it provides opportunity to reap the cost–benefits of increasing the productive machine hours for equipment used in an enterprise (Cundiff, 1996). It also provides the potential to greatly reduce the infrastructural (and energy) demands of biofuel systems. Thus, the disadvantages of opening the harvest window (in terms of reduced feedstock quality, greater nutrient inputs, or greater harvest energy) may be more than offset by advantages to the system as a whole.

Of course, the value of such feedstock management strategies must also be weighed in the context of the conversion process—which in some ways may be the final arbiter. On the one hand, systems utilizing combustion or pyrolysis/gasification for conversion may be less sensitive to the variable nature of feedstocks that would be harvested over the course of a year, but at the same time they may be less efficient or more subject to fouling when processing biomass harvested during the growing season because of its greater concentrations of minerals in the feedstock. This also induces greater regulatory hurdles if combusting green tissues that have high N and sulfur concentrations leads to elevated NO_X and SO_X emissions. In contrast, liquid fuel systems may not face such challenges in terms of emissions or system fouling. However, the sensitivity of microbes and enzymes to small changes in feedstock constituents would likely mean that the biochemical conversion systems would have to be "recalibrated" to address the changing nature of the feed supply.

Successful assemblage of a biofuel system is unlikely if the enterprise is viewed as a sandwiching together of each maximized system component. Biofuel systems must be viewed as integrated wholes with each component optimized for the greatest systemic outcomes. Of course, that may lead us to two fundamentally deeper questions: what systems are we truly talking about, and what outcomes do we want from them? Consider, for example, the results of a modeling effort by Egbendewe-Mondzozo *et al.* (2011), who studied the performance of ethanol production systems based on multiple cropping systems in several subwatersheds in Michigan. The authors concluded that economic performance was greater with more centralized biofuel processing systems, but decentralized preprocessing plants led to greater environmental sustainability.

While we acknowledge the limited treatment of these issues, the reader should recognize both the complicated interactions that occur throughout a biomass-to-bioenergy system and how its function relates to a greater set of environmental, economic, social, and policy issues (Fike *et al.*, 2007; Ng *et al.*, 2011). Determining the costs and benefits of the different options for bioenergy systems can be challenging, particularly when those doing the assessment place different weights on the value of different outcomes. Moreover, maximizing the system may not be the ultimate goal as biofuel production is viewed in a larger, sustainability context. We will touch on that topic in the next section.

The Bigger Picture: Sustainability Issues for Herbaceous Energy Systems

Several sustainability issues need addressing when considering herbaceous energy crop production. Water, species diversity, land use, and climate impacts are among the many aspects of energy cropping that need assessment as these systems are deployed. As many reviews and several subsequent chapters of this book have been devoted to the environmental impacts of energy cropping, it is not our goal to cover these issues in detail. Rather, we aim to emphasize several of the salient points about the traditional measures of "sustainability" as they relate to the crops and systems we have discussed.

Water

The long-term impacts of converting lands currently in traditional row cropping systems to bioenergy crops may have profound effects both on water quantity and quality, but the outcomes of such changes are not completely certain. We consider the move from first- to second-generation biofuels as a step toward greater sustainability (Fike *et al.*, 2007) in terms of reduced erosion and nutrient runoff, but several studies suggest that reduced soil losses and greater water quality may not be the whole of the story. Although perennial energy crops typically are expected to reduce erosion and nutrient runoff or leaching, some have modeled greater N loading to surface waters (Wu and Liu, 2012) based on the assumption that fertilized acres will expand as energy crops displace "native grasslands." In contrast, Ng *et al.* (2010) reported that the greater the fraction of land converted to miscanthus, the greater the expected benefit to water quality in terms of reduced NO_3-N loading.

Regardless of which assumptions are borne out, perennial energy crops do use more water than maize and existing pastures or grasslands (Stephens *et al.*, 2001; VanLoocke *et al.*, 2010; Le *et al.*, 2011; Wu and Liu, 2012). *M.* × *giganteus* has greater root water capture capacity relative to other energy crops (Monti and Zatta, 2009), and work in the United Kingdom suggests that the large evapotranspiration potential of *M.* × *giganteus*, coupled with water extraction capacity to depths of 2–3 m could reduce "hydrologically effective rainfall" around 140–180 mm each year (Stephens *et al.*, 2001). Modeling efforts indicate that increased plantings of miscanthus or switchgrass could raise the demands on available water supplies and reduce soil moisture storage (VanLoocke *et al.*, 2010; Le *et al.*, 2011; Wu and Liu, 2012). However, substantial impacts might not be felt until miscanthus fractional cover reached >25% in dry regions and >50% across the whole of the Midwest (VanLoocke *et al.*, 2010).

Although improvements in water quality often are seen as a collateral benefit of deploying energy cropping systems, this could be a primary objective in some scenarios. Certainly, giant reed is being tested or deployed for phytoremediation efforts in constructed wetlands (Sagehashi *et al.*, 2011; Calheiros *et al.*, 2012), but much larger-scale applications have been suggested. For example, harvesting invasives in wetland areas has been posited as an effective way to recapture nutrients lost from agriculture fields (Jakubowski *et al.*, 2010) while also supplying large quantities of biomass for bioenergy production. While much work would need to be done to bring such a system to fruition, it could be a "triple play" strategy by also helping to shift wetland systems to more desirable vegetation.

Issues of clean and abundant water do not simply reside in the domain of agronomy/crop production systems. Water is the key to some industrial processes as well, and water demands have constrained the growth of the ethanol industry in the past because of its high water input per unit of product. Water supply limitations have potential to challenge the industry in future as well, and likely will be greatest for biorefineries sited near urban centers or in regions with limited precipitation.

Diversity and Wildlife

Plans to convert large expanses of land to monoculture plantings of any species can be disconcerting given the potential negative consequences to ecosystem diversity and wildlife habitat, especially in sensitive sites such as wetlands. These concerns are greater for many of the species of interest because they are not native to Europe or North America, the two regions where the biofuel crops predominantly are being studied. Uncertainties about the impact of energy crops reflect the relatively short time and small spatial scales associated with their use—and perhaps some experience of their invasive potential.

There are different takes on the diversity impacts of biofuel plantings, and these certainly vary by species. For example, in addition to their potential invasivity, giant reed and reed canary grass have strong potential for negative effects on species diversity in stream communities (Cushman and Gaffney, 2010; Spyreas *et al.*, 2010). Thus, there is little in the literature to speak positively of these species in this regard. Rather, much of the ecology literature surrounding giant reed is related to restoring stream habitats dominated by the species. Unfortunately, both are fierce competitors, and resistance to (re)invasion might not be sustainable over time (Quinn and Holt, 2009).

In contrast, and despite the fact that it is grown as a monoculture, several studies have given *M.* × *giganteus* rather favorable ratings for supporting wildlife diversity. The lack of annual tillage, low inputs of pesticides, and the infrequent winter season harvest regime have been considered beneficial for an array of invertebrates, mammals, and birds (Felten and Emmerling, 2011). Headland spaces around these crops also are considered exceptional habitats for many desired plant and animal species (Sage *et al.*, 2008). However, because crop structure plays a key role in its suitability as habitat for different species, miscanthus might not be acceptable for some grassland birds, particularly as stands age and thicken (Semere and Slater, 2007; Bellamy *et al.*, 2009).

Planting native species such as switchgrass has been proposed as a method of increasing grassland bird abundance, but more recently there is recognition that large, monotypic stands may prove less than ideal. In particular, concern has been voiced about the quality of highly productive upland switchgrass ecotypes as habitat for ground-nesting birds, because the stand densities that are suitable for productive energy cropping systems may prove impenetrable, especially for fledglings. This issue—and the opportunity to meet the needs of different avian fauna—might be addressed with varied harvest timings (Roth *et al.*, 2005) that would allow for different biomass structures. In fields sown to lowland ecotypes, however, the potential habitat limitations for ground-nesting birds also may be addressed through planting management. Because lowlands maintain more bunch structure and are not negatively affected by row planting widths of up to about 90 cm (Ma *et al.*, 2001; Foster *et al.*, 2012), wide-row planting configurations could meet the needs of both the wildlife and biomass producer communities.

Faunal species diversity in these systems can further be increased and the nature of the diversity improved—for example, by increasing invertebrate pest predators—by planting and managing for diverse floristic mixtures (Robertson *et al.*, 2011; Werling *et al.*, 2011; Robertson *et al.*, 2012). However, we have noted the debate about biomass productivity relative to plant species diversity and have also, albeit briefly, noted the potential drag that diversified feedstocks can represent for the current biomass conversion platforms.

Thus we return to an earlier theme: for bioenergy systems to meet as many of society's desired production and sustainability goals as possible, they must be implemented holistically. In this way management becomes an optimization process, accounting for the different (and sometimes seemingly conflicting) needs and outcomes of the system. As biofuels go "prime time," greater efforts at landscape planning may be useful to avoid the potential consequences of letting the "horse out of the barn"—or the "plant out of the field" (Haughton *et al.*, 2009). Sustainability frameworks can be used to identify lands where energy crops are environmentally suitable for planting and to identify appropriate indicators of biodiversity (such as butterfly abundance (Haughton *et al.*, 2009). While such approaches may instrumental for landscape planners, the ability to create (or impose?) large-scale (e.g., watershed-level) land management plans will take greater coordination and cooperation between producers and governments.

Marginal Land and Indirect Land-use Change

Much of the touted promise of energy crop production is embedded in the idea that the species of choice can be suitable for use on marginal land. This has an implied message that "marginal" lands already are of limited function and productivity, and that by planting energy crops on these sites, society will achieve multiple positive outcomes. However, this idea may run up against the reality that marginal lands are not well defined, may serve productive purposes that have positive economic outcomes (as, for example, livestock production), or may truly be unsuited for cropping due to soil factors such as erosivity or poor drainage and steepness of slope (Gopalakrishnan *et al.*, 2011). Thus, developing metrics or frameworks that identify multiple criteria for classifying marginal-but-suitable sites for biofuel production—and which can calculate tradeoffs between existing land uses and management practices—can improve the overall sustainability of biofuel implementation strategies (Gopalakrishnan *et al.*, 2011).

While such analyses may prove useful at regional scales, there are broader questions of bioenergy production effects on direct and indirect land-use change that may have global implications. For example, an analysis by Fargione *et al.*, (2008) suggested that large carbon debts would arise as a function of land conversion for biofuel production; this paper seems to have struck a nerve as the subsequent debate has been fierce. The issues within this topic are abundant—biofuel systems as GHG emitting versus GHG sparing; food versus fuel; developed world versus underdeveloped world—and it is well beyond the scope of this chapter to flesh them out. Creutzig *et al.* (2012) acknowledged that reconciling the different views and assessments has not occurred and is "a fundamental prerequisite for future bioenergy assessment." However, to truly understand the environmental impacts of these systems via land-use change will require global frameworks that "provide a globally accepted" GHG accounting which can be applied from fine to large spatial resolution (Smith *et al.*, 2012).

Energy and Carbon Balances

A few researchers have explored the energy dynamics of giant reed production and largely have come to positive conclusions about the feedstock's viability as an energy crop (Angelini *et al.*, 2005; Angelini *et al.*, 2009; Mantineo *et al.*, 2009). Energy balance responses have had a rather broad range, particularly because of establishment year inputs and irrigation requirements. Mantineo *et al.* (2009) suggested the added costs of irrigation and fertility can be justified from an energy balance standpoint, however, and the five-year average net energy yield was ~419 GJ ha^{-1} when averaged over N and irrigation rates. Although yield response to N and irrigation were positive, the questions of GHG balance and return on investment for these added inputs were not addressed, although our reading of the data suggests the law of diminishing returns still applies. Given the plant's massive rhizomes and root system (Figure 4.5), there is surprisingly little information about carbon sequestration for this species.

Net energy yield for miscanthus (~296 GJ ha^{-1}) was not as great as for giant reed in the Mantineo study, but rates as high as 590 GJ ha^{-1} have been reported (Lewandowski and Schmidt, 2006). In studies of carbon sequestration, *M.* × *giganteus* has fared well when compared with arable energy crops (Drewer *et al.*, 2012). This is largely a function of high yields coupled with low fertility demands as discussed previously. Carbon sequestration potential also is high, but this must be put in context, as the ability to sequester carbon will primarily be a function of previous land use. Several researchers have seen the benefits of displacing agricultural crops with dedicated *M.* × *giganteus* energy crops (St. Clair *et al.*, 2008; Dondini *et al.*, 2009; Hillier *et al.*, 2009; Zeri *et al.*, 2011; Don *et al.*, 2012), but the feasibility of

Figure 4.5. Top, left to right: roots and rhizome systems for upland and lowland switchgrass ecotypes. Bottom, left to right: roots and rhizome systems for *Miscanthus* × *giganteus* and giant reed.

converting the production of agricultural enterprises from groceries to "grassoline" is questioned in light of the food versus fuel debate. There is some question in the literature about the carbon accounting attendant to converting grasslands to *M.* × *giganteus* energy cropping. Some data suggest these conversions will provide no benefits (and perhaps negative impacts) for carbon sequestration at least in the short term (Don *et al.*, 2012; Zimmermann *et al.*, 2012). Others have suggested there will be a payback for such conversions but that it might take 30 years for a grassland-to-energy plantation conversion to reach carbon neutrality (Hughes *et al.*, 2010; Donnelly *et al.*, 2011). As we noted earlier, this does not bode well if expected stand life is no more than 20 or 25 years.

Although Schmer *et al.* (2008) reported much lower net energy yields for switchgrass (60 GJ ha^{-1}), it is important to note that these results are estimated from 10 fields across a wide geographic range under dryland management. Traditional agricultural equipment also was used to harvest the biomass, and yields averaged 5.2–11.1 Mg ha^{-1}. In terms of carbon balances, similar general findings as with miscanthus are reported for switchgrass: converting croplands to switchgrass is a net carbon sink, while converting grasslands to switchgrass may present no gains (or worse) without added nutrient inputs (Chamberlain *et al.*, 2011; Monti *et al.*, 2012). We leave this in rather general terms as many of the changes reported in the literature are derived from modeled results; enduring outcomes are not well quantified and will require longer monitoring (Zeri *et al.*, 2011). We also note that targeting "marginal" grasslands for bioenergy cropping systems might not bear the sustainability gains that are commonly assumed.

Additional Impacts of Climate Change: C_4 versus C_3 species

A general scan of potential grass bioenergy crops (Table 4.2) reveals that most species of interest use one of the C_4 photosynthetic pathways. This is the case because of the high productivity and greater water and nutrient use efficiencies associated with this photomechanism. The C_4 photosynthetic syndrome has evolved many times in grasses (and other species) in response to conditions such as lower atmospheric CO_2 concentrations, warmer temperatures, and low precipitation (Edwards *et al.*, 2010). It is interesting, and perhaps not a little ironic, to think that humans may take advantage of this adaptation in part to mitigate the effects of human-induced climate changes. However, questions and uncertainties about the impacts of climate change have resulted in some debate about the utility and sustained use of C_4 species for this purpose into the future.

One key attribute of the C_4 photomechanism is the advantage it provides under low atmospheric CO_2 concentrations. In low CO_2 environments, the enzyme systems and leaf architecture of C_4 plants allow for the efficient capture of CO_2 at temperatures higher than optimum for C_3 species. Assuming present trajectories for CO_2 emissions, current atmospheric CO_2 concentrations (about 391 ppm) are predicted to reach 500 ppm by 2050 (IPCC, 2007). In this scenario, C_3 species would be favored because the competitive abilities of C_4 species (e.g., greater tolerance of higher atmospheric temperatures and greater nutrient use efficiencies) would prove less of an advantage in a CO_2-enriched atmosphere. Climate change effects may play out in several ways given the potential interactions of resources such as N, CO_2, precipitation, and temperature (see Shaw *et al.*, 2002) along with alterations to current seasonal temperature and precipitation regimes. This abridged and greatly simplified account of the complex interactions of plants and their environs should give some sense of the uncertainties and of the difficulty of making predictions about feedstock selections into the future.

As an aside, we note that both regardless and independent of climate change effects, C_3 species are likely to predominate—or be the only grass in town—at greater latitude. For example, the grass-based bioenergy cropping systems of Scandinavia have largely utilized reed canary grass because the species combines relatively high productivity with a capacity to accommodate the lower temperatures and shorter seasons typical of these high-latitude regions. Currently, the growing environment in such locations precludes the use of C_4 species—and will continue to do so unless and until breeding efforts or genetic manipulation can be deployed to increase the geographic range of C_4 plants.

Waste (or Co-product) Streams

Managing—or even utilizing—waste streams from bioenergy facilities have got little attention compared with other sustainability issues attendant to biofuel production systems. On the contrary, many bioenergy systems are seen as providing disposal solutions for an array of agricultural and municipal wastes. However, as these systems come online and expand, they too will create unique waste streams that present potential environmental risks (Menetrez, 2010)—or opportunities.

In biochemical process systems, the waste waters from alcohol distillation are the largest pollutant load, but these have potential use for generating biogas to power the system (Arias-Polo, 2011). Also, recycled process water used in biorefining can have high levels of metals such as manganese or zinc and the ability to capture and remove such contaminants will reduce environmental impacts by increasing water reuse (Palumbo *et al.*, 2012).

Waste products will be produced not only in alcohol conversion pathways. Residues from biogas systems may present a disposal issue (Odlare *et al.*, 2012), and pyrolytic process technologies are likely to generate oil–water mixtures that will need to be processed or eliminated during the conversion process (Cornelissen *et al.*, 2009). Gasification systems may generate the least waste, as only minerals and silicates which make up the ash remain after processing. Regardless of conversion system, high priority should be placed on efficiently capturing and returning the nutrients in waste streams to production fields in order to tighten nutrient cycles, reduce fertilizer input needs, and increase system sustainability (Arthurson, 2009).

To conclude, we make one note about biochar, a co-product of pyrolysis systems. Biochar has value as an agricultural or horticultural amendment or for industrial processes that utilize high-grade coking products. Currently biochar may be more valuable than the oils coming from pyrolysis systems. A tremendous body of literature surrounding biochars has arisen in the past decade, and some scientists suggest that pyrolysis systems have tremendous potential to improve sustainability because biochars can be used to sequester carbon in soils and this provides the added benefit of greater soil quality and consequent increased plant productivity (Laird, 2008). The feasibility of this idea remains in question, however, as at this date pyrolysis chars are selling for as much as \$1000 Mg^{-1}, and application rates to achieve sizeable production increases will probably not be less than 10 Mg ha^{-1}.

Life-cycle Models and Challenges to Determining Sustainability

Despite their potential, there still are many questions about the sustainability and ecological benefits of herbaceous energy cropping systems. These issues progress all the way from the field scale—What to plant? How to manage it?—to a global systems impact level—What is the CO_2 or GHG emissions balance? What is the energy return? The ability to aggregate and synthesize information across the supply chain to understand the impacts at fine and coarse spatial scales is one of the challenges we face. Current methodologies rely heavily on life-cycle assessment (LCA) as a tool for analyzing energy balance, GHG, and other environmental impacts. However, the values for LCA models vary significantly—even for a given biofuel species—due to the different assumptions and datasets used by modelers. Figure 4.6 gives us another way to conceive of this set of issues and interacting factors.

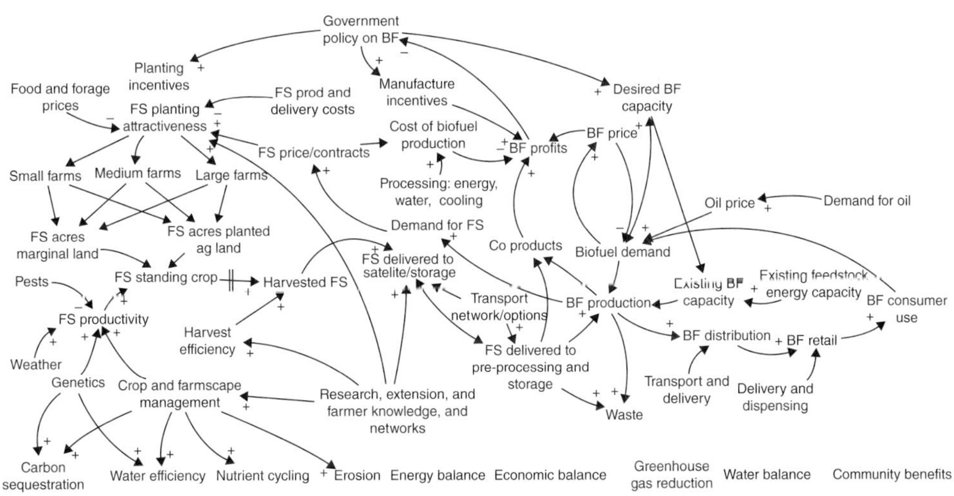

Figure 4.6. Simplified causal loop diagram of the biofuel supply chain. (Created by Steven C. Hodges). Note the level of model complexity even without addressing the linkages to sustainability metrics.

Discrepancies in model outcomes arise from the LCA components—the system boundaries and life-cycle inventories—used in the modeling effort. System boundaries can be defined by spatial, temporal, and process boundaries in order to suit the goal and scope of a given model. Having more uniform system boundaries would decrease the variation among models and provide a more realistic basis for outcome comparisons. However, setting the constraints on the spatial boundary may be more difficult, as this can progress from local to watershed to regional, national, or even global boundaries, as mentioned previously. Locally based LCA has benefit for specific goals for a given community. These models should be based on—and produce—a more precise dataset for life-cycle inventory than large-scale models that must subsume greater "data diversity" with respect to climate, weather, soil properties, topography, transport networks, etc. Thus discrepancies in estimates of energy balance and GHG emissions arise from the good or not-so-good assumptions inherent in the modeling effort. Another factor in the variation among models lies in conducting life-cycle inventories that use complex models in which the components are estimated with some uncertainty. Therefore, finding standardized methods and metrics for the various conditions and locations, time scales, and logistical and economic aspects of these systems remains a key challenge—and great need—in order to increase our understanding of system sustainability.

Some Concluding Thoughts

We have addressed in some detail what makes for an ideal perennial biofuel feedstock and have spent a fair number of words discussing the potential and limitations of three of the day's foremost perennial feedstock candidates for warm to temperate zones. We have also spent a fair amount of discussion advocating that bioenergy systems will have the greatest opportunity for achieving sustainability if they are conceived and implemented with a view toward the "greatest good." The implementation of bioenergy systems should be an optimization process rather than an economic maximization exercise. The former approach has the best chance for success at the multiple levels of interplay that will occur as these systems are developed—the latter may just be another form of mining.

We have spoken little about economics and bioenergy production. Uncertainties surrounding the value of fossil fuels, new cost-sparing advances in production, logistics and conversion, process technologies, and changes in carbon emission policies and regulations all "cloud the crystal ball." With any potential set of conditions, the reality is that producers must recognize bioenergy cropping as a rewarding enterprise relative to their existing farm operations before they will engage in the endeavor—at least economics has been the primary motivating factor in cropping decisions historically. Whether bioenergy systems can be economically feasible depends both on internal factors (e.g., costs of production) and external factors (e.g., energy prices or climate change mitigation strategies). Environmental policy and production standards may also play roles in valorizing alternative energy markets as society begins working to deal with threats of climate change, species diversity loss, etc. The outcome at the farm level is uncertain, but such factors and their associated production incentives (or disincentives) could result in very different paths taken on the road to energy cropping implementation.

Of course, the idea that a bioenergy system would exist at all is predicated on a level of societal acceptance that may or may not exist at present. Garnering societal acceptance will be critical for policy makers, businesses, producers, and community leaders who would strive to move the industry forward. At the local, producer community level, residents must feel that the effort is a plus, that it provides economic reward, that it is not a hazard, and that it fosters community development and revitalization.

In the bigger scheme of things, bioenergy systems must be viewed by society as worthwhile and without significant risk to human or ecosystem health. Balancing a growing global population's demands for food, feed, fiber, and fuel in ways that provide high standards of living for Earth's inhabitants without degrading Earth's ecosystems is a daunting task. This will be true both in terms of the science and the policy, but the fact that there is such interest in these systems gives some reason for optimism.

References

Abbassi, B., Al-Zboon, K., Radaideh, J., and Wahbeh, A. (2011) Using constructed wetlands to improve drainage water quality from hydroponics farms. *Irrigation and Drainage* 60: 370–380.

Adamski, D., Johnson, P.J., Boe, A.A., Bradshaw, J.D., and Pultyniewicz, A. (2010) Descriptions of life-stages of Blastobasis repartella (Lepidoptera: Gelechioidea: Coleophoridae: Blastobasinae) and observations on its biology in switchgrass. *Zootaxa* 2656: 41–54.

Adler, P.R., Sanderson, M.A., Boateng, A.A., Weimer, P.J., and Jung, H.J.G. (2006) Biomass yield and biofuel quality of switchgrass harvested in fall or spring. *Agronomy Journal* 98: 1518–1525.

Adler, P.R., Sanderson, M.A., Weimer, P.J., and Vogel, K.P. (2009) Plant species composition and biofuel yields of conservation grasslands. *Ecological Applications* 19: 2202–2209.

Ahmad, R., Liow, P.S., Spencer, D.F., and Jasieniuk, M. (2008) Molecular evidence for a single genetic clone of invasive Arundo donax in the United States. *Aquatic Botany* 88: 113–120.

Ahonsi, M.O., Agindotan, B.O., Gray, M.E., and Bradley, C.A. (2011) First report of basal stem rot and foliar blight caused by Pythium sylvaticum on Miscanthus sinensis in Illinois. *Plant Disease* 95: 616.

Ahonsi, M.O., Agindotan, B.O., Williams, D.W., *et al.* (2010) First report of Pithomyces chartarum causing a leaf blight of Miscanthus × giganteus in Kentucky. *Plant Disease* 94: 480–481.

Aliscioni, S.S., Giussani, L.M., Zuloaga, F.O., and Kellogg, E.A. (2003) A molecular phylogeny of *Panicum* (Poaceae: Paniceae): tests of monophyly and phylogenetic placement within the Panicoideae. *American Journal of Botany* 90: 796–821.

Allison, G.G., Morris, C., Lister, S.J., *et al.* (2012) Effect of nitrogen fertiliser application on cell wall composition in switchgrass and reed canary grass. *Biomass and Bioenergy* 40: 19–26.

Amalraj, V.A. and Balasundaram, N. (2006) On the taxonomy of the members of Saccharum complex. *Genetic Resources and Crop Evolution* 53: 35–41.

An, G., Miyakawa, S., Kawahara, A., Osaki, M., and Ezawa, T. (2008) Community structure of arbuscular mycorrhizal fungi associated with pioneer grass species Miscanthus sinensis in acid sulfate soils: habitat segregation along pH gradients. *Soil Science and Plant Nutrition* 54: 517–528.

Angelini, L.G., Ceccarini, L., and Bonari, E. (2005) Biomass yield and energy balance of giant reed (Arundo donax L.) cropped in central Italy as related to different management practices. *European Journal of Agronomy* 22: 375–389.

Angelini, L.G., Ceccarini, L., Nassi o di Nasso, N., and Bonari, E. (2009) Comparison of *Arundo donax L.* and Miscanthus × giganteus in a long-term field experiment in Central

Italy: analysis of productive characteristics and energy balance. *Biomass and Bioenergy* 33: 635–643.

Anonymous. (2008) Optimization of switchgrass management for commercial fuel pellet production, in *Weeds on the Horizon-Biomass Energy From Silvergrass, Environmental Protection* (ed. T. Panorama), Vol. 2012. Ministry of Foreign Affairs, Republic of China (Taiwan).

Arias-Polo, G.N. (2011) Sustainable energy proposal ethanol distilleries. *ICIDCA Sobre los Derivados de la Caña de Azúcar* 45: 7–14.

Arthurson, V. (2009) Closing the global energy and nutrient cycles through application of biogas residue to agricultural land—potential benefits and drawback. *Energies* 2: 226–242.

Atkinson, C.J. (2009) Establishing perennial grass energy crops in the UK: a review of current propagation options for Miscanthus. *Biomass and Bioenergy* 33: 752–759.

Barrett, S.W., Swetnam T.W., and Baker, W.L. (2005) Indian fire use: deflating the legend. *Fire Management Today* 65: 31–34.

Beale, C.V. and Long, S.P. (1995) Can perennial C4 grasses attain high efficiencies of radiant energy conversion in cool climates? *Plant, Cell and Environment* 18: 641–650.

Beaty, E.R., Engel, J.L., and Powell, J.D. (1978) Tiller development and growth in switchgrass. *Journal of Range Management* 31: 361–365.

Beccari, G., Covarelli, L., Balmas, V., and Tosi, L. (2010) First report of Miscanthus × giganteus rhizome rot caused by Fusarium avenaceum, Fusarium oxysporum and Mucor hiemalis. *Australasian Plant Disease Notes* 5: 28–29.

Belesky, D.P. and Fedders, J.M. (1995) Warm-season grass productivity and growth rate as influenced by canopy management. *Agronomy Journal* 87: 42–48.

Bellamy, P.E., Croxton, P.J., Heard, M.S., *et al.* (2009) The impact of growing miscanthus for biomass on farmland bird populations. *Biomass and Bioenergy* 33: 191–199.

Boe, A. and Gagné, R.J. (2011) A new species of gall midge (Diptera: Cecidomyiidae) infesting switchgrass in the Northern Great Plains. *BioEnergy Research* 4: 77–84.

Boland, J.M. (2006) The importance of layering in the rapid spread of *Arundo donax* (giant reed). *Madroño* 53: 303–312.

Bonanno, G. (2012) *Arundo donax* as a potential biomonitor of trace element contamination in water and sediment. *Ecotoxicology and Environmental Safety* 80: 20–27.

Bow, J.R., Muir, J.P., Weindorf, D.C., Rosiere, R.E., and Butler, T.J. (2008) Integration of cool-season annual legumes and dairy manure compost with switchgrass. *Crop Science* 48: 1621–1628.

Bradshaw, J.D., Prasifka J.R., Steffey, K.L., and Gray, M.E. (2010) First report of field populations of two potential aphid pests of the bioenergy crop Miscanthus × giganteus. *Florida Entomologist* 93: 135–137.

Bullard, M.J. (1996) The agronomy of Miscanthus. *Landwards* 51: 12–15.

Bullard, M.J., Christian, D.G., Knight, J.D., Lainsbury, M.A., and Parker, S.R. (2001) Biomass and Energy Crops II, University of York, York, UK, 18–21 December 2001. Aspects of Applied Biology vi + , 65th edn. 374p.

Cadoux, S., Riche, A.B., Yates, N.E., and Machet, J.M. (2012) Nutrient requirements of Miscanthus × giganteus: conclusions from a review of published studies. *Biomass and Bioenergy* 38: 14–22.

Calheiros, C.S.C., Quitério, P.V.B., Silva, G., *et al.* (2012) Use of constructed wetland systems with Arundo and Sarcocornia for polishing high salinity tannery wastewater. *Journal of Environmental Management* 95: 66–71.

Calheiros, C.S.C., Teixeira, A., Pires, C., *et al.* (2010) Bacterial community dynamics in horizontal flow constructed wetlands with different plants for high salinity industrial wastewater polishing. *Water Research (Oxford)* 44: 5032–5038.

Carpenter, J.E. (2011) Impact of GM crops on biodiversity. *GM Crops* 2: 7–23.

Carris, L.M., Castlebury, L.A., and Zale, J. (2008) First report of Tilletia pulcherrima bunt on switchgrass (*Panicum virgatum*) in Texas. *Plant Disease* 92: 1707.

Carver, P., Hocking, T.J., and Bullard, M.J. (2001) Photosynthetic responses of Miscanthus ecotypes. *Aspects of Applied Biology* 65: 215–222.

Casler, M.D. (2005) Ecotypic variation among switchgrass populations from the Northern USA. *Crop Science* 45: 388–398.

Casler, M.D. and Boe, A.R. (2003) Cultivar x environment interactions in switchgrass. *Crop Science* 43: 2226–2233.

Cassida, K.A., Kirkpatrick T.L., Robbins, R.T., Muir, J.P., Venuto, B.C., and Hussey, M.A. (2005a) Plant-parasitic nematodes associated with switchgrass (*Panicum virgatum L.*) grown for biofuel in the South Central United States. *Nematropica* 35: 1–10.

Cassida, K.A., Muir, J.P., Hussey, M.A., Read, J.C., Venuto, B.C., and Ocumpaugh, W.R. (2005b) Biomass yield and stand characteristics of switchgrass in South Central U.S. environments. *Crop Science* 45: 673–681.

Cavallaro, V., Tringali, S., and Patanè, C. (2011) Large-scale in vitro propagation of giant reed (*Arundo donax L.*), a promising biomass species. *Journal of Horticultural Science and Biotechnology* 86: 452–456.

Ceotto, E. and Candilo M.D. (2010) Shoot cuttings propagation of giant reed (*Arundo donax L.*) in water and moist soil: the path forward? *Biomass and Bioenergy* 34: 1614–1623.

Chamberlain, J.F., Miller, S.A., and Frederick, J.R. (2011) Using DAYCENT to quantify on-farm GHG emissions and N dynamics of land use conversion to N-managed switchgrass in the Southern U.S. *Agriculture, Ecosystems and Environment* 141: 332–341.

Chen, B., Lai, H., Lee, D., and Juang, K. (2011) Using chemical fractionation to evaluate the phytoextraction of cadmium by switchgrass from Cd-contaminated soils. *Ecotoxicology* 20: 409–418.

Cherubini, F., Jungmeier, G., Wellisch, M., *et al.* (2009) Toward a common classification approach for biorefinery systems. *Biofuels, Bioproducts and Biorefining* 3: 534–546.

Chou, C. (2009) Miscanthus plants used as an alternative biofuel material: the basic studies on ecology and molecular evolution. *Renewable Energy* 34: 1908–1912.

Christian, D.G., Lamptey, J.N.L., Forde, S.M.D., and Plumb, R.T. (1994) First report of barley yellow dwarf luteovirus on Miscanthus in the United Kingdom. *European Journal of Plant Pathology* 100: 167–170.

Christian, D.G., Riche, A.B., and Yates, N.E. (2008) Growth, yield and mineral content of Miscanthus × giganteus grown as a biofuel for 14 successive harvests. *Industrial Crops And Products* 28: 320–327.

Clark, R.B. (2002) Differences among mycorrhizal fungi for mineral uptake per root length of switchgrass grown in acidic soil. *Journal of Plant Nutrition* 25: 1753–1772.

Clark, R.B., Zeto, S.K., and Zobel, R.W. (1999a) Arbuscular mycorrhizal fungal isolate effectiveness on growth and root colonization of Panicum virgatum in acidic soil. *Soil Biology and Biochemistry* 31: 1757–1763.

Clark, R.B., Zobel, R.W., and Zeto, S.K. (1999b) Effects of mycorrhizal fungus isolates on mineral acquisition by Panicum virgatum in acidic soil. *Mycorrhiza* 9: 167–176.

Clifton-Brown, J., Robson, P., Sanderson, R., Hastings, A., Valentine, J., and Donnison, I. (2011) Thermal requirements for seed germination in Miscanthus compared with

Switchgrass (*Panicum virgatum*), Reed canary grass (*Phalaris arundinaceae*), maize (*Zea mays*) and perennial ryegrass (*Lolium perenne*). *GCB Bioenergy* 3: 375–386.

Clifton-Brown, J.C., Breuer, J., and Jones, M.B. (2007) Carbon mitigation by the energy crop, Miscanthus. *Global Change Biology* 13: 2296–2307.

Clifton-Brown, J.C., Jones, M.B., and Breuer, J. (2001a) Yield performance of M. × giganteus during a 10 year field trial in Ireland. *Aspects of Applied Biology* 65: 153–160.

Clifton-Brown, J.C. and Lewandowski, I. (2000) Water use efficiency and biomass partitioning of three different Miscanthus genotypes with limited and unlimited water supply. *Annals of Botany* 86: 191–200.

Clifton-Brown, J.C. and Lewandowski, I. (2002) Screening Miscanthus genotypes in field trials to optimise biomass yield and quality in Southern Germany. *European Journal of Agronomy* 16: 97–110.

Clifton-Brown, J.C., Lewandowski, I., Andersson, B., *et al.* (2001b) Performance of 15 Miscanthus genotypes at five sites in Europe. *Agronomy Journal* 93: 1013–1019.

Clifton-Brown, J.C., Lewandowski, I., Bangerth, F., and Jones, M.B. (2002) Comparative responses to water stress in stay-green, rapid- and slow senescing genotypes of the biomass crop, Miscanthus. *New Phytologist* 154: 335–345.

Cornelissen, T., Jans, M., Stals, M., *et al.* (2009) Flash co-pyrolysis of biomass: the influence of biopolymers. *Journal of Analytical and Applied Prolysis* 85: 87–97.

Cornelius, D.R. and Johnston, C.O. (1941) Differences in plant type and reaction to rust among several collections of *Panicum virgatum* L. *Journal of American Society of Agronomy* 33: 115–124.

Cortés, E., Goolsby, J.A., Moran, P.J., and Marcos-García, M.A. (2011a) The effect of the armored scale, *Rhizaspidiotus donacis* (Hemiptera: Diaspididae), on shoot growth of the invasive plant *Arundo donax* (Poaceae: Arundinoideae). *Biocontrol Science and Technology* 21: 535–545.

Cortés, E., Kirk, A.A., Goolsby, J.A., Moran, P.J., Racelis, A.E., and Marcos-García, M.A. (2011b) Impact of the Arundo scale *Rhizaspidiotus donacis* (Hemiptera: Diaspididae) on the weight of *Arundo donax* (Poaceae: Arundinoideae) rhizomes in Languedoc southern France and Mediterranean Spain. *Biocontrol Science and Technology* 21: 1369–1373.

Cortese, L.M., Honig, J., Miller, C., and Bonos, S.A. (2010) Genetic diversity of twelve switchgrass populations using molecular and morphological markers. *BioEnergy Research* 3: 262–271.

Creutzig, F., Popp, A., Plevin, R., Luderer, G., Minx, J., and Edenhofer, O. (2012) Reconciling top-down and bottom-up modelling on future bioenergy deployment. *Nature Climate Change* 2: 320–327.

Cundiff, J.S. (1996) Simulation of five large round bale harvesting systems for biomass. *Bioresource Technology* 56: 77–82.

Cundiff, J.S., Fike, J.H., Parrish, D.J., and Alwang, J. (2009) Logistic constraints in developing dedicated large-scale bioenergy systems in the Southeastern United States. *Journal of Environmental Engineering* 135: 1086–1096.

Curley, E.M., O'Flynn, M.G., and McDonnell, K.P. (2010) Nitrate leaching losses from Miscanthus × giganteus impact on groundwater quality. *Journal of Agronomy* 8: 31–36.

Cushman, J.H. and Gaffney, K.A. (2010) Community-level consequences of invasion: impacts of exotic clonal plants on riparian vegetation. *Biological Invasions* 12: 2765–2776.

Das, M.K., Fuentes, R.G., and Taliaferro, C.M. (2004) Genetic variability and trait relationships in switchgrass. *Crop Science* 44: 443–448.

Davies, M.J., Longbottom, H., and Atkinson, C.J. (2011) Changes in duration of rhizome cold storage and manipulation of the growing environment to promote field establishment of *Miscanthus giganteus*. *Biomass and Bioenergy* 35: 4268–4279.

Davis, J.J. (1914) The Oat Aphis. United States Department of Agriculture Bulletin, 16p.

Davis, S.C., Parton, W.J., Dohleman, F.G., *et al.* (2010) Comparative biogeochemical cycles of bioenergy crops reveal nitrogen-fixation and low greenhouse gas emissions in a Miscanthus × giganteus agro-ecosystem. *Ecosystems* 13: 144–156.

DeHaan, L.R., Weisberg, S., Tilman, D., and Fornara, D. (2010) Agricultural and biofuel implications of a species diversity experiment with native perennial grassland plants. *Agriculture, Ecosystems and Environment* 137: 33–38.

DiTomaso, J.M., Barney, J.N., and Fox, A.M. (2007) Biofuel feedstocks: the risk of future invasions. CAST (Council for Agricultural Science and Technology) Commentary, 7p.

Dohleman, F.G., Heaton, E.A., Arundale, R.A., and Long, S.P. (2012) Seasonal dynamics of above- and below-ground biomass and nitrogen partitioning in Miscanthus × giganteus and Panicum virgatum across three growing seasons. *GCB Bioenergy* 4: 534–544.

Dohleman, F.G. and Long, S.P. (2009) More productive than maize in the Midwest: how does Miscanthus do it? *Plant Physiology* 150: 2104–2115.

Don, A., Osborne, B., Hastings, A., *et al.* (2012) Land-use change to bioenergy production in Europe: implications for the greenhouse gas balance and soil carbon. *GCB Bioenergy* 4: 372–391.

Donald, C.M. (1968) The breeding of crop ideotypes. *Euphytica* 17: 385–403.

Dondini, M., Groenigen, K.J.V., Galdo, I.D., and Jones, M.B. (2009) Carbon sequestration under Miscanthus: a study of 13C distribution in soil aggregates. *GCB Bioenergy* 1: 321–330.

Donnelly, A., Styles, D., Fitzgerald, J., and Finnan, J. (2011) A proposed framework for determining the environmental impact of replacing agricultural grassland with Miscanthus in Ireland. *GCB Bioenergy* 3: 247–263.

Dove, M.R. (1983) Theories of swidden agriculture, and the political economy of ignorance. *Agroforestry Systems* 1: 85–99.

Dowd, P.F. and Johnson, E.T. (2009) Differential resistance of switchgrass *Panicum virgatum L.* lines to fall armyworms *Spodoptera frugiperda* (J.E. Smith). *Genetic Resources and Crop Evolution* 56: 1077–1089.

Drewer, J., Finch, J.W., Lloyd, C.R., Baggs, E.M., and Skiba, U. (2012) How do soil emissions of N2O, CH4 and CO2 from perennial bioenergy crops differ from arable annual crops? *GCB Bioenergy* 4: 408–419.

Eckert, B., Weber, O.B., Kirchhof, G., Halbritter, A., Stoffels, M., and Hartmann, A. (2001) Azospirillum doebereinerae sp. nov., a nitrogen-fixing bacterium associated with the C4-grass Miscanthus. *International Journal of Systematic and Evolutionary Microbiology* 51: 17–26.

Edwards, E.J., Osborne, C.P., Strömberg, C.A.E., and Smith, S.A. (2010) The origins of C4 grasslands: integrating evolutionary and ecosystem science. *Science (Washington)* 328: 587–591.

Egbendewe-Mondzozo, A., Swinton, S.M., Bals, B.D., and Dale, B.E. (2011) Can dispersed biomass processing protect the environment and cover the bottom line for biofuel? East Lansing: Michigan State University, Department of Agricultural Economics, 38p.

Ercoli, L., Mariotti, M., Masoni, A., and Bonari, E. (1999) Effect of irrigation and nitrogen fertilization on biomass yield and efficiency of energy use in crop production of Miscanthus. *Field Crops Research* 63: 3–11.

Esbroeck, G.A.V., Hussey, M.A., and Sanderson, M.A. (2003) Variation between Alamo and Cave-in-Rock switchgrass in response to photoperiod extension. *Crop Science* 43: 639–643.

Evers, G.W. and Butler, T.J. (2000) Switchgrass establishment on coastal plain soil. Proceedings/Reports of the American Forage and Grassland Council, 37th North American Alfalfa Improvement Conference, Madison, Wisconsin, July 16–19, 2000.

Ezaki, B., Nagao, E., Yamamoto, Y., Nakashima, S., and Enomoto, T. (2008) Wild plants, Andropogon virginicus L. and Miscanthus sinensis Anders, are tolerant to multiple stresses including aluminum, heavy metals and oxidative stresses. *Plant Cell Reports* 27: 951–961.

Farage, P.K., Blowers, D., Long, S.P., and Baker, N.R., (2006) Low growth temperatures modify the efficiency of light use by photosystem II for CO2 assimilation in leaves of two chilling-tolerant C4 species, Cyperus longus L. and Miscanthus × giganteus. *Plant, Cell and Environment* 29: 720–728.

Fargione, J., Hill, J., Tilman, D., Polasky, S., and Hawthorne, P. (2008) Land clearing and the biofuel carbon debt. *Science (Washington)* 319: 1235–1238.

Fargione, J.E., Cooper, T.R., Flaspohler, D.J., *et al.* (2009) Bioenergy and wildlife: threats and opportunities for grassland conservation. *Bioscience* 59: 767–777.

Farr, D.F. and Rossman, A.Y. (2012) *Fungal Databases, Systematic Mycology and Microbiology Laboratory, Agricultural Research Service, USDA*, http://nt.ars-grin.gov/fungaldatabases (accessed 11 July 2012).

Farr, D.F., Rossman, A.Y., Palm, M.E., and McCray, E.B. (2004) *Fungal databases, Systematic Botany and Mycology Laboratory, Agricultural Research Service, USDA*. Available at: http://nt.ars-grin.gov/fungaldatabases (accessed 8 March 2004).

Felten, D. and Emmerling, C. (2011) Effects of bioenergy crop cultivation on earthworm communities—a comparative study of perennial (Miscanthus) and annual crops with consideration of graded land-use intensity. *Applied Soil Ecology* 49: 167–177.

Fike, J.H., Parrish, D.J., Alwang, J., and Cundiff, J.S. (2007) Challenges for deploying dedicated, large-scale, bioenergy systems in the USA. *CAB Reviews: Perspectives in Agriculture, Veterinary Science, Nutrition and Natural Resources*. doi:10.1079/PAVSNNR20072064.

Fike, J.H., Parrish D.J., Wolf, D.D., *et al.* (2006a) Long-term yield potential of switchgrass-for-biofuel systems. *Biomass and Bioenergy* 30: 198–206.

Fike, J.H., Parrish, D.J., Wolf, D.D., *et al.* (2006b) Switchgrass production for the upper southeastern USA: influence of cultivar and cutting frequency on biomass yields. *Biomass and Bioenergy* 30: 207–213.

Foster, J.L., Guretzky, J.A., Huo, C., Kering, M.K., and Butler, T.J. (2012) Effects of row spacing, seeding rates, and planting date on establishment and biomass of switchgrass. *Crop Science*. doi:10.2135.cropsci2012.03.0171.

Garnock Jones, P.J. (1978) Plant communities on Lakeba and southern Vanua Balavu, Lau group, Fiji. *Journal of the Royal Society of New Zealand* 15: 95–117.

Garrett, K.A., Dendy, S.P., Power, A.G., Blaisdell, G.K., Alexander, H.M., and McCarron, J.K. (2004) Barley yellow dwarf disease in natural populations of dominant tallgrass prairie species in Kansas. *Plant Disease* 88: 574.

Garten, C.T., Jr., Brice, D.J., Castro, H.F., *et al.* (2011) Response of "Alamo" switchgrass tissue chemistry and biomass to nitrogen fertilization in West Tennessee, USA. *Agriculture, Ecosystems and Environment* 140: 289–297.

Garten, C.T., Jr., Smith, J.L., Tyler, D.D., *et al.* (2010) Intra-annual changes in biomass, carbon, and nitrogen dynamics at 4-year old switchgrass field trials in west Tennessee, USA. *Agriculture, Ecosystems & Environment* 136: 177–184.

Gauder, M., Graeff-Hönninger, S., Lewandowski, I., and Claupein, W. (2012) Long-term yield and performance of 15 different Miscanthus genotypes in southwest Germany. *Annals Of Applied Biology* 160: 126–136.

Ge, Y., Zhang, C., Jiang, Y., *et al.* (2011) Soil microbial abundances and enzyme activities in different rhizospheres in an integrated vertical flow constructed wetland. *CLEAN - Soil, Air, Water* 39: 206–211.

Ghimire, S.R. and Craven, K.D. (2011) Enhancement of switchgrass (*Panicum virgatum L.*) biomass production under drought conditions by the ectomycorrhizal fungus Sebacina vermifera. *Applied and Environmental Microbiology* 77: 7063–7067.

Głowacka, K. (2011) A review of the genetic study of the energy crop Miscanthus. *Biomass and Bioenergy* 35: 2445–2454.

Gloyna, K., Thieme, T., and Zellner, M. (2011) Miscanthus, a host for larvae of a European population of Diabrotica v. virgifera. *Journal of Applied Entomology* 135: 780–785.

Goolsby, J.A., Kirk, A.A., Moran, P.J., *et al.* (2011) Establishment of the armored scale, Rhizaspidiotus donacis, a biological control agent of *Arundo donax*. *Southwestern Entomologist* 36: 373–374.

Gopalakrishnan, G., Negri, M.C., and Snyder, S.W. (2011) A novel framework to classify marginal land for sustainable biomass feedstock production. *Journal Of Environmental Quality* 40: 1593–1600.

Gravert, C.E., Tiffany, L.H., and Munkvold, G.P. (2000) Outbreak of smut caused by Tilletia maclagani on cultivated switchgrass in Iowa. *Plant Disease* 84: 596.

Gubišová, M., Gubiš, J., Žofajová, A., Mihálik, D., and Kraic, J. (2012) Enhanced in vitro propagation of Miscanthus × giganteus. *Industrial Crops And Products* 41: 279–282.

Guretzky, J.A., Biermacher, J.T., Cook, B.J., Kering, M.K., and Mosali, J. (2011) Switchgrass for forage and bioenergy: harvest and nitrogen rate effects on biomass yields and nutrient composition. *Plant Soil* 339: 69–81.

Hanson, J.D. and Johnson, H.A. (2005) Germination of switchgrass under various temperature and pH regimes. *Seed Technology* 27: 203–210.

Haughton, A.J., Bond, A.J., Lovett, A.A., *et al.* (2009) A novel, integrated approach to assessing social, economic and environmental implications of changing rural land-use: a case study of perennial biomass crops. *Journal of Applied Ecology* 46: 315–322.

Heaton, E.A., Dohleman, F.G., Miguez, F., *et al.* (2010) Miscanthus: a promising biomass crop. *Advances in Botanical Research* 56: 75–137.

Heggenstaller, A.H., Moore, K.J., Liebman, M., and Anex, R.P. (2009) Nitrogen influences biomass and nutrient partitioning by perennial, warm-season grasses. *Agronomy Journal* 101: 1363–1371.

Hess, J.R., Wright, C.T., and Kenney, K.L. (2007) Cellulosic biomass feedstocks and logistics for ethanol production. *Biofuels, Bioproducts and Biorefining* 1: 181–190.

Hickman, G.C., VanLoocke, A., Dohleman, F.G., and Bernacchi, C.J. (2010) A comparison of canopy evapotranspiration for maize and two perennial grasses identified as potential bioenergy crops. *GCB Bioenergy* 2: 157–168.

Hillier, J., Whittaker, C., Dailey, G., *et al.* (2009) Greenhouse gas emissions from four bioenergy crops in England and Wales: integrating spatial estimates of yield and soil carbon balance in life cycle analyses. *GCB Bioenergy* 1: 267–281.

Hirsch, R.L., TeBeest, D.O., Bluhm, B.H., and West, C.P. (2010) First report of rust caused by Puccinia emaculata on switchgrass in Arkansas. *Plant Disease* 94: 381.

Hitchcock, A.S. (1935) *Manual of the Grasses of the United States*, United States Department of Agriculture, Washington, DC.

Hodkinson, T.R., Chase, M.W., Lledó, M.D., Salamin, N., and Renvoize, S.A. (2002) Phylogenetics of Miscanthus, Saccharum and related genera (Saccharinae, Andropogoneae, Poaceae) based on DNA sequences from ITS nuclear ribosomal DNA and plastid trnL intron and trnL-F intergenic spacers. *Journal of Plant Research* 115: 381–392.

Hodkinson, T.R. and Renvoize, S. (2001) Nomenclature of Miscanthus × giganteus (Poaceae). *Kew Bulletin* 56: 759–760.

Hopkins, A.A., Vogel, K.P., Moore, K.J., Johnson, K.D., and Carlson, I.T. (1995a) Genotype effects and genotype by environment interactions for traits of elite switchgrass populations. *Crop Science* 35: 125–132.

Hopkins, A.A., Vogel, K.P., Moore, K.J., Johnson, K.D., and Carlson, I.T. (1995b) Genotypic variability and genotype x environment interactions among switchgrass accessions from the midwestern USA. *Crop Science* 35: 565–571.

Huang, S.X., Su, X.J., Haselkorn, R., and Gornicki, P. (2003) Evolution of switchgrass (*Panicum virgatum L.*) based on sequences of the nuclear gene encoding plastid acetyl-CoA carboxylase. *Plant Science* 164: 43–49.

Huggett, A.J., Leather, S.R., and Walters, K.F.A. (1999) Suitability of the biomass crop *Miscanthus sinensis* as a host for the aphids *Rhopalosiphum padi (L.)* and *Rhopalosiphum maidis (F.)*, and its susceptibility to the plant luteovirus Barley Yellow Dwarf virus. *Agricultural and Forest Entomology* 1: 143–149.

Hughes, J.K., Lloyd, A.J., Huntingford, C., Finch, J.W., and Harding, R.J. (2010) The impact of extensive planting of Miscanthus as an energy crop on future CO2 atmospheric concentrations. *GCB Bioenergy* 2: 79–88.

Hultquist, S.J., Vogel, K.P., Lee, D.J., Arumuganathan, K., and Kaeppler, S. (1996) Chloroplast DNA and nuclear DNA content variations among cultivars of switchgrass, Panicum virgatum L. *Crop Science* 36: 049–1052.

Hurej, M. and Twardowski, J. (2009) Phyrophagous insects on *Miscanthus giganteus* (Miscanthus × giganteus L.). *Progress in Plant Protection* 49: 1183–1186.

Idris, S.M., Jones, P.L., Salzman, S.A., Croatto, G., and Allinson, G. (2012a) Evaluation of the giant reed (*Arundo donax*) in horizontal subsurface flow wetlands for the treatment of dairy processing factory wastewater. *Environmental Science and Pollution Research* 19: 3525–3537.

Idris, S.M., Jones, P.L., Salzman, S.A., Croatto, G., and Allinson, G. (2012b) Evaluation of the giant reed (*Arundo donax*) in horizontal subsurface flow wetlands for the treatment of recirculating aquaculture system effluent. *Environmental Science and Pollution Research* 19: 1159–1170.

IPCC. (2007) Climate Change 2007: Synthesis Report, in R.K. Pachauri and A. Reisinger (eds), Contribution of Working Groups I, II and III to the Fourth Assessment Report of the Intergovernmental Panel on Climate Change [Core Writing Team, IPCC, Geneva, Switzerland], 104p.

IPCC. (2011) IPCC Special Report on Renewable Energy Sources and Climate Change Mitigation, in *Prepared by Working Group III of the Intergovernmental Panel on Climate Change* (eds O. Edenhofer, R. Pichs-Madruga, Y. Sokona, K. Seyboth, P. Matschoss, S. Kadner, T. Zwickel, P. Eickemeier, G. Hansen, S. Schlömer, and C. von Stechow), Cambridge University Press, Cambridge, UK and New York, USA, 1075p.

Jakubowski, A.R., Casler, M.D., and Jackson, R.D. (2010) The benefits of harvesting wetland invaders for cellulosic biofuel: an ecosystem services perspective. *Restoration Ecology* 18: 789–795.

Johnson, M., Dudley, T., and Burns, C. (2006) Seed production in *Arundo donax*? in Cal-IPC News, Fall edition, 2p.

Johnson, P.C., Clementson, C.L., Mathanker, S.K., Grift, T.E., and Hansen, A.C. (2012) Cutting energy characteristics of Miscanthus × giganteus stems with varying oblique angle and cutting speed. *Biosystems Engineering* 112: 42–48.

Jones, M.B. and M. Walsh. (2007) *Miscanthus for Energy and Fiber*. Earthscan, Sterling, Virginia, USA.

Jung, J.Y. and Lal, R. (2011) Impacts of nitrogen fertilization on biomass production of switchgrass (*Panicum virgatum L.*) and changes in soil organic carbon in Ohio. *Geoderma* 166: 145–152.

Karpiscak, M.M., Gerba, C.P., Watt, P.M., Foster, K.E., and Falabi, J.A. (1996) Multi-species plant systems for wastewater quality improvements and habitat enhancement. *Water Science and Technology* 33: 231–236.

Ker, K., Fyles, J.W., Driscoll, B.T., Seguin, P., and Smith, D. (2010) Development of a Low-Input Switchgrass-Microbe System to Improve Energy Ratio and Production of Biofuels. *ASA–CSSA–SSSA 31 October–4 November 2010 International Annual Meetings*, Long Beach, California.

Ker, K., Seguin, P., Driscoll, B.T., Fyles J.W., and Smith, D.L. (2012) Switchgrass establishment and seeding year production can be improved by inoculation with rhizosphere endophytes. *Biomass and Bioenergy*. http://dx.doi.org/10.1016/i.biomioe2012.09.031.

Kering, M.K., Biermacher, J.T., Butler, T.J., Mosali, J., and Guretzky, J.A. (2012a) Biomass yield and nutrient responses of switchgrass to phosphorus application. *BioEnergy Research* 5: 71–78.

Kering, M.K., Butler, T.J., Biermacher, J.T., and Guretzky, J.A. (2012b) Biomass yield and nutrient removal rates of perennial grasses under nitrogen fertilization. *BioEnergy Research* 5: 61–70.

Khudamrongsawat, J., Tayyar, R., and Holt, J.S. (2004) Genetic diversity of giant reed (*Arundo donax*) in the Santa Ana River, California. *Weed Science* 52: 395–405.

Kim, C.S., Zhang, D., Auckland, S.A., *et al.* (2012) SSR-based genetic maps of Miscanthus sinensis and M. sacchariflorus, and their comparison to sorghum. *TAG Theoretical and Applied Genetics* 124: 1325–1338.

Kim, H.S., Zhang G.R., Juvik, J.A., and Widholm, J.M. (2010) Miscanthus × giganteus plant regeneration: effect of callus types, ages and culture methods on regeneration competence. *GCB Bioenergy* 2: 192–200.

Kirchhof, G., Eckert, B., Stoffels, M., Baldani, J.I., Reis, V.M., and Hartmann, A. (2001) *Herbaspirillum frisingense* sp. nov., a new nitrogen-fixing bacterial species that occurs in C4-fibre plants. *International Journal of Systematic and Evolutionary Microbiology* 51: 157–168.

Knoll, J.E., Anderson, W.F., Strickland, T.C., Hubbard, R.K., and Malik, R. (2012) Low-input production of biomass from perennial grasses in the Coastal Plain of Georgia, USA. *BioEnergy Research* 5: 206–214.

Koslowsky, S.D. and Boerner, R.E.J. (1989) Interactive effects of aluminium, phosphorus and mycorrhizae on growth and nutrient uptake of Panicum virgatum L. (Poaceae). *Environmental Pollution* 61: 107–125.

Kromdijk, J., Schepers, H.E., Albanito, F., *et al.* (2008) Bundle sheath leakiness and light limitation during C4 leaf and canopy CO2 uptake. *Plant Physiology* 148: 2144–2155.

Krupinsky, J., Berdahl, J.D., Schoch, C.L., and Rossman, A.Y. (2004) Leaf spot on switch grass (*Panicum virgatum*), symptoms of a new disease caused by Bipolaris oryzae. *Canadian Journal Of Plant Pathology-Revue Canadienne De Phytopathologie* 26: 371–378.

Laird, D.A. (2008) The charcoal vision: a win-win-win scenario for simultaneously producing bioenergy, permanently sequestering carbon, while improving soil and water quality. *Agronomy Journal* 100: 178–181.

Lal, R. (2004) Carbon sequestration in dryland ecosystems. *Environmental Management* 33: 528–544.

LaMondia, J.A. (1996) Response of additional herbaceous perennial ornamentals to Meloidogyne hapla. *Journal of Nematology* 28: 636–638.

Layton, C.N. and Bergstrom, G.C. (2011) Outbreaks of smut caused by Tilletia maclaganii on switchgrass in New York and Pennsylvania. *Plant Disease* 95: 1587.

Le, P.V.V., Kumar, P., and Drewry, D.T. (2011) Implications for the hydrologic cycle under climate change due to the expansion of bioenergy crops in the Midwestern United States. *Proceedings of the National Academy of Sciences of the United States of America* 108: 15085–15090.

Lee, D.K., Owens, V.N., and Doolittle, J.J. (2007) Switchgrass and soil carbon sequestration response to ammonium nitrate, manure, and harvest frequency on conservation reserve program land. *Agronomy Journal* 99: 462–468.

Lemus, R., Parrish, D.J., and Abaye, O. (2008) Nitrogen-use dynamics in switchgrass grown for biomass. *BioEnergy Research* 1: 153–162.

Lewandowski, I., Clifton-Brown J.C., Scurlock, J.M.O., and Huisman, W. (2000) Miscanthus: European experience with a novel energy crop. *Biomass and Bioenergy* 19: 209–227.

Lewandowski, I. and Schmidt, U. (2006) Nitrogen, energy and land use efficiencies of miscanthus, reed canary grass and triticale as determined by the boundary line approach. *Agriculture, Ecosystems and Environment* 112: 335–346.

Lewandowski, I., Scurlock J.M.O., Lindvall, E., and Christou, M. (2003) The development and current status of perennial rhizomatous grasses as energy crops in the US and Europe. *Biomass and Bioenergy* 25: 335–361.

Ma, N., Yokoyama, K., and Marumoto, T. (2007) Effect of peat on mycorrhizal colonization and effectiveness of the arbuscular mycorrhizal fungus Gigaspora margarita. *Soil Science and Plant Nutrition* 53: 744–752.

Ma, Z., Wood, C.W., and Bransby, D.I. (2001) Impact of row spacing, nitrogen rate, and time on carbon partitioning of switchgrass. *Biomass and Bioenergy* 20: 413–419.

Madakadze, I.C., Coulman, B.E., McElroy, A.R., Stewart, K.A., and Smith, D.L. (1998) Evaluation of selected warm-season grasses for biomass production in areas with a short growing season. *Bioresource Technology* 65: 1–12.

Madakadze, I.C., Stewart, K.A., Peterson, P.R., Coulman B.E., and Smith, D.L. (1999) Cutting frequency and nitrogen fertilization effects on yield and nitrogen concentration of switchgrass in a short season area. *Crop Science* 39: 552–557.

Mantineo, M., D'Agosta, G.M., Copani, V., Patanè, C., and Cosentino, S.L. (2009) Biomass yield and energy balance of three perennial crops for energy use in the semi-arid Mediterranean environment. *Field Crops Research* 114: 204–213.

Mariani, C., Cabrini, R., Danin, A., *et al.* (2010) Origin, diffusion and reproduction of the giant reed (*Arundo donax L.*): a promising weedy energy crop. *Annals Of Applied Biology* 157: 191–202.

Maughan, M., Bollero, G., Lee, D.K., *et al.* (2012) Miscanthus × giganteus productivity: the effects of management in different environments. *GCB Bioenergy* 4: 253–265.

McLaughlin, S.B. and Kszos L.A. (2005) Development of switchgrass (*Panicum virgatum*) as a bioenergy feedstock in the United States. *Biomass and Bioenergy* 28: 515–535.

McMillan, C. (1959) The role of ecotypic variation in the distribution of the central grassland of North America. *Ecological Monographs* 29: 285–308.

Mekete, T., Gray, M.E., and Niblack T.L. (2009) Distribution, morphological description, and molecular characterization of Xiphinema and Longidorous spp. associated with plants (*Miscanthus* spp. and *Panicum virgatum*) used for biofuels. *GCB Bioenergy* 1: 257–266.

Mekete, T., Reynolds, K., Lopez-Nicora, H.D., Gray, M.E., and Niblack, T.L. (2011a) Distribution and diversity of root-lession nematode (*Pratylenchus* spp.) associated with Miscanthus × giganteus and Panicum virgatum used for biofuels, and species identification in a multiplex polymerase chain reaction. *Nematology* 13: 673–686.

Mekete, T., Reynolds, K., Lopez-Nicora, H.D., Gray, M.E., and Niblack, T.L. (2011b) Plant-parasitic nematodes are potential pathogens of Miscanthus × giganteus and Panicum virgatum used for biofuels. *Plant Disease* 95: 413–418.

Menetrez, M.Y. (2010) The potential environmental impact of waste from cellulosic ethanol production. *Journal of the Air and Waste Management Association* 60: 245–250.

Miao, Z.W., Shastri, Y., Grift, T.E., Hansen, A.C., and Ting, K.C. (2012) Lignocellulosic biomass feedstock transportation alternatives, logistics, equipment configurations, and modeling. *Biofuels, Bioproducts & Biorefining* 6: 351–362.

Miguez, F.E., Villamil, M.B., Long S.P., and Bollero, G.A. (2008) Meta-analysis of the effects of management factors on Miscanthus × giganteus growth and biomass production. *Agricultural and Forest Meteorology* 148: 1280–1292.

Missaoui, A.M., Paterson, A.H., and Bouton, J.H. (2006) Molecular markers for the classification of switchgrass (*Panicum virgatum L.*) germplasm and to assess genetic diversity in three synthetic switchgrass populations. *Genetic Resources and Crop Evolution* 53: 1291–1302.

Mitchell, R.B. and Schmer, M.R. (2012) Switchgrass harvest and storage, in *Switchgrass, Green Energy and Technology* (ed. A. Monti), Springer-Verlag, London, pp. 113–127.

Mitchell, R., Vogel, K.P., and Sarath, G. (2008) Managing and enhancing switchgrass as a bioenergy feedstock. *Biofuels, Bioproducts and Biorefining* 2: 530–539.

Monteiro, R.A., Balsanelli, E., Wassem, R., *et al.* (2012) Herbaspirillum-plant interactions: microscopical, histological and molecular aspects. *Plant and Soil* 356: 175–196.

Monti, A., Barbanti, L., Zatta, A., and Zegada-Lizarazu, W. (2012) The contribution of switchgrass in reducing GHG emissions. *GCB Bioenergy* 4: 420–434.

Monti, A. and Zatta, A. (2009) Root distribution and soil moisture retrieval in perennial and annual energy crops in Northern Italy. *Agriculture, Ecosystems and Environment* 132: 252–259.

Moore, G.W., Watts, D.A., and Goolsby, J.A. (2010) Ecophysiological responses of giant reed (Arundo donax) to herbivory. *Invasive Plant Science and Management* 3:521–529.

Moser, L.E. and Vogel, K.P. (1995) Switchgrass, Big Bluestem, and Indiangrass, in *Forages: An Introduction to Grassland Agriculture*, 5th edn.,Vol. 1 (eds R.F. Barnes, D.A. Miller, and C.J. Nelson), Iowa State University Press, Ames, pp. 409–420.

Muir, J.P., Sanderson, M.A., Ocumpaugh, W.R., Jones, R.M., and Reed, R.L. (2001) Biomass production of 'Alamo' switchgrass in response to nitrogen, phosphorus, and row spacing. *Agronomy Journal* 93: 896–901.

Murphy, D., Krentler, J.G., and Georg, H. (1995) Rapidly renewable raw materials for use in building. *Landtechnik* 50: 342–355.

Murphy, I.J. and Coats, J.R. (2011) The capacity of switchgrass (*Panicum virgatum*) to degrade atrazine in a phytoremediation setting. *Environmental Toxicology and Chemistry* 30: 715–722.

Nabity, P.D., Zangerl, A.R., Berenbaum, M.R., and DeLucia, E.H. (2011) Bioenergy crops Miscanthus × giganteus and Panicum virgatum reduce growth and survivorship of Spodoptera frugiperda (Lepidoptera: Noctuidae). *Journal of Economic Entomology* 104: 459–464.

Ng, T.L., Cai, X.M., and Ouyang, Y.F. (2011) Some implications of biofuel development for engineering infrastructures in the United States. *Biofuels, Bioproducts and Biorefining* 5: 581–592.

Ng, T.L., Eheart J.W., Cai, X.M., and Miguez, F. (2010) Modeling miscanthus in the Soil and Water Assessment Tool (SWAT) to simulate its water quality effects as a bioenergy crop. *Environmental Science and Technology* 44: 7138–7144.

Nixon, P.M.I., Boocock, H., and Bullard, M.J. (2001) An evaluation of planting options for miscanthus. *Aspects of Applied Biology* 65: 123–130.

Nolan, A., McDonnell, K., McSiurtain, M., Carroll J.P., Finnan, J., and Rice, B. (2009) Conservation of miscanthus in bale form. *Biosystems Engineering* 104: 345–352.

Odlare, M., Abubaker, J., Lindmark, J., Pell, M., Thorin, E., and Nehrenheim, E. (2012) Emissions of N_2O and CH_4 from agricultural soils amended with two types of biogas residues. *Biomass and Bioenergy* 44: 112–116.

O'Neill, N.R. and Farr, D.F. (1996) Miscanthus blight, a new foliar disease of ornamental grasses and sugarcane incited by Leptosphaeria sp. and its anamorphic state Stagonospora sp. *Plant Disease* 80: 980–987.

Ochatt, S.J., Atif, R.M., Patatochatt, E.M., Jacas, L., and Conreux, C. (2010) Competence versus recalcitrance for in vitro regeneration. 38: 102–108.

Orth, W.D. and Klug, E. (1995) Winter kill of new Miscanthus plantings on light soils - also a technological problem. p. 265–F268. Garmisch-Partenkirchen. Vorträge zum Generalthema des 107. VDLUFA-Kongresses vom 18.-23.9.1994 in Garmisch-Partenkirchen: Grünland als Produktionsstandort und Landschaftselement, 1995.

Palumbo, A.J., Taylor, S.C., Addison, S.L., Slade, A.H., and Glover, C.N. (2012) Metal biosorption in lignocellulosic biofuel biorefinery effluent: an initial step towards sustainability of water resources. *Journal of Industrial Microbiology and Biotechnology* 39: 1345–1356.

Pan, Y., Xie, Y., Chen, X., and Li, F. (2012a) Effects of flooding and sedimentation on the growth and physiology of two emergent macrophytes from Dongting Lake wetlands. *Aquatic Botany* 100: 35–40.

Pan, Y., Xie, Y., Li, F., and Pan, B. (2012b) Morphological and physiological responses to burial depth and sediment type in the wetland macrophyte Miscanthus sacchariflorus. *Fundamental and Applied Limnology* 180: 271–277.

Papazoglou, E.G. (2009) Heavy metal allocation in giant reed plants irrigated with metalliferous water. *Fresenius Environmental Bulletin* 18: 166–174.

Papazoglou, E.G., Karantounias G.A., Vemmos, S.N., and Bouranis, D.L. (2005) Photosynthesis and growth responses of giant reed (*Arundo donax L.*) to the heavy metals Cd and Ni. *Environmental International* 31: 243–249.

Papazoglou, E.G., Serelis, K.G., and Bouranis, D.L. (2007) Impact of high cadmium and nickel soil concentration on selected physiological parameters of *Arundo donax L. European Journal of Soil Biology* 43: 207–215.

Parrish, D.J., Casler, M.D., and Monti, A. (2012) The evolution of switchgrass as an energy crop, in *Switchgrass: A Valuable Biomass Crop for Energy* (ed. A. Monti), Springer, Dordrecht, The Netherlands, pp. 1–28.

Parrish, D.J. and Fike, J.H. (2005) The biology and agronomy of switchgrass for biofuels. *Critical Reviews in Plant Sciences* 24: 423–459.

Parrish, D.J. and Fike, J.H. (2009) Growth and production of herbaceous energy crops, in *Soils, Plant Growth and Crop Production, Encyclopedia of Life Support Systems (EOLSS), Developed under the Auspices of the UNESCO* (ed. W.H. Verheye), Eolss Publishers, Oxford, UK, 31p. Available at: http://www.eolss.net (accessed 8 October 2012).

Parrish, D.J., Fike, J.H., Samson, R., and Bransby, D.I. (2008) Establishing and managing switchgrass as an energy crop. *Forage and Grazinglands*. doi:10.1094/FG-2008-0220-01-RV.

Parrish, D.J., Gammon, T.C., and Graves, B. (1985) Production of fermentables and biomass by six temperate fuelcrops. *Energy in Agriculture* 4: 319–330.

Perlack, R.D., Wright, L.L., Turhollow, A.F., Graham R.L., Stokes, B.J., and Erbach, D.C. (2005) Biomass as Feedstock for a Bioenergy and Bioproducts Industry: The Technical Feasibility of a Billion Ton Annual Supply. (Oak Ridge National Laboratory, Oak Ridge, TN) ORNL/TM-2005/66.

Perdue, R.E., Jr. (1958) *Arundo donax* source of musical reeds and industrial cellulose. *Economic Botany* 12: 368–404.

Perrin, R., Vogel, K., Schmer, M., and Mitchell, R. (2008) Farm-scale production cost of switchgrass for biomass. *BioEnergy Research* 1: 91–97.

Picasso, V.D., Brummer, E.C., Liebman, M., Dixon, P.M., and Wilsey, B.J. (2011) Diverse perennial crop mixtures sustain higher productivity over time based on ecological complementarity. *Renewable Agriculture and Food Systems* 26: 317–327.

Płazek, A., Dubert, F., Janowiak, F., Krepski, T., and Tatrzańska, M. (2011) Plant age and in vitro or in vivo propagation considerably affect cold tolerance of Miscanthus × giganteus. *European Journal of Agronomy* 34: 163–171.

Poenaru, I.C., Voicu, G., Paraschiv, G., Moiceanu, G., and Voicu, P. (2012) Study regarding choosing the optimum rhizome planting equipment. *Actual Tasks on Agricultural Engineering: 40th International Symposium on Agricultural Engineering*, Opatija, Croatia, pp. 283–292.

Porter, C.L., Jr. (1966) An analysis of variation between upland and lowland switchgrass *Panicum virgatum* L. in central Oklahoma. *Ecology* 47: 980–992.

Prasifka, J.R., Bradshaw, J.D., Boe, A.A., Lee, D.K., Adamski, D., and Gray, M.E. (2010) Symptoms, distribution and abundance of the stem-boring caterpillar, Blastobasis repartella (Dietz), in switchgrass. *BioEnergy Research* 3: 238–242.

Prasifka, J.R., Bradshaw, J.D., and Gray, M.E. (2012) Potential biomass reductions to Miscanthus × giganteus by stem-boring caterpillars. *Environmental Entomology* 41: 865–871.

Prasifka, J.R., Bradshaw, J.D., Lee, S.T., and Gray, M.E. (2011) Relative feeding and development of armyworm on switchgrass and corn, and its potential effects on switchgrass grown for biomass. *Journal of Economic Entomology* 104: 1561–1567.

Prasifka, J.R., Bradshaw, J.D., Meagher, R.L., Nagoshi, R.N., Steffey, K.L., and Gray, M.E. (2009) Development and feeding of fall armyworm on Miscanthus × giganteus and switchgrass. *Journal of Economic Entomology* 102: 2154–2159.

Pude, R. (2003) New cultivation methods of Miscanthus in Europe. *Berichte über Landwirtschaft* 81: 405–415.

Pude, R., Treseler, C.H., and Noga, G. (2004) Morphological, chemical and technical parameters of Miscanthus genotypes. *Journal of Applied Botany and Food Quality* 78: 58–63.

Pude, R., Treseler, C.H., Trettin, R., and Noga, G. (2005) Suitability of Miscanthus genotypes for lightweight concrete. *Bodenkultur* 56: 61–69.

Pyter, R.J., Dohleman, F.G., and Voigt, T.B. (2010) Effects of rhizome size, depth of planting and cold storage on Miscanthus × giganteus establishment in the Midwestern USA. *Biomass and Bioenergy* 34: 1466–1470.

Qin, X., Xie, Y., and Chen, X. (2010) Comparative study on the aerenchyma of four dominant wetland plants in Dongting Lake. *Journal of Wuhan Botanical Research* 28: 400–405.

Quinn, L.D. and Holt, J.S. (2009) Restoration for resistance to invasion by giant reed (*Arundo donax*). *Invasive Plant Science and Management* 2: 279–291.

Racelis, A.E., Goolsby, J.A., Penk, R., Jones, W.K., and Roland, T.J. (2010) Development of an inundative, aerial release technique for the arundo wasp, biological control agent of the invasive *Arundo donax* L. *Southwestern Entomologist* 35: 495–501.

Raintree, J.B. and Warner, K. (1986) Agroforestry pathways for the intensification of shifting cultivation. *Agroforestry Systems* 4: 39–54.

Resop, J.P., Cundiff, J.S., and Heatwole, C.D. (2011) Spatial analysis to site satellite storage locations for herbaceous biomass in the piedmont of the Southeast. *Applied Engineering in Agriculture* 27: 25–32.

Riggs, P.J., Moritz, R.L., Chelius, M.K., *et al.* (2002) Isolation and characterization of diazotrophic endophytes from grasses and their effects on plant growth. Nitrogen fixation: global perspectives. Proceedings of the 13th International Congress on Nitrogen Fixation, Hamilton, Ontario, Canada, 2–7 July 2001. pp. 263–267.

Robertson, B.A., Doran, P.J., Loomis, L.R., Robertson, J.R., and Schemske, D.W. (2011) Perennial biomass feedstocks enhance avian diversity. *GCB Bioenergy* 3: 235–246.

Robertson, B.A., Porter, C., Landis, D.A., and Schemske, D.W. (2012) Agroenergy crops influence the diversity, biomass, and guild structure of terrestrial arthropod communities. *BioEnergy Research* 5: 179–188.

Rooney, W.L., Blumenthal, J., Bean, B., and Mullet, J.E. (2007) Designing sorghum as a dedicated bioenergy feedstock. *Biofuels, Bioproducts and Biorefining* 1: 147–157.

Rossa, B., Tuffers, A.V., Naidoo, G., and von Willert, D.J. (1998) *Arundo donax L.*(Poaceae)—a C3 species with unusually high photosynthetic capacity. *Botanica Acta* 111: 216–21.

Roth, A.M., Sample, D.W., Ribic, C.A., Paine, L., Undersander, D.J., and Bartelt, G.A. (2005) Grassland bird response to harvesting switchgrass as a biomass energy crop. *Biomass and Bioenergy* 28: 490–498.

Sage, R., Waltola, G., Cunningham, M., *et al.* (2008) Headlands around SRC plantations have potential to provide new habitats for plants and butterflies on farmland. *Aspects of Applied Biology* 90: 303–309.

Sagehashi, M., Liu, C., Fujii, T., *et al.* (2011) Cadmium removal by the hydroponic culture of giant reed (*Arundo donax*) and its concentration in the plant. *Journal of Water and Environment Technology* 9: 121–127.

Saltonstall, K., Lambert, A., and Meyerson, L.A. (2010) Genetics and reproduction of common (Phragmites australis) and giant reed (*Arundo donax*). *Invasive Plant Science and Management* 3: 495–505.

Sanderson, M.A., Egg, R.P., and Wiselogel, A.E. (1997) Biomass losses during harvest and storage of switchgrass. *Biomass and Bioenergy* 12: 107–114.

Sanderson, M.A., Read, J.C., and Reed, R.L. (1999) Harvest management of switchgrass for biomass feedstock and forage production. *Agronomy Journal* 91: 5–10.

Sanderson, M.A. and Reed, R.L. (2000) Switchgrass growth and development: water, nitrogen, and plant density effects. *Journal of Range Management* 53: 221–227.

Sartori, F., Lal, R., Ebinger, M.H., and Parrish, D.J. (2006) Potential soil carbon sequestration and CO2 offset by dedicated energy crops in the USA. *Critical Reviews in Plant Sciences* 25: 441–472.

Scally, L., Waldren, S., Hodkinson, T.R., and Jones, M.B. (2001) Morphological and molecular systematics of the genus Miscanthus. *Aspects of Applied Biology* 65: 231–237.

Schaeffer, S., Baxendale, F., Heng-Moss, T., *et al.* (2011) Characterization of the arthropod community associated with switchgrass (Poales: Poaceae) in Nebraska. *Journal Of The Kansas Entomological Society* 84: 87–104.

Schmer, M.R., Vogel, K.P., Mitchell, R.B., Moser, L.E., Eskridge, K.M., and Perrin, R.K. (2006) Establishment stand thresholds for switchgrass grown as a bioenergy crop. *Crop Science* 46: 157–161.

Schmer, M.R., Vogel, K.P., Mitchell, R.B., and Perrin, R.K. (2008) Net energy of cellulosic ethanol from switchgrass. *Proceedings of the National Academy of Sciences of the United States of America* 105: 464–469.

Schrotenboer, A.C., Allen, M.S., and Malmstrom, C.M. (2011) Modification of native grasses for biofuel production may increase virus susceptibility. *GCB Bioenergy* 3: 360–374.

Seawright, E.K., Rister, M.E., Lacewell, R.D., *et al.* (2009) Economic implications for the biological control of *Arundo donax*: Rio Grande Basin. *Southwestern Entomologist* 34: 377–394.

Semere, T. and Slater, F.M. (2007) Ground flora, small mammal and bird species diversity in miscanthus (Miscanthus × giganteus) and reed canary-grass (*Phalaris arundinacea*) fields. *Biomass and Bioenergy* 31: 20–29.

Sharma, K.P., Kushwaha, S.P.S., and Gopal, B. (1998) A comparative study of stand structure and standing crops of two wetland species, *Arundo donax* and *Phragmites karka*, and primary production in *Arundo donax* with observations on the effect of clipping. *Tropical Ecology* 39: 3–14.

Shaw, M.R., Zavaleta, E.S., Chiariello, N.R., Cleland, E.E., Mooney, H.A., and Field, C.B. (2002) Grassland responses to global environmental changes suppressed by elevated CO2. *Science (Washington)* 298: 1987–1990.

Sill, W.H. and Pickett, R.C. (1957) A new virus disease of Switchgrass, *Panicum virgatum L. Plant Disease Reporter* 41: 241–249.

Smart, A.J., Moser, L.E., and Vogel, K.P. (2004) Morphological characteristics of big bluestem and switchgrass plants divergently selected for seedling tiller number. *Crop Science* 44: 607–613.

Smith, K.A. and Searchinger, T.D. (2012) Crop-based biofuels and associated environmental concerns. *GCB Bioenergy* 4: 479–484.

Smith, P., Davies, C.A., Ogle, S., *et al.* (2012) Towards an integrated global framework to assess the impacts of land use and management change on soil carbon: current capability and future vision. *Global Change Biology* 18: 2089–2101.

Smith, R. and Slater, F.M. (2010) The effects of organic and inorganic fertilizer applications to Miscanthus × giganteus, *Arundo donax* and Phalaris arundinacea, when grown as energy crops in Wales, UK. *GCB Bioenergy* 2: 169–179.

Sobral, B.W.S., Braga, D.P.V., LaHood, E.S., and Keim, P. (1994) Phylogenetic analysis of chloroplast restriction enzyme site mutations in the Saccharinae Griseb. subtribe of the Andropogoneae Dumort. tribe. *Theoretical and Applied Genetics* 87: 843–853.

Spencer, D.F., Ksander G.G., and Whitehand, L.C. (2005) Spatial and temporal variation in RGR and leaf quality of a clonal riparian plant: *Arundo donax*. *Aquatic Botany* 81: 27–36.

Spencer, J.L. and Raghu, S. (2009) Refuge or reservoir? The potential impacts of the biofuel crop Miscanthus × giganteus on a major pest of maize. *PLoS ONE*: e8336.

Springer, T.L., Aiken, G.E., and McNew, R.W. (2001) Combining ability of binary mixtures of native, warm-season grasses and legumes. *Crop Science* 41: 818–823.

Spyreas, G., Wilm, B.W., Plocher, A.E., *et al.* (2010) Biological consequences of invasion by reed canary grass (*Phalaris arundinacea*). *Biological Invasions* 12: 1253–1267.

St. Clair, S., Hillier, J., and Smith, P. (2008) Estimating the pre-harvest greenhouse gas costs of energy crop production. *Biomass and Bioenergy* 32: 442–452.

Stephens, W., Hess, T.M., and Knox, J.W. (2001) The effect of energy crops on hydrology. *Aspects of Applied Biology* 65: 101–108.

Stewart, J.R., Toma, Y., Fernández, F.G., Nishiwaki, A., Yamada, T., and Bollero, G. (2009) The ecology and agronomy of Miscanthus sinensis, a species important to bioenergy crop development, in its native range in Japan: a review. *GCB Bioenergy* 1: 126–153.

Stout, W.L. and Jung, G.A. (1995) Biomass and nitrogen accumulation in switchgrass: effects of soil and environment. *Agronomy Journal* 87: 663–669.

Sudha, S. and Vasudevan, N. (2009) Constructed wetlands for treating wastewater from crocodile farm. *Journal of Ecotoxicology and Environmental Monitoring* 19: 277–284.

Sun, Q., Lin, Q., Yi, Z., Yang, Z., and Zhou, F.A.S. (2010) A taxonomic revision of Miscanthus s.l. (Poaceae) from China. *Botanical Journal of the Linnean Society* 164: 178–220.

Thelemann, R., Johnson, G., Sheaffer, C., Banerjee, S., Cai, H.W., and Wyse, D. (2010) The effect of landscape position on biomass crop yield. *Agronomy Journal* 102: 513–522.

Thomsen, P.M., Brummer, E.C., Shriver, J., and Munkvold, G.P. (2008) Biomass yield reductions in switchgrass due to smut caused by *Tilletia maclaganii*. *Plant Health Progress*. doi:101094/PHP-2008-0317-01-RS.

Tilman, D., Reich, P.B., and Knops, J.M.H. (2006) Biodiversity and ecosystem stability in a decade-long grassland experiment. *Nature (London)* 441: 629–632.

Tjepkema, J. (1975) Nitrogenase activity in the rhizosphere of Panicum virgatum. *Soil Biology and Biochemistry* 7: 179–180.

Tomaso-Peterson, M. and Balbalian, C.J. (2010) First report of Bipolaris oryzae causing leaf spot of switchgrass in Mississippi. *Plant Disease* 94: 643.

Triplett, J.K., Wang Y.J., Zhong, J.S., and Kellogg, E.A. (2012) Five nuclear loci resolve the polyploid history of switchgrass (*Panicum virgatum L.*) and relatives. *PLoS ONE* 7: e38702.

Truman, C.C., Potter, T.L., and Nuti, R.C. (2012) Quantifying variable rainfall intensity events on seasonal runoff and sediment losses from strip and conventionally tilled peanuts. *Peanut Science* 39: 62–68.

Tzanakakis, V.A., Paranychianakis, N.V., Londra, P.A., and Angelakis, A.N. (2011) Effluent application to the land: changes in soil properties and treatment potential. *Ecological Engineering* 37: 1757–1764.

VanLoocke, A., Bernacchi, C.J., and Twine, T.E. (2010) The impacts of Miscanthus × giganteus production on the Midwest US hydrologic cycle. *GCB Bioenergy* 2: 180–191.

VanLoocke, A., Twine, T.E., Zeri, M., and Bernacchi, C.J. (2012) A regional comparison of water use efficiency for miscanthus, switchgrass and maize. *Agricultural and Forest Meteorology* 164: 82–95.

Virtue, J.G., Reynolds, T., Malone, J., Preston, C., and Williams, C. (2010) *Managing the weed risk of cultivated Arundo donax L*, New Zealand Plant Protection Society, Hastings, pp. 176–179.

Vogel, K.P. (1996) Energy production from forages (or American agriculture-back to the future). *Journal of Soil and Water Conservation* 51: 137–139.

Vogel, K.P. (2000) Improving warm-season forage grasses using selection, breeding, and biotechnology. Native warm-season grasses: research trends and issues. Proceedings of the Native Warm-Season Grass Conference and Expo, Des Moines, IA, USA, 12–13 September 1996, pp. 83–106.

Vogel, K.P., Brejda, J.J., Walters, D.T., and Buxton, D.R. (2002) Switchgrass biomass production in the Midwest USA: harvest and nitrogen management. *Agronomy Journal* 94: 413–420.

Vogel, K.P. (2004) Switchgrass. in *Warm-season (C4) grasses* (eds L.E. Moser, L. Sollenberger, and B. Burson). Agronomy Monograph 45, ASA, CSSA, and SSSA, Madison, Wisconsin. pp. 561–588.

Voigt, T.B., Lee, D.K., and Kling, G.J. (2012) Perennial herbaceous crops with potential for biofuel production in the temperate regions of the USA. *CAB Reviews* 7: 1–13.

Vu, A.L., Dee, M.M., Gwinn, K.D., and Ownley, B.H. (2011) First report of spot blotch and common root rot caused by *Bipolaris sorokiniana* on switchgrass in Tennessee. *Plant Disease* 95: 1195.

Walkden, H.H. (1943) Cutworm and armyworm populations in pasture grasses, waste lands, and forage crops. *Journal of Economic Entomology* 36: 376–81.

Wang, D., Portis, A.R., Jr., Moose, S.P., and Long, S.P. (2008) Cool C4 photosynthesis: pyruvate Pi dikinase expression and activity corresponds to the exceptional cold tolerance of carbon assimilation in Miscanthus × giganteus. *Plant Physiology* 148: 557–567.

Wang, Y.S., Pao, S.T., Kwei, C.M., Chan, G.S., and Chu, H.F. (1961) Studies on the sorghum aphid, *Aphis sacchari* Zehntner. *Acta Entomologica Sinica* 10: 363–380.

Waramit, N., Moore, K.J., and Heggenstaller, A.H. (2011) Composition of native warm-season grasses for bioenergy production in response to nitrogen fertilization rate and harvest date. *Agronomy Journal* 103: 655–662.

Watts, D.A. and Moore, G.W. (2011) Water-use dynamics of an invasive reed, *Arundo donax*, from leaf to stand. *Wetlands* 31: 725–734.

Waxman, K.D. and Bergstrom, G.C. (2011) First report of a leaf spot caused by *Bipolaris oryzae* on switchgrass in New York. *Plant Disease* 95: 1192.

Wen, Z.Y., Ignosh, J., Parrish, D., Stowe, J., and Jones, B. (2009) Identifying farmers' interest in growing switchgrass for bioenergy in southern Virginia. *Journal of Extension* 47: 5RIB7.

Werling, B.P., Meehan, T.D., Robertson, B.A., Gratton, C., and Landis, D.A. (2011) Biocontrol potential varies with changes in biofuel-crop plant communities and landscape perenniality. *GCB Bioenergy* 3: 347–359.

Widmer, T.L. and Kirk, A.A. (2005) Nigrospora oryzae associated with shoot tip death of *Arundo donax*. Proceedings of the 13th EWRS Symposium, Bari, Italy, 19–23 June 2005.

Wilkins, C. (1997) The uptake of copper, arsenic and zinc by Miscanthus - environmental implications for use as an energy crop. *Aspects of Applied Biology* 49: 335–340.

Williams, C.M.J., Biswas, T.K., Harris, P.L., and Heading, S. (2008) Use of giant reed to treat wastewater for resource recycling in South Australia. *Acta horticulturae* 792: 701–707.

Wu, Y.P. and Liu S.G. (2012) Impacts of biofuels production alternatives on water quantity and quality in the Iowa River Basin. *Biomass and Bioenergy* 36: 182–191.

Yamashita, S., Nonaka, N., Namba, S., Doi, Y., and Yora, K. (1985) Miscanthus streak virus, a geminivirus in Miscanthus sacchariflorus Benth. et Hook. *Annals of the Phytopathological Society of Japan* 51: 582–590.

Yan, J., Chen, W., Luo, F., *et al.* (2012) Variability and adaptability of Miscanthus species evaluated for energy crop domestication. *GCB Bioenergy* 4: 49–60.

Young, H.A., Lanzatella, C.L., Sarath, G., and Tobias, C.M. (2011) Chloroplast genome variation in upland and lowland switchgrass. *PLoS ONE*: e23980.

Zalapa, J.E., Price, D.L., Kaeppler, S.M., Tobias, C.M., Okada, M., and Casler, M.D. (2011) Hierarchical classification of switchgrass genotypes using SSR and chloroplast sequences: ecotypes, ploidies, gene pools, and cultivars. *Theoretical and Applied Genetics* 122: 805–817.

Zale, J., Freshour, L., Agarwal, S., *et al.* (2008) First report of rust on switchgrass (*Panicum virgatum*) caused by *Puccinia emaculata* in Tennessee. *Plant Disease* 92: 1710.

Zeri, M., Anderson-Teixeira, K., Hickman, G., Masters, M., DeLucia, E., and Bernacchi, C.J. (2011) Carbon exchange by establishing biofuel crops in Central Illinois. *Agriculture, Ecosystems & Environment* 144: 319–329.

Zhang, Y., Zalapa, J.E., Jakubowski, A.R., *et al.* (2011a) Natural hybrids and gene flow between upland and lowland switchgrass. *Crop Science* 51: 2626–2641.

Zhang, Y., Zalapa, J.E., Jakubowski, A.R., *et al.* (2011b) Post-glacial evolution of Panicum virgatum: centers of diversity and gene pools revealed by SSR markers and cpDNA sequences. *Genetica* 139: 933–948.

Zimmermann, J., Dauber, J., and Jones, M.B. (2012) Soil carbon sequestration during the establishment phase of Miscanthus × giganteus: a regional-scale study on commercial farms using 13C natural abundance. *GCB Bioenergy* 4: 453–461.

Zub, H.W. and Brancourt-Hulmel, M. (2010) Agronomic and physiological performances of different species of Miscanthus, a major energy crop. A review. *Agronomy for Sustainable Development* 30: 201–214.

Chapter 5

Sustainable Oil Crops Production

C. Eynck[1], D. Shrestha[2], J. Vollmann[3], K.C. Falk[4], W. Friedt[5], H.P. Singh[6], and E. Obeng[6]

[1]*Linnaeus Plant Sciences Inc., Saskatoon, Canada*
[2]*University of Idaho, Moscow, ID, USA*
[3]*University of Natural Resources and Life Sciences, Vienna, Tulln, Austria*
[4]*Agriculture and Agri-Food Canada, Saskatoon Research Centre, Saskatoon, Canada*
[5]*Justus Liebig University Giessen, Giessen, Germany*
[6]*Fort Valley State University, Fort Valley, GA, USA*

Introduction

Plant oils are major agricultural commodities with a production volume of 129 M t annually (Carlsson, 2009). Historically, they have been used in human and animal nutrition and to a lesser extent for nonedible applications, such as lubricants and detergents (Lu *et al.*, 2011). Recently, however, due to high energy prices and a greater awareness of the environmental consequences of fossil fuels, plant oils have increasingly been used for the production of biodiesel, a renewable liquid fuel consisting of fatty acid methyl or ethyl esters (Durrett *et al.*, 2008).

Presently, the international biodiesel market is dominated by rapeseed oil, followed by soybean and palm oil. However, other crop platforms such as camelina and the recently emerged Ethiopian mustard hold promise as new biodiesel feedstocks. Here, we present an overview on sustainability considerations associated with each of these crops.

Soybean (*Glycine max*)

Soybean (*Glycine max* [L.] Merr.) is the leading oil crop worldwide with an annual production of over 260 M t in the year 2010. Due to growing market demand, the global soybean acreage has more than doubled over the period of the last 30 years from about 50.6 M ha in 1980 to 102.5 M ha in 2010 (FAOSTAT, 2012). The United States, Brazil, Argentina, India, China, and Paraguay are the main soybean producers, while only 2% of the soybean acreage is in Europe. Since soybean is of subtropical origin, low temperature and/or lack of precipitation are the two most significant factors limiting production, particularly in Europe, Africa, and Australia. On average, the soybean seed contains 20% oil and 40% protein with a typical oil composition of 54% linoleic, 22% oleic, 10% linolenic, and 10% palmitic acid (Cober *et al.*, 2009). Soybean is a member of the legume family capable of biological dinitrogen fixation

Biofuel Crop Sustainability, First Edition. Edited by Bharat P. Singh.

in symbiosis with rhizobial bacteria (*Bradyrhizobium japonicum* [Kirchner] Jordan) present in soybean root nodules; as a consequence, soybean oil content and oil yield per hectare are rather low when compared to other oilseeds. However, as soybean is high in protein and soybean meal is the major protein source for livestock feeding due to its favorable content of the essential amino acids lysine and tryptophan, soybean oil is also available in abundance and can be utilized both for human consumption as well as for biofuel production depending on market needs.

Until recently, most plant breeding research on soybean oil was devoted to optimizing the fatty acid composition for human consumption and food processing. In particular, selection for reduced saturated fatty acid content (palmitic and stearic acids) or for low linolenic acid was carried out in order to minimize health risks associated with either saturated fats or trans-fatty acid formation (Cober *et al.*, 2009). Moreover, mutant lines and genetic stocks of soybeans are available with considerable differences in individual fatty acid concentration for particular end uses. Additionally, high oleic acid (about 90% C18:1) soybeans have recently been created through a genetic engineering approach. A high oleic soybean oil may possess a better biodiesel quality, and the oxidative stability is greatly improved due to a reduced level of polyunsaturated fatty acids, which is crucial particularly for lubricants used as motor or hydraulic oil (Cahoon *et al.*, 2009). Breeding for increased oil content of the soybean seed is possible at the expense of protein content. However, as both oil and protein are important constituents in trade, the total commercial value of a high oil soybean crop might be lower than the value of a conventional crop; therefore, economic models based on oil and protein prices have been proposed as a selection index for breeding of high value soybeans (Leffel, 1990).

Sustainability in Soybean Production

GM Soybean

Genetically modified soybean (GM soybean) was introduced in 1996 and is predominantly grown in the major soybean producing regions both in North and South America. At present, about 73% of the global soybean acreage is planted with GM soybeans (James, 2012) which exhibit transgenic resistance traits against nonselective glyphosate- or glufosinate-based herbicides. This permitted crop management changes toward minimum or no-tillage as well as other management and weed control changes (Bertram and Pedersen, 2004; Heatherly *et al.*, 2002). A study on the sustainability of the US soybean production system (Heatherly *et al.*, 2009) revealed a 93% decrease in soil erosion with better preservation of top soil, 70% decrease in herbicide runoff, reduced soil moisture loss, decreased wind erosion, 80% less phosphorus contamination of surface water, and an over 50% reduction in fuel use associated with GM soybean and conservation tillage as compared to the conventional soybean production system. This represents a significant benefit of the GM no-tillage soybean production system on the individual field level. Apart from reduced energy costs, conservation tillage in soybean–corn cropping systems resulted in an increase of soil organic carbon content in long-term experiments (Varvel and Wilhelm, 2010). Heatherly *et al.* (2009) emphasized the significant reduction in the global warming potential of no-tillage soybeans over conventional tillage practices, which is a consequence of reduced carbon dioxide emission due to lower fuel consumption and increased carbon storage in soil organic matter. In addition, soybean grain yield and biological nitrogen fixation were increased under no-tillage as compared to a conventional tillage system (Zotarelli *et al.*, 2012), which might be due to the negative impacts of disrupting soil aggregates during tillage, apart from other effects.

Due to its economic and environmental benefits, conservation tillage in association with planting herbicide-resistant GM soybeans was rapidly adopted in the main soybean growing regions, particularly in Brazil and the United States (Heatherly *et al.*, 2009; Zotarelli *et al.*, 2012). As a consequence, the widespread use of glyphosate herbicides promoted the evolution of herbicide-resistant weeds (Baucom and Mauricio, 2004; CAST, 2012). This is considered a major constraint on the sustainability of the GM soybean no-tillage production system (Christoffoleti *et al.*, 2008; Shaw *et al.*, 2011) both in North and South America, as the incidence of herbicide-resistant weeds could require termination of conservation tillage or application of additional herbicides. The natural selection of glyphosate-resistant weed populations has probably been accelerated by the fact that, apart from soybean, the glyphosate-resistant crop technology is presently also utilized in maize, rapeseed, cotton, and sugar beet (Shaw *et al.*, 2011). Moreover, high seed multiplication rates of weed species and a predominantly single gene inheritance of glyphosate resistance contributed to a rapid spread of resistant weeds. Management practices have, therefore, been developed for controlling herbicide-resistant weed populations without the need for totally abandoning conservation tillage. Depending on the weed and crop species, strategies include changes in herbicide rotations, adapted herbicide tank mixtures, mechanical weed control procedures, cultural strategies such as variation of planting time, seeding rate, planting configuration, cultivar choice, mulching, cover crops, or optimized nutrient management; furthermore, diversifying a corn–soybean rotation into a more complex rotation (e.g., maize, soybean, small grain, forage) would dilute selection pressure on specific weeds (CAST, 2012). However, all strategies of controlling herbicide-resistant weeds are more labor and cost intensive than the no-tillage system, and they might partly oppose soil conservation or other environmental targets. Other threats of soybean sustainability frequently discussed in relation with GM soybean cultivation such as biodiversity issues and ecosystem effects will be covered in the paragraph on land-use change below.

Soybean Dinitrogen Fixation

Mineral nitrogen fertilizer is the major external input in oil crops such as rapeseed or sunflower, while symbiotic dinitrogen fixation has been credited as major advantage of soybean both economically and in terms of environmental impact. Thus, although soybean is low in net energy production per unit area as compared to other biofuel producing crops, soybean biodiesel is outstanding in both nitrogen-use efficiency (NUE) and reduction of greenhouse gas (GHG) emissions due to the low rates of nitrogen and pesticide applications required (de Vries *et al.*, 2010). Up to 80% of nitrogen uptake of a well-nodulated soybean crop may be derived from atmospheric N_2, but lower values (40–60%) were reported from temperate regions (Schweiger *et al.*, 2012) or from high-yield environments with additional nitrogen application (Salvagiotti *et al.*, 2009). The ratio of energy output to energy input (energy return on investment, EROI) is a helpful indicator of energy efficiency of biofuels. For maize-based ethanol the EROI is rather low ranging from 1.1:1 to 1.65:1 depending on different technology assumptions, while for cellulose-based ethanol EROI ratios can range from 4.4:1 to 11:1; for soybean- and rapeseed-based biodiesel, EROI values of about 3.7:1 were reported (Solomon, 2010). For different soybean growing regions of Brazil, better EROI values for GM soybean as compared to conventional production were found due to lower labor, diesel, and herbicide inputs associated with the GM soybean production system, and high EROI was generally observed for less intensified crop management systems (Romanelli *et al.*, 2012). As dinitrogen fixation contributes to an advantage both in EROI and sustainability of soybean as compared to other oilseeds, management and optimization of symbiotic dinitrogen fixation appears particularly important. This is also emphasized by the results of a meta-analysis on N uptake and N_2 fixation (Salvagiotti *et al.*, 2008), which revealed a negative N balance for 80% of over 600 data sets due to the high N export rates from the field through the seed harvested. Apart from direct energy return and given that the soybean nitrogen balance is positive, a desirable crop rotation effect of soybean dinitrogen fixation on succeeding crops has to be taken into account as an additional sustainability feature as well.

Land-use Change

While GM soybean production has advantages on the farm level due to the beneficial effects of adopting conservation tillage, the sustainability of GM soybean has severely been questioned on larger scale and ecological level, as it permits the expansion of cropping acreages associated with reduced biodiversity, land-use changes, and various economic and social effects. In the Cerrado and Amazon regions of Brazil, most deforestation has been claimed to be due to cattle grazing land with about 6% attributed to soybean biodiesel directly; however, soybean expansion to grazing area would probably cause further indirect deforestation elsewhere (Lima *et al.*, 2011). Although the carbon debt due to land-use change for biofuel production is much lower for soybean than for oil palm based on differences in land-use change assumptions (Achten and Verchot, 2011), a multi-scale integrated assessment of biodiesel production from soybean points to the rather low energy output to input ratio and the small amount of net energy obtained from soybean (Borzoni, 2011). As most Brazilian GHG emissions are from deforestation and changes in land use rather than from fossil energy consumption, Borzoni (2011) considered soybean biodiesel use in Brazil as highly questionable. In addition, biofuel exports from lands of high carbon stock or high diversity would clearly not meet the sustainability criteria of the European Union renewable energy directive of 2008 (Zahniser, 2010). Reijnders and Huijbregts (2008) also emphasized that life-cycle emissions of biodiesel from Brazilian soybeans or European rapeseed are higher than for fossil diesel but could be reduced by improving crop N efficiency or increasing soil carbon content through conservation tillage. Similar conclusions were also drawn for soybean-based biodiesel produced in Argentina (Panichelli *et al.*, 2009).

Based on the critical impact of expanding soybean production on rural communities, agricultural ecosystems, and biodiversity, initiatives for certified sustainable production have been formed such as the Round Table on Responsible Soy (RTRS) association. A list of certification standards for soybean production has been established (RTRS, 2010). The absence of quantitative empirical data on the various side effects of intensified soybean production in local contexts and the lack of life-cycle analyses particularly for biofuel production are impediments in political decision making toward more sustainable soybean production.

Rapeseed (*Brassica napus*)

Besides soybean as the world's dominating oilseed crop, oilseed rape, a member of the *Brassicaceae*, is the second most important oilseed of the world. The world's total acreage of *Brassica. napus* oilseed rape/canola along with related oilseed brassicas, that is, turnip rape (*Brassica rapa*) and mustard species, is about 30 M ha, with a current production of around 50 M t. The major growing countries are China (6.6–7.4 M ha in 2008–2010), India (5.5–6.3 M ha, mainly mustard), Canada (6.4–6.5 M ha), France (1.4–1.5 M ha), Germany

Table 5.1. Annual rapeseed yield (t ha^{-1}) in Canada, China, and the major European growing countries France, Germany, and the United Kingdom, 2008–2010.

Country	2008	2009	2010
Canada	1.95	1.95	1.82
China	1.84	1.88	1.78
France	3.32	3.77	3.29
Germany	3.76	4.29	3.90
UK	3.30	3.35	3.47

Source: http://faostat.fao.org/site/339/default.aspx

(1.4–1.5 M ha) and Australia (1.7 M ha). Other countries with significant production levels of up to 1 M ha are the Ukraine, Poland, the United Kingdom, Russia, the United States, and Pakistan (FAOSTAT, 2012). In Europe, oilseed rape is mostly grown as a winter crop on a total acreage of 8.2 M ha. To date, most oilseed rape is produced within the EU member states, that is, 6.1–6.9 M ha in 2008–2010, with 18.9–21.4 M t seed. The highest seed and oil yields are harvested in Central and Western Europe, where the annual yield of winter crops varies between 3.3 and 4.3 t ha^{-1} (Table 5.1). Nevertheless, a great potential is seen for increasing production in Eastern Europe depending on climatic conditions and the economic situation. The yield potential in other major rapeseed growing regions of the world is much lower (Table 5.1). This is either due to the harsh environment and short growing season of spring canola in Western Canada, or to the small-scale production of alternative rapeseed types as a second crop in Central China. The prospects of enhancing yield under such production conditions seem to be rather limited.

Breeding and Variety Types

As a facultative outcrossing species, oilseed rape crops can either represent open pollinated (OP) varieties or hybrids. In all major growing areas including Europe, an increasing proportion of the registered cultivars represent single-cross or F_1 hybrids versus OP line varieties. In central Europe, winter oilseed rape generates the highest seed and oil yields in comparison to other oil crops; therefore, it is the most important oilseed in this region. In Germany, average farm yields currently vary around 3.5 t ha^{-1}, with a peak of 4 t ha^{-1} in 2004 (Statistisches Bundesamt, www.destatis.de). A total of 87 oilseed rape cultivars are currently listed in Germany, 77 of which are winter types (Anonymous, 2012). Two of the winter cultivars are low glucosinolate but high erucic (0+) types, while all others represent 00 types (zero erucic acid, low seed glucosinolates (equivalent to canola quality). Since the introduction of the first restored winter oilseed rape hybrid variety in 1995 (Paulmann and Frauen, 1997), the proportion of hybrids has grown steadily, with 43 current hybrids comprising almost 50% of the registered winter oilseed rape varieties in Germany (Anonymous, 2012). In terms of cultivation, there is a trend toward an increased use of hybrids in all major canola and oilseed rape growing areas worldwide. For example, according to breeders' information, more than 70% of the rapeseed acreage in Germany today represents hybrid crops (Frauen, Norddeutsche Pflanzenzucht Hans Georg Lembke KG, personal communication). Under real farm conditions, hybrid varieties tend to generally out-yield line (OP) varieties; the yield advantage of hybrids has been estimated at 8–16%, depending on the yield (Christen and Friedt, 2007). The newest generation of winter oilseed rape hybrids displays improved yield performance and stability, and also achieves high and stable oil contents (Anonymous, 2012). New hybrid varieties such as "Avatar" and

"Raptor" are the basis for enhanced farm yields and profitable rapeseed production. Other major advantages of hybrids versus lines are (i) their faster and more vigorous development in fall leading to a better crop establishment before winter, (ii) their better nutrient uptake ability resulting in more economic application of fertilizers, and (iii) their generally better winter survival and stronger growth in spring.

Oil Quality and its Improvement

The seed oil of 00-quality oilseed rape/canola is widely used both for human nutrition and as an important renewable resource for nonfood purposes. Almost all of the oilseed rape production in Europe is of 00-quality. A relatively small amount of high erucic acid rapeseed (HEAR, also called "traditional rapeseed") is still grown in Canada and Europe for industrial end uses. Due to its favorable fatty acid composition, the oil extracted from current 00 rapeseed varieties is particularly suitable for the production of biodiesel (rapeseed methyl ester, RME). Its relatively high mono-unsaturated fatty acid (oleic acid, 18:1) content together with comparatively low contents of poly-unsaturated fatty acids (linoleic acid, 18:2; linolenic acid, 18:3) are determinants of high RME quality. The quality of biodiesel has originally been characterized by the German standard E DIN 51606 which later has been replaced by the European standard for biodiesel (DIN EN 14214).

The achievable biodiesel yield from a given amount of seed depends primarily on its oil content. Further progress in oil content and yield can be expected from breeding in the future. However, a large number of genes or quantitative trait loci (QTL) are involved in seed oil biosynthesis, and a strong environmental modification of seed oil content has been observed (Nesi *et al.*, 2008). Therefore, further investigations are essential to obtain deeper insight into this complex trait in order to further improve oil content in high-yielding cultivars. In this context, the identification and utilization of genes contributing to oil content via genetic analysis in rapeseed, for example, comparative QTL mapping in different genetic backgrounds will help to identify gene loci with a key function on this complex trait by knowledge-based breeding approaches. For example, in a comparative study homeologous genomic regions involved in oil content in different genetic backgrounds could be identified (Delourme *et al.*, 2006), and novel alleles were found in individual genotypes. Recently, Zhu *et al.* (2012) identified 39 genes which were differentially expressed in rapeseed lines differing in a QTL influencing oil content. Among these genes, six were differentially expressed regardless of temperature, indicating the major relevance for oil content. Such research work is a basis for the marker-assisted combination of favorable alleles at different genetic loci to increase seed oil content in modern oilseed rape by breeding.

Broadening Genetic Diversity for Breeding of Oilseed Rape

It has been shown many times and is generally accepted that genetic diversity within a species is not only nice to have but also is valuable and desirable. It is also a prerequisite for achieving high crop yield of hybrid cultivars via heterosis. Genetic variance can be created by sexual hybridization (crossing) between released cultivars, adapted genotypes of landraces, or even alien exotic germplasm. In addition, entirely new variation can be created by interspecific or intergeneric hybridization within the *Brassicaceae* family. Since rapeseed is a natural hybrid between *Brassica oleracea* and *Brassica rapa*, completely novel types of rapeseed can be created by interspecific hybridization between selected genotypes of the two parental species. It has been demonstrated that the progeny of such interspecific hybrids ("resynthetic [RS] rapeseed") give rise to completely new material which is clearly discriminated from commercial oilseed rape cultivars by multivariate analysis. At the same time, different interspecific families show obvious genetic distances and form separate subclusters within the interspecific progenies. The development of inbred lines derived from such wide crosses can lead to new test hybrids (F_1) with a high-yield potential (Friedt *et al.*, 2004; Gehringer *et al.*, 2007; Seyis *et al.*, 2003, 2006). Therefore, the use of novel RS material for the development of parents for new F_1 combinations can lead to new hybrid varieties which out-yield classical and modern variety types regarding both seed and oil production.

Sustainability Aspects of Rapeseed and Biodiesel Production

Environment friendly and sustainable land use and agricultural production is characterized by many components. One of the major factors is the diversity of crops and agricultural production system. Nowadays, the production systems in many countries, particularly in the so-called developed world, are characterized by a small number of plant species which are grown in short rotations on relatively large areas as compared to earlier times. Consequently, few (cash) crops dominate the landscape; for example, only four crop plants, that is, wheat, maize, barley, and oilseed rape, cover about 80% of the arable land in Germany with oilseed rape having stabilized its position as the major noncereal crop during the last decade. For ecological but also economic reasons it is necessary to maintain a minimum acreage of leaf crops such as rapeseed in cereal-dominated agricultural systems in order to reduce negative effects of short rotations or monoculture, for example, increased disease and pest damage. On the other hand, rapeseed itself is also subject to many pests and diseases potentially causing severe damage to the crop: insect pests like pollen beetles can induce the loss of flower buds, and fungal pathogens such as blackleg (*Leptosphaeria maculans*), sclerotinia stem rot (*Sclerotinia sclerotiorum*), and clubroot (*Plasmodiophora brassicae*) affect the productivity and seed yield of oilseed rape crops. Whereas the resistance of rapeseed cultivars against diseases has been substantially improved during the last decades, resistance against insect pests has not been achieved yet. But basic studies have recently been initiated to develop resistant plants by gene technology for example, via RNAi (Krause *et al.*, unpublished data).

The Renewable Energy Directive (RED) of the European Union, which was introduced in 2009, requires that until 2017, GHG emissions from the production and use of biofuels for transport are to be at least 35% lower than those from fossil fuels; thereafter, 50% lower. According to the commission's studies, rapeseed oil meets the RED requirement, delivering GHG cuts of at least 38% compared with conventional fuels. Regarding the sustainability of rapeseed and RME production and use, a detailed LCA study of biodiesel has recently been carried out by Herrmann *et al.* (2012). Based on today's climate change potential from the production and use of biodiesel of 57 kg CO_2 equivalent/1000 km (petrochemical diesel: 214 kg CO_2 equivalent/1000 km), the authors have assessed the specific environmental impacts from the production and use of biodiesel as it is today, based on rapeseed oil and different types of alcohols, and using technologies that are currently available or will be available shortly. Their modeling of the LCA is based on a specific Danish biodiesel production facility and different options have been applied for the simulation of environmental improvement of production methods. All relevant process stages were included, such as rapeseed production including carbon sequestration and N_2O balances as well as transportation of products used during the life cycle of biodiesel (functional unit "1000 km transportation for a standard passenger car"). Based on their analysis, the authors recommend that additional options and incentives regarding the better use of rape straw be investigated, particularly considering carbon sequestration issues (from the perspective of potential climate change) of using bioalcohol instead of petrochemical alcohol for the transesterification process (Herrmann *et al.*, 2012).

Potentials of Variety and Crop Improvement

Studies on the future crop potential have to be based on the current knowledge and available material, that is, modern rapeseed varieties. There is no doubt that the yield potential of rapeseed or canola is much higher than the average farm yield obtained until now. Many field trials as well as practical observations and experience of breeders and farmers show that the seed yield potential of modern winter oilseed rape cultivars is certainly higher than 6 t/ha. While maximum yield requires optimum fertilization and other plant treatments, reasonable yields can still be achieved at reduced N-fertilization levels and suboptimal field conditions (i.e., soil quality). Whereas the variety type, that is, OP or hybrid, is less important at marginal sites or low-input conditions, maximal seed and oil yield under high-fertility conditions tends to be more easily and safely achieved with hybrids than with OP varieties. While emissions due to N leaching under low-input conditions will be naturally low, such emissions can be avoided under high-input conditions by highly productive varieties which combine a strong nutrient uptake ability due to an efficient root system and a pronounced NUE or translocation efficiency.

There is no doubt that it is possible to develop new rapeseed varieties with improved NUE. Thus, F_1 hybrids tended to have the highest yield in replicated multi-environment field experiments in comparison to other variety types under all nitrogen regimes studied, including zero mineral N-fertilization. Nevertheless, some new experimental hybrids (Test-Hy) were very close to the commercial hybrids under different fertilization levels. Therefore, it is deduced that not only the nutrient uptake but also the translocation and metabolism efficiency of rapeseed can be improved by breeding, making use of extremely broad genetic material. In a recent study, Ulas *et al.* (2012) have attributed genotypic differences of nitrogen efficiency to root-growth characteristics. When comparing N efficient cv. "Apex" with cv. "Capitol" they found a higher root density and more living fine roots in the former variety. The authors concluded that genotypes which invest particularly into root growth in the vegetative stage may be more N efficient than others. By using such genotypes for improving N efficiency in farmers' fields, the sustainability of rapeseed cultivation and its use for biofuel production would be greatly enhanced. Furthermore, oilseed rape is known to have beneficial effects as a breaking crop in cereal-dominated rotations. For example, in an extensive evaluation of more than 700 Austrian cases, Vollmann (2001) has found grain yield advantages in winter wheat and winter barley grown in rotation with oilseed rape: when wheat and barley were planted after rapeseed, they yielded 615 kg ha^{-1} (wheat) and 430 kg ha^{-1} (barley) more than after a cereal pre-crop (http://ipp.boku.ac.at/pz/oilseeds/raps2001/). Results of various German studies show even higher effects of rapeseed cultivation on the grain yield of subsequent wheat crops ranging from 7 to 20 dt ha^{-1} (details under www.ufop.de/agrar). These findings are basically in agreement with other observations and farmers' experience and indicate that oilseed rape is an indispensable field crop in agricultural farm systems otherwise dominated by cereals as is usually the case in many parts of the world today.

Conclusions

Today, oilseed rape is a favorite plant for basic and breeding research, taking advantage of its close relationship to major model plants such as *Arabidopsis thaliana* and *B. rapa*. After the genome of these two species has been fully sequenced earlier, the sequence of the *B. napus* genome is expected to be available in the near future. This is going to boost gene discovery, the elucidation of genetic networks and the control of major characteristics of rapeseed including complex agronomic traits such as oil content and seed yield. In addition to its outstanding importance as a vegetable oil source, oilseed rape is an important crop for biofuel production today and is expected to be so in the future because of the following reasons: (1) Rapeseed is an important dicotyledonous crop plant with nutrient requirement, etc. differing from monocots (grasses); therefore, it is grown as a major alternative to cereals in many parts of the world, particularly Canada, China, and Europe. (2) Oilseed rape is characterized by a high oil content and valuable oil quality, that is, fatty acid pattern, which allows its alternative use as a food, feed, or fuel; this gives the farmer the freedom to produce for different purposes and supply separate markets and industrial production chains. (3) Besides seed oil as valuable major compound, the rapeseeds are also rich in protein which determines the high nutritional value of rapeseed meals and cakes as a feed for farm animal (cf. Wittkop *et al.*, 2009). (4) The species *B. napus* comprises a broad morpho-physiological diversity including spring and winter growth types, oil as well as forage types, OP, and hybrid cultivars, etc. (5) Therefore, the rapeseed plant has a high potential for further improvements regarding yield and quality; novel germplasm can be developed by interspecific and intergeneric hybridization as a basis for the creation of better breeding lines and hybrid cultivars. This potential of enhancing seed and oil yield is often underestimated in respective studies on the energy costs and balance. Taking this potential into account will allow making use of the great agronomical and industrial advantages of the rapeseed plant.

Ethiopian Mustard (*Brassica carinata*)

Origin

Brassica carinata, commonly known as Ethiopian or Abyssinian mustard, is an amphidiploid (BBCC, $2n = 34$) formed through the interspecific hybridization of *Brassica nigra* L. (BB, $2n = 16$) and *B. oleracea* L. (CC, $2n = 18$). It is a highly productive, pest-tolerant oilseed commonly grown in the central highlands of Ethiopia at altitudes of 2200 to 2800 m above sea level (Belayneh *et al.*, 1983). In Ethiopia, the crop is typically grown in the rainy season from late June to early December where it is grown for its edible leaves and seed oil. The oil, however, is of low quality (high erucic acid) and, therefore, typically blended with other oils. Also, the meal is high in allyl glucosinolates (sinigrin) and is typically discarded after extraction or fed in mixtures with other meal feedstocks. Until recently, there was little or no commercial production of this species outside of Ethiopia or neighboring countries.

Agronomics

Ethiopian mustard is highly heat and drought tolerant (Cohen and Knowles, 1983; Ferreres *et al.*, 1983), has good resistance to blackleg disease (Gugel *et al.*, 1990), resistance to aphids and flea beetles (Bayeh and Gebre Medhin, 1992), relatively large seed size (Getinet *et al.*, 1996) and some germplasms are resistant to alternaria black spot (Yitbarek, 1992). Getinet *et al.* (1996) were likely the first to show that it could be grown successfully in Western Canada. Their research was based on 11 Ethiopian accessions grown over 2 years in central Saskatchewan. Maturity of the accessions, relative to the early maturing *B. napus* cv. Westar, ranged from 1 week to just over 3 weeks later. Seed yield of the accessions was similar to Westar in the first year and lower in the second. They concluded that this species could help diversify and stabilize canola production in Western Canada, particularly in heat- and drought-prone areas.

Research conducted by Pan *et al.* (2011) on water stress and nitrogen (N) effects on photosynthesis and growth indicated that *B. carinata* maximized the utilization efficiency of the most limiting resource by developing physiological adaptations such as changes in root and leaf development. They also found that seed yield was strongly related to instantaneous NUE. If water was not limiting, increasing N up to 100 kg ha^{-1} significantly increased seed yield but maximum yields could be obtained with only 50 kg ha^{-1} N. In another study by Pan *et al.* (2011), where the effect of cultivar, seeding rate, and N was tested, a number of interesting observations were made. First, maximum seed yields were obtained over a wide range of seeding rates suggesting that this species has high plasticity under low plant populations, that is, when plant populations were low, branching was high. This concurs with work by Punia *et al.* (2001) who observed plant populations as low as 22 plants m^{-2} under semi-arid conditions. They also observed a linear response to applied N up to 150 kg ha^{-1} N but this resulted in a decrease in oil content, and corresponding increase in protein content. The response to N is in agreement with other studies (Thakral *et al.*, 1997; Kaur and Sidhu, 2004).

Quality and Uses of Oil, Meal, and Whole Plant

A market niche for Canadian canola and other oilseeds' oil in the emerging bioeconomy is for its use in biodiesel production.

Oil extracted from mustard seed is typically high in erucic acid so it is generally not suited to human consumption. For example, fatty acid profiles of 66 accessions investigated by Warwick *et al.* (2007) were observed to be high in erucic acid (C22:1; 30.9–45.7%) with approximately 5.1–11.6% oleic (C18:1), 13.7–18.9% linoleic (C18:2), and 10.2–16.0% linolenic (C18:3) acids).

The biofuel industry is rapidly growing because of increasing energy demand and diminishing petroleum reserves on a global scale (Jessup, 2009). With the introduction of Bill C-33, the Canadian government legislated a mandatory 2% renewable content in diesel and heating oil by 2012. To meet these targets, biodiesel production in Canada is predicted to rise. While rapeseed (*B. napus*) oil is the main feedstock for the synthesis of European biodiesel (Vicente *et al.*, 2006), oil from another brassicaceous species such as Ethiopian mustard are also used as biodiesel feedstocks.

Mustard meal (i.e., *Brassica juncea*) has high glucosinolate content and is used for the production and processing of condiment mustard. Glucosinolates in the mustard meal give mustard its characteristic pungency and spiciness. A recent and growing market for mustard-derived biopesticides based on the biocidal properties of the breakdown products of glucosinolates, called isothiocyanates, is already producing products with significant value from conventional varieties of mustard. Only certain types of glucosinolates, such as allyl glucosinolates, form isothiocyanates (AITC) with biocidal activity. The primary glucosinolate (>95%) in both condiment *B. juncea* and *B. carinata* is allyl isothiocyante, or sinigrin (2-propenyl). These isothiocyanates are biocidal to a wide range of soil organisms including nematodes, bacteria, and fungi (Brown and Morra, 1997; Sarwar *et al.*, 1998; Barrau *et al.*, 2009; Furlan *et al.*, 2010), and have also been shown to have anti-carcinogenic properties (Mithen *et al.*, 2000). A process for producing mustard-seed-derived isothiocyanates was developed at Agriculture and Agri-Food Canada in Guelph by Dr. Mikio Chiba, and commercialized by MPT (Mustard Products and Technologies), Saskatoon, Saskatchewan. Similar work has been reported in Europe by Palmieri (1999), where novel and low-cost extraction of glucosinolates can be utilized in protecting fruit in storage. Since meal from current Ethiopian mustard varieties cannot be used directly in meal preparations, this value-added use will surely improve the economics of growing this species as a sustainable biorefinery platform where its primary use is fuel.

Ethiopian Mustard Breeding

The Saskatoon Research Centre of Agriculture & Agri-Food Canada (AAFC-SRC) initiated a breeding program in the mid-1990s to develop early maturing *B. carinata* strains. After approximately 15 years of breeding for earliness, several promising strains have been developed from these initial populations and the first strain AAC A100 was commercialized in 2012. Approximately 2600 hectares of this variety were grown in 2012 (Males, Agrisoma Biosciences Inc., personal communication). In addition to early maturity, breeding lines with high oil content (>45%), high protein content (>35% on a whole seed basis), relatively large seed size (1000 seed weight > 3 g), low fiber content, and both low and high erucic acid contents have been developed.

Sustainability Considerations

Sustainable agriculture or production of food/fuel and novel biomaterials is becoming more commonplace and offers great opportunities. The move toward a biobased economy will clearly benefit the entire value chain if species such as Ethiopian mustard are developed that require fewer inputs (i.e., fertilizer, pesticides, water) while producing higher seed and more biomass. These so-called bioplatforms or biorefineries could be used to create renewable materials such as biopolymers, lubricants, soaps etc. thereby reducing the need for products derived from fossil fuels. Whole plant utilization is key to this process where the extracted oil, for example, is used as a feedstock for fuel, lubricants, and the production of biopolymers, while the meal is used as food for aquaculture once anti-nutritives such as AITC are removed or used as a potent biopesticide.

Fahd *et al.* (2012) provide an excellent example of total crop utilization on marginal land in Southern Europe. Their work focused on bioenergy and biochemicals from *B. carinata* (i.e., a nonfood crop) with two hypotheses: (i) cropping for bioenergy (biodiesel + heat) and (ii) bioenergy and biomaterials. Their assessment was made using SUMMA (SUstainability Multi-scale Multi-method Approach) which considers a number of factors/elements to develop a comprehensive understanding of the process under investigation (Fahd *et al.*, 2012). They conclude that the actual growing of the crop and not the processing of the crop is the limiting factor that makes the process uneconomical. The inputs exceed the net gain in energy. One of the major factors is the poor productivity of this species (and others) on marginal lands. The low- and high-input regimes investigated were based on seed yields of 1.2 t ha^{-1} and 1.7 t ha^{-1}, respectively. The scenario would improve, therefore, if crop productivity increased, the cost of fossil fuel rose dramatically (above €130–€140), labor and service costs were reduced by half, and/or alternative crops such as soy or sunflower were used. However, these will likely have higher-input costs, which would negate any positive effect.

Ethiopian mustard is of great interest as a biofuel feedstock due to its potentially better agronomic performance in certain areas of Spain (Bouaid *et al.*, 2005; Vicente *et al.* 2006;

Martinez-Lozano *et al.*, 2009), Italy (Cardone *et al.*, 2002, 2003) and most recently in Western Canada and parts of the United States (Males, Agrisoma Biosciences Inc., personal communication). In all instances, these areas are not suitable for the cultivation of oilseed *B. napus* but may not necessarily be in marginal areas. Cardone *et al.* (2003), for example, found that *B. carinata* was more productive than rapeseed under poor soil conditions, low cropping systems (inputs), and semi-arid environments.

Research on the production of biodiesel from Ethiopian mustard in Europe is well documented, including studies on agronomics, chemistry of production, and economics. The EU, in particular Spain, has looked at a number of field crops or "energy" crops as sources of biomass such as rapeseed, cardoon (*Cynara cardunculus*), sweet sorghum (*Sorghym bicolour* L.) among others (Martinez-Lozano *et al.*, 2009). Their conclusion, for the parameters considered, was that using *B. carinata* only as a source of biomass for the production of electricity was not economical. However, if higher yields and lower costs of production were possible, the economics may improve. The issue of productivity that is, seed yield, can be addressed through traditional breeding and/or the development of hybrids. OP strains developed and tested by the author (K.C. Falk) over the past 5 years average over 2 t ha^{-1} seed. The development of a pollination control system is also a promising avenue for improving yields over OP strains. AAFC-SRC is currently field testing a pollination control system based on cytoplasmic male sterility.

B. carinata is a unique study subject in that it is being considered not only for its oil as a source of fuel, but also for its great biomass potential. Approximately 33% and 22% of its dry matter consists of cellulose and hemicellulose, respectively (González-García *et al.*, 2009). In this case study, the E85 blend had the least GHG emissions. Regardless of the blend, ethanol-based fuels reduced global warming, GHG emissions, and the use of fossil-based fuels. However, fossil fuel has less environmental impact relative to photochemical oxidants, acidification, and eutrophication. Most of these negative effects are due to agricultural inputs. González-García *et al.* (2010) studied the environmental performance of a number of crops as lignocellulosic feedstocks in the production of ethanol. In this instance, second-generation feedstocks such as stems, leaves, and/or chaff following harvest were compared for flax, hemp, poplar, Ethiopian mustard, and alfalfa. In contrast, first-generation feedstocks refer to biofuel from starch or sugars, or biodiesel from extracted plant oils. Their LCA focused on the four so-called impact categories: global warming, photochemical oxidant formation, acidification, and eutrophication (the reader is referred to their publication for a detailed explanation of each). Although there was not any single clear winner, Ethiopian mustard fared very well. When compared to conventional automotive fuel, the life-cycle GHG emissions were 145% lower for E85 blends derived from this species; it also had the best rating for eutrophying emissions. The major contributor to eutrophying emissions are nitrogen emissions to air and water from agricultural fertilizers. Ethiopian mustard fared well in this category owing to its high biomass production and the fact that the allocation factor was 100%.

The development of a sustainable oil-based feedstock depends on a number of factors including, but not limited to, choice of platform specie(s), quality of oil, and availability at low cost (Carlsson, 2009) and the environmental impact should also be minimal. Furthermore, the final product, that is, oil, should have a broad range of industrial applications and preferably be confined to nonfood uses. In his review, he mentions several potential "technical oil" platforms, among them *B. carinata.* Many of its virtues are highlighted, that is, agronomic performance under adverse conditions, high seed yield, disease and pest resistance as well as the fact that Ethiopian mustard is amenable to transformation and, therefore, suitable as a molecular platform.

The topic of growing crops specifically for energy, whether liquid fuel or as a source of biomass for the production of electricity, in rotation with food crops has been reviewed by Zegada-Lizarazu and Monti (2011). The authors focus their review on annual crops, which are most often grown in rotation. Clearly, one of the greatest benefits to including new crops, alongside food crops such as rapeseed and sunflower, is as a break for the control of pests (insects and diseases) and to improve soil fertility and tilth.

It is important to develop nonfood biorefineries or bioplatform crops so as not to create a situation where food uses compete with fuel. Nonedible, high erucic acid oil from Ethiopian mustard, for example, would clearly fall into this category. Not only is the oil industrial, the remaining post-harvest biomass has little or no feed value and can, therefore, be used directly in the production of ethanol or perhaps generation of electricity. It is worthwhile to note that fuel-acres could still displace food-acres in areas where producers see an economic benefit to growing the former (and are paid a premium to do so). The creation of bioprocessing industries has benefits to producers, local economies, the environment, and may have a positive impact on managing crop residues. Clearly the latter would require close monitoring so as not to adversely affect soil tilth, particularly, in areas with low organic matter or soils subject to erosion.

Improvements in crop yields, whether for biomass or seed + biomass is critical to the economics of growing sustainable field crops in the emerging bioeconomy. Advances in genetics/breeding, species diversity, and crop management practices will reduce the cost of production and stabilize yields. Whole plant or total crop utilization coupled with newer processing technologies (i.e., ethanol production, biodiesel, or jet fuel) and crop management strategies is paramount to developing a sustainable biorefinery industry.

Camelina (*Camelina sativa*)

Camelina (*Camelina sativa* L. Crantz), also known as false flax, gold of pleasure, or German sesame, is an ancient oilseed that has been cultivated since prehistoric times with archaeological records reaching as far back as into the New Stone Age (Bouby, 1998; Knörzer, 1978). Findings of mostly carbonized seeds and capsules indicate that the Southeastern Europe and Southwestern Asian steppe regions are most likely the centre of origin of the species (Knörzer, 1978; Zohary and Hopf, 2000).

While the seeds of camelina constituted an important part of the human diet in the Bronze and Iron Ages in most parts of Europe (Hjelmquist, 1979), its importance as a food crop declined during the Middle Ages (Hjelmquist, 1979; Knörzer, 1978). Later, it was primarily the seed oil that was utilized, either to a limited extent as edible oil, as lamp oil preceding the discovery of gas and electricity, or in the production of soaps and coatings. In the middle of the twentieth century, camelina acreages almost completely dwindled away with the introduction of higher-yielding winter rapeseed varieties; it was grown only sporadically, such as to compensate for an out-wintered rapeseed crop. According to Crowley (1999), the lower cost of hydrogenating rapeseed oil and a highly limited market for camelina oil were the major reasons for the lack of interest in the crop.

With the colonization through European settlers, camelina was brought into the New World, where early on its numerous positive agronomic attributes, including a high-yield potential, early maturity, low fertilizer requirements, frost and drought tolerance, and resistance to a number of insect pests were praised (Porcher, 1863). Later, trials in both the United States and Canada corroborated these early findings (Robinson, 1987; Plessers *et al.*, 1962); yet, despite numerous favorable accounts of camelina's agronomic characteristics, until lately camelina has for the most part only been grown on a trial basis. Similar to the situation in Europe, the lack of knowledge of potential large-scale applications for camelina oil has hampered the establishment of camelina as a crop in North America. However, with the rise in petroleum prices and an increasing demand for renewable fuels, camelina has gained renewed interest as an oilseed crop in Europe (Zubr, 1997; Hebard, 1998; Leonard, 1998; Bonjean and Le Goffic, 1999), Australia (Francis and Campbell, 2003) and, particularly, North America (Putnam *et al.*, 1993) with great potential as a low-input source of biofuel and feedstock for the oleochemical industry.

Camelina is a member of the *Brassica* (*Brassicaceae*) family. It is a fast growing, spring or winter annual that is primarily self-pollinated (Walsh *et al.*, 2012). The rather small seed (average thousand seed weight: 1 g) contains 30–46% oil (Vollmann *et al.*, 2007) with a typical composition of 25–42% α-linolenic, 13–21% linoleic, 14–20% oleic, 12–18% gondoic (eicosenoic), and 2–4% erucic acid (Budin *et al.,* 1995; Zubr, 1997). The high content of poly-unsaturated fatty acids renders camelina oil suboptimal for many bio-based industrial applications; therefore, camelina lines with a high content of mono-unsaturated fatty acids along with lines expressing modified fatty acids for particular industrial end uses have been developed through genetic modification (D. Puttick, Linnaeus Plant Sciences Inc., personal communication). The ease of transformation of camelina, a simple floral dip procedure similar to that used for the model plant *A. thaliana* (Lu and Kang, 2008; Liu *et al.*, 2012) places camelina among the most attractive oilseed candidates for a biotechnology platform (Collins-Silva *et al.*, 2011; Snapp and Lu, 2012).

After extraction of the seed oil, a meal of high nutritional value remains, with a protein fraction rich in essential amino acids and a low content of antinutritives, including glucosinolates (Zubr, 1997; Zubr and Matthäus, 2002). In addition, camelina meal contains bio-active compounds such as flavonoids and phenolic products (Cherian, 2012). Camelina meal is approved as animal feed component in Europe (Act [EU] Nr. 575/2011) and the U.S. Food and Drug Administration (FDA) approved the use of camelina meal in feed rations given to poultry, broiler, beef (each up to 10%), and swine (up to 2%). In Canada, camelina meal does not have the "Generally Regarded as Safe" (GRAS) status yet and, therefore, cannot be used as a feed ingredient.

Much of the research on camelina crop production and the performance of camelina biofuel is conducted in the United States (Pavlista *et al.*, 2012; Gesch and Cermak, 2011) and Canada (Gugel and Falk, 2006; Johnson *et al.*, 2010; Blackshaw *et al.*, 2011); due to this, the following discussion, with the exception of the section on camelina in organic agriculture, will focus on sustainability aspects of camelina cultivation for biofuel production in North America. However, results obtained here may be transferred to other regions with similar climatic and geographical conditions.

Camelina Production and Sustainability

Camelina as a Low-input Crop

Feedstock costs can comprise up to 80% or more of biofuel production costs (Demirbas, 2006), therefore, apart from an environmental benefit, lower agricultural production inputs render a dedicated biofuel crop also more attractive economically. It is generally accepted that camelina has lower fertilizer, water, and pesticide requirements than other traditional commodity crops like canola or sunflower (Moser, 2010). As mentioned earlier in this chapter, nitrogen (N) fertilizer is the most important agricultural input in oil crops. For optimum production of camelina, nitrogen requirements range from 80–100 kg ha^{-1} combined soil and fertilizer NO_3-N (Jackson, 2008) to 90–100 kg ha^{-1} applied N (Budin *et al.*, 1995; Johnson *et al.*, 2010). This is lower than for most other *Brassica* species, such as alba mustard (*Sinapis alba*), spring canola, and both juncea mustard and juncea canola (*B. juncea*) (Johnson *et al.*, 2010; Gan *et al.*, 2007; Grant and Bailey, 1993).

Numerous times camelina has been described as well adapted to drought conditions (Porcher, 1863; Plessers *et al.*, 1962; Robinson, 1987; Putnam *et al.*, 1993; Zubr, 1997; Blackshaw *et al.*, 2011). In support of this in some cases rather anecdotal notion, recently a number of consecutive studies in the irrigated semi-arid environment of Arizona revealed that total evapotranspiration, which equates to the minimum water requirement of a crop, is markedly less for camelina than that typically needed by grain and vegetable crops or that needed by *Lesquerella fendleri*, an industrial oilseed crop native to the semi-arid Southwestern United States (French *et al.*, 2009; Hunsaker *et al.*, 1998, 2011, 2012). This makes camelina a suitable alternative in areas where long-lasting droughts, urbanization, and associated competition for water supplies hamper the successful cultivation of crops such as wheat or cotton (French *et al.*, 2009).

Camelina is resistant to blackleg (*L. maculans*) (Salisbury, 1987) and alternaria blight (*Alternaria brassicae*) (Conn *et al.*, 1988), two diseases that can severely affect canola production. Including camelina in a canola rotation might, therefore, contribute to the mitigation of these diseases (Kutcher *et al.*, 2012; Zelmer and McVetty, 2009). Camelina is also resistant to many insect pests including crucifer flea beetle (*Phyllotreta cruciferae*) (Pachagounder *et al.*, 1998), cabbage root fly (*Delia brassicae* Wiedemann) (Finch, 1978), diamondback moth (*Plutella xylostella*) (Deng *et al.*, 2004), and mustard sawfly (*Athalia proxima*) (Singh and Sachan, 1997). Equipped with resistance to many pests and diseases, camelina requires neither seed treatment nor late season insecticide treatment which is a significant economic and environmental advantage compared to other oilseed crops. While camelina is susceptible to a number of diseases that are common for *Brassica* crops (reviewed in Séguin-Swartz *et al.*, 2009), with sclerotinia stem rot (*Sclerotina sclerotiorum*), aster yellows (*Candidatus* Phytoplasma asteris), and downy mildew (*Hyaloperonospora camelinae*) being the most important diseases in camelina production in North America, usually none of them occur to a degree that would justify fungicide application. In addition, genotypes resistant to aster yellows (Olivier, unpublished data), downy mildew (Vollmann *et al.*, 2001; K.C. Falk, AAFC-SRC, personal communication), and sclerotinia stem rot (Eynck *et al.*, 2012) have been identified, rendering the development of cultivars resistant to these diseases readily available.

Land-use Change Considerations for Camelina Production in Conventional Agriculture

For a biofuel to be sustainable it is of utmost importance that it is derived from feedstocks produced with no or only little competition with food production and with no or only minor GHG emissions due to land-use changes (Tilman *et al.*, 2009). Camelina may offer several scenarios in which these issues can be at least partially avoided. As a crop that is early maturing, adapted to low rainfall conditions, and well suited for no-till management, camelina has great potential to replace summer fallow acreages (Moser, 2010; Shonnard *et al.*, 2010). Fallow is the practice of keeping land out of production during a growing season in order to provide extra soil moisture and nutrients for the following crop. Wheat–summer fallow is a rotation traditionally associated with dryland farming (production of crops in areas with low levels of natural rainfall without irrigation; www.globalwaterforum.org/tools/glossary/) and also the prevailing cropping system in the US Great Plains, extending from the Canadian border in the north to the southern part of Texas and bounded by the Rocky Mountains on the west and higher rainfall zones to the east (Hansen *et al.*, 2012). While summer fallow has been traditionally implemented to minimize the risk of crop failures in an environment with low or infrequent precipitation, extensive fallow has also been associated with a number of sustainability issues, such as poor precipitation use efficiency, intensified soil erosion by wind and/or water, decreased soil organic matter, and uncertain economic returns (Black *et al.*, 1981; Campbell *et al.*, 1990). With its low moisture and nutrient requirements, camelina offers a suitable alternative to summer fallow in both spring and winter wheat rotations. It provides several benefits to the subsequent wheat crop, including reduced soil erosion, increased soil organic matter, and reduction of wheat pests and diseases. Furthermore, Shonnard *et al.* (2010) argue that with the incorporation of camelina into the rotation, land can be returned to wheat with moisture and nutrient levels similar to those after a fallow year, resulting not only in equal yield improvements but also additional revenue for the farmer. Although similar effects have been described for other fallow replacement crops (Pikul *et al.*, 1997), the hypothesis that growing camelina instead of summer fallow has only little impact on soil water and nutrient levels yet needs to be corroborated; therefore, further research is required to elucidate if and to what extent inclusion of camelina improves the overall productivity and sustainability of the traditional wheat–summer fallow rotation.

Although both spring and winter forms of camelina exist, winter types are not commonly grown and camelina crops seeded in fall are in most cases spring types that over-winter. However, recently there is a growing interest in winter camelina as well (Gesch and Cermak, 2011; Gesch and Archer, 2013). While winter camelina is seeded in fall with the establishment of a rosette before winter, spring-type camelina can, alternatively to spring seeding, be sown on frozen ground in late November to early December. Since the seedlings are able to cope with several cycles of freeze–thaw, the seeds start to germinate early in spring at temperatures near 0 °C, well before the emergence of weeds, a feature that increases the competitiveness of the crop against weeds (Putnam *et al.*, 1993; Zubr, 1997) and supersedes herbicide applications (Gesch and Cermak, 2011), which further decreases the environmental impact of camelina production. Fall or winter seeding of camelina may be particularly utile in the dryland regions of the Northern Great Plains where the crop takes advantage of soil moisture that accumulates during the cooler spring months (Pilgeram *et al.*, 2007). Further, an over-wintering camelina crop contributes to the prevention of soil erosion (Lal *et al.*, 1991), uptake of excess N (Staver and Brinsfield, 1998), and provides habitat for small wild life (Tillman *et al.*, 2004). Another advantage of fall-seeded camelina is that it may be harvested early enough to allow the production of a second crop (double cropping). In a study on double cropping of winter camelina with short season soybean and sunflower, Gesch and Archer (2013) found the yields of the double cropped soybean and sunflower to be slightly below that of the corresponding monocrop which is to be expected due to seeding dates later than optimal. However, net returns and combined oil yields of the double cropping system were comparable to those of conventionally managed soybean or sunflower monocrops. According to the authors, this may present at least a partial solution to the food versus fuel debate (Gesch and Archer, 2013).

Camelina is an extremely hardy plant that can be grown on land that does not support the profitable production of major commodity crops. It was successfully grown on soils reclaimed from mine spoil (Petrikova *et al.*, 1996) and on once-irrigated, abandoned farmland (http://westernfarmpress.com/markets/camelina-offers-hope-california-biofuel-crop).

In all the above-described scenarios the production of camelina as a biofuel feedstock is possible without the conversion of native ecosystems into agricultural lands and with no or minor reduction of acreages of arable land available for food production. Nevertheless, further research is needed to determine the economic viability of production systems including camelina.

Camelina in Mixed Cropping

Mixed cropping is the practice of growing two or more crops in the same field during a growing season with the aim to use resources more efficiently and minimize the production risk compared to monocropping. It is frequently used in organic agriculture and other areas with low external inputs in order to increase productivity, yield security, and product diversity (Jensen, 2006; Paulsen, 2011). Similar to double cropping, mixed cropping systems in which food and energy crops are grown simultaneously represent a land-use option where biofuel production does not compete with food production and does not result in land-use changes (Tilman *et al*., 2009).

Mixed cropping systems of peas (*Pisum sativum*), lupins (*Lupinus angustifolius*), and spring wheat (*Triticum aestivum)* grown together with camelina have been proposed in Europe with the rationale to not only produce food but at the same time biofuel to be used directly on the producing farm in the form of straight vegetable oil (Paulsen, 2007, 2008, 2011; Paulsen et al., 2011). Mixtures with camelina resulted in an overall yield gain (main crop + camelina) and the LER values (land equivalent ratio, the sum of the fractions of the intercropped yields divided by the sole-crop yields) indicated increases in resource efficiency. Concordantly, the yields of the main crops in mixed cropping systems with camelina surpassed those in the monocrop, meaning parallel to the production of biofuel even increases in yields for food production are possible. Mixtures including camelina resulted in a significant overproduction of vegetable oil and could, therefore, contribute significantly to the on-farm fuel demand of the whole crop rotation without competition to food production and with low environmental costs (Paulsen *et al.,* 2011). Next to elevated land-use efficiency values, further arguments for mixed cropping with camelina are increased control of annual weeds (Saucke and Ackermann, 2006), reduced lodging, low production costs, and good eco-balances (Paulsen, 2007). Thus, LCAs for mixed cropping systems with camelina showed low energy and environmental costs (Sergis-Christian and Brouwers, 2005; Paulsen, 2007). Furthermore, CO_2 emissions from the production of camelina oil were estimated to be only one-third of those from organically produced rapeseed oil (Sergis-Christian and Brouwers, 2005). The overall reduction of GHG emissions has been shown to be over-proportional in comparison to possible yield losses in the main crop and higher than the additional GHG emissions from the mixed cropping system (Paulsen, 2008). Similar to conventional production of camelina biofuel, if the meal were to be used as animal feed or as additional source of biofuel, the reduction in GHG emissions would increase even further (Paulsen, 2008). Additionally, setting up local crushing facilities could contribute to stimulating the economy in rural communities help to diversify the range of agricultural products and would meet the requirement of close local energy cycles in organic agriculture (Stotz and Remmele, 2005; Paulsen, 2008).

Oil Palm (*Elaeis guineensis*)

Oil palm differs from most other oil crops tapped for biodiesel production in that, because of being a tropical perennial tree, its production is confined to developing countries with high population density and intense pressure on land. Biofuel feedstocks, like palm oil, that are also used for food, must first of all satisfactorily prove that fuel is not produced at the expense

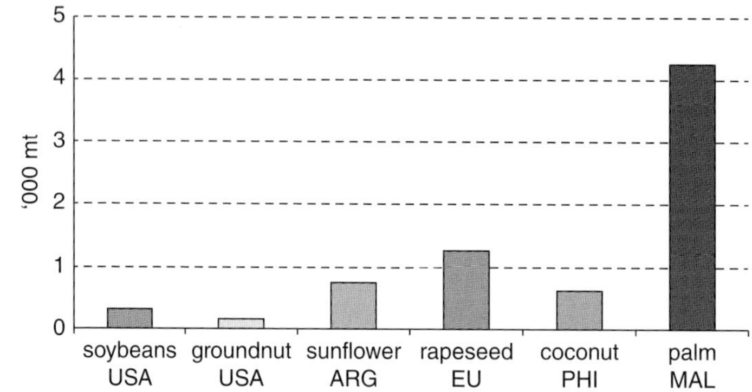

Figure 5.1. Oilcrop yields (Cf. FAO, 2006).

of food. Also, all rural enterprises in developing countries must focus on poverty alleviation and job creation. Palm oil, even before being considered a feedstock for biodiesel has been enmeshed in controversies surrounding its role in deforestation, aggravation of climate change, and destruction of biodiversity. Thus, appropriate prevention mechanisms must be put in place to ensure forests are not cleared to establish new plantations to produce additional oil for biodiesel. Low input and high yields are basic criteria for any preferred biofuel crop; oil palm also must pass this test. As use of oil palm for biodiesel is still not in significant proportion to its total production, most of the issues concerning oil palm have their origin in pre-biodiesel period, but need to be understood in order to develop a future course that is sustainable for the oil palm industry for all purposes.

Food versus Fuel Dilemma

From Figure 5.1, it is clear that because of much higher yield, palm oil plantations can relieve enormous amount of land devoted to producing other oil crops. Roughly, 0.26 ha of oil palm yields 1 t of oil compared to 2.22 ha of soybean, 2 ha of sunflower, and 1.52 ha of rapeseed (MPOC, 2006). Lee and Tan (2011), citing 2007 data, point out that 10.6 Mha was needed to produce 38.5 M t of palm oil while 94.2 M ha of soybean produced 36.9 M t of soya oil. Thus, oil palm is roughly 10 times more efficient in land utilization compared to soybean for oil synthesis.

Except for coconut, a considerable portion of the land areas used for other oil crops lies in South America or temperate countries where the pressure on land is substantially lower than in the main oil palm producing countries of tropical Southeast Asia. But in the present day global commercial trade environment, it does not matter where which crop is produced, market determines the final commodity destination. Thus, the land relieved from oilseed crop production can be beneficially utilized meeting the often decried shortfalls in staple food crop supply. Moreover, since oil seed crops are produced on the prime agricultural land, the land saved from oilseed crop production can be directly transferred to food grain production.

Oil palm plantations can also help diversify food choices in the countries of production through companion farming activities (Malaysian Palm Oil Board, 2007). Livestock are often raised on oil palm plantations and grazed on the ground vegetation. This has added advantages

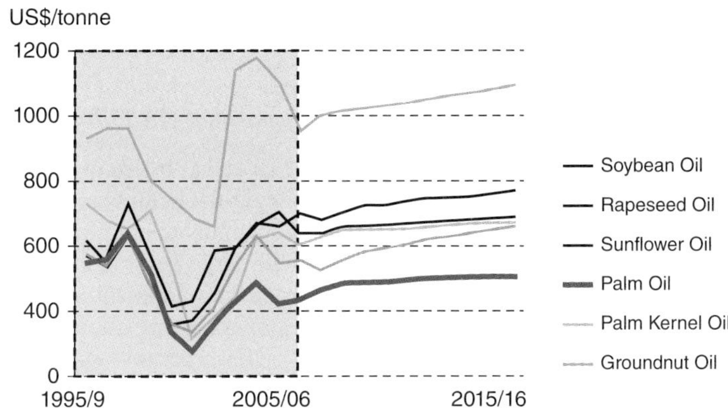

Figure 5.2. Vegetable oil past prices and projections (Cf. FAO, 2006).

of keeping weeds under control while enriching soil by manure. Immature plantations are also intercropped with short-term crops like banana, pineapple, and sugarcane.

The food use of palm oil is mostly in cooking and confectionary. The affordability of palm oil has not changed over the years as is evidenced from Figure 5.2. The preference for palm oil for cooking is not universal because of its high content of saturated fatty acids. In western countries, its use as food is mainly in confectionary. This puts a limit on the demand for palm oil. Thus, in reality, the oil palm industry needs new consumption outlets for its product especially taking into account that several additional countries with favorable climate have also established and plan to expand palm oil plantations.

How much of the oil palm production will be diverted to biodiesel in the producing countries is difficult to predict. The producing countries can develop a flexible policy for biodiesel production similar to the one Brazil has on sugarcane ethanol, and let the market determine the utilization of palm oil. The EU is the main biodiesel consumer in the world. However, most of EU biodiesel has been produced from domestically grown rapeseed. Whether the EU will

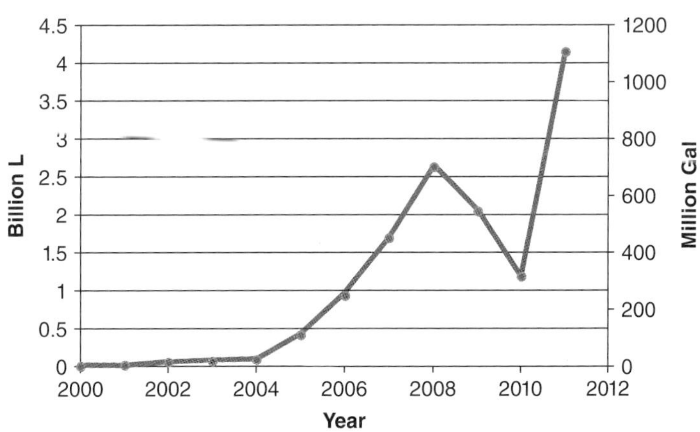

Figure 5.3. Soybean biodiesel production in USA (*Source*: NBB, 2012).

import palm oil in substantial quantity for biodiesel is uncertain and depends upon regulatory developments.

Rural Employment and Economic Impacts of Expanded Oil Palm Cultivation

Oil palm cultivation is carried out on an industrial scale as well as by small farmers. According to World Growth (2010), small land holders owned 43% of total palm plantations in Indonesia and 40% in Malaysia. Rist *et al.* (2010) stated that cooperatives are a meaningful solution to problems small farmers may face in competition with large commercial plantations. The Malaysian Government created an agency (Federal Land Development Agency) in 1956 to help small farmers in aggregating holdings into estates of economic size and provide needed management and infrastructure support (Basiron, 2007). This program succeeded in improving income of farmers as they were now able to raise more economically viable crops.

Lam *et al.* (2009) report that the cost of producing palm oil at US\$228 t^{-1} is substantially lower than US\$400 t^{-1} and US\$648 t^{-1}, for soybean and rapeseed, respectively. This enables both large as well as small growers to receive higher return from oil palm compared to annual oilseed crops.

Several reports suggest that the palm oil industry brings much needed jobs to rural communities. Kui (2008) stated that the oil palm industry in Malaysia provided employment to more than 570 000 people, out of which 405 000 were engaged in the cultivation of the crop. Two important components of commercial oil palm plantations, raising of seedlings and harvesting, continue to be performed mainly by manual labor. It is estimated that these activities create 1 job for every 2.3 ha (UNEP, 2011). A typical 2000 ha palm oil plantation in Malaysia employs a manger, three assistant mangers, and nine field staff in addition to manual workers for various field tasks (Basiron, 2007).

On the other side of the equation, disregard of land tenure systems of local communities during acquisition of large blocks of land for plantation corporations, low wages to workers and unsafe working conditions are often cited concerns as palm oil industry expands. Sustainable certification systems criteria have recently been put in place which include safeguard to protect vulnerable communities from land tenure abuses (RSPO, 2007). Albán and Cárdenas (2007) suggest that profit sharing together with improvements in living and working conditions could improve the satisfaction level of palm oil industry employees.

Land-use Change due to Oil Palm

Oil palm is adapted geographically to the equatorial region also home to ecologically critical rain forests. To confound the equation, this area is also heavily populated with continued high birth rate. Agriculture is the main livelihood and thus relied upon for feeding and providing livelihood to this fast multiplying human mass. Land area of the region is inelastic, but is absolute for crop production. Severe shortage of farm land tempts farmers to resort to clearing of forests to get additional area for raising crops. Some, like Butler (2011), contend that oil palm plantations actually help arrest destruction of tropical rainforests as higher return from land and labor provides a disincentive toward the alternative of cattle rearing that requires much more land for grazing purposes.

Commercial interests with high profit motivation have attributed a great deal to the onslaught on the primary tropical forests. The forest tree species provide valuable quality timber and powerful companies have used their influence and monetary strength to get access to logging. To aggravate the situation further, they have shown wanton disregard to the ecology of these regions and left the land barren to be eroded and degraded. Deforestation and subsequent land change in Malaysia and Indonesia, the two countries with maximum land under oil palm plantation, thus, must be viewed taking into account these existing ground realities.

Simultaneously, though, a sizeable amount of primary forests in Indonesia were also cleared for establishing oil palm plantation under the licenses issued by the Indonesian Government. Many plantation companies possess the right under these agreements to open further large areas of forests. However, under the revised Indonesian Government policy, new plantations can only be established on previously cleared lands. The new policy may not be favorable to prior license holders as the new land offering may not be contiguous to the current holdings and the lands offered in exchange may not be as fertile as newly cleared forests. How these companies will come to terms with the new forest preservation policy remains to be seen.

Both Malaysia and Indonesia have now developed policy guidelines to manage oil palm expansion. In Malaysia, by law no nonfarming land can be newly brought under oil palm cultivation. Malaysia has 20% of its land area designated for agriculture while reserving 60% for forest. This compares favorably to only 25% under forest in Europe. The other major oil palm producing country, Indonesia, has also promulgated laws setting aside 25% land to forest and restricting of new oil palm plantation to set aside land for cultivation (World Growth, 2010).

Oil palm producing countries also have developed schemes to undo some of the harm done previously by restoring degraded lands to crop production. Tree crops are better suited to planting on undulating eroded lands of these countries. They have mapped eroded and poor fertility soils which they intend to bring back to high value use by establishing new oil palm plantations on them.

A sufficient knowledge base is also available to implement measures such as contouring to prevent soil from future erosion and use of legumes and crop waste to enrich the soil to ensure long-term sustainability of oil palm production systems on eroded soils. Additional land demands to 2020 for oil palm plantations range from 1 to 28 Mha in Indonesia and 0.06 to 5 Mha in Malaysia. Indonesia has enough degraded land to meet full demand while Malaysia can meet the demand at the lower end of projection (Wicke *et al.*, 2011).

The emphasis is also shifting from bringing more acreage under oil palm to increasing production from existing plantations. Yield of oil palm has remained stagnate since 1970s at approximately 17 t ha^{-1} fruit bunches, equivalent to 3.8 t ha^{-1} crude palm oil (FAO, 2008). This is not to say that improved planting materials with higher-yielding potentials have not been developed during the period. The average commercially achievable oil palm yield potential over a production cycle of 25 years is at least 23 t ha^{-1} of fruit or >5 t ha^{-1} crude palm oil (Corley and Tinker, 2007). Difficulties encountered in improving yields can be attributed to the expansion of oil palm cultivation to less suited soils and climates. Lot of research attention is being focused on developing best management practices for oil palm cultivation under these new conditions.

There appears to be adequate amounts of lands available outside of forests to accommodate all future projection of oil palm expansion in different tropical countries. To illustrate this point, Corley (2009) cites statistics on oil palm plantation prospective lands for three countries. Indonesia is said to possess 8 Mha of Imperata grassland (Garrity *et al.*, 1996), Brazil 15.5 Mha of "fallow" land in the Carajas region of Para state alone (da Costa, 2004), and Colombia 49 Mha of permanent pasture land similar to that previously planted under oil palm (Pagiola *et al.*, 2004). The Brazilian Government has plans to increase its share of global palm oil market but aims to establish the plantations under the Program for Sustainable Production of Oil Palm which limits the expansion to degraded lands only. Brazilian policy makers appear to be committed to the commitment the country made at the 2009 United Nations Climate Change Conference in Copenhagen to reduce deforestation by 80% by 2020 (Regalado, 2011; Tollefson, 2011).

Oil Palm Greenhouse Gas Footprint

It has been reported that 55–60% of oil palm plantations between 1990 and 2005 in Malaysia and Indonesia were established by clearing of virgin forest lands (Koh and Wilcove, 2008). Part of the forests cleared were peat swamps, which serve as reservoir of carbon and their clear felling and drainage lead to vast amounts of carbon loss through oxidation. Koh *et al.* (2011) report that one-tenth of the total forest area cleared (880,000 ha or 6 per cent of total peat land area) in the Malaysian peninsula was peat land swamp. This clearance resulted in the loss of approximately 4.6 Mg of carbon from peat oxidation and the loss of carbon sequestration service to the tune of approximately 660 000 Mg of carbon annually (Koh *et al.*, 2011).

There are many estimates available of the ecosystem carbon payback time (ECPT) for oil palm plantation on various types of lands. Fargione *et al.* (2008) calculated it to be 423 years for peat, and 86 years for lowland tropical rainforests. Danielsen *et al.* (2008) reported an ECPT of 692 years for peat, 93 years for other forests, and 10 years for degraded grassland while Gibbs *et al.* (2008) gave ECPT values of about 918 years for tropical peat, 90 years for forest, 30 years for degraded forest, 8 years for woody savanna, and <10 years for grassland and cropped lands. Syahrinudin (2005) is of the opinion that oil palm becomes a valuable carbon sink much sooner on grassland and degraded forest land. This perception is based on the estimate of total carbon stocks in grassland at 5.9–7.6 t ha^{-1} and carbon accumulation equivalent to approximately 13 t ha^{-1} yr^{-1}, from the fifth year onward in oil palm. Thus, it is very clear that a palm oil industry based on deforestation is unsustainable, but in contrast, quite sustainable on marginal lands. Fortunately, the palm oil industry is now more regulated and monitored than in periods of unrestricted expansion. The main oil palm producing countries have further pledged to restrict new plantations to agricultural and degraded lands.

GHG footprint generated in oil palm cultivation on agricultural lands is from the burning of fossil fuels needed to provide energy for planting, cultivation, harvesting, post harvesting, transportation, and processing. On the other hand, by being a perennial evergreen, the oil palm plant assimilates carbon throughout the year and thus serves as carbon sink. Tan *et al.* (2009)observed that oil palm was more an efficient CO_2 assimilator than the tropical rain forest; the amount of annual net CO_2 assimilated by oil palm and rain forest amounted to 64.5 and 42.4 t ha^{-1}, respectively.

As for the fossil fuel part of the GHG emission, Lam *et al.* (2009) noted that the energy consumption to energy generation ratios for oil palm, soybean, and rapeseed were 1:10, 1:2.5, and 1:3, respectively, showing a clear superiority of oil palm in energy efficiency.

Adaptation, Growth Habit, Morphological and Physiological Attributes of Importance to Sustainability

Positive Traits

The perennial growth habit of oil palm is an advantage over the annual growth habit of most other biofuel candidate oil crops. Oil palm is usually grown for 25–30 years before replanting. This results in considerable savings in labor and fossil energy costs compared to yearly planting and cultivation required for annual oilseed crops and in less danger of soil erosion associated with tillage operations. The evergreen nature provides the plant with a year-round canopy thereby enabling it to continually intercept incoming radiation for CO_2 assimilation. According to Tan *et al.* (2009), oil palm is more efficient than the rain forest in biomass accumulation, with annual rates of the two at 8.3 t and 5.8 t, respectively. In reality, oil palm is a forest species domesticated for its high biomass accumulation and its storage into desirable oil entities (Basiron, 2007).

The oil palm plant canopy architecture is close to ideal. It is formed in a way to intercept maximum amount of incident sunlight while simultaneously ensuring light interspersion throughout the canopy. Firstly, the leaf is divided into leaflets, which gives frequently spaced gaps in the leaf spread. The leaflets are narrow (~6 cm) and long (up to 1.3 m) but numerous (200–300 per mature leaf) (Corley and Tinker, 2007), a setup allowing high leaf area but sufficient gap to allow downward light travel. Further, the linear leaflets are arranged alternately on each side of the leaf stalk, essentially forming a light sieve. New leaves emerge from apical meristems (and not axially), rendering them fully exposed to the sunlight. The top leaves are narrow and upright, thus capturing solar radiation efficiently but at the same time leaving gaps for the downward light movement. The angle of leaf rachis to stem increases sequentially from top to bottom ensuring light transmissibility among canopy layers. The rachis becomes almost horizontal at the bottom to cover the ground fully preventing loss of light to the soil. The canopy also has a built-in self-pruning system, thereby saving unnecessary upkeep costs of nonfunctional leaves and confining the crown to 30–50 leaves.

The biomass accumulation capability of oil palm is especially impressive considering the fact that it has a C_3 photosynthetic apparatus. Dufrene and Saugier (1993) recorded oil palm photosynthetic rate approximating 25 μmol m^{-2} s^{-1} at light saturation, a very high value for a perennial tree-like species. Squire and Corley (1987) found that at the commercial planting density of 130–150 palms ha^{-1}, oil palm attained full canopy cover with a leaf area index of around 6. Additionally, by year 10, the plantation intercepted 96% of the photosynthetically active radiation (PAR) with a mean lifetime interception value of 88%.

The oil palm begins fruiting during the third year of planting, gradually increasing to reach optimal yield by the tenth year and continuing economic fruit bearing up to 20–25 years (FAO, 2002). Oil palm produces fruit bunches throughout the year thereby providing continuous yield. The fruit bunches are harvested at a frequency of 10–15 days. A single bunch may contain 1500 plum size fruits weighing up to 40 kg (Lee and Tan, 2011). Both hard kernel and fleshy mesocarp surrounding of the fruit are filled with oil. The palm pulp oil (PO) is the major portion of the oil yield and superior to palm kernel oil (PKO) in nutritional value as it is rich in carotenoids and tocopherols. The PO portion of palm oil is refined to edible oil and its use can further expand into production of biodiesel.

Wahid *et al.* (2004) describe the biomass accumulation and partitioning pattern in oil palm plantings under optimal environment and best management practices. They estimated the standing biomass of a 9-year-old oil palm planting amounting to 56 t ha^{-1} and annual total dry matter production equaling 36.7 t ha^{-1}. A substantial portion of the synthesized biomass was translocated to the reproductive organ, as indicated by a bunch index (fruit bunch dry weight/annual total dry matter production) of 0.46. The harvest index, which in oil palm is expressed as palm oil/annual total dry matter production, equaled 0.185, equal to an oil yield of 6.8 t ha^{-1} ignoring the kernel oil.

Taking into account the mutually antagonistic relationships between yield components, Breure (2003) considered 10–11 t ha^{-1} yr^{-1} as the realistic yield potential for oil palm. These values are similar to those variously reported for breeder's materials (Rajanaidu and Jalani, 1990; Lee and Toh, 1991). However, in commercial plantings, the yields obtained are in the range of 5–7 t ha^{-1} yr^{-1} (Henson, 1991).

Limitations

Oil palm has its origin in the tropical rainforest region of West Africa. As such, it has two basic climatic requirements: high temperature and high rainfall. It is not sufficient to have adequate soil moisture, but the atmospheric humidity also must remain high for oil palm to achieve optimal photosynthetic capacity (Wahid *et al.*, 2004). To meet these rigid climatic conditions, oil palm cultivation is mostly confined within 10° latitude from equator with minimal annual rainfall of 1600 mm yr^{-1} (Corley and Tinker, 2007). Thus, oil palm production is more restrictive geographically than other oil crops.

The oil palm plant prioritizes assimilates to vegetative parts over developing fruits under limited soil moisture conditions (Corley *et al.*, 1971). This arrangement assures continuous formation of new leaves that provide axil for inflorescence development but hamper filling of current fruit bunches resulting in reduced yields under drought. Although both male and female flowers are present on the same plant in oil palm, because of the monoecious nature of inflorescence, transfer of pollen for fertilization depends on insects which sometimes can become a limiting factor.

Nutrient and Other Management Practices in Oil Palm Production

As oil palm is a heavy producer, it naturally also needs high amounts of mineral nutrients. Tarmizi (2000) estimated that a supplementation of 0.5 to 1.1 kg/plant/yr of N, 0.7 to 1.1 kg/plant/yr of P_2O_5, and 0.5 to 2 kg/plant/yr of K_2O were needed for optimum palm growth on Malaysian soils.

Organic matter from different oil palm operations is mostly returned back to plantation to enrich the soil. Empty fruit bunches after oil extraction are good mulch to add organic matter to the soil and improve soil moisture retention. The residual ash after burning of fruit bunch fiber for energy generation is rich in potassium and often used as fertilizer. Pruned palm fronds are also good mulching material. Hamdan *et al.* (1998) reported that 60 t ha^{-1} yr^{-1} of empty fruit bunch without any inorganic fertilizer could support oil palm growth. But usually application is made at a lower rate and supplemented with inorganic fertilizers. This practice reduced fertilizer cost of immature palms by up to 58% and of mature palms by 5%. The fertilization rate is determined based on foliar and rachis analysis to make the applications based on soil needs.

Tan *et al.* (2009) report that oil palm plantations are increasingly incorporating best management practices in their cultivation routine. Examples of such methods are no burning during land clearing of old trees to plant new, and decreased reliance on pesticides through integrated pest management (IPM). Rat control with the help of barn owl is one of these IPM practices. The rhinoceros beetle, *Oryctes rhinoceros*, is an economically damaging pest of oil palm. Use of two pathogens of rhinoceros beetle, entomophagous fungus *Metarhizium anisopliae* and the *Oryctes* virus appear to provide an effective control against it (Murphy, 2007).

Uses for Palm Biomass and Palm Oil Mill Effluents

On average, only 10% of the total dry biomass produced by oil palm is utilized for oil synthesis (Basiron, 2005). Oil palm is a source of biomass supply all along its fruiting years and again when cleared for replanting. At the palm oil mill, 70–75% of the fresh fruit bunch is left as solid residue after oil extraction. These consist of empty fruit bunches, palm fiber, and palm kernel shell. The empty fruit bunches are often returned to the oil plantation for use as mulch. According to Choo *et al.* (2012), fiber and shell can generate adequate electricity to meet the total energy demand of the palm oil mill. Fiber has also been shown to be a good raw material for the paper industry. The kernel shell can also be turned into kernel meal used for animal feed. Chan (2009) estimated that 10.4 dry t ha^{-1} of biomass is also produced annually from pruned fronds which again can be used beneficially for plantation mulching.

Chan *et al.* (1980) estimated that based on 126 plants ha^{-1}, 75.4 dry t of trunks and 14.5 dry t of fronds are removed during land clearing for replanting. Basing their calculation on 5% plantation being cleared and replanted every year, Choo *et al.* (2012) estimated that 17.3 million dry t of trunk and 3.3 million dry t of frond biomass are available annually in Malaysia.

Aside from biodiesel and electricity generation by combustion, the biomass and palm oil mill effluents can be useful in generating energy in other forms. Empty fruit bunches are subjected to high pressure and temperature to form compressed, high density briquettes which can be used for household fuel or to feed biomass-fired boilers. These bunches can also be fed into a gasification system at temperatures of 700°–1000 °C to yield producer gases (SYNGAS) comprising of CO, hydrogen, and CH_4 (Choo *et al.*, 2012). Powdered empty fruit bunches have been utilized to produce bio-oil by pyrolysis at 300–700 °C (Mohamad *et al.*, 2009).

For every t of fresh fruit bunch processed, about 0.65 m^3 of palm oil mill effluent is generated (Choo *et al.*, 2012). This effluent is high in organic matter and thus has high biochemical oxygen demand (BOD). This discharge has to be decomposed to bring the BOD level to an acceptable level before discharge to land or water. This is done by anaerobic digestion which releases approximately 28 m^3 of biogas containing about 60–70% methane and 30–40% CO_2 for every cubic meter of effluent treated (Choo *et al.*, 2012). The biogas can be utilized for generating heat or electricity.

Recent Research Achievements and Current and Future Initiatives

Lee and Tan (2011) cite several recent research accomplishments to improve oil palm production and yield. The oil palm DNA has been decoded enabling scientists to develop clones with greater biomass accumulation and oil partitioning traits. Several new clones also bear fruit early to accelerate economic returns. Some of the Nigerian introductions by Malaysian Palm Oil Board possess reduced height genes; their introductions into commercial cultivars have produced shorter trees that are easier to harvest and can be planted at higher densities.

In palm oil the ratio of unsaturated to saturated fatty acid is approximately 1. Due to the high percentage of saturated fatty acids, oil-palm-based biodiesel has a high cloud point creating difficulty in engine ignition at low temperatures. Unsaturated fatty acids are also desirable for cooking oil. Thus, research utilizing genetic engineering is underway to develop cultivars with a lower percentage of saturated palmitic acid and increasing the percentage of mono-unsaturated fatty acids like oleic acid (Pinzi *et al.*, 2009).

There are certain additional research areas that deserve due attention. There is a large gap between the potential yielding ability of improved oil palm cultivars and their yield in commercial plantations. This indicates deficiencies in the current technologies for the oil palm production systems. Furthermore, as new plantations are established on marginal soils, new production systems will be needed that remediate growth impediments due to limited soil nutrient availability or low soil moisture holding capacity of these soils. Strikingly, very little attention has been paid to study oil palm roots such as with regard to root size, water and nutrient uptake abilities, and root turnover rates. The importance of oil palm roots for the plant upkeep will increase as the oil palm expansion will expose plants to climates of occasional droughts and soils with limiting fertilities.

Life-cycle Analysis of Biofuel Production from Oil Crops

Life-cycle analysis (LCA) is a cradle-to-grave analysis to quantify the energy and environmental impacts of making a product. Energy life-cycle analysis (ELCA) provides a systematic methodology to quantify the total energy coming into a system from various sources and the overall energy efficiency of processes involved in producing the final product. ELCA accounts for the amount of energy used in the production of a biofuel and compares it to the amount of energy contained in it (Morris, 2005).

Environmental LCA on the other hand provides an accounting for the environmental impact of a product. At present, the Renewable Fuel Standard (RFS), the renewable fuel volume mandate in the United States implemented by the United States Environmental Protection Agency (EPA, 2010b), requires an environmental LCA and a report on life cycle GHG emissions for biofuels. This allows the quantification of the reduction of GHG emissions of biofuels compared to the use of petroleum fuel. Therefore, most of the environmental LCA in the literature focus on GHG emission reduction.

In general, there are two types of LCA: (1) attributional LCA and (2) consequential LCA. Attributional LCA (ALCA) provides information about the impacts of the processes used to produce (and consume and dispose of) a product, but does not consider indirect effects arising from changes in the output of a product. Consequential LCA (CLCA), on the other hand, provides information about the consequences of changes in the level of output (and consumption and disposal) of a product, including effects both inside and outside the life cycle of a product (Brander *et al.*, 2009). CLCA models the causal relationships originating from the decision to change the output of the product and, therefore, seeks to inform policy makers on the broader impacts of policies which are intended to change levels of production (Brander *et al.*, 2009). Although a CLCA aims at including all other systems that are affected by a biofuel system, it is practically impossible to trace every possible interaction. As John Muir, a Scottish–American naturalist and preservationist once said "When one tugs at a single thing in nature, he finds it is attached to the rest of the world." This statement correctly characterizes the difficulty in conducting CLCA. For one, the data available for modeling holds large uncertainties and secondly, the CLCA model may not be able to account for all relations among various systems in the world.

Despite this difficulty and the uncertainty and incompleteness of available data, the Energy Independence and Security Act (EISA) (EISA, 2007) mandates to include a CLCA in biofuel LCA. However, it is important to keep in mind that the science of CLCA is still evolving and that its results need to be interpreted with caution.

Energy Life-cycle Analysis

Much of the attention directed toward biofuels is focused on the fact that they are renewable compared to their petroleum fuel counterparts (EPA, 2002; Knothe *et al.*, 2010). A widely adapted measure of the efficiency of biofuels is the Net Energy Ratio (NER) which is the ratio of energy output from biodiesel when used as fuel to the biodiesel share of input energy that was used in the fuel production cycle:

$$\mathrm{NER} = \frac{\text{Renewable Fuel Energy Output}}{\text{Biodiesel share of Energy Input}}.$$

During biodiesel production, the input energy consists mainly of fossil fuels such as coal, natural gas, diesel, and gasoline. It is important to distinguish the difference between total energy input and biodiesel share of energy input. For example, as soybean provides both meal and oil, it is important to properly allocate the total energy spent during soybean production among these two products.

Soybean Biodiesel

Several LCA studies on soybean biodiesel revealed that the energy return from soybean biodiesel is increasing over time because of increasing yields without significant increases in agricultural inputs. Thus, the NER of soybean biodiesel was reported to be 2.51 in 1994 (Ahmed *et al.*, 1994). In 1998, the National Renewable Energy Laboratory (NREL) published a detailed energy and environmental LCA study and found the NER to be 4.21 (Sheehan *et al.*, 1998). Analyses similar to NREL's report with updated data found the NER to be 4.56 in 2003 (Pradhan *et al.*, 2009) and 5.54 in 2006 (Pradhan *et al.*, 2011). This shows a clear trend toward increasing returns in biodiesel energy per unit of fossil fuel expended in production.

Compared to ethanol from corn which was reported to possess an NER of less than 2 (Shapouri *et al.*, 2006), soybean biodiesel offers a significantly higher return on fossil energy investment (see also the term "energy return on investment" [EROI] referred to earlier in this chapter). However, even though soybean biodiesel returned 5.5 times more energy than the amount of fossil fuel expended, it still shows a lower return than conventional diesel. Conventional diesel fuel returns 6 times more energy than the energy spent on mining, processing, and transportation.

To be economically competitive with regular diesel, the NER of a biodiesel needs to be at least equally high. However, an NER of 6 is only a starting point for biodiesel to compete with petroleum diesel at the price point. The turnaround time for biodiesel from placing the soybean seed into the ground to turning the oil into biodiesel is 1 year. Compared to this, it takes on average 30 days to process crude oil from well to pump. This presents an economic disadvantage for soybean biodiesel: the same return on investment is generated later than for conventional diesel. In addition, since the primary use of soybean oil is for food, the price of oil fluctuates significantly depending on the amount of available surplus and predictions regarding future market trends. As a result, without some sort of market regulations or subsidies, the soybean biodiesel industry operates under great uncertainty.

Since 2004, the US soybean biodiesel industry experienced quite extreme fluctuations in production volume because of uncertainties in government subsidies and mandates to use biodiesel such as RFS2 (Figure 5.3) (Also see "Biodiesel Tax Incentive Reform and Extension Act," 2009; http://www.govtrack.us/congress/bills/111/s1589).

The first comprehensive life-cycle inventory (LCI) quantifying and comparing energy and emissions associated with the production and use of soybean biodiesel with that of petroleum diesel was completed by NREL (Sheehan *et al.*, 1998). The study investigated the emissions associated with soybean cropping, transport, crushing, oil transesterification, biodiesel transport, and the use of biodiesel in vehicles. The results of the study showed that the use of soybean biodiesel reduces the GHG emissions by 78.5% compared to the use of petroleum diesel, which was subsequently used as an industry standard. The number also served as a basis for the mandatory GHG reduction threshold for biomass-based diesel established by the EISA in 2007. Accordingly, in order to be classified as biomass-based diesel, a renewable fuel must reduce life-cycle GHG emissions by 50% compared to the 2005 baseline diesel fuel that it replaces (EPA, 2010a). Biomass-based diesel that meets the 50% reduction criteria qualifies as advanced biofuel in the United States.

In 2008, researchers included the indirect land-use change in the calculation of total GHG emission when corn ethanol was used as biofuel (Searchinger *et al.*, 2008). They showed that with the inclusion of indirect land-use changes, corn ethanol produces 93% more GHG emissions and that the breakeven time for corn ethanol to neutralize the increased carbon emissions is 167 years. This was based on the assumption that for every bushel of corn used to produce corn ethanol, the replacement bushel was associated with deforestation of rainforest, hence releasing huge amounts of GHG.

Although there has been substantial rebuttal to the assumptions made by Searchinger *et al.* and the certainty of the used data has been questioned (Kim *et al.*, 2009), this study generated great interest. When the US Environmental Protection Agency (EPA) conducted a second analysis on indirect land-use changes from soybean biodiesel it found that with the inclusion of indirect land-use changes soybean biodiesel reduces GHG emissions by on average 57% compared to conventional diesel. This result was significantly lower than the previously reported 78.5% reduction but still met the minimum 50% reduction to qualify as an advanced biofuel.

Rapeseed, Camelina, and Ethiopian Mustard Biodiesel

Although soybean oil is the major feedstock for biodiesel in the United States, it is not the major feedstock in other parts of the world. Rapeseed oil is the major source of biofuel in Europe and contributes about 85% of the oil for biodiesel production worldwide, followed by soybean oil, and palm oil (Mittelbach and Remschmidt, 2005). According to the European Biodiesel Board (EBB, 2011), European (EU-27) production of rapeseed biodiesel reached 2.8 billion gallons [5.7 M t] compared to US canola biodiesel production of 1.1 billion gallons [4.2 billion L] in 2010. Germany is the largest producer of rapeseed biodiesel among EU countries accounting for about 30% of the total European rapeseed biodiesel production.

Unlike soybean, which is primarily grown for meal, canola and camelina are grown primarily for oil. Rapeseed cultivars with low erucic acid content in the oil and low glucosinolate content in the meal are referred to as "canola" in North America and Australia and as "double low rapeseed" in Europe (Rakow, 2012). High erucic acid, low glucosinolate rapeseed (HEAR) cultivars contain over 50% erucic acid and are grown mainly for industrial purposes (Zelmer and McVetty, 2009). Apart from the fatty acid profile of the oil, canola or double low rapeseed and HEAR rapeseed cultivars are very similar in terms of agronomics and processing; therefore, the LCA results from one crop can be applied to the other.

Over 90% of canola that is grown in the United States is grown on the Northern Great Plains of North Dakota and Minnesota. In Canada, canola acreage is continuously increasing, mainly due to the replacement of summer fallow acreages (CCC, 2010). With a value of 3.93, the NER for canola biodiesel is slightly lower than for soybean biodiesel (5.5). In a study conducted by the Canola Council of Canada (CCC) in 2010, the use of rapeseed biodiesel resulted in GHG emission reductions of 90% when compared to petroleum diesel; however, this study did not consider indirect land-use changes.

Research in camelina is not as advanced as in rapeseed/canola and hence field data to conduct a LCA are limited. Krohn and Fripp (2012) found that biodiesel from camelina grown as niche crop reduces the GHG emission by 40–60%. Shonnard *et al.* (2010) reported camelina derived green diesel reduced GHG emission by 80%. This shows a considerable variation in results. The author's own analysis with data from Pacific Northwest shows that the NER of camelina biodiesel is about 3.8 and the total GHG reduction is 73.3% compared to petroleum diesel (unpublished data). Similar to camelina, Ethiopian mustard (*B. carinata*) is a relatively new crop with potential as a biodiesel feedstock.

Oil Palm

In East Asian countries, palm oil is the major feedstock for biodiesel. In the period from 1995–2010, palm oil production more than tripled to 46.7 M t, with Indonesia (47%) and Malaysia (38%) as the main producers. With about 4000 kg of oil ha^{-1} (about 500 gal of biodiesel per acre), palm oil yields are significantly higher than that of, for example, soybean (375 kg ha^{-1} [about 47 gal of biodiesel per acre]) (Hansen, 2007).

Without consideration of indirect land-use changes, the GHG reduction from palm oil biodiesel was shown to be 76% (Silalertruksa *et al.*, 2012) and the NER 3.58 (Pleanjai and Gheewala, 2009). However, after inclusion of indirect land-use changes, palm oil biodiesel showed a much lower GHG reduction. A LCA conducted by the US EPA determined that biodiesel produced from palm oil does not meet the minimum 20% GHG emission reduction required by the federal RFS (RFS2).

Conclusion

While soybean oil is the main source of oil for biodiesel production in the United States, rapeseed oil is the main feedstock for biodiesel in Europe. Palm oil is the main oil crop for biodiesel in tropical regions of the world. Camelina has a great potential to be grown as a replacement crop for summer fallow which renders the inclusion of direct or indirect land-use changes in the LCA for this crop unnecessary. In fact, a credit could be applied for the amount of soybean or canola meal that is replaced by camelina meal. Camelina, however, similar to Ethiopian mustard, is in its initial research and developmental stage and more data is needed for a more robust LCA analysis.

LCA shows that biodiesel from soybean and rapeseed reduces GHG emission by more than 50% compared to petroleum diesel, the threshold set by US EPA for biofuels to be qualified as advanced biofuel. The biodiesel from palm oil failed to make the list. This US EPA estimation for palm oil, though, is based on land-use change values which are not applicable to degraded or idle lands that have been set aside by the producing countries for establishing oil palm plantations to produce the needed feedstock.

References

Achten, W.M.J. and Verchot, L.V. (2011) Implications of biodiesel-induced land-use changes for CO_2 emissions: case studies in tropical America, Africa, and Southeast Asia. *Ecology and Society* 14: 14. http://dx.doi.org/10.5751/ES-04403-160414 (accessed 21 September 2012).

Ahmed, I., Decker, J., and Morris, D. (1994) *How Much Energy Does it Take to Make a Gallon of Soydiesel*? National Soydiesel Development Board, Jefferson City, MO.

Albán, M. and Cárdenas, H. (2007) Biofuels trade and sustainable development: the case of Ecuador's palm oil diesel. Ecuador, International Institute for Environment and Development, 60p.

Anonymous. (2012) Beschreibende Sortenliste 2012. Bundessortenamt, Hannover, Germany (in German).

Barrau, C., Porras, M., Romero, E., *et al.* (2009) Brassica carinata for control of Phytophthora spp. in strawberry field crops. *Revista de Ciencias Agrarias* 32: 135–138.

Basiron, Y. (2007) Palm oil production through sustainable plantations. *European Journal of Lipid Science and Technology* 109: 289–295.

Baucom, R.S. and Mauricio, R. (2004) Fitness costs and benefits of novel herbicide tolerance in a noxious weed. *Proceedings of the National Academy of Sciences of the United States of America* 101: 13386–13390.

Basiron, Y. (2005) Biofuel: an alternative fuel in the Malaysian scenario. Malaysian palm oil board bulletin/palm oil developments, 32/Information Series, 2005.

Bayeh, M. and Gebre Medhin, T. (1992) Insect pests of noug, linseed and Brassica in Ethiopia, in *Oilseeds Research and Development in Ethiopia. Proceedings of the First National Oilseeds Workshop*, Addis Ababa, Ethiopia, 3–5 December 1991, pp. 174–178.

Belayneh, H., Riley, K.W., Tadesse, N., and Alemaw, G. (1983) The response of three oilseed *Brassica spp.* to different planting dates and seed rates in the highlands of Ethiopia. *Ethiopian Journal of Agricultural Science* 5: 22–31.

Bertram, M.G. and Pedersen, P. (2004) Adjusting management practices using glyphosate-resistant soybean cultivars. *Agronomy Journal* 96: 462–468.

Black, A.L., Brown, P.L., Halvorson, A.D., and Siddoway, F.H. (1981) Dryland cropping strategies for efficient water-use to control to control saline seeps in the northern Great Plains, U.S.A. *Agricultural Water Management* 4: 295–311.

Blackshaw, R.E., Johnson, E.N., Gan, Y., *et al.* (2011) Alternative oilseed crops for biodiesel feedstock on the Canadian prairies. *Canadian Journal of Plant Science* 91: 889–896.

Bonjean, A. and Le Goffic, F. (1999) La cameline—*Camelina sativa* (L.) Crantz: une opportunité pour l'agriculture et l'industrie européenne. [Camelina – *Camelina sativa* (L.) Crantz: an opportunity for European agriculture and industry].Oléagineux, corps gras, lipides (OCL) 6: 28–34 (in French, English summary).

Borzoni, M. (2011) Multi-scale integrated assessment of soybean biodiesel in Brazil. *Ecological Economics* 70: 2028–2038.

Bouaid, A., Diaz, Y., Martinez, M., and Aracil, J. (2005) Pilot plant studies of biodiesel production using Brassica carinata as raw material. *Catalysis Today* 106: 193–196.

Bouby, L. (1998) Two early finds of gold-of-pleasure (*Camelina* sp.) in middle Neolithic and Chalcolithic sites in western France. *Antiquity* 72: 391–398.

Brander, M., Tipper, R., Hutchison, C., and Davis, G. (2009) Consequential and Attributional Approaches to LCA: a Guide to Policy Makers with Specific Reference to Greenhouse Gas LCA of Biofuels. TP-090403-A. Ecometrica press. [online] Available at: http://d3u3pjcknor73l.cloudfront.net/assets/media/pdf/approachesto_LCA3_technical.pdf (accessed 20 September 2012).

Breure, K. (2003) The search for yield in oil palm: basic principles, in *The Oil Palm. Management for Large and Sustainable Yields*, (eds T. Fairhurst and R. Hardter), Potash & Phosphate Institute/Potash Institute of Canada and International Potash Institute, pp. 59–98.

Brown, P.D. and Morra, M.J. (1997) Control of soil-borne plant pests using glucosinolate-containing plants. *Advances in Agronomy* 61: 167–231.

Budin, J.T., Breene, W.M., and Putnam, D.H. (1995) Some compositional properties of camelina (*Camelina sativa* L. Crantz) seeds and oils. *Journal of the American Oilchemists Society (JAOCS)* 72: 309–315.

Butler, R. (2011) In Brazil, palm oil plantations could help preserve Amazon. Available at: http://e360.yale.edu/content/feature.msp?id=2415 (accessed 19 November 2012).

Cahoon, E.B., Clemente, T.E., Damude, H.G., and Kinney, A.J. (2009) Modifying vegetable oils for food and non-food purposes, in *Oil Crops, Series: Handbook of Plant Breeding*, (eds J. Vollmann and I. Rajcan), Vol. 4, Springer, New York, pp. 31–56.

Campbell, C.A., Zentner, R.P., Janzen, H.H., and Bowren, K.E. (1990) Crop rotation studies on the Canadian prairies, Research Branch, Agriculture Canada, Ottawa, ON, Canada, Publication 1841/E, 133p.

Cardone, M., Mazzonicini, M., Menini, S., *et al.* (2003) Brassica carinata as an alternative oil crop for the production of biodiesel in Italy: agronomic evaluation, fuel production by transesterification and characterization. *Biomass and Bioenergy* 25: 623–636.

Cardone, M., Prati, M.V., Rocco, V., Seggiani, M., Senatore, A., and Vitolo, S. (2002) Brassica carinata as an alternative oil crop for the production of biodiesel in Italy: engine performance and regulated and unregulated exhaust emissions. *Environmental Science and Technology* 36: 4656–4662.

Carlsson, A.S. (2009) Plant oils as feedstock alternatives to petroleum—A short survey of potential oil crop platforms. *Biochimie* 91: 665–670.

CAST. (2012) Herbicide-resistant weeds threaten soil conservation gains: finding a balance for soil and farm sustainability, Issue paper 49, Council for Agricultural Science and Technology (CAST), Ames, Iowa, February 2012.

CCC. (2010) Life Cycle Analysis of Canola Biodiesel. Canola Council of Canada. [online] Available at: http://canola.ab.ca/uploads/biodiesel/Canola_Lifecycle_Analysis.pdf (accessed 22 September 2012).

Chan, K.W. (2009) Biomass production and uses in oil palm industry, in *Sustainable Production of Palm Oil: A Malaysian Perspective*, (eds G. Singh, K.H. Lim, L. Teo, and K.W. Chan), Malaysian Palm Oil Association, pp. 133–161.

Chan, K.W., Watson, I., and Lim, K.C. (1980) Use of oil palm waste material for increased production. Proceedings of the conference of soil science and agricultural development in Malaysia, Malaysian Society of Soil Science, Kuala Lumpur, Malaysia, pp. 214–241.

Cherian, G. (2012) Camelina sativa in poultry diets: Opportunities and challenges, in *Biofuel Co-products as Livestock Feed: Opportunities and Challenges*, (ed. H.P.S. Makkar), FAO, Rome. Available at: http://www.fao.org/docrep/016/i3009e/i3009e00.htm (accessed 2 October 2012).

Choo, Y.M., Yung, C-L., and Ma, A-N. (2012) Oil Palm, in *Handbook of Bioenergy Plants*, (eds C. Kole, C.P. Joshi, and D.R. Shonnard), CRC Press, Boca Raton, FL, pp. 433–451.

Christen, O. and Friedt, W. (eds) (2007) Winterraps—Das Handbuch für Profis, DLG-Verlags-GmbH, Frankfurt, Germany, 323p (in German).

Christoffoleti, P.J., Galli, A.J.B., Carvalho, S.J.P., *et al.* (2008) Glyphosate sustainability in South American cropping systems. *Pest Management Science* 64: 422–427.

Cober, E.R., Cianzio, S.R., Pantalone, V.R., and Rajcan, I. (2009) Soybean, in *Oil Crops*, (eds J. Vollmann and I. Rajcan), Springer, New York, pp. 57–90.

Cohen, B.M. and Knowles, P.F. (1983) Evaluation of *Brassica* species in California, Proceedings of the 6th International Rapeseed Congress, Paris, France, pp. 282–285.

Collins Silva, J.E., Lu, C., and Cahoon, E.B. (2011) Camelina—a designer biotech oilseed crop. *Inform* 22: 610–613.

Conn, K.L., Tewari, J.P., and Dahiya, J.S. (1988) Resistance to *Alternaria brassicae* and phytoalexin-elicitation in rapeseed and other crucifers. *Plant Science* 56: 21–25.

Corley, R.H.V. (2009) How much palm oil do we need? *Environmental Science and Policy* 13: 134–139.

Corley, R.H.V., Gray, B.S., and Ng, S.K. (1971) Productivity of the oil palm (Eleais guineensis Jacq.) in Malaysia. *Experimental Agriculture* 7: 129–136.

Corley, R.H.V. and Tinker, R.H.V. (2007) *The Oil Palm*, Blackwell Publishing, Oxford.

Crowley, J.G. (1999) Evaluation of Camelina sativa as an alternative oilseed crop, Teagasc, Dublin, Ireland. ISBN 1841700495, 11p.

da Costa, R.C. (2004) Potential for producing bio-fuel in the Amazon deforested areas. *Biomass and Bioenergy* 26: 405–415.

Danielsen, F., Beukema, H., Burgess, N., Parish, F., *et al.* (2008) Biofuel plantations on forested lands: double jeopardy for biodiversity and climate. *Conservation Biology* 23: 348–358.

Delourme, R., Falentin, C., Huteau, V., *et al.* (2006) Genetic control of oil content in oilseed rape (*Brassica napus* L.). *Theoretical and Applied Genetics* 113: 1331–1345.

Demirbas, A. (2006) Biodiesel production via non-catalytic SCF method and biodiesel fuel characteristics. *Energy Conversion and Management* 47: 2271–2282.

Deng, S., Yun, G., Zhang, Q., Xu, H., and Cai, Q. (2004) Effect of false flax (*Camelina sativa*) on larval feeding and adult behavioral response of the diamondback moth (*Plutella xylostella*). *Acta Entomologica Sinica* 47: 474–478.

de Vries, S.C., van de Ven, G.W.J., van Ittersum, M.K., and Giller, K.E. (2010) Resource use efficiency and environmental performance of nine major biofuel crops, processed by first-generation conversion techniques. *Biomass and Bioenergy* 34: 588–601.

Dufrene, E. and Saugier, B. (1993) Gas exchange of oil palm in relation to light, vapour pressure deficit, temperature and leaf age. *Functional Ecology* 7: 97–104.

Durrett, T.P., Benning, C., and Ohlrogge, J. (2008) Plant triacylglycerols as feedstocks for the production of biofuels. *The Plant Journal* 54: 593–607.

EBB. (2011) The EU biodiesel industry statistics. European Biodiesel Board. Available at: http://www.ebb-eu.org/stats.php (accessed 13 February 2012).

EISA. (2007) Title II: Energy security through increased production of biofuels, in *Energy Independence and Security Act of 2007*: *One Hundred Tenth Congress of the United States of America*.

EPA (U.S. Environmental Protection Agency). (2002) A comprehensive analysis of biodiesel impacts on exhaust emissions; draft technical report. EPA420-P-02-001. Available at: http://www.epa.gov/otaq/models/analysis/biodsl/p02001.pdf (accessed 12 July 2010).

EPA (U.S. Environmental Protection Agency). (2010a) EPA finalizes regulations for the National Renewable Fuel Standard Program for 2010 and beyond. EPA-420-F-10-007. Available at: http://www.epa.gov/otaq/renewablefuels/420f10007.pdf (accessed 23 September 2012).

EPA (U.S. Environmental Protection Agency). (2010b) Renewable fuel standard program (RFS2) regulatory impact analysis, US Environmental Protection Agency, Washington, DC. EPA-420-R-10-006. Available at: http://nepis.epa.gov/EPA/html/DLwait.htm?url=/Adobe/PDF/P1006DXP.PDF (accessed 12 September 2012).

Eynck, C., Séguin-Swartz, G., Clarke, W.E., and Parkin, I.A.P. (2012) Monolignol biosynthesis is associated with resistance to *Sclerotinia sclerotiorum* in *Camelina sativa*. *Molecular Plant Pathology* 13: 887–899.

Fahd, S., Fiorentino, G., Mellino, S., and Ulgiati, S. (2012) Cropping bioenergy and biomaterials in marginal land: The added value of the biorefinery concept. *Energy* 37: 79–93.

FAO. (2002) Small-scale palm oil processing in Africa. Rome, Italy, FAO Agricultural Services Bulletin 148. ISSN 1010-1365.

FAO. (2006) Biofuels and commodity markets—Palm oil focus, Commodities and Trade Division, FAO, Rome.

FAO. (2008). In: FAOSTAT Online Statistical Service. Food and Agriculture Organization of the United Nations. Available at: http://faostat.fao.org (accessed 11 October 2012).

FAOSTAT. (2012) FAO Statistical Database, FAO, Rome. Available at: http://faostat.fao.org (accessed 2 August 2012).

Fargione, J., Hill, J., Tilman, D., Polasky, S., and Hawthome, P. (2008) Land clearing and the biofuel debt. *Science* 319: 1235–1238.

Ferreres, E., Fernandez, M., Minguez, I., and Dominguez, J. (1983) Productivity of B. juncea and B. carinata in relation to rapeseed B. napus. Agronomic studies, Proc. 6th Int. Rapeseed Congress, Paris, France, pp. 293–299.

Finch, S. (1978) Volatile plant chemicals and their effect on host plant finding by the cabbage root fly (*Delia brassicae*). *Entomologia Experimentalis et Applicata* 25: 150–159.

Francis, C.M. and Campbell. (2003) New high quality oil seed for temperate and tropical Australia, RIRDC (Rural Industries Research and Development Corporation), 27p. Publication No. 03/045. ISBN 0642586136.

French, A.N., Hunsaker, D., Thorp, K., and Clarke, T. (2009) Evapotranspiration over a camelina crop at Maricopa, Arizona. *Industrial Crops and Products* 29: 289–300.

Friedt, W., Seyis, F., Snowdon R.J., and Lühs, W. (2004) Resynthetic rapeseed germplasm as a source of novel genetic variation for breeding of *Brassica napus. GCIRC* (Groupe consultatif international de recherche sur le colza) *Bulletin* 21: 38–46.

Furlan, L., Bonetto, C., Finotto, A., *et al.* (2010) The efficacy of biofumigant meals and plants to control wireworm populations. *Industrial Crops and Products* 31: 245–254.

Gan, Y., Malhi, S.S., Brandt, S., Katepa-Mupondwa, F., and Kutcher, H.R. (2007) *Brassica juncea* canola in the Northern Great Plains: responses to diverse environments and nitrogen fertilization. *Agronomy Journal* 99: 1208–1218.

Garrity, D.P., Soekardi, M., Van Noordwijk, M., *et al.* (1996) The Imperata grasslands of tropical Asia: area, distribution, and typology. *Agroforestry Systems* 36: 3–29.

Gehringer, A., Snowdon, R.J., Spiller, T.H., Basunanda, P., and Friedt, W. (2007) New oilseed rape (*Brassica napus*) hybrids with high levels of heterosis for seed yield under oligotrophic conditions. *Breeding Science* 57: 315–320.

Gesch, R.W. and Archer, D.W. (2013) Double-cropping with winter camelina in the northern Corn Belt to produce fuel and food. *Industrial Crops and Products* 44: 718–725.

Gesch, R.W. and Cermak, S.C. (2011) Sowing date and tillage effects on fall-seeded camelina in the northern corn belt. *Agronomy Journal* 103: 980–987.

Getinet, A., Rakow, G., and Downey, R.K. (1996) Agronomic performance and seed quality of Ethiopian mustard in Saskatchewan. *Canadian Journal of Plant Science* 76: 387–392.

Gibbs, H., Johnston, M., Foley, J., *et al.* (2008) Carbon payback times for crop-based biofuel expansion in the tropics: the effects of changing yield and technology. *Environmental Research Letters* 3: 1–10.

González-García, S., Gasol, C.M., Gabarrell, X., Rieradevall, J., Moreira, M.T., and Feijoo, G. (2009) Environmental aspects of ethanol-based fuels from *Brassica carinata*: a case study of second generation ethanol. *Renewable and Sustainable Energy Reviews* 13: 2613–2620.

González-García, S., Moreira, T.M., and Feijoo, G. (2010) Comparative environmental performance of lignocellulosic ethanol from different feedstocks. *Renewable and Sustainable Energy Reviews* 14: 2077–2085.

Grant, C.A. and Bailey, L.D. (1993) Fertility management in canola production. *Canadian Journal of Plant Science* 73: 651–670.

Gugel, R.K. and Falk, K.C. (2006) Agronomic and seed quality evaluation of *Camelina sativa* in western Canada. *Canadian Journal of Plant Science* 86: 1047–1058.

Gugel, R.K., Séguin-Swartz, G., and Petrie, A. (1990) Pathogenicity of three isolates of Leptosphaeria maculans on Brassica species and other crucifers. *Canadian Journal of Plant Pathology* 12: 75–82.

Hamdan, A.B., Ahmad, T.M., and Mohd, T.D. (1998) Empty fruit bunch mulching and nitrogen fertilizer amendment: The resultant effect on oil palm performance and soil properties. *PORIM Bulletin* 3: 1–14.

Hansen, N.C., Allen, B.L., Baumhardt, R.L., and Lyon, D.J. (2012) Research achievements and adoption of no-till, dryland cropping in the semi-arid U.S. Great Plains. *Field Crops Research* 132: 196–203.

Hansen, S. (2007) Feasibility study of performing a life cycle assessment on crude palm oil production in Malaysia. *The International Journal of Life Cycle Assessment* 12: 50–58.

Heatherly, L., Dorrance, A., Hoeft, R., *et al.* (2009) Sustainability of U.S. soybean production: conventional, transgenic, and organic production systems, Special Publication No. 30, Council for Agricultural Science and Technology, Ames, IA.

Heatherly, L.G., Elmore, C.D., and Spurlock, S.R. (2002) Weed management systems for conventional and glyphosphate-resistant soybean with and without irrigation. *Agronomy Journal* 94: 1419–1428.

Hebard, A. (1998) *Camelina sativa*—a pleasurable experience or another false hope? *Lipid Technology* 10: 81–83.

Henson, I.E. (1991) Estimating potential productivity of oil palm, in *Proceedings of the 1990 ISOPB International Workshop on Yield Potential in the Oil Palm,* (eds V. Rao, I.E. Henson, and N. Rajanaidu), Palm Oil Research Institute of Malaysia, Kuala Lumpur, pp. 98–108.

Herrmann, I.T., Jørgensen, A., Bruun, S., and Hauschild, M.Z. (2012) Potential for optimized production and use of rapeseed biodiesel. Based on a comprehensive real-time LCA case study in Denmark with multiple pathways. *The International Journal of Life Cycle Assessment*, doi:10.1007/s11367-012-0486-8.

Hjelmquist, H. (1979) Beiträge zur Kenntnis der prähistorischen Nutzpflanzen in Schweden. *Opera Botanica* 47: 34–57 (in German).

Hunsaker, D.J., French, A.N., Clarke, T.R., and El-Shikha, D.M. (2011) Water use, crop coefficients, and irrigation management criteria for camelina production in arid regions. *Irrigation Science* 29: 27–43.

Hunsaker, D.J., French, A.N., and Thorp, K.R. (2012) Camelina water use and seed yield response to irrigation scheduling in an arid environment. *Irrigation Science*, doi:10.1007/s00271-012-0368-7.

Hunsaker, D.J., Nakayama, F., Dierig, D., and Alexander, W. (1998) Lesquerella seed production: water requirement and management. *Industrial Crops and Products* 8: 167–182.

Jackson, G.D. (2008) Response of camelina to nitrogen, phosphorus, and sulfur. Fertilizer Fact 49, Montana State University Extension Service, 2p.

James, C. (2012) Global status of commercialized biotech/GM crops: 2011, Brief 43, International Service for the Acquisition of Agri-biotech Applications. Available at: http://www.isaaa.org (accessed 4 August 2012).

Jensen, E.S. (2006) Intercropping of cereals and grain legumes for increased production, weed control, improved product quality and prevention of N losses in European organic farming systems. Available at: http://www.coreorganic.org/library/EU_folder/intercrop.pdf (accessed 23 September 2012).

Jessup, R.W. (2009) Development and status of dedicated energy crops in the United States. *In Vitro Cellular and Developmental Biology* 45: 282–290.

Johnson, E.N., Falk, K., Klein-Gebbinck, H., *et al.* (2010) Agronomy of *Camelina sativa* and *Brassica carinata*. Final Report. Available at AAFC Scott Research Farm, Box 10, Scott, SK, S0K 4A0, Canada.

Kaur, P. and Sidhu, M.S. (2004) Effect of sowing date, nitrogen level and row spacing on the growth and yield of African sarson (Brassica carinata A. Braun). *Journal of Research Punjab Agricultural University* 41: 27–34.

Kim, H., Kim, S., and Dale, B.E. (2009) Biofuels, land use change, and greenhouse gas emissions: Some unexplored variables. *Environmental Science and Technology* 43: 961–967.

Knörzer, K.H. (1978) Entwicklung und ausbreitung des Leindotters (*Camelina sativa* s.l.) [Evolution and spreading of gold of pleasure (*Camelina sativa* s.l.).]. *Berichte der Deutschen Botanischen Gesellschaft* 91: 187–195 (in German, English summary).

Knothe, G., Krahl, J., and Van Gerpen, J.H. (2010) *The Biodiesel Handbook*, 2nd edn, AOCS Press, Champaign, IL.

Koh, L.P., Miettinen, J., Liew, S.C., and Ghazoul, J. (2011) Remotely sensed evidence of tropical peatland conversion to oil palm. *Proceedings of the National Academy of Sciences of the United States of America* 108: 5127–5132.

Koh, L.P. and Wilcove, D.S. (2008) Is oil palm agriculture really destroying tropical biodiversity? *Conservation Letters* 1: 60–64.

Krohn, B.J. and Fripp, M. (2012) A life cycle assessment of biodiesel derived from the "niche filling" energy crop camelina in the USA. *Applied Energy* 92: 92–98.

Kui, D.P.C. (2008) Malaysia's oil palm-hallmark of sustainable development. Global Oils and Fats Business Magazine, Vol. 5, Issue No. 4.

Kutcher, H.R., Eynck, C., *et al.* (2012) Disease implications of canola-intensive crop rotations, in *Topics in Canadian weed Science: New Crops and Crops with Second-generation Traits.* (eds H.J. Beckie and L.M. Hall), Canadian Weed Science Society, Pinawa, Manitoba, Canada, pp. 49–66.

Lal, R., Regnier, E., Eckert, D.J., Edwards, W.M., and Hammond, R. (1991) Expectations of cover crops for sustainable agriculture, in *Cover Crops for Clean Water. Proceedings of the International Conference, Jackson, TN, 9–11 April 1991*, (ed. W.L. Hargrove), Soil and Water Conservation Society of America, Ankeny, Iowa, pp. 1–11.

Lam, M.K., Tan, K.T., Lee, K.T, and Mohammed, A.R. (2009) Malaysian palm oil: Surviving the food versus fuel debate for sustainable future. *Renewable Energy and Sustainable Energy Reviews* 13: 1464–1466.

Lee, K.T. and Tan, K.T. (2011) Palm oil as an energy crop, in *Energy Crops*, (eds N.G. Halford and A. Karp), Royal Society of Chemistry, Cambridge, pp. 187–195.

Lee, C.H. and Toh, P.Y. (1991) Yield performance of Golden Hope OPRS D x P planting materials. *The Planter* 67: 317–324.

Leffel, R.C. (1990) Economic models and breeding strategies for soybean improvement. *Journal of Production Agriculture* 3: 582–586.

Leonard, E.C. (1998) Camelina oil: α-linolenic source. *Inform* 9: 830–838.

Lima, M., Skutsch, M., and de Medeiros Costa, G. (2011) Deforestation and the social impacts of soy for biodiesel: Perspectives of farmers in the south Brazilian Amazon. *Ecology and Society* 16: 4.

Liu, X., Brost, J., Hutcheon, C., *et al.* (2012) Tranformation of the oilseed crop *Camelina sativa* by Agrobacterium-mediated floral dip and simple large-scale screening of tranformants. *In Vitro Cellular and Developmental Biology—Plant* 48: 462–468.

Lu, C. and Kang, J. (2008) Generation of transgenic plants of a potential oilseed crop *Camelina sativa* by *Agrobacterium*-mediated transformation. *Plant Cell Reports* 27: 273–278.

Lu, C., Napier, J.A., Clemente, T.E., and Cahoon, E.B. (2011) New frontiers in oilseed biotechnology: meeting the global demand for vegetable oils for food, feed, biofuel, and industrial applications. *Current Opinion in Biotechnology* 22: 252–259.

Malaysian Palm Oil Board (2007) MPOB report. Available at: http://econ.mpob.gov.my/economy/overview07.htm (accessed January 2013).

Martinez-Lozano, S., Gasol, C.M., Rigola, M., *et al.* (2009) Feasibility assessment of Brassica carinata bioenergy systems in Southern Europe. *Renewable Energy* 34: 2528–2535.

Mithen, R.F., Dekker, M., Verkerk, R., Rabot, S., and Johnson, I.T. (2000) The nutritional significance, biosynthesis and bioavailability of glucosinolates in human foods. *Journal of the Science of Food and Agriculture* 80: 967–984.

Mittelbach, M. and Remschmidt, C. (2005) *Biodiesel, the Comprehensive Handbook*, 2nd edn, Boersedruck Ges. m.b.H, Vienna.

Mohamad, A.S., Nor, K.A.B., and Chow, M.C. (2009) Optimization of pyrolysis of oil palm empty fruit bunches. *Journal of Oil Palm Research* 21: 653–658.

Morris, D. (2005) The carbohydrate economy, biofuels and the net energy debate, Institute for Local Self-Reliance, Minneapolis, MN. Available at: http://www.ilsr.org/carbohydrate-economy-biofuels-and-net-energy-debate/ (accessed 4 September 2012).

Moser, B.R. (2010) Camelina (*Camelina sativa* L.) oils as a biofuels feedstock: Golden opportunity or false hope? *Lipid Technology* 22: 270–273.

MPOC. (2006) *Oil Palm- Tree of Life*, Malaysian Oil Palm Council, Selangor, Malaysia.

Murphy, D.J. (2007) Future prospects for oil palm in the 21st century: Biological and related challenges. *European Journal of Lipid Science and Technology* 109: 296–306.

NBB. (2012) US Biodiesel Production Statistics. National Biodiesel Board. [online] Available at: http://www.biodiesel.org/production/production-statistics (accessed 2 September 2012).

Nesi, N., Delourme, R., Bregeon, M., Falentin, C., and Renard, M. (2008) Genetic and molecular approaches to improve nutritional value of *Brassica napus* L. seed. *Comptes Rendus Biologies* 331: 763–771.

Pachagounder, P., Lamb, R.J., and Bodnaryk, R.P. (1998) Resistance to the flea beetle *Phyllotreta cruciferae* (Coleoptera: Chrysomelidae) in false flax, *Camelina sativa* (Brassicaceae). *Canadian Entomologist* 130: 235–240.

Pagiola, S., Agostini, P., Gobbi, J., *et al.* (2004) Paying for biodiversity conservation services in agricultural landscapes. Rome, FAO Environment Dept Paper 96, 32p.

Palmieri, S. (1999) Glucosinolates as a source for novel bioactive formulations, in *Proceedings of the 6th Symposium of Industrial Crops and Products*, Bonn, Germany, pp. 65–75.

Panichelli, L., Dauriat, A., and Gnansounou, E. (2009) Life cycle assessment of soybean-based biodiesel in Argentina for export. *International Journal of Life Cycle Assessment* 14: 144–159.

Pan, X., Lada, R.R., Caldwell, C.D., and Falk, K.C. (2011) Water-stress and N-nutrition effects on photosynthesis and growth of Brassica carinata. *Photosynthetica* 49: 309–315.

Paulmann, W. and Frauen, M. (1997) The firstly restored F1-hybrids of winter oilseed rape. *GCIRC Bulletin* 13: 24–25.

Paulsen, H.M. (2007) Mischfruchtanbausysteme mit Ölpflanzen im ökologischen Landbau. 1. Ertragsstruktur des Mischfruchtanbaus von Leguminosen oder Sommerweizen mit Leindotter (*Camelina sativa* L. Crantz) [Organic mixed cropping systems with oilseeds. 1. Yields of mixed cropping systems of legumes or spring wheat with false flax (*Camelina sativa* L. Crantz.)]. *Landbauforschung Völkenrode* 57: 107–117 (in German, English summary).

Paulsen, H.M. (2008) Enhancing GHG balances in organic farms by integration of new bioenergy crop contents. Available at: http://www.orgprints.org/17119/ (accessed 4 August 2012).

Paulsen, H.M. (2011) Improving green-house gas balances of organic farms by the use of straight vegetable oil from mixed cropping as farm own fuel and its competition to food production. *Landbauforschung—vTI Agriculture and Forestry Research* 3: 209–216.

Paulsen, H.M., Wichmann, V., Shuemann, U., and Richter, B. (2011) Use of straight vegetable oil mixtures of rape and camelina as on farm fuels in agriculture. *Biomass and Bioenergy* 35: 4015–4024.

Pavlista, A.D., Baltensperger, D.D., Isbell, T.A., and Hergert, G.W. (2012) Comparative growth of spring-planted canola, brown mustard and camelina. *Industrial Crops and Products* 36: 9–13.

Petrikova, V., Vana, J., and Ustjak, S. (1996) Production and utilization of industrial and energy crops on reclaimed soils. *Metod Zemědělské Praxi* 17: 1–24 (in Czech, English summary).

Pikul Jr., J.L., Aase, J.K., and Cochram, V.L. (1997) Lentil green manure as fallow replacement in the semiarid northern Great Plains. *Agronomy Journal* 89: 867–874.

Pilgeram, A.L., Sands, D.C., Boss, D., *et al.* (2007) *Camelina sativa*, a Montana omega-3 fatty acid and fuel crop, in *Issues in New Crops and New Uses*, (eds J. Janick and A. Whipkey), ASHS Press, Alexandria, VA, pp. 129–131.

Pinzi, S., Garcia, I.L., Lopez-Gimenez, F.J., *et al.* (2009) The ideal vegetable oil-based biodiesel composition: A review of social, economical and technical implications. *Energy and Fuels* 23: 2325–2341.

Pleanjai, S. and Gheewala, S.H. (2009) Full chain energy analysis of biodiesel production from palm oil in Thailand. *Applied Energy* 86: S209–S214.

Plessers, A.G., McGregor, W.G., Carson, R.B., and Nakoneshny, W. (1962) Species trials with oilseed plants. II. Camelina. *Canadian Journal of Plant Science* 42: 452–459.

Porcher, F.P. (1863) Resources of the Southern fields and forests, medical, economical, and agricultural. Being also a Medical Botany of the Confederate States; with practical information on the useful properties of the trees, plants, and shrubs. Steam- Power Press of Evans and Cogswell, No. 3 Broad Street, Charleston, VA. xxv + 601 pp., ill. Call number 3041 Conf. (Rare Book Collection, University of North Carolina at Chapel Hill). Available at: http://docsouth.unc.edu/imls/porcher/porcher.html (accessed 8 August 2012).

Pradhan, A., Shrestha, D.S., McAloon, A.J., *et al.* (2009) Energy life cycle assessment of soybean biodiesel, Agricultural Economic Report Number 845, US Department of Agriculture.

Pradhan, A., Shrestha, D.S., McAloon, A.J., Yee, W.C., Haas, M.J., and Duffield, J. (2011) Energy life-cycle assessment of soybean biodiesel revisited. *Transactions of the ASABE (American Society of Agricultural and Biological Engineers)* 54: 1031–1039.

Punia, S.S., Chahar, S., and Agarwal, S.K. (2001) Influence of crop geometry and nitrogen on seed yield and yield attributes of Ethiopian mustard (*Brassica carinata*). *Indian Journal of Agronomy* 46: 732–735.

Putnam, D.H., Budin, J.T., Field, L.A., and Breene, W.M. (1993) Camelina: a promising low input oilseed, in *New Crops* (eds J. Janick and J.E. Simon), Wiley, New York, pp. 314–322.

Rajanaidu, N. and Jalani, S. (1990) Performance of D X P planting material in various parts of the world, in *Proceedings on Worldwide Performance of D x P Oil Palm Planting Materials, Clones and Interspecific Hybrids* (eds N. Rajanaidu and B.S. Jalani), Palm Oil Research Institute of Malaysia, Kuala Lumpur, pp. 1–34.

Rakow, G. (2012) Classical genetics and traditional breeding, in *Genetics, Genomics and Breeding of Oilseed Brassicas* (eds D. Edwards, J. Batley, I. Parkin, and C. Kole), CRC Press, New York, pp. 73–84.

Regalado, A. (2011) Furor over proposed Brazilian Forest Law. *Science Insider* http://news.sciencemag.org/scienceinsider/2011/05/furor-over-proposed-brazilian.html (accessed 11 October 2012).

Reijnders, L. and Huijbregts, M.A.J. (2008) Biogenic greenhouse gas emissions linked to the life cycles of biodiesel derived from European rapeseed and Brazilian soybeans. *Journal of Cleaner Production* 16: 1943–1948.

Robinson, R.G. (1987) Camelina: a useful research crop and a potential oilseed crop. University of Minnesota Agricultural Experiment Station Technical Bulletin 579 (Item No. AD-SB-3275), pp. 1–12.

Rist, L., Feintrenie, L., and Levang, P. (2010) The livelihood impacts of oil palm: smallhodlers in Indonesia. *Biodiversity and Conservation* 19: 1009–1024.

Romanelli, T.L., Nardi, H.D.S., and Saad, F.A. (2012) Material embodiment and energy flows as efficiency indicators of soybean (*Glycine max*) production in Brazil. *Eng. Agric. Jaboticabal* 32: 261–270.

RSPO. (2007) RSPO certification systems. Roundtable on Sustainable Palm Oil, Zurich.

RTRS (Round Table on Responsible Soy). (2010) RTRS standard for responsible soy production. Version 1.0, RTRS document of 10 June 2010. Available at: http://www.responsiblesoy.org/ (accessed 16 August 2012).

Salisbury, P.A. (1987) Blackleg resistance in weedy crucifers. *Eucarpia Cruciferae Newsletter* 12: 90.

Salvagiotti, F., Cassman, K.G., Specht, J.E., Waters, D.T., Weiss, A., and Dobermann, A. (2008) Nitrogen uptake, fixation and response to fertilizer N in soybeans: a review. *Field Crops Research* 108: 1–13.

Salvagiotti, F., Specht, J.E., Cassman, K.G., Walters, D.T., Weiss, A., and Dobermann, A. (2009) Growth and nitrogen fixation in high-yielding soybean: impact of nitrogen fertilization. *Agronomy Journal* 101: 958–970.

Sarwar, M., Kirkegaard, J.A., Wong, P.T.W., and Desmarchelier, J.M. (1998) Biofumigation potential of brassicas. III In-vitro toxicity of isothiocyanates to soil-borne fungal pathogens. *Plant and Soil* 201: 103–112.

Saucke, H. and Ackermann, K. (2006) Weed suppression in mixed cropped grain peas and false flax (*Camelina sativa*). *Weed Research* 46: 453–461.

Schweiger, P., Hofer, M., Hartl, W., Wanek, W., and Vollmann, J. (2012) N_2 fixation by organically grown soybean in Central Europe: Method of quantification and agronomic effects. *European Journal of Agronomy* 41: 11–17.

Searchinger, T., Heimlich, R., Houghton, R., *et al.* (2008) Use of US croplands for biofuels increases greenhouse gases through emissions from land-use change. *Science* 319: 1238–1240.

Séguin-Swartz, G., Eynck, C., Gugel, R.K., *et al.* (2009) Diseases of *Camelina sativa* (false flax). *Canadian Journal of Plant Pathology* 31: 375–386.

Sergis-Christian, L. and Brouwers, J. (2005) Dezentral hergestelltes, kaltgepresstes Pflanzenöl im ökologischen Vergleich mit Dieselkraftstoff. Witzenhausen: AG Land-Regionalentwicklung, Arbeitsergebnisse SH3, 67p (in German).

Seyis, F., Friedt, W., and Lühs, W. (2006) Yield of *Brassica napus* L. hybrids developed using resynthesized rapeseed material. *Field Crops Research* 96: 176–180.

Seyis, F., Snowdon, R.J., Lühs, W., and Friedt, W. (2003) Molecular characterisation of novel resynthesised rapeseed (*Brassica napus* L.) lines and analysis of their genetic diversity in comparison to spring rapeseed cultivars. *Plant Breeding* 122: 473–478.

Shapouri, H., Wang, M., and Duffield, J. (2006) Chapter 5: Net energy balance and fuel-cycle analysis, in *Renewable Based Technology* (eds J. Dewulf and H. Van Langenhove), John Wiley and Sons, New York, pp. 73–86.

Shaw, D.R., Owen, M.D.K., Dixon, P.M., *et al.* (2011) Benchmark study on glyphosate-resistant cropping systems in the United States. Part 1: Introduction to 2006–2008. *Pest Management Science* 67: 741–746.

Sheehan, J., Camobreco, V., Duffield, J., Graboski, M., and Shapouri, H. (1998) Life cycle inventory of biodiesel and petroleum diesel for use in an urban bus, NREL/SR-580-24089, National Renewable Energy Laboratory, U.S. Department of Energy, Golden, Colorado.

Shonnard, D.R., Williams, L, and Kalnes, T.N. (2010) Camelina-derived jet fuel and diesel: sustainable advanced fuels. *Environmental Progress and Sustainable Energy* 29: 382–392.

Silalertruksa, T., Bonnet, S., and Gheewala, S.H. (2012) Life cycle costing and externalities of palm oil biodiesel in Thailand. *Journal of Cleaner Production* 28: 225–232.

Singh, S.P. and Sachan, G.C. (1997) Effect of different temperatures and host plants on the developmental behaviour of mustard sawfly, *Athalia proxima*. *Indian Journal of Entomology* 59: 34–40.

Snapp, A.R. and Lu, C. (2012) Engineering industrial fatty acids in oilseeds. *Frontiers in Biology*, doi:10.1007/s11515-012-1228-9.

Solomon, B.D. (2010) Biofuels and sustainability. *Annals of the New York Academy of Sciences* 1185: 119–134.

Squire, G. and Corley, R.H.V. (1987) Oil palm, in *Tree crop physiology* (eds M.R. Sethuraj and A.S. Raghavendra), Amsterdam, Elsevier Science, pp. 141–167.

Staver, K.W. and Brinsfield, R.B. (1998) Using cereal grain winter cover crops to reduce groundwater nitrate contamination in the mid-Atlantic coastal plain. *Journal of Soil and Water Conservation* 53: 230–240.

Stotz, K. and Remmele, E. (2005) Daten und Fakten zur dezentralen Ölgewinnung in Deutschland. *Straubing* 2005: 59. Berichte aus dem TFZ 3 (in German).

Syahrinudin (2005) The potential of oil palm and forest plantations for carbon sequestration on degraded land in Indonesia. PhD, Ecology and development series, Cuvillier, Gèottingen.

Tan, K.T., Lee, K.T., Mohamed, A.R., and Bhatia, S. (2009) Palm oil: addressing issues and towards sustainable development. *Renewable and Sustainable Energy Reviews* 13: 420–427.

Tarmizi, A.M. (2000) Nutritional requirements and efficiency of fertilizer use in Malaysia, in *Advances in Oil Palm Research* (eds B. Yuso B.S. Jalani, and K.W. Chan), Kula Lumpur, Malaysia, Malaysian Palm Oil Board. Vol. 1, pp. 411–440.

Thakral, S.K., Singh, B.P., and Faroda, A.S. (1997) Effect of soil moisture and fertility management on LAI, LAD and CGR of Brassica species in Aridisols. *Annals of Agricultural Research* 18: 115–117.

Tillman, G., Schomberg, H., Phatak, S., *et al.* (2004) Influence of cover crops on insect pests and predators in conservation tillage cotton. *Journal of Economic Entomology* 97: 1217–1232.

Tilman, D., Socolow, R., Foley, J.A., *et al.* (2009) Beneficial biofuels—the food, energy, and environment trilemma. *Science* 325: 271–272.

Tollefson, J. (2011) Brazil revisits forest code. *Nature* 476: 259–260.

Ulas, A., Schulte auf'm Erley, G., Kamh, M., Wiesler, F., and Horst, W.J. (2012) Root-growth characteristics contributing to genotypic variation in nitrogen efficiency of oilseed rape. *Journal of Plant Nutrition and Soil Science* 175: 489–498.

Varvel, G.E. and Wilhelm, W.W. (2010) Long-term soil organic carbon as affected by tillage and cropping systems. *Soil Science Society of America Journal* 74: 915–921.

Vicente, G., Martinez, M., and Aracil, J. (2006) A comparative study of vegetable oils for biodiesel production in Spain. *Energy and Fuels* 20: 394–398.

Vollmann, J. (2001) Vorfruchtwert und Produktionstechnik von Winterraps in Österreich - eine empirische Untersuchung. [online] Available at: http://plantbreeding.boku.ac.at/oilcrops/raps2001/index.html (accessed 2 October 2012).

Vollmann, J., Moritz, T., Kargl, C., Baumgartner, S., and Wagentristl, H. (2007) Agronomic evaluation of camelina genotypes selected for seed quality characteristics. *Industrial Crops and Products* 26: 270–277.

Vollmann, J., Steinkellner, S., and Glauninger, J. (2001) Variation in resistance of camelina (*Camelina sativa* [L.] Crtz.) to downy mildew (*Peronospora camelinae* Gäum.). *Journal of Phytopathology* 149: 129–133.

Wahid, M.B., Abdullah, S.N.A., and Henson, I.E. (2004) Oil palm-achievements and potential. Proceedings of the 4th International Crop Science Congress, 26 September to 1 October 2004, Brisbane, Australia. Published on CDROM.

Wicke, B., Sikkema, R., Dornburg, V., and Faaij, A.P.C. (2011) Exploring land use changes and the role of palm oil production in Indonesia and Malaysia. *Land Use Policy* 28: 193–206.

Walsh, K.D., Puttick, D.M., Hills, M.J., *et al.* (2012) First report of outcrossing rates in camelina [*Camelina sativa* (L.) Crantz], a potential platform for bioindustrial oils. *Canadian Journal of Plant Science* 92: 681–685.

Warwick, P.D., Coleman, J.L., Hackley, P.C., *et al.* (2007) USGS assessment of undiscovered oil and gas resources in Cenozoic strata of the U.S. Gulf of Mexico Coastal Plain and State Waters. Proceedings of the 27th Annual GCSSEPM Bob F. Perkins Research Conference, The Paleogene of the Gulf of Mexico and Caribbean Basins: Processes, Events, and Petroleum Systems, December 2–5, 2007, Houston, Texas, p. 2–44, CDROM.

Wittkop, B., Snowdon, J., and Friedt, W. (2009) Status and perspectives of breeding for enhanced yield and quality of oilseed crops for Europe. *Euphytica* 170: 131–140.

World Growth. (2010) Facts on Palm Oil. World Growth, Melbourne, Australia.

WWF. (2011). Palm oil: environmental impacts. World Wildlife Fund (accessed 9 November 2012).

Yitbarek, S. (1992) Pathological research on noug, linseed, gomenzer and rapeseed in Ethiopia, in *OilseedsResearch and Development in Ethiopia, Proceedings of the First National Oilseeds Workshop, 3–5 December 1991, Addis Ababa, Ethiopia*.

Zahniser, S. (2010) EU environmental sustainability requirements and Brazilian biofuel exports. *EuroChoices* 9: 28–34.

Zegada-Lizarazu, W. and Monti, A. (2011) Energy crops in rotation. A review. *Biomass and Bioenergy* 35: 12–25.

Zelmer, C.D. and McVetty, P.B.E. (2009) Industrial Products, in *Biology and Breeding of Crucifers* (ed. S.K. Gupta), CRC Press, Taylor and Francis Group, Boca Raton, London, New York, pp. 343–359.

Zhu, Y., Cao, Z., Xu, F., *et al.* (2012) Analysis of gene expression profiles of two near-isogenic lines differing at a QTL region affecting oil content at high temperatures during seed maturation in oilseed rape (*Brassica napus* L.). *Theoretical and Applied Genetics* 124: 515–531.

Zohary, D. and Hopf, M. (2000) *Domestication of Plants in the Old World*, Oxford University Press, New York.

Zotarelli, L., Zatorre, N.P., Boddey, R.M., *et al.* (2012) Influence of no-tillage and frequency of a green manure legume in crop rotations for balancing N outputs and preserving soil organic C stocks. *Field Crops Research* 132: 185–195.

Zubr, J. (1997) Oil-seed crop: Camelina sativa. *Industrial Crops and Products* 6: 113–119.

Zubr, J. and Matthäus, B. (2002) Effects of growth conditions on fatty acids and tocopherols in *Camelina sativa* oil. *Industrial Crops and Products* 15: 155–162.

Chapter 6

Short-rotation Woody Crop Biomass Production for Bioenergy

L.C. Kiser[1] and T.R. Fox[2]

[1] *Abraham Baldwin Agricultural College, Tifton, GA, USA*
[2] *Virginia Polytechnic Institute and State University, Blacksburg, VA, USA*

Introduction

Woody biomass is used to provide bioenergy source in the form of fuel pellets that displace fossil fuels in the production of electricity and heat and also in the form of cellulosic ethanol liquid fuel. Woody biomass for bioenergy can be generated by growing short-rotation woody crops (SRWCs) or by using products of conventional forestry. This chapter focuses on the sustainability of SRWC systems in the United States. SRWC systems are defined as a silvicultural system based upon short clear-cutting cycles, generally between 1 and 15 years, employing intensive cultural techniques such as fertilization, irrigation, and weed control, and utilizing genetically superior planting material (Drew *et al.*, 1987).

In the United States, the U.S. Department of Energy (US DOE) examined the potential for annual production of 1 billion dry tonnes of biomass for bioenergy to displace approximately 30% of fossil fuel use. The updated report, U.S. Billion-Ton Update: Biomass Supply for a Bioenergy and Bioproducts Industry was released in August 2011. The efficacy of biomass production with SRWC systems was evaluated under baseline (1% yield increase annually) and high-yield (2–4% yield increase annually) production scenarios. At baseline production, it was estimated that annual production of 1 billion dry tonnes of biomass could occur by 2030 (Figure 6.1).

Under the high-yield production scenario, this goal was estimated to be obtained by 2022. Assumptions for the baseline scenario were that energy crop production began in 2014 and energy crops contributed 400 million dry tonnes annually by 2030. Of these 400 million dry tonnes, SRWC accounted for 126 million dry tonnes (Figure 6.1). Unlike the use of crop and traditional forest harvesting residues where land-use changes are not required, biomass production with SRWC systems requires changing existing agricultural or forest land to SRWC production. Assuming average operational yields in SRWC systems range from 5 to 10 dry tonnes ha^{-1} yr^{-1}, meeting the DOE's production goal will require an extensive land-use conversion. In order to produce 126 million dry tonnes of biomass at 5 dry tonnes ha^{-1} yr^{-1},

Biofuel Crop Sustainability, First Edition. Edited by Bharat P. Singh.

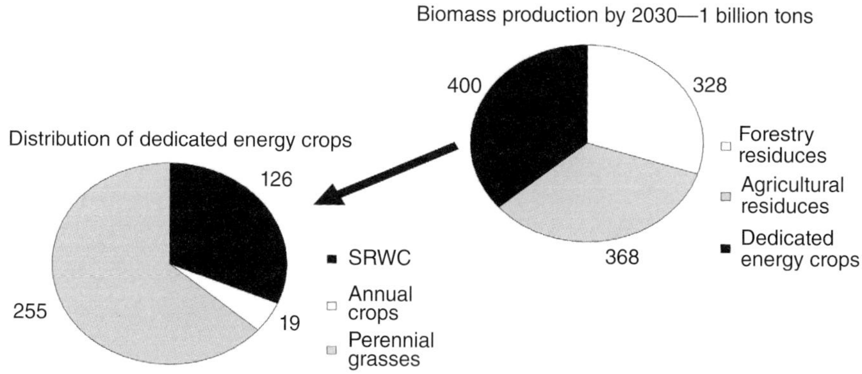

Figure 6.1. Biomass production goals reported in the U.S. Department of Energy's Billion-Ton Update report from August 2011. Values (dry metric tonnes yr^{-1}) based on baseline production which assumes a 1% annual yield increase starting in 2012. By 2030, SRWC are anticipated to contribute 126 million dry tonnes annually. Adapted from U.S. Department of Energy's Billion-Ton Update (2011).

25.2 Mha of land must be converted solely to use as SRWC systems. To put this in perspective, this is an area the size of the state of Wyoming. Doubling production to 10 dry tonnes ha^{-1} yr^{-1} will still require an area the size of the state of Mississippi to be converted to SRWC systems. This suggests that sustainability of SRWC systems is affected by issues of scale.

Bioenergy production with SRWC systems has been successfully implemented in Sweden with 16,000 ha planted representing 0.5% of arable land (Mola-Yudego, 2010). While production cost reductions have contributed, initial implementation in Sweden was mostly due to taxes on fossil fuels equivalent to the price of oil and four times the price of coal (Gustavsson, 1997). Unlike Sweden, the United States has an abundant indigenous supply of coal and implementation of SRWC systems for production of bioenergy has yet to be widely implemented due to costs approximately double that of coal (2.60–3 US$ GJ^{-1} vs. 1.40–1.90 US$ GJ^{-1}) (Abrahamson *et al.*, 1998; Volk *et al.*, 2006). This has led to considerations of co-firing woody biomass with coal in New York (Tharakan *et al.*, 2005). Similar to Sweden, further implementation of SRWC systems for bioenergy in the United States may require incentives such as taxes and subsidies that increase the cost competitiveness of SRWC biomass (Tharakan *et al.*, 2005; Volk *et al.*, 2006). Without these price guarantees, landowners have to assume too great a risk of a profitable return on their investment in an SRWC biomass bioenergy system. This suggests that sustainability of SRWC systems is further affected by economic issues.

Biomass yields with SRWC systems can be improved with proper site selection and silvicultural treatments such as species and provenance choice, fertilization, irrigation, tillage, and weed control (Mead, 2005). Environmental sustainability issues that arise from the intensive silvicultural inputs used in SRWC systems are typically more similar to those in agriculture than in traditional forestry operations. These systems are highly mechanized and require intensive silvicultural inputs to achieve desired yields. Environmental sustainability is defined in the broadest context as using the forest or land in a way where future generations can derive the same benefits as the present generation (Andersson *et al.*, 2000). Sustainable management maintains the ecological capacity of the site for tree growth over time (Evans, 1997) by avoiding degradation of soil nutrient and water supplies (Andersson *et al.*, 2000; Reijnders, 2006).

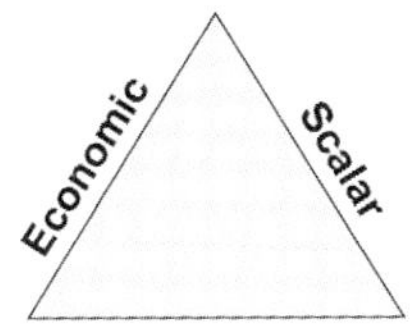

Figure 6.2. Three primary factors affecting the sustainability of SRWC systems for biomass bioenergy production.

Managing for sustainability requires an understanding of the ecological controls on production and the implementation of silvicultural strategies that work in harmony with ecosystem processes (Kimmins, 1997). In order to be environmentally sustainable, SRWC systems must also be optimized to achieve high yields at the lowest cost while minimizing pests, diseases, and damage to the environment (Mitchell, 1992).

We propose that the overall sustainability of bioenergy production with SRWC systems is affected by three primary sustainability factors: scalar, economic, and environmental (Figure 6.2).

These factors are intimately interconnected. For example, increasing the intensity of silvicultural inputs reduces the amount of land required to meet the desired yields producing a positive effect (scalar). However, this may also have negative effects by increasing production costs (economic) and soil erosion/nitrate leaching (environmental). The relationship of these factors suggests that SRWC systems must strike the right balance to meet production goals cost efficiently while minimizing negative environmental impacts. It is our estimation that the current yield per unit area must be dramatically increased in order for SRWC to be economically viable and also to reduce the amount of land required to produce the necessary tonnage. The purpose of this chapter was to review the research conducted on improvements in SRWC yield and any potential negative environmental impacts that arise from the intensive silvicultural techniques used to achieve higher yields. This allowed us to investigate the efficacy of SRWC for bioenergy production in the context of the production goals of the US DOE, which is to produce 126 million dry tonnes by 2022 (2–4% annual yield increase) or 2030 (1% annual yield increase). We focused on species currently most scrutinized for use in SRWC biomass bioenergy systems—shrub willow (*Salix spp.*), poplar (*Populus spp.*), and pine (*Pinus spp.*). Typical SRWC production systems and their relative advantages are discussed followed by evaluations of yield potentials. We then summarized potential environmental sustainability issues that arise from the intensity of silvicultural inputs. Then, we assessed the potential of SRWC systems for bioenergy production in the context of our definition of sustainability (Figure 6.2). Lastly, we briefly summarized the potential of shrub willow and poplar for phytoremediation.

Shrub Willow (*Salix spp.*)

Shrub willow is considered ideal for the silvicultural techniques used in SRWC systems due to very fast growth rates, ease of clonal propagation from cuttings, and an easily exploitable genetic diversity (Dickmann, 2006). Its main advantage over pine is its ability to resprout after cutting, known as coppice regeneration (Dickmann, 2006). The use of shrub willow for SRWC systems has been extensively evaluated over the past two decades with research centered in Northeastern Canada, Sweden, and the United States. Information has been provided on sustainability issues (Abrahamson *et al.*, 1998), cutting cycle and spacing effects on biomass yield (Kopp *et al.*, 1997), genetic improvements (Smart *et al.*, 2005), and nutrient management (Adegbidi *et al.*, 2001, 2003; Aronsson and Bergstrom, 2001; Park *et al.*, 2004, 2005; Quaye *et al.*, 2011). Shrub willow commonly hybridized and evaluated for use in SRWC bioenergy coppice systems includes the North American species *Salix eriocephala* Michx., *Salix discolor* Muhl., and *Salix* × *rubra* Huds., the European species *Salix viminalis* L., *Sarracenia purpurea* L., *Salix alba* L., and *Salix dasyclados* Wimm., and the Asian species *Salix miyabeana* Seemen.

In general, little agreement exists on whether shrub willow SRWC systems should be established on degraded or high-quality sites. Dickmann (2006) suggests that poor or marginal sites are unsuitable for SRWC systems due to lack of return on the intensive silvicultural investment arguing instead that the best available agricultural and forested sites are more suitable for SRWC. In contrast, others argue that SRWC systems have more beneficial effects on the soil and environment when grown on marginal lands (infertile, compacted, eroded soils) compared to croplands and natural forests (Reijnders, 2006; Blanco-Canqui, 2010).

Abrahamson *et al.* (1998) propose that shrub willow SRWC systems have been designed to maintain soil productivity and enhance nutrient cycling conferring advantages relative to agricultural crops. Abrahamson *et al.* (1998) summarized research conducted to refine shrub willow SRWC systems and listed the following operational recommendations and assumptions: harvesting in the winter, after translocation of nutrients to the root system, maintains soil productivity and enhances nutrient cycling; winter harvesting facilitates maximum coppice regeneration; shrub willow biomass crops promote stable nutrient cycling, soil biological activity, and organic matter accumulation; frequent harvesting will result in a removal of nutrients from the site requiring fertilization; fertilization during the growing season will facilitate crop nutrient assimilation, minimizing nutrient losses from runoff and leaching; clones with high nutrient-use efficiency (NUE) will permit planting on sites with marginal nutrient pools and reduce the need for fertilization; and lastly, waste organic material fertilizers improve site soil fertility, efficiency, and economic feasibility.

Shrub Willow (Salix spp.) Silvicultural Inputs and Yield

Shrub willow biomass yields can vary widely depending upon inherent site resource availability, clone selection, planting density, and cutting cycles. The intensity of silvicultural prescriptions such as weed control, fertilization, and irrigation also has a large effect on yield. Establishment of a shrub willow SRWC system requires intensive site preparation. This includes preplanting, application of herbicides, and tillage (Kopp *et al.*, 1997). Clonal stem cuttings are selected based on their site adaptation. Breeding programs hope to increase yields and optimize wood chemistry thereby lowering production and pretreatment costs (Smart *et al.*, 2005). Long-term breeding programs on *S. viminalis* have produced yield increases of 12–67% in Sweden and 8–143% in the United Kingdom (Volk *et al.*, 2006). However, plantings of the Swedish variety of this clone in the United States were severely damaged by the potato leafhopper (*Empoasca fabae*) (Volk *et al.*, 2006). Volk *et al.* (2006) suggest that a long-term breeding program is likely to produce similar yield increases with species adapted for growth in the United States.

A set of initial operational recommendations for shrub willow SRWC systems was provided by academic and government entities in New York, Canada, and Sweden in the early to mid-1990s (Abrahamson *et al.*, 1998). Briefly, operational recommendations were to plant shrub

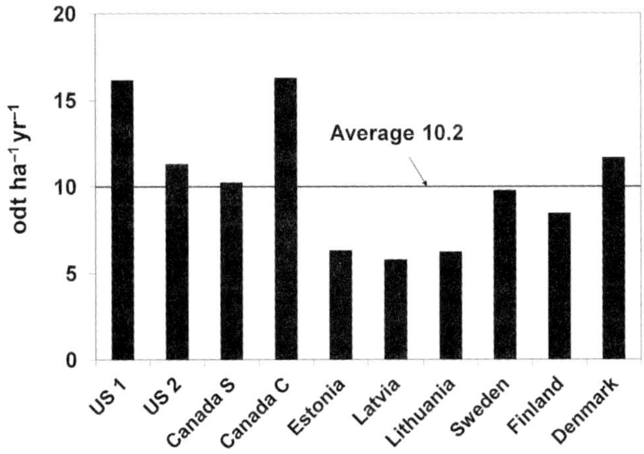

Figure 6.3. Potential annual willow (*Salix spp.*) yield in oven-dry metric tonnes ha^{-1} yr^{-1}. The yield for US 1 represents mean of biennial and triennial harvest cycles and intensive fertilization and irrigation, adapted from Kopp *et al.* (1997). The yield for US 2 represents mean of inorganically and organically fertilized plots, adapted from Adegbidi *et al.*, 2003. Canada yield represents mean of unfertilized and fertilized sandy (S) and clayey (C) sites, adapted from Labrecque and Teodorescu (2003). Yields from European countries, derived from model predictions of yield producing capabilities of the top 25% growers, adapted from Mola-Yudego (2010).

willow clones in a "double row" system of 15 320 tree ha^{-1} (6200 tree ac^{-1}) using agricultural practices, harvest after the first growing season to promote multiple-stem coppice regrowth, and harvest during the winter on a 3–4-year rotation for up to 7 rotations. Crop requirements include moderately productive soils, ≤8% slope, and weed control during the first 2 years. Yields are anticipated to range from 11–18 dry tonnes ha^{-1} yr^{-1}.

Data from studies in the United States (Kopp *et al.*, 1997; Adegbidi *et al.*, 2003), Canada (Labrecque and Teodorescu, 2003), and Europe (Mola-Yudego, 2010) indicate an average yield of 10.2 dry tonnes ha^{-1} yr^{-1} (Figure 6.3) which is below the anticipated yield range of 11–18 dry tonnes ha^{-1} yr^{-1} suggested by Abrahamson *et al.* (1998).

However, it is important to note that this data set represents a large range in inherent site resource availability and silvicultural intensity. In planting densities ranging from 15 152 to 111 111 thousand cuttings ha^{-1}, triennial rather than biennial or annual cutting cycles greatly increased the yield (Kopp *et al.*, 1997). Similar results were found for two studies conducted in Sweden suggesting a triennial cutting cycle to be optimal (Kopp *et al.*, 1997). However, Kopp *et al.* (1997) also suggest that interactions between shrub willow clones and site conditions make it difficult to identify optimal spacing and cutting cycle combinations for all situations. In a study where 10 shrub willow clones were grown with the same planting densities (18 000 cuttings ha^{-1}), the yield ranged from a low of 6.2 dry tonnes ha^{-1} yr^{-1} for a *Salix interior* × *eriocephala* clone to a high of 16.9 dry tonnes ha^{-1} yr^{-1} for a *S. miyabeana* clone (Labrecque and Teodorescu, 2005). Within *S. miyabeana* clones, the yield ranged from 9.4 to 16.9 dry tonnes ha^{-1} yr^{-1}. Results of this study suggest high-yield variability among shrub willow species and clones within a species.

We used a study conducted by Labrecque and Teodorescu (2003) on abandoned farmland in Quebec, Canada, as a case study in the evaluation of sustainability of shrub willow SRWC

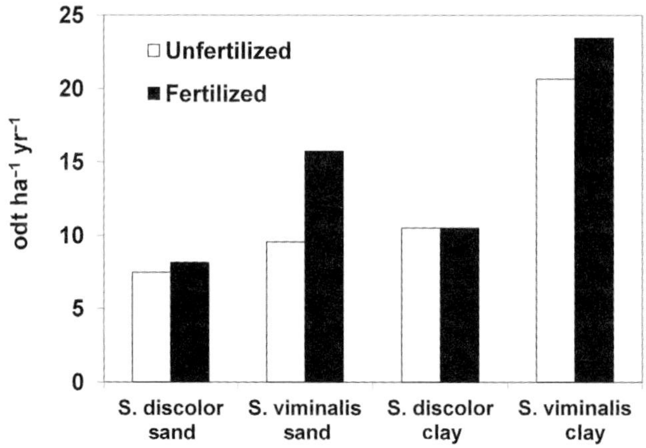

Figure 6.4. Yields in oven-dry metric tonnes ha^{-1} yr^{-1} of two willow species growing on sandy and clayey textured soils in Quebec, Canada. Planting density was approximately 22 000 cuttings ha^{-1}. Fertilization was conducted with a single application of wastewater sludge equating to 273, 100, 214, and 50 kg ha^{-1} of total-N, available-N, total-P, and total-K, respectively. Adapted from Labrecque and Teodorescu (2003).

systems since it simultaneously provided information on yield and fertilizer nutrient capture of two shrub willow species (*S. discolor*, *S. viminalis*) growing on degraded sandy and clayey textured agricultural soils. Planting density was approximately 22 000 cuttings ha^{-1}. After one growing season, the trees were cut early in the winter to produce coppice regeneration. In the following spring, wastewater sludge was applied in a single application. Nutrient applications were 273, 214, and 50 kg ha^{-1} of total-N, total-P, and total-K, respectively. Available-N was anticipated to be approximately 100 kg ha^{-1}. Plants were harvested after the third growing season. Site nutrient removal by the unfertilized crop varied little across species and soil type, with the exception of *S. viminalis* growing on the clayey soil where removal of N, P, and K was two to three times higher. The response to fertilization varied with species and soil texture. *S. discolor* exhibited little to no growth response to fertilization on either soil type (Figure 6.4).

While *S. viminalis* exhibited a strong growth response to fertilization on the sandy soil, this response was lower on the clayey soil. On both soil types, *S. discolor* was much less effective at capturing nutrients added through fertilization than *S. viminalis* (Figure 6.5).

Since this was an application of organic nutrients, the nutrient of most interest here in terms of potential site export from leaching/runoff was available-N. *S. viminalis* captured at least 100% of available-N while *S. discolor* captured from 0% to 34% (Figure 6.5). Specifically, results of this study indicate that *S. viminalis* SRWC systems are likely to be more sustainable in terms of site N export, due to effective capture of N added through fertilization. In general terms, results of this study indicate N leaching/runoff rates at each fertilizer application can be highly variable, ranging from 0 to 100 kg N ha^{-1}, depending on choice of species and soil texture. Furthermore, decisions regarding fertilization rates need to be specific to species and site conditions in terms of nutrient and water availability. For example, shrub willow growth in experimental SRWC plots planted on a relatively fertile soil in New York, a Glossoboric Hapludalf, was not increased by fertilization with wood ash (Park *et al.*, 2005) or paper sludge (Quaye *et al.*, 2011).

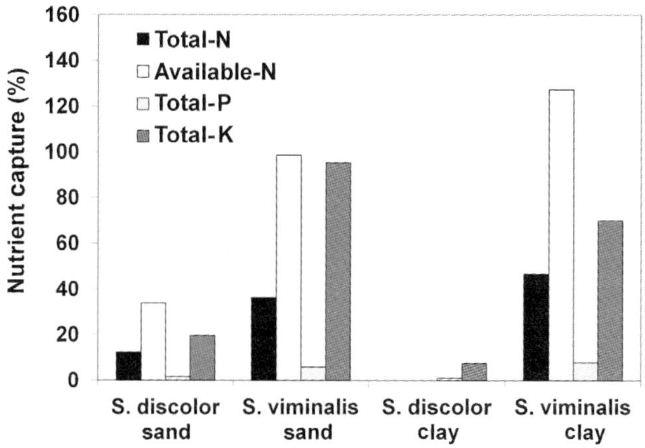

Figure 6.5. Fertilizer nutrient capture (plant nutrient content/fertilizer nutrient content) of two willow species growing on sandy and clayey textured soils in Quebec, Canada. Planting density was approximately 22 000 cuttings ha^{-1}. Fertilization was conducted with a single application of wastewater sludge equating to 273, 100, 214, and 50 kg ha^{-1} of total-N, available-N, total-P, and total-K, respectively. Values greater than 100% assumed to result from soil nutrient supply. Adapted from Labrecque and Teodorescu (2003).

Adegbidi *et al.* (2001) determined fertilization effects on *S. alba*, *S. alba* var. sanguinea, *S.* × *rubra*, *S. purpurea*, and *S. dasyclados* growing on the same New York site in the studies by Park *et al.* (2005) and Quaye *et al.* (2011). Planting density was 107 600 cuttings ha^{-1}. Fertilization was conducted with an approximate annual application of 336, 112, and 224 kg ha^{-1} of N, P, and K, respectively. Fertilizer sources included ammonium nitrate, urea, triple superphosphate, and muriate of potash. During the growing season, irrigation supplied 2–6 cm water week^{-1}. Shrub willow biomass was harvested annually after leaf fall. The highest yields were achieved with unfertilized and fertilized *S. purpurea* and *S. dasyclados* and ranged from 9.6 to 12.5 dry tonnes ha^{-1} yr^{-1}. *S. alba* exhibited the highest growth response to fertilization (+2.8 dry tonnes ha^{-1} yr^{-1}) followed by *S. alba* var. sanguinea (+2 dry tonnes ha^{-1} yr^{-1}), *S. dasyclados* (+1.7 dry tonnes ha^{-1} yr^{-1}), and *S.* × *rubra* (+1.6 dry tonnes ha^{-1} yr^{-1}). *S. purpurea* exhibited essentially no growth response to fertilization (+0.2 dry tonnes ha^{-1} yr^{-1}), possibly indicating no fertilizer uptake. Yields of unfertilized *S. purpurea* and *S. dasyclados* were 1.5× to 1.8× greater, respectively, than *S. alba* var. sanguinea and *S.* × *rubra*. Results indicate highly variable species-dependent yields and growth responses to fertilization. Results of this study further suggest that fertilizer uptake and potential leaching/runoff rates are also highly variable and species dependent.

Nitrate (NO_3^-) leaching was assessed in an experimental setting where *S. viminalis* clones were planted in fiberglass cylinders with a volume of 1.2 m^3 (Aronsson and Bergstrom, 2001). We present data only from cylinders containing a postglacial, alluvial sandy soil. Nitrate leaching data from cylinders containing a postglacial, alluvial clayey soil was reportedly skewed by preferential flowpaths in the cylinders (Aronsson and Bergstrom, 2001). Fertilizer was applied in irrigation water at low and high rates for 3 years. The low fertilization rate was equivalent to 110 kg N ha^{-1} in year 1 and 122 kg N ha^{-1} in years 2 and 3. The high fertilization rate was equivalent to 220 kg N ha^{-1} in year 1 and 244 kg N ha^{-1} in years 2 and 3.

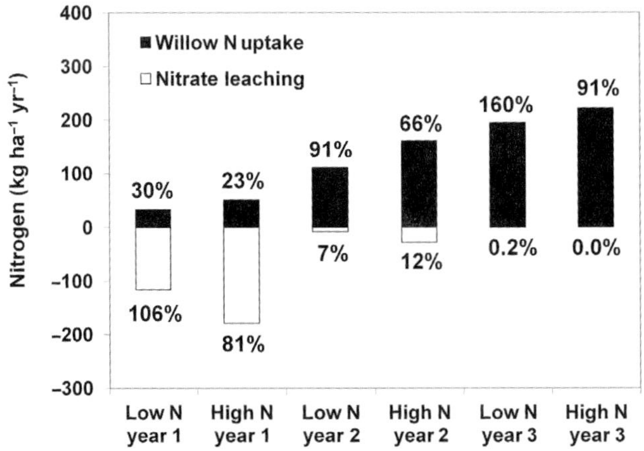

Figure 6.6. Willow (*Salix viminalis*) nitrogen (N) uptake and nitrate (NO_3^-) leaching in 1.2 m^3 cylinders filled with a sandy soil. Soluble fertilizer was applied in 650 mm of irrigation water annually at low (110–122 kg ha^{-1}) and high (220–244 kg ha^{-1}) rates. Numbers above and below columns represent willow N uptake and nitrate leaching, respectively, as a percent of N added through fertilization. Values greater than 100% assume supply from soil N. Adapted from Aronsson and Bergstrom (2001).

Approximately 650 mm of irrigation water was supplied annually. In the first year, NO_3^- leaching was 106% and 81% of N added at the low and high rate, respectively (Figure 6.6).

Also in the first year, shrub willow N uptake was 30% and 23% of N added at the low and high rate, respectively. Nitrate leaching sharply declined in the second year and was near zero by the third year while shrub willow N uptake exhibited the opposite trend. The trend in NO_3^- leaching and shrub willow N uptake mirrored the trend in annual yield which averaged only 1.8 dry tonnes ha^{-1} in the first year and sharply increased to an average of 11.9 dry tonnes ha^{-1} in the second year. Results of this study suggest that fertilization of shrub willow should not be conducted during the first year due to little effect on growth arising from limited fertilizer N uptake, $\leq$30% of added N, and high rates of NO_3^- leaching, $\geq$80% of added N. However, fertilization can be conducted by the second year as demonstrated by the near 700% increase in yield, with substantially greater uptake and less leaching of N.

Adegbidi *et al.* (2001) compared the NUE of five shrub willow species with and without fertilization; the species included *S. alba, S. alba* var. sanguinea, *S.* × *rubra, S. purpurea,* and *S. dasyclados*. Planting density was 107 600 cuttings ha^{-1}. Fertilization was conducted with an approximate annual application of 336, 112, and 224 kg ha^{-1} of N, P, and potassium (K), respectively. Fertilizer sources included ammonium nitrate, urea, triple superphosphate, and muriate of potash. During the growing season, irrigation supplied 2–6 cm water $week^{-1}$. Shrub willow biomass was harvested annually after leaf fall. Fertilization decreased the NUE of N in all species and of P and K in some cases. Nitrogen NUE tended to be the highest in *S. purpurea,* and *S. dasyclados*. Phosphorus NUE was slightly higher in *S. dasyclados* than all other species; however, the NUE of K was clearly the highest.

Adegbidi *et al.* (2001) also assessed the effects of planting density, cutting cycle, and irrigation on the NUE of N, P, and K in another experiment which compared only *S. alba* and *S. dasyclados*. Fertilization was conducted with an approximate annual application of 224, 112, and 224 kg ha^{-1} of N, P, and K, respectively. Fertilizer sources included ammonium nitrate, urea, triple superphosphate, and muriate of potash. During the growing season, irrigation maintained soil moisture close to field capacity. Planting density did not have a large effect on the NUE of N, P, and K; however, NUE tended to be higher at the two lowest planting densities. In addition, the NUE of *S. dasyclados* was higher than *S. alba* at all planting densities. Cutting cycle increased the NUE of N, P, and K of both species in the order of triennial > biennial > annual. Again, the NUE of *S. dasyclados* was higher than *S. alba* at all cutting cycles. Irrigation increased the NUE of N for both species and was the highest in *S. dasyclados*. Irrigation increased the NUE of P noticeably in *S. dasyclados* but not in *S. alba*. Irrigation slightly decreased the NUE of K in both species.

In the study by Adegbidi *et al.* (2001), optimal NUE was achieved when *S. dasyclados* was planted at $\leq$36 960 cuttings ha^{-1}, harvested on a triennial cycle, and irrigated. While the NUE of *S. alba* was suboptimal compared to *S. dasyclados*, the optimal NUE response of this species did however, follow the same trend as *S. dasyclados*. The remaining species were ranked according to their NUE as *S. purpurea* > *S. alba* > *S. alba* var. sanguinea > *S.* × *rubra*. Results of this study indicate that the NUE of N, P, and K can be increased with species choice by an average of 1.4×, 1.5×, and 2.6×, respectively. Within a species, in the case of *S. alba* and *S. dasyclados*, harvesting on a triennial rather than annual cutting cycle can improve the NUE of N, P, and K by an average of 1.7×, 1.7×, and 1.6×, respectively. Also within a species, irrigation can increase the NUE of N and P by an average of 1.4× and 1.1×, respectively.

Poplar (*Populus spp.*)

Poplar offers advantages similar to those of shrub willow in the production of bioenergy from SRWC systems—fast growth, ease of clonal propagation from cuttings, an exploitable genetic diversity, and coppice regeneration (Dickmann, 2006). Initial research in poplar SRWC systems focused on both yield potential and sustainability issues while the majority of later research has focused primarily on genetic improvements, clonal selection, and physiological characteristics as related to yields. Later research has also provided information on the potential to use poplar for phytoremediation. Environmental sustainability research was provided on nitrate leaching by McLaughlin *et al.* (1985) and soil and organic matter loss by Jug *et al.* (1999). Research has also been conducted on potential yields (Hansen, 1991; Fisher *et al.*, 2005; Aylott *et al.*, 2008; Christersson, 2010; Johansson and Karacic, 2011), species/hybrid effects on yields (Herve and Ceulemans, 1996; Laureysens *et al.*, 2003; Bungart and Huttl, 2004; Coyle *et al.*, 2006a), and potential pest problems (Nordman *et al.*, 2005; Coyle *et al.*, 2006b). Poplar commonly hybridized and evaluated for use in SRWC systems includes the North American species *Populus deltoides* Marsh., *Populus balsamifera* L., *Populus trichocarpa* Torr. & A. Gray, and *Populus tremuloides* Michaux, the European species *Populus nigra* L., *Populus tremula* L., and *P opulus alba* L., and the Asian species *P. maximowiczii* A. Henry and *Populus tomentosa* Carriere.

Similar to shrub willow, recommendations have been made to establish poplar SRWC systems on degraded agricultural land (Blanco-Canqui, 2010) with the assumption that soil organic matter can be increased from litter and root inputs. Mixed results have been reported with Hansen (1993) and Jug *et al.* (1999) finding increases in mineral soil C and Sartori *et al.* (2007a) finding no change. Kauter *et al.* (2003) argue that poplars are relatively demanding requiring good-quality sites with at least 350 mm precipitation during the growing season and deep, fertile soils to achieve the desired yields. Dickmann (2006) also suggests that degraded sites are unsuitable for poplar SRWC systems due to lack of return on the intensive silvicultural investment.

The most important drivers of poplar SRWC system productivity have been suggested to be proper species/hybrid selection for site conditions and pest/disease resistance (Hansen, 1991) and weed control at establishment (Hansen, 1991; Kauter *et al.*, 2003). Typical operational procedures may include tillage, fertilization, and irrigation (Sartori *et al.*, 2007a). Rotation lengths for poplar range from 3–5 years in coppice regeneration to 5–15 years in more traditional single-stem plantations (Hansen, 1991). Planting densities range from 1500 cuttings ha^{-1} (Sartori *et al.*, 2007a) in more traditional plantation designs to 10 000 cuttings ha^{-1} (Al Afas *et al.*, 2005) in double-row systems that more closely resemble willow SRWC systems. Thinning is sometimes conducted at higher planting densities to provide a supply of biomass at intermediate points in the rotation.

Poplar (Populus spp.) Silvicultural Inputs and Yield

According to Hansen (1991), early poplar yields differed considerably among experimental small plots and commercial plantations. Experimental small plot yields often exceeded 15 dry tonnes ha^{-1} yr^{-1} while yields in commercial plantations were $\leq$5 dry tonnes ha^{-1} yr^{-1} (Hansen, 1991). This discrepancy in production resulted from less intensive management (genetic improvements, fertilization, and irrigation) in commercial plantations. Poplar yields in experimental small plots also varied by type of planting and intensity of silvicultural inputs averaging 11.4 dry tonnes ha^{-1} yr^{-1} across sites in North America, Europe, and South Korea (Figure 6.7).

The yield from coppice regeneration was higher than that from the first planting of cuttings and suckers. The addition of intensive silviculture techniques (SRWC) further increased the yield of first plantings and coppice regeneration. According to Fisher *et al.* (2005), current commercial production potential varies widely even with intensive silvicultural inputs, ranges

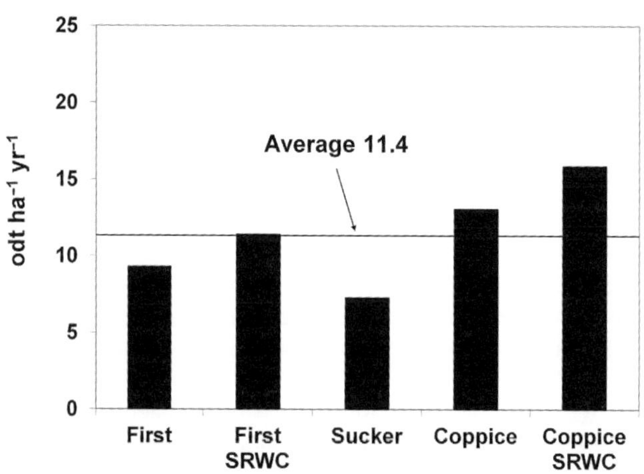

Figure 6.7. Annual poplar (*Populus spp.*) yields in oven-dry metric tonnes ha^{-1} yr^{-1} on experimental small plots under different planting/regeneration systems. First and Sucker are plantings of stem and root sucker cuttings, respectively. Coppice refers to regenerated root stools. SRWC is intensive silviculture, fertilization, and irrigation. Adapted from Hansen (1991).

from 1.2 to 16.1 dry tonnes ha^{-1} yr^{-1} across the site fertility gradient, and is below the yield range reported for experimental small plots by Hansen (1991). Current commercial production potential on agricultural land in Sweden is approximately half that of experimental small plot yields at 7 dry tonnes ha^{-1} yr^{-1} (Christersson, 2010; Johansson and Karacic, 2011). Models have predicted similar potential commercial yields in the United Kingdom (Aylott *et al.*, 2008). Model predictions of poplar yield across a site-quality gradient in Eastern Europe and Asia suggest that potential yields are most strongly influenced by inherent site characteristics, nutrient and water availability as influenced by soil and climate, and intensive silvicultural inputs may result in small rather than large yield increases on poor-quality sites (Fisher *et al.*, 2005).

While coppicing of poplar has been found to result in higher yields (Hansen, 1991), this response is inconsistent, varying among species/hybrid and coppicing year (Herve and Ceulemans, 1996; Laureysens *et al.*, 2003). On a landfill site near Antwerp, Belgium, coppicing of poplar was conducted after 1 year of growth and stool mortality rates ranged from 0% to 57% (Laureysens *et al.*, 2003). This response was highly variable among species/hybrid and specific clones within a species/hybrid. In general, stool mortality was less when coppicing after 2, 3, and 4 years of growth since total stool mortality was only increased to 10–65% (Laureysens *et al.*, 2003). Stool mortality was the lowest in a native *P. nigra* clone and the highest in a *P. trichocarpa* × *deltoides* clone (Laureysens *et al.*, 2003). Mortality also varies greatly in noncoppiced poplar SRWC systems. Coyle *et al.* (2006a) evaluated the yield and survival of poplar clones growing on sandy bottomland and upland sites in South Carolina. Fertilization and irrigation was conducted to provide optimal growing conditions. Clones of *P. deltoides*, *P. deltoides* × *maximowiczii*, *P. nigra* × *maximowiczii*, *P. deltoides* × *nigra*, *P. trichocarpa* × *deltoides*, and *P. trichocarpa* × *nigra* were planted at densities of 5900 and 1600 cuttings ha^{-1} on the bottomland and upland sites, respectively. Survival ranged from 7 to 99% and hybrid clones exhibited the highest survival rates on both sites. Survival was higher on the bottomland sites, attributed to higher water availability. However, survival of some clones was higher on the upland site, due to flooding at the bottomland site. Results of the study by Laureysens *et al.* (2003) suggest that coppicing should take place after at least 2 years of growth to minimize mortality. Results of the studies by Laureysens *et al.* (2003) and Coyle *et al.* (2006a) suggest that mortality is also highly dependent upon species/hybrid selection and their specific site adaptations.

Bungart and Huttl (2004) conducted a study on a lignite coal mining substrate in the Lusatian region in Germany, evaluating the efficacy of establishing a poplar SRWC system on a surface coal mine spoil. The substrate was composed of clayey and sandy overburden that was free of lignite and pyrite. The substrate was tilled to a depth of 25 cm to homogenize and promote establishment. Eight poplar species/hybrids were planted with a density of 8333 cuttings ha^{-1} and fertilized with 338, 150, and 275 kg ha^{-1} of N, P, and K, respectively, over three applications. They found relatively low yields on this poor-quality site. Yields varied across the eight poplar species/hybrids averaging 4.7 dry tonnes ha^{-1} yr^{-1} (Figure 6.8).

However, *P. trichocarpa* × *deltoides* clones did achieve yields of 7 dry tonnes ha^{-1} yr^{-1} suggesting species/hybrid selection can have a positive effect on yields on poor-quality sites.

Heilman and Norby (1998) evaluated the NUE of various hybrid poplars (*P.* hybrids), *P. trichocarpa*, and *P. deltoides* across typical yield potential ranges. NUE of *P. deltoides* and *P. trichocarpa* did not vary among low- and high-yield potentials. In contrast, in the high-yield scenario, *P.* hybrids exhibited enhanced NUE for N and Ca. In general, NUE was the highest in *P. trichocarpa*.

The effects of fertilization and herbaceous ground cover on yield and nitrate (NO_3^-) leaching were assessed in a poplar SRWC system by McLaughlin *et al.* (1985). Cuttings of *P. deltoides* ×

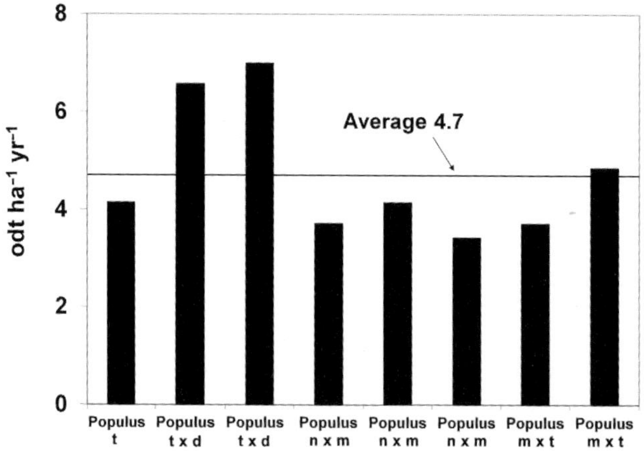

Figure 6.8. Annual poplar (*Populus spp.*) yields in oven-dry metric tonnes ha^{-1} yr^{-1} on a clayey-sandy lignite coal mining substrate. Planting density was 8333 cuttings ha^{-1}. Trees were fertilized with a total of 338, 150, and 275 kg ha^{-1} of N, P, and K, respectively, over three applications. Species/hybrids were *Populus trichocarpa* (t), *P. trichocarpa* × *deltoides* (t × d), *P. nigra* × *maximowiczii* (n × m), and *P. maximowiczii* × *trichocarpa* (m × t). Adapted from Bungartt and Huttl (2004).

trichocarpa were planted at a density of 10 000 cuttings ha^{-1}. The soil was a mixed, frigid, coarse loamy Alfic Haplorthod that was previously cropped with potatoes for 50 consecutive years. Cuttings were planted in three treatments, (1) bare soil, (2) birdsfoot trefoil (*Lotus corniculatus* L.) ground cover, and (3) native weed ground cover, in order to assess the effects of fertilization and cover type on growth and nitrate leaching. Bare soil was maintained with herbicides. Nitrogen fertilizer was applied as ammonium nitrate (NH_4-NO_3) at an annual rate of 56 kg ha^{-1} in both June and July. Nitrate leaching was measured during the growing season with porous cup lysimeters at 1.2 m in depth. In the first two growing seasons, nitrate concentration was the highest in the unfertilized and fertilized bare soil treatments (Figure 6.9).

By the third growing season, nitrate leaching was similar among treatments. Height was similar among treatments in the first growing season. In the second growing season, fertilization began to increase height most noticeably in the birdsfoot trefoil ground cover treatment. By the third growing season, height was highest in the fertilized bare soil treatment followed by the fertilized birdsfoot trefoil and native weed treatments. Results indicated that fertilization of poplar cuttings was not necessary until the third growing season since in the first 2 years, growth was not dramatically increased while nitrate leaching was increased. Furthermore, ground cover was effective at reducing nitrate leaching without a large reduction in poplar growth. This indicates that establishment of ground cover in poplar SRWC systems can maximize fertilizer nutrient capture and reduce nitrate leaching. In addition, ground cover will likely significantly reduce erosion relative to bare soil, further contributing to sustainability of these systems.

Pine (*Pinus spp.*)

Pine is a viable option for SRWC systems due to the extent of the already well-established plantation land base, 50 years of research and development (R&D), and demonstrated higher

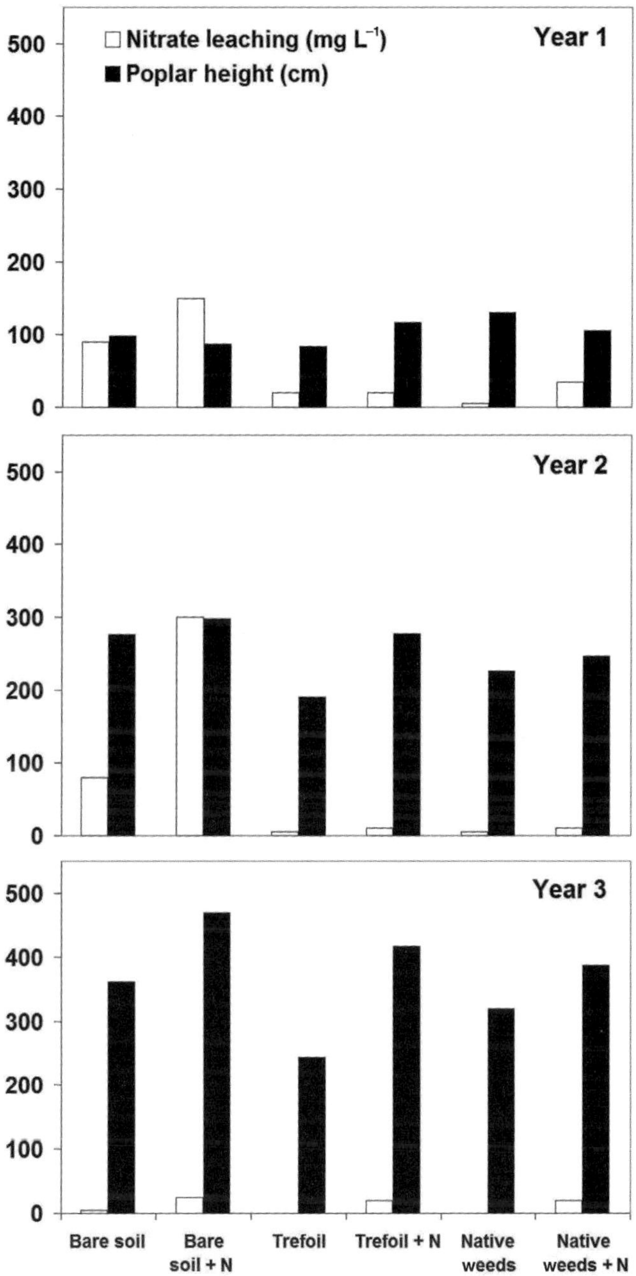

Figure 6.9. Poplar (*Populus deltoides* × *trichocarpa*) height (cm) and nitrate (mg L^{-1}) leaching in the first three growing seasons. Nitrate leaching was reported as the highest concentration detected with porous cup lysimeters at 1.2 m in depth during each growing season. Planting density was 10 000 cuttings ha^{-1}. Treatments were unfertilized and fertilized bare soil, birdsfoot trefoil ground cover, and native weed ground cover. Ammonium nitrate (NH_4-NO_3) was applied at an annual rate of 56 kg ha^{-1} in both June and July. Adapted from McLaughlin *et al.* (1985).

productivity than hardwoods on many sites (Kline and Coleman, 2010). Pine SRWC systems differ from those of shrub willow and poplar due to a lack of coppice regeneration. Instead, seedlings must be planted to establish each single-stem rotation resulting in higher cost. In particular, loblolly pine (*Pinus taeda* L.) shows promise for SRWC systems due to outstanding growth rates and already highly advanced genetic improvements (Dickmann, 2006). Therefore, we focused primarily on the potential for loblolly pine SRWC systems. However, it is important to note that radiata pine (*P. radiata* D. Don) also shows promise for use in SRWC systems (Acuna *et al.*, 2010).

Loblolly pine is the most widely planted tree in the United States (Dickmann, 2006). This species is, however, confined to warm temperate climates such as that found in the southern United States. Currently, there are 13 Mha of pine plantations in the southern United States comprised primarily loblolly pine (Munsell and Fox, 2010) with the remainder comprised slash pine (*P. elliottii* Englemn.). This encompasses the area of land needed to meet the SRWC portion of the DOE's Billion-Ton goal to annually produce 126 million dry tonnes of biomass assuming a production rate of 10 dry tonnes ha^{-1} yr^{-1}. However, this assumes that all loblolly pine production would be used for bioenergy which is not a viable option, considering the importance of loblolly pine plantation production to pulp and lumber products industries. However, utilizing some portion of the already well-established loblolly pine plantation infrastructure for SRWC biomass bioenergy systems would eliminate the need to convert a proportional quantity of agricultural or natural forest land area thereby easing the burden of establishing these systems. In addition, due to the decades of research dedicated to refining loblolly pine plantation productivity and sustainability, these plantations arguably have the highest short-term potential for successful widespread biomass bioenergy production with SRWC systems.

Extensive research has been conducted to increase loblolly pine yields and productivity has increased fourfold since the 1950s (Munsell and Fox, 2010) from a combination of genetic improvements and intensive silvicultural techniques such as site preparation, fertilization, and control of competing vegetation (Allen, 1987; Fox *et al.*, 2007a, 2007b). Research has been conducted to improve nutrient management (Birk and Vitousek, 1986; Allen, 1987; Albaugh *et al.*, 2004, 2008). Research has also been conducted on the potential for these intensive silvicultural techniques to degrade the soil resource (Vitousek *et al.*, 1992; Gresham, 2002; Johnston and Crossley, 2002; Field *et al.*, 2003; Grace, 2004) and the result in site nutrient export (Field *et al.*, 2003; Lee and Jose, 2005).

In the southern United States, loblolly pine plantations have been extensively planted on degraded former agricultural soils that were abandoned between the late 1800s and mid-1900s (Fox *et al.*, 2007b). Loblolly pine plantations are also often located on relatively infertile sandy soils typical to the southern US Coastal Plain region (Albaugh *et al.*, 2009). Therefore, unlike shrub willow and poplar there is little argument whether loblolly pine SRWC systems should be established on poor- or high-quality sites since pines grow relatively well on poor-quality sites and silvicultural advancements have aided in generally overcoming inherent site fertility limitations. Establishment of loblolly pine plantations on degraded former agricultural soils has not been found to increase mineral soil C (Johnson *et al.*, 2003; Sartori *et al.*, 2007b). However, increases in mineral soil N and P have been observed following fertilization suggesting the potential for enhancement of soil fertility (Pritchett and Comerford, 1982; Gresham, 2002; Will *et al.*, 2006, Everett and Palm-Leis, 2009). While there is concern that continuous short rotations and intensive silvicultural techniques will degrade the soil over time in loblolly pine plantations, only windrowing is widely thought to have lasting negative impacts (Johnston and Crossley, 2002).

Typical loblolly pine plantation operational procedures begin with site preparation prior to planting which may include chopping and burning of the existing vegetation and tillage or disking of the soil. This is conducted to provide for ease of planting and to control competing vegetation. Competing vegetation is also often controlled with herbicides until canopy closure. Site preparation was initially developed, mostly by trial and error rather than formal R&D, to recreate old-field conditions occurring on abandoned agricultural land where growth and seedling survival was observed to be the highest (Fox *et al.*, 2007a). Planting densities can range from approximately 400 to 2000 trees ha^{-1} with mid-rotation thinning conducted in the higher densities. Fertilization at planting is primarily conducted only on P-deficient sites (Fox *et al.*, 2007a). More often, operational fertilization is conducted with a single mid-rotation application of 225 kg N ha^{-1} and 28 kg P ha^{-1} (Fox *et al.*, 2007a). Rotation lengths vary typically ranging from 15 to 20 years for pulpwood and >25 years for sawtimber.

Pine (Pinus spp.) Silvicultural Inputs and Yield

Current loblolly pine plantation yields throughout the southern United States average approximately 6 dry tonnes ha^{-1} yr^{-1} (Fox *et al.*, 2007a). The use of intensive silviculture in these plantations has the potential to more than double yields to 13 dry tonnes ha^{-1} yr^{-1} (Fox *et al.*, 2007a). Control of competing vegetation (weed control), fertilization, and their combination are the intensive silvicultural techniques typically used to achieve higher loblolly pine yields (Jokela, 2004). Jokela (2004) reported loblolly pine growing on a Spodosol in central Florida that yielded 4 dry tonnes ha^{-1} yr^{-1} without weed control or fertilization. Rotation-length intensive weed control and fertilization increased the yield to 9.3 and 10.7 dry tonnes ha^{-1} yr^{-1}, respectively. The combination of both weed control and fertilization further increased the yield to 11.7 dry tonnes ha^{-1} yr^{-1}. Maximizing loblolly pine yields is conducted by maintaining optimal foliar nutrition which increases leaf area to the point where light capture and photosynthesis are maximized (Albaugh *et al.*, 1998). Optimal loblolly pine foliar concentrations for the three primary plant nutrients are 1.2% N, 0.12% P, and 0.30% K (Jokela, 2004).

Jokela *et al.* (2004) reported the effects of intensive silvicultural techniques on loblolly pine yield from seven long-term studies in the southern United States. This study provided a valuable synthesis of information on loblolly pine yield potentials across a diverse array of soil–climate combinations as influenced by silvicultural treatments. Sites were located in North Carolina (NC), Georgia (GA), Florida (FL), Louisiana (LA), and Oklahoma (OK). The NC site was located on a somewhat excessively drained Psammentic Hapludult; fertilization incrementally applied a total of 902, 157, and 337 kg ha^{-1} of N, P, and K, respectively, and irrigation maintained soil water content >40% of field capacity during the growing season. The FL site was located on a somewhat poorly drained Ultic Alaquod and fertilization incrementally applied a total of 360, 143, and 317 kg ha^{-1} of N, P, and K, respectively. In GA, a wet and a dry site were evaluated. The wet GA site was located on a somewhat poorly drained Arenic Paleaquult. The dry GA site was located on a well-drained Grossarenic Plinthic Paleudult. At both sites, fertilization incrementally applied a total of 1255, 196, and 164 kg ha^{-1} of N, P, and K, respectively. The LA site was located on a moderately well drained Plinthaquic Paleudult and fertilization incrementally applied a total of 335, 200, and 50 kg ha^{-1} of N, P, and K, respectively. The OK site was located on a somewhat poorly drained Aquic Paleudult and no fertilization was conducted. Fertilization rates at the FL and LA sites were closer to operational rates compared to the very high N additions at the NC and GA sites. Fertilization was combined with control of competing vegetation (weed control) which was generally conducting with mechanical site preparation prior to planting and with herbicides following

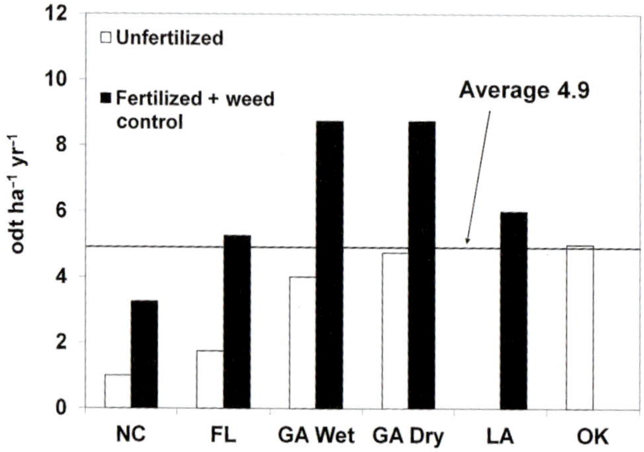

Figure 6.10. Annual loblolly pine (*Pinus taeda* L.) yields in oven-dry metric tonnes ha^{-1} yr^{-1} in unfertilized and fertilized + weed control treatments across an array of sites in the southern US representative of the region's range in soil–climate conditions. Annual yields estimated from standing biomass after 15 years of growth. Adapted from Jokela *et al.* (2004).

planting. Without fertilization, the yield ranged from a low of 1 dry tonnes ha^{-1} yr^{-1} at the NC site to a high of 5 dry tonnes ha^{-1} yr^{-1} at the OK site (Figure 6.10).

The combination of fertilization and weed control at the least doubled yield at the NC (3.3 dry tonnes ha^{-1} yr^{-1}), FL (5.3 dry tonnes ha^{-1} yr^{-1}), and GA (8.8 dry tonnes ha^{-1} yr^{-1}) sites. Within treatments, the yield did not differ among the wet and dry GA sites. Overall, results suggest dramatic variation in loblolly pine yield across sites. In addition, regardless of site-specific yields, fertilization and weed control tended to at least double yields. Lastly, high inherent site fertility can produce yields equivalent to or exceeding yields achieved with fertilization and weed control such as observed at the OK site.

The effects of planting density and genetics on loblolly pine yield were examined by Land *et al.* (2004) in a study conducted in east-central Mississippi (MS). Eight open-pollinated loblolly pine families from NC were planted at densities of 1076, 1682, and 4306 trees ha^{-1}. The soil was a Glossic Fragiudult. Average annual yield at stand age 0–5 years ranged from 0.8 to 3.2 dry tonnes ha^{-1} yr^{-1} from the lowest to the highest planting density (Figure 6.11).

Average annual yield at stand age 0–9 years ranged from 5.8 to 10.7 dry tonnes ha^{-1} yr^{-1} from the lowest to the highest planting density. Average annual yield reached a maximum at stand age 0–13 years and ranged from 9.8 to 13.1 dry tonnes ha^{-1} yr^{-1} from the lowest to the highest planting density. Average annual yield at stand age 0–17 in the 1682 and 4306 trees ha^{-1} planting densities leveled off and decreased, respectively, below that at stand age 0–13 years indicating annual growth increment reached a maximum at stand age 13. However, in the lowest planting density of 1076 trees ha^{-1}, growth increment continued to increase. Average annual yield at stand age 0–17 ranged from 10.4 to 11.3 dry tonnes ha^{-1} yr^{-1} from the lowest to the highest planting density. In the fast growth families, average annual yield was higher than in the slow growth families and did not appear to be affected by crown size (Figure 6.12).

Also in the fast growth families, average annual yield reached a maximum at stand age 0–13 years and then began to decrease. In the slow growth families, the small crown type

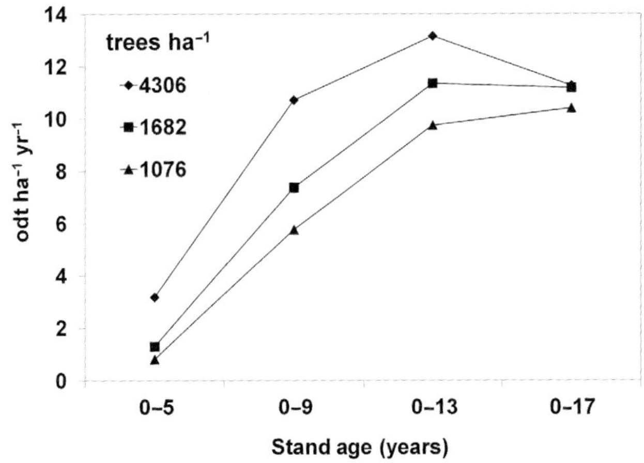

Figure 6.11. Average annual loblolly pine (*Pinus taeda* L.) yields in oven-dry metric tonnes ha^{-1} yr^{-1} at three planting densities and four age classes. Values represent eight open-pollinated NC families grown in east-central MS. Adapted from Land *et al.* (2004).

exhibited lower average annual yield and this yield declined after stand age 13. In contrast, the large crown type in the slow growth family exhibited higher average annual yield and this yield leveled off rather than decreased after stand age 12. Results indicated that genetics can have a significant effect on yields. Results of this study also indicate that annual yield is very low from stand age 0–5 years. After the trees become well established and the canopy closes after stand age 5, annual yield greatly accelerates and reaches a maximum at approximately stand age 13. As one would anticipate, the highest planting density of 4306 trees ha^{-1} produced the

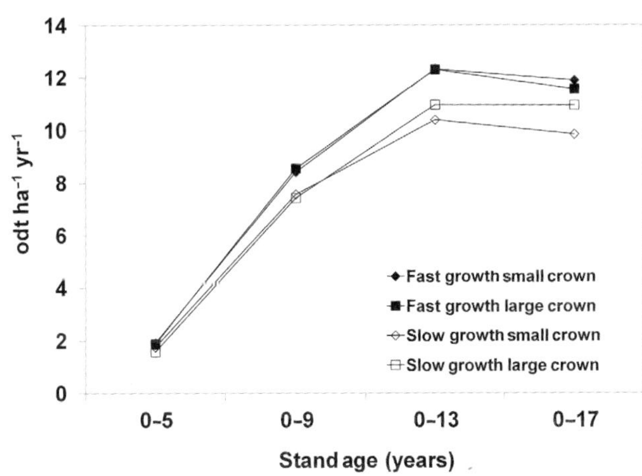

Figure 6.12. Annual loblolly pine (*Pinus taeda* L.) yields in oven-dry metric tonnes ha^{-1} yr^{-1} for eight open-pollinated NC families expressing different growth rate/crown size combinations grown in east-central MS. Each family category represents the mean of two families. Values are averaged across three planting densities (1076, 1682, and 4306 trees ha^{-1}). Adapted from Land *et al.* (2004).

highest total yield in the shortest amount of time. The total yield in the highest planting density at age 9 was nearly double that of the two lower planting densities. However, due to growth dynamics represented in Figure 6.12, harvesting at stand age 17 results in nearly equivalent yields among all planting densities.

Li *et al.* (1991) conducted a study that examined the N NUE of 23 open-pollinated loblolly pine families from the NC and Virginia (VA) Coastal Plain. An N fertilizer solution was applied to 1-month-old seedlings grown in a greenhouse at low (5 ppm) and high (50 ppm) rates for 20 weeks. Nitrogen NUE at the low N fertilizer rate was higher and exhibited greater genetic variability ranging from 120 to 210 kg biomass kg N^{-1} compared with the 20–30 kg biomass kg N^{-1} range of the high fertilizer N rate. This large genetic variation and high heritability in N NUE traits suggested a strong potential for genetic improvement in NUE (Li *et al.*, 1991). Results also suggested that some families were more highly adapted to growing on N poor sites indicating proper family selection according to site conditions to be a key consideration. Typical NUEs for mature loblolly pine stands (17–21 years) range from 220 to 370 kg biomass kg N^{-1} and 4300–6000 kg biomass kg P^{-1} (Dicus and Dean, 2001; Albaugh *et al.*, 2004, 2008).

Potential nitrate (NO_3^-) leaching was assessed in an annually fertilized 7-year-old loblolly pine plantation growing on a well-drained, sandy loam Rhodic Paleudult in Northwest FL (Lee and Jose, 2005). Trees were planted at a density of 1400 trees ha^{-1}. Soluble N fertilizer was applied with a drip irrigation system at rates of 0, 56, 112, and 224 kg N ha^{-1} yr^{-1}. Nitrate leaching was measured with tension lysimeters placed at a maximum depth of 120 cm. According to Lee and Jose (2005), tree growth was increased by the addition of 56 kg N ha^{-1} yr^{-1} (Figure 6.13).

However, additional amounts of N fertilizer did not increase tree growth above the 56 kg N ha^{-1} yr^{-1} rate while greatly increasing nitrate leaching to 145 and 260 kg NO_3^- ha^{-1} yr^{-1} in the 112 and 224 kg N ha^{-1} yr^{-1} fertilization rates, respectively.

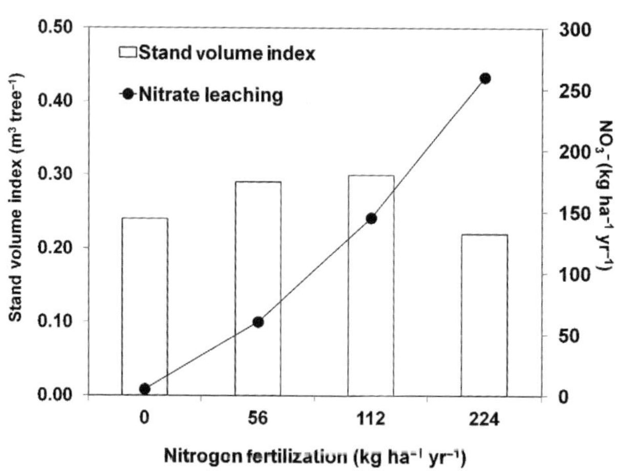

Figure 6.13. Nitrate (NO_3^-) leaching and stand volume index (m^3 $tree^{-1}$) along a gradient of N fertilization in 7-year-old loblolly pine planted at a density of 1400 trees ha^{-1} in Northwest FL. Soluble N fertilizer was applied with a drip irrigation system. Reported NO_3^- leaching (kg ha^{-1} yr^{-1}) estimated from NO_3^- concentrations detected at 120 cm in depth with tension lysimeters. Adapted from Lee and Jose (2005).

Results of this study indicated that care must be taken when choosing fertilization rates since nitrate leaching can be increased without benefits to tree growth.

Environmental Sustainability Issues

The intensity of silvicultural practices can have a positive effect by increasing yields. Yield increases are required to make SRWC biomass bioenergy cost competitive with fossil fuel energy sources and to also reduce the amount of land required to produce the necessary tonnage. However, negative environmental effects of SRWC systems also arise from the intensity of silvicultural practices, resulting in a trade-off between yield increases and environmental impacts. We identified factors with the highest potential to negatively impact the environmental sustainability of SRWC systems as tillage and erosion, herbicides, nutrient management, pests and pathogens, and soil compaction.

Tillage and Erosion

One of the proclaimed advantages of SRWC systems is the establishment on marginal lands leading to improvements in soil quality over time (Blanco-Canqui, 2010). This is suggested to occur due to high inputs from litter and root production (Abrahamson *et al.*, 1998). Mann and Tolbert (2000) argue that improvements to marginal lands are inconsistent and largely based on short-term results. The use of waste organic fertilizers, lime-stabilized sewage sludge, and composted poultry manure, has been found to increase soil organic matter and nitrogen (N) content while increasing yields similarly to urea (Adegbidi *et al.*, 2003). However, soil erosion has also occurred due to tillage required at establishment (Abrahamson *et al.*, 1998). When erosion occurs, the assertion that these systems increase or even maintain soil quality is contradicted due to loss of soil from the site negating accumulations of organic matter. Erosion represents possibly the greatest threat to long-term soil productivity due to loss of organic matter and the nutrient-rich soil surface (Lal, 1998). However, since tillage is only required at establishment erosion is likely limited to the establishment year. Therefore, erosion should be less in coppice regenerated systems compared to annually tilled systems. Erosion rates on tilled land increase dramatically as slope increases with typical values of 10 Mg ha^{-1} yr^{-1} (0–2% slope), 35 Mg ha^{-1} yr^{-1} (2–6% slope), and >100 Mg ha^{-1} yr^{-1} (6–12% slope) (Pimentel and Krummel, 1987). Therefore, location of SRWC systems is a key consideration. As a result of erosion during the establishment phase of shrub willow SRWC systems (every 20–30 years), research was conducted on mitigation strategies and found that the use of cover crops and changes in the timing of tillage-reduced erosion (Volk *et al.*, 2006). The most effective strategy was to establish a rye cover crop after plowing and disking in the fall, kill it with post-emergence herbicide in the spring, and use pre-emergence herbicides to control new weed growth (Volk *et al.*, 2006). Best management practices (BMPs) are well established and have been shown to be effective at minimizing erosion and sedimentation of surface waters. Due to tillage in SRWC systems resembling agriculture, effective agricultural BMPs such as grassed waterways, riparian buffers, conservation tillage, contour cropping, and cover cropping should be implemented (Logan, 1993).

Herbicides

Establishment of bioenergy SRWC systems requires intensive site preparation including application of herbicides (Haywood, 1988; Hansen, 1991; Kopp *et al.*, 1997; Kauter *et al.*, 2003). In the absence of mechanical control, herbicides are required to control weeds that can suppress plantings to the extent that replanting is required or biomass production is greatly decreased. Since SRWC systems can mimic large-scale agriculture, there are potential concerns with herbicides and their decomposition products entering groundwater similar to findings in large-scale agricultural landscapes (Barbash *et al.*, 2001; Spalding *et al.*, 2003; Sanchis *et al.*, 2012). Exposure to these chemicals can have carcinogenic and endocrine system disrupting effects (Ackerman, 2007). This potentially raises concerns for SRWC systems as the frequency of herbicides in groundwater increases proportionally with herbicide use (Barbash *et al.*, 2001). The judicious use of these chemicals by following label restrictions and using BMPs has been recommended to minimize water contamination (Abrahamson *et al.*, 1998). Baker and Mickelson (1994) made a number of recommendations to minimize water contamination with herbicides including reducing the application rate with banding or soil incorporation, using riparian buffers to capture surface runoff, timing the application relative to expected intense storms, and avoiding their use if weed infestation is below the economic effect level.

Nutrient Management

The application of fertilizers, from both inorganic and organic sources, is necessary in some cases to achieve the desired yields in bioenergy SRWC systems. Fertilization may also be necessary to replenish nutrients removed from the soil in the harvested biomass. Intensive harvesting of all aboveground biomass on short rotations can potentially result in a depletion rate of available soil nutrients that exceeds the rate of resupply from weathering of soil minerals, organic matter mineralization, and atmospheric deposition. If this occurs, it could impact the sustainability of SRWC systems since inherent site quality would decrease over time. This decrease in inherent site quality would require a proportional increase in fertilizer applications to maintain the desired yield levels, increasing costs and the potential for negative environmental impacts. Genetic variation in nutrient accumulation and site removal at harvest has been found (Adegbidi *et al.*, 2001) suggesting proper clonal selection to be one mitigation strategy. Abrahamson *et al.* (1998) suggest that harvesting in the winter, after leaf fall and translocation of nutrients from the stem to the root system, can minimize site nutrient removals. Since nutrients in the stem are concentrated in the bark and phloem rather than in the stem wood, on-site debarking and dispersal of bark across the site has been found effective at reducing site nutrient removals in intensively managed *Eucalyptus spp.* plantations (Goncalves *et al.*, 2004). This technique could also be effective at minimizing site nutrient removals in single-stem poplar and pine SRWC systems. Assessing the effects of SRWC systems on the soil resource over time would also prove valuable. Monitoring of soil productivity could be conducted with the sustainability monitoring process developed by Burger and Kelting (1999). To briefly summarize their process, soil productivity baseline conditions are established. Then, soil factors that affect biomass productivity such as nutrient availability, porosity, and water holding capacity are monitored over time to assess changes that arise from management and harvesting activities. This allows inferences to be made on the long-term effects of specific SRWC system operations on soil productivity and biomass yield across sites with similar soils. Then, site- and soil-specific management and harvesting activities can be developed that minimize the long-term negative impacts on productivity.

Fertilization has the potential to negatively impact the sustainability of SRWC systems when added nutrients are not captured by the crop but instead leach to groundwater or run off to surface waters. Intensive bioenergy SRWC systems can resemble intensive agriculture where studies have found only 30–50% of applied N fertilizer and 45% of phosphorus (P) fertilizer was taken up by crops (Tilman *et al.*, 2002). In the study by Labrecque and Teodorescu

(2003), uptake of available-N from wastewater sludge by shrub willow was estimated to range from 0% to 100%. When plants do not utilize nutrients supplied with fertilization, leaching and runoff of N and P can negatively impact sustainability through eutrophication of surface waters and the associated negative ecological impacts (Tilman *et al.*, 2002). The potential for fertilization of bioenergy SRWC systems to negatively impact sustainability through this mechanism creates the need for fertilization regimes and BMPs that minimize nutrient losses. Fertilization at rates based on nutrient demand of the crop can increase yields while minimizing negative environmental impacts. Research has shown that waiting until the trees are established, typically $\geq$3 years after planting of shrub willow and poplar cuttings, minimizes fertilizer N loss from nitrate leaching (McLaughlin *et al.*, 1985; Aronsson and Bergstrom, 2001). Nitrate leaching in loblolly pine is also minimized by carefully matching fertilizer rates to tree demand (Lee and Jose, 2005). In addition, herbaceous cover crops have been found effective at capturing excess fertilizer N while not drastically reducing tree growth (McLaughlin *et al.*, 1985). Nutrient losses can be also minimized with species/clone selection and silvicultural techniques with the highest NUE (Li *et al.*, 1991; Heilman and Norby, 1998; Singh, 1998; Adegbidi *et al.*, 2001) thereby decreasing the amount of applied nutrients necessary to achieve the desired yields.

Pests and Pathogens

Damage of SRWC systems by pests and pathogens has been found to be problematic in specific instances indicating a potential for negative impacts on sustainability. Plantings of Swedish shrub willow varieties of *S. viminalis* in the United States have incurred severe damage from potato leafhoppers (*Empoasca fabae* Uhler) (Volk *et al.*, 2006). In a study conducted in New York, *S. alba* incurred pest damage from rabbits and potato leafhoppers to the extent that it was effectively eliminated from the experiment (Kopp *et al.*, 1997). The highest planting density of 111 111 cuttings ha^{-1} led to the highest damage from rabbits and was attributed to smaller stem diameters in the highest planting density (Kopp *et al.*, 1997). Also in the study by Kopp *et al.* (1997), mortality of shrub willow stools from stem canker fungus (*Cryptodiaporthe salicella* Hoehn.) was increased in the highest planting density. Astrom and Ramstedt (1994) found this fungus to readily infect shrub willow and identified it as the most important canker pathogen in Swedish biomass shrub willow SRWC systems. Leaf rust diseases (*Melampsora spp.*) can also be problematic for which planting of resistant shrub willow varieties is the most practical control (Royle and Ostry, 1995). Recommendations to minimize damage from pests and pathogens include lowering planting density and eliminating cover in the case of rabbits (Kopp *et al.*, 1997) and to avoid water and nutrient stress in the case of fungal infection with *Cryptodiaporthe salicella* (Astrom and Ramstedt, 1994). Strategies for minimizing insect damage include planting more resistant *S. purpurea*, *S. eriocephala*, and *S. eriocephala* × *interior* varieties (Nordman *et al.*, 2005) and avoiding dense, contiguous monoculture plantings (Dickmann, 2006; Dalin *et al.*, 2009).

Poplar is generally more resistant to defoliators than willow (Nordman *et al.*, 2005) but is also susceptible to pest damage to the extent that yields and sustainability can be negatively affected. In the United States, major poplar pests include the cottonwood leaf beetle (*Chrysomela scripta* F.), the cottonwood leaf curl mite (*Tetra lobulifera* (Keifer)), and leaf rust (*Melampsora medusae* Thuem.) (Coyle *et al.*, 2006b). Susceptibility to these pests varies widely among poplar clones and proper clonal selection was reported to be an effective control method (Coyle *et al.*, 2006b). Nordman *et al.* (2005) also suggest genetics to be a key resistance determining factor.

Loblolly pine plantations can be damaged by fusiform rust (*Cronartium fusiforme* Hedg. & Hunt ex Cumm.) and bark beetles (*Dendroctonus spp.*). Loblolly pine susceptibility to fusiform rust was found to vary considerably among 55 controlled pollinated and 48 open-pollinated families (Kinloch and Stonecypher, 1969) suggesting control by proper genetic selection. Bark beetles can decimate loblolly pine stands and outbreaks can be prevented or minimized by thinning (Allen *et al.*, 2005). Matson *et al.* (1987) suggest that thinning reduces loblolly pine susceptibility to bark beetles by increasing resin flow rates, starch concentrations in phloem, and bark and phloem thickness.

Soil Compaction

SRWC systems use intensive management practices requiring the use of machinery with the potential to compact the soil. Compaction increases soil bulk density and reduces porosity which in turn, reduces the capacity of the soil to hold water and for root growth. Compaction may also reduce water infiltration to the extent where ponding occurs and erosion rates are accelerated. Compaction can be alleviated with subsoil ripping (Mann and Tolbert, 2000). However, since this is both energy intensive and costly (Mann and Tolbert, 2000), avoiding or minimizing compaction is the best alternative. Based on the number of machinery passes over time, potential compaction in SRWC systems should be lower in magnitude than in annually harvested agricultural systems, but higher in magnitude than in traditional forestry systems with harvest cycles $\geq$20 years.

The effect of soil compaction by harvesting machinery on shrub willow (*S. viminalis*) yield was conducted in the United Kingdom (Souch *et al.*, 2004). Compaction was restricted to U-shaped zones extending 0.4 m deep and 0.3 m beyond each rut created by the loader tires. In these zones, root length density was 14% lower. However, in the noncompacted surface soil around these zones, root length density was higher than in the control treatment suggesting compensation for restricted rooting in the compacted zones. In spite of this compensatory mechanism, the yield was reduced by compaction, decreasing from 34.4 (control) to 32.2 (moderate compaction) to 30.4 (heavy compaction) dry tonnes ha^{-1} in clay loam and from 13.7 (control) to 13.6 (moderate compaction) to 12.1 (heavy compaction) dry tonnes ha^{-1} in sandy loam. While compaction was restricted to areas surrounding wheel ruts, over time an ever increasing area may become compacted leaving less space for compensatory root growth. Alleviating compaction in rutted areas may require costly tillage techniques such as subsoil ripping.

Bioenergy Potential, Production, and Economics

Sustainability of SRWC systems is affected by three primary factors—environmental, economic, and scalar (Figure 6.2). The current yield potential of SRWC systems with typical operational costs and procedures across the inherent site fertility gradient appears to range from 5 to 10 dry tonnes ha^{-1} yr^{-1}. Selection of the best genetics and intensive silvicultural practices such as optimal fertilization, irrigation, and weed control have shown the potential to double or triple yields of all species (Hansen, 1991; Kopp *et al.*, 1997; Labrecque and Teodorescu, 2003; Bungart and Huttl, 2004; Jokela, 2004; Fox *et al.*, 2007b). Yield increases can proportionally reduce the area of land required to meet production goals. However, intensive silvicultural practices may raise production costs to the level where SRWC systems may

Table 6.1. Hypothetical production scenarios for shrub willow, poplar, and loblolly pine SRWC systems.

Production scenario	Site fertility	Silvicultural input and relative cost	Relative environmental impact	Yield (dry tonnes ha^{-1} yr^{-1})		
				Shrub willow	Poplar	Loblolly pine
1	Low	Low	Low	5	5	5
2	Low	High	High	10	10	10
3	High	Low	Low	15	10	10
4	High	High	High	20	15	15

become economically unviable. In addition, these intensive silvicultural practices tend to increase the negative environmental impacts of SRWC systems.

The US DOE has a goal to produce 1 billion dry tonnes of biomass for bioenergy by 2030 (US DOE Billion-Ton Update, 2011). SRWC are expected to contribute 126 million. In the United States, the contribution of SRWC biomass to the energy production goals of the DOE was considered feasible at supply prices of 50–60 US$ dry tonne^{-1} (US DOE Billion-Ton Update, 2011). At a supply price of 40 US$ dry tonne^{-1}, the contribution of biomass produced by SRWC systems to energy production in the United States was estimated to be zero (US DOE Billion-Ton Update, 2011). We assessed four hypothetical production scenarios representing two levels of inherent site fertility and two levels of silvicultural input intensity and cost (Table 6.1).

We assumed potential annual yield ranges of 5–15 dry tonnes ha^{-1} yr^{-1} for poplar and loblolly pine and 5–20 dry tonnes ha^{-1} yr^{-1} for shrub willow. Input costs (US$) included all establishment and rotation-length silvicultural inputs and were derived from values reported by Kline and Coleman (2010) for poplar and loblolly pine and from Dawson (2007) for shrub willow (Table 6.2).

We used average United States farm fertilizer prices from March 2011 reported by the U.S. Department of Agriculture's Economic Research Service (urea 526 US$ tonne^{-1}; triple superphosphate 633 US$ tonne^{-1}; potassium chloride 601 US$ tonne^{-1}). The net present value (NPV) was determined across a stumpage price range of 5–40 US$ dry tonne^{-1} using a discount rate of 8% and rotation lengths of 4 years for shrub willow and 8 years for poplar and loblolly pine. Since shrub willow and poplar SRWC systems utilize coppice regeneration following initial establishment, NPV values for these species represent averages over five rotations (one establishment rotation and four coppice regenerated rotations) whereas in noncoppicing loblolly pine establishment costs are incurred in each rotation.

Based on the production scenarios listed in Table 6.1 and described in detail in Table 6.2, shrub willow SRWC systems on high fertility sites became profitable at stumpage prices of 20 US$ dry tonne^{-1} in scenarios 3 (low input, 15 dry tonnes ha^{-1} yr^{-1}) and 4 (high input, 20 dry tonnes ha^{-1} yr^{-1}) (Figure 6.14).

On low fertility sites, shrub willow SRWC systems never became profitable in scenario 1 (low input, 5 dry tonnes ha^{-1} yr^{-1}) but became profitable in scenario 2 (high input, 10 dry tonnes ha^{-1} yr^{-1}) at stumpage prices of 40 US$ dry tonne^{-1}. At the maximum stumpage price of 40 US$ dry tonne^{-1}, NPV was −80, 161, 1190, and 1431 US$ ha^{-1} for scenarios 1, 2, 3, and 4, respectively.

Results for poplar SRWC systems were similar to those for shrub willow with no profitability in the low fertility site scenario 1 (low input, 5 dry tonnes ha^{-1} yr^{-1}) (Figure 6.15).

In the low fertility site scenario 2 (high input, 10 dry tonnes ha^{-1} yr^{-1}), profitability was achieved at stumpage prices of 35 US$ dry tonne^{-1}. On high fertility sites, scenarios 3

Table 6.2. Rotation cost estimates for establishment and silvicultural inputs in poplar (Kline and Coleman, 2010), loblolly pine (Kline and Coleman, 2010), and shrub willow (Dawson, 2007) SRWC systems.

		Poplar				Loblolly pine	
		First rotation Cost (US$ ha^{-1})		Coppice rotation Cost (US$ ha^{-1})		All rotations Cost (US$ ha^{-1})	
	Establishment/ silvicultural input	Low input	High input	Low input	High input	Low input	High input
Year 1	Site preparation	618	618			278	278
	Cuttings/seedlings	216	216			74	74
	Planting	112	112			112	112
	Herbicide/pesticide	321	321			130	130
Year 2	Herbicide/pesticide		172				
Year 3							
Year 4	Fertilization (kg ha^{-1}) 225 N, 45 P, 100 K		426		426		426
Year 5							
Year 6							
Year 7							
Year 8	Total cost	1267	1865	0	426	594	1020
		Shrub willow					
		First rotation Cost (US$ ha^{-1})		Coppice rotation Cost (US$ ha^{-1})			
	Establishment/ silvicultural input	Low input	High input	Low input	High input		
Year 1	Site preparation	237	237				
	Cuttings	1482	1482				
	Planting	1283	1283				
	Herbicide/pesticide	499	499				

(low input, 10 dry tonnes ha^{-1} yr^{-1}) and 4 (high input, 15 dry tonnes ha^{-1} yr^{-1}) became profitable at stumpage prices of 25 US$ dry tonne^{-1}, a higher stumpage price than shrub willow. At the maximum stumpage price of 40 US$ dry tonne^{-1}, NPV was −148, 250, 716, and 1115 US$ ha^{-1} for scenarios 1, 2, 3, and 4, respectively.

Unlike shrub willow and poplar, all loblolly pine SRWC system production scenarios became profitable (Figure 6.16) due to these systems having the lowest establishment costs.

On the low fertility sites, scenarios 1 (low input, 5 dry tonnes ha^{-1} yr^{-1}) and 2 (high input, 10 dry tonnes ha^{-1} yr^{-1}) became profitable at stumpage prices of 30 and 25 US$ dry tonne^{-1}, respectively. On the high fertility sites, scenarios 3 (low input, 10 dry tonnes ha^{-1} yr^{-1}) and 4 (high input, 15 dry tonnes ha^{-1} yr^{-1}) became profitable at stumpage prices of 15 US$ dry tonne^{-1}. At the maximum stumpage price of 40 US$ dry tonne^{-1}, NPV was 271, 798, 1136, and 1662 US$ ha^{-1} for scenarios 1, 2, 3, and 4, respectively.

Our NPV analysis enabled us to make some generalizations regarding the efficacy of SRWC systems for biomass bioenergy production. Shrub willow and poplar SRWC systems appeared to benefit the most from establishment on higher fertility sites where higher production (dry tonnes ha^{-1} yr^{-1}) is achieved due to inherent site characteristics. Furthermore, results indicated that in order to be profitable these systems must be established on sites that can produce at

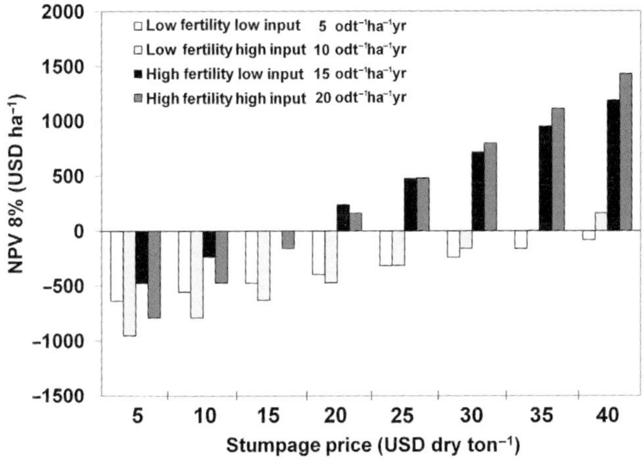

Figure 6.14. Analysis of shrub willow (*Salix spp.*) biomass stumpage price ranging from 5 to 40 US$ dry tonne^{-1} on 4-year rotation net present value (NPV) at 8% under four SRWC production scenarios representing low and high site fertility and silvicultural inputs/costs. Values represent averages over five rotations (one establishment rotation and four coppice regenerated rotations).

least 10 dry tonnes ha^{-1} yr^{-1}. Loblolly pine SRWC systems were profitable at all sites and production levels. However, the lowest level of production (5 dry tonnes ha^{-1} yr^{-1}) also required the highest stumpage price to become profitable. Results for all species indicated that further investment in intensive silviculture such as fertilization has the potential to lower the breakeven stumpage price on both low and high fertility sites. Furthermore, investment in intensive silviculture results in the highest return on investment provided that the projected

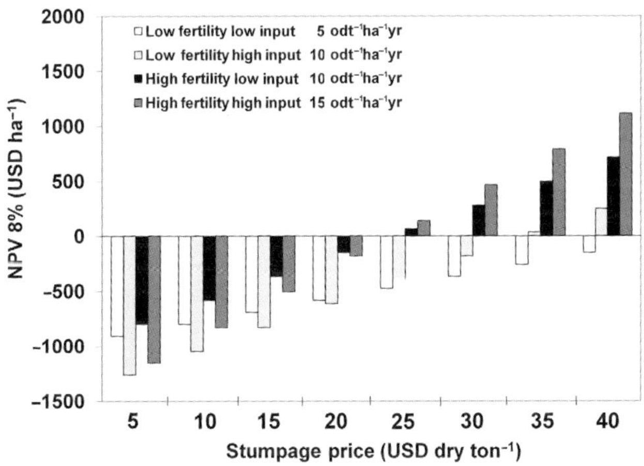

Figure 6.15. Analysis of poplar (*Populus spp.*) biomass stumpage price ranging from 5 to 40 US$ dry tonne^{-1} on 8-year rotation net present value (NPV) at 8% under four SRWC production scenarios representing low and high site fertility and silvicultural inputs/costs. Values represent averages over five rotations (one establishment rotation and four coppice regenerated rotations).

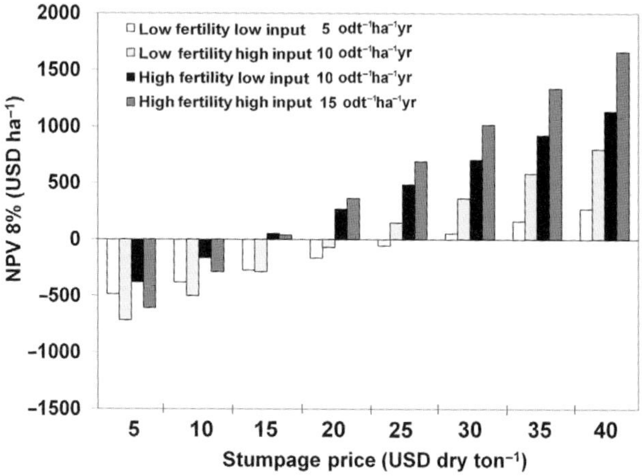

Figure 6.16. Analysis of loblolly pine (*Pinus taeda* L.) biomass stumpage price ranging from 5 to 40 US$ dry tonne^{-1} on 8-year rotation net present value (NPV) at 8% under four SRWC production scenarios representing low and high site fertility and silvicultural inputs/costs.

yield increases are achieved. This indicates that intensive silviculture produces a positive economic effect. When intensive silviculture increases yields, the area of land needed to produce 126 million dry tonnes of biomass is reduced proportionally producing a positive scalar effect. However, intensive silviculture can also have negative environmental effects through degradation of the soil resource and contamination of ground and surface water with herbicides and excess nutrients. When the soil is degraded, intensive silvicultural practices can result in yield declines over time rather than increases. Negative environmental impacts of SRWC systems can also potentially negate the assumed positive attributes of phasing out fossil fuels. Both SRWC systems and conventional forestry can be used to produce essentially the same product, woody biomass for bioenergy in the form of fuel pellets and cellulosic ethanol. Biomass production with SRWC systems may result in negative environmental impacts similar to those found in agricultural systems as a result of tillage, fertilization, and herbicide/pesticide application. In contrast, in conventional forestry where management intensity is much lower or even nonexistent negative environmental impacts are also much lower, mirroring the level of management intensity. Therefore, the production goal of 126 million dry tonnes produced solely by SRWC systems set by the US DOE may be difficult to realize according to our definition of sustainability. In addition, it may be unrealistic to convert an area of land ranging from 12.6 to 25.2 Mha to SRWC systems. Instead, it may be necessary for conventional forestry to contribute a greater proportion to the overall biomass production goal of 1 billion dry tonnes. Using SRWC systems to produce biomass for bioenergy may not be universally applicable across the gradient of inherent site fertility but instead can be implemented in the most sustainable manner on high fertility sites where high productivity can be achieved with minimal silvicultural inputs and associated negative environmental impacts.

Phytoremediation Potential

Phytoremediation is the use of plants to decontaminate soils and waste materials such as sewage sludge and industrial waste by taking up metals and organic contaminants and accumulating them in their tissues. Then, plants can be harvested allowing for proper disposal and removal of contaminants from the site. Plants most desirable for phytoremediation are known as hyperaccumulators that refers to their relatively high ability to remove contaminants from their growth medium and accumulate them in their tissues.

Shrub willow is suggested to function well in phytoremediation due to adaptability to harsh growing conditions including seasonal flooding (Smart *et al.*, 2005), extensive fibrous root systems allowing for efficient capture of contaminants (Kuzovkina and Quigley, 2005), and the ability to hyperaccumulate heavy metal (Smart *et al.*, 2005; Volk *et al.*, 2006) and organic (Kuzovkina and Quigley, 2005; Volk *et al.*, 2006) contaminants. Willow plantings function as riparian buffers by capturing the runoff of excess nutrients such as N and P from agricultural areas (Smart *et al.*, 2005), minimizing eutrophication of surface waters. Willow also removes contaminants from the environment by accumulating heavy metals from wastewater; then, they are burned and the heavy metals are disposed of by scrubbing the combustion gases and proper handling of ashes (Perttu and Kowalik, 1997). Research on the use of shrub willow for phytoremediation in Sweden revealed it to be useful for treating municipal wastewater (Perttu and Kowalik, 1997; Borjesson and Berndes, 2006). The application of wastewater to willow increased yields by 30–100% (Borjesson and Berndes, 2006). Irrigation of willow with municipal wastewater has also, however, been found to increase the occurrence of infection with rust (*Melampsora epitea*) disease (Toome *et al.*, 2009) leading to a need for control methods. Research is also being conducted in the United States to assess the use of willow to control site water problems in regards to controlling water flowing through contaminated soils (high pH and salts) and sites with contaminated groundwater chemistry (petroleum, industrial sites) (Volk *et al.*, 2006).

Poplar is also a hyperaccumulator of heavy metals and significant genetic variation in this ability has been observed in a number of studies (Sebastiani *et al.*, 2004; Laureysens *et al.*, 2004; Zalesny *et al.*, 2007, 2009). On an old household waste disposal site, poplar was found to be particularly efficient at taking up cadmium (Cd), zinc (Zn), and aluminum (Al) (Laureysens *et al.*, 2004). This study also found that metal concentrations were the highest in senescing leaves potentially allowing for annual contaminant removal. Sebastiani *et al.* (2004) found that industrial organic waste applied to poplar did not exert toxic effects and that Zn, copper (Cu), and chromium (Cr) concentrations consistently differed among poplar clones across a gradient of waste application. Poplar also shows potential for use in riparian buffer strips which capture excess runoff of excess nutrients from agricultural areas. Haycock and Pinay (1993) suggest that poplar riparian buffers are more effective than grass at retaining nitrate (NO_3^-) during winter months due to higher contributions of C aiding soil microbes in nitrate reduction.

Conclusions

Woody biomass is used to provide a bioenergy source in the form of fuel pellets that displace fossil fuels in the production of electricity and heat and also in the form of cellulosic ethanol liquid fuel. SRWC are anticipated by the U.S. Department of Energy to contribute 126 million dry tonnes annually to the production goal of 1 billion dry tonnes. Our evaluation of the efficacy of this SRWC contribution indicated that yields (dry tonnes ha^{-1}) must be increased over current levels in order to produce a positive scalar effect. Yields must be doubled or even tripled from current average operational yields in order to reduce the area of land needed to meet production goals to a reasonable level. Research has shown that this can be achieved in experimental small plots with intensive silvicultural inputs. However, this also results in higher costs producing a negative economic effect. In addition, intensive silviculture has been shown to produce negative environmental effects in some cases. In order for SRWC biomass bioenergy production to be successful in the near term, production goals must be set to reasonable levels within the scope of the three primary factors we have identified—environmental, economic, and scalar. The most successful implementation of shrub willow, poplar, and pine SRWC production systems will occur in locations where relatively high yields can be achieved with the minimum of intensive silvicultural inputs and associated negative environmental effects. In the context of our review of the literature, SRWC biomass production with pine shows the most immediate promise due to relatively high yields on infertile sites. Currently, shrub willow and poplar SRWC biomass production requires more fertile sites. This trend will likely continue until genetic improvements are more fully developed dramatically increasing shrub willow and poplar yield on relatively infertile sites.

References

Abrahamson, L.P., Robison, D.J., Volk, T.A., *et al.* (1998) Sustainability and environmental issues associated with shrub willow bioenergy development in New York (U.S.A.). *Biomass and Bioenergy* 15: 17–22.

Ackerman, F. (2007) The economics of atrazine. *International Journal of Occupational and Environmental Health* 13: 441–449.

Acuna, E., Espinosa, M., Cancino, J., Rubilar, R., and Munoz, F. (2010) Estimating the bioenergy potential of *Pinus radiata* plantations in Chile. *Ciencia E Investigación Agraria* 37: 93–102.

Adegbidi, H.G., Briggs, R.D., Volk, T.A., White, E.H., and Abrahamson, L.P. (2003) Effect of organic amendments and slow-release nitrogen fertilizer on shrub willow biomass production and soil chemical characteristics. *Biomass and Bioenergy* 25: 389–398.

Adegbidi, H.G., Volk, T.A., White, E.H., Abrahamson, L.P., Briggs, R.D., and Bickelhaupt, D.H. (2001) Biomass and nutrient removal by shrub willow clones in experimental bioenergy plantations in New York State. *Biomass and Bioenergy* 20: 399–411.

Al Afas, N., Pellis, A., Niinemets, U., and Ceulemans, R. (2005) Growth and production of a short rotation coppice culture of poplar. II. Clonal and year-to-year differences in leaf and petiole characteristics and stand leaf area index. *Biomass and Bioenergy* 28: 536–547.

Albaugh, T.J., Allen, H.L., Dougherty, P.M., and Johnsen, K.H. (2004) Long term growth responses of loblolly pine to optimal nutrient and water resource availability. *Forest Ecology and Management* 192: 3–19.

Albaugh, T.J., Allen, H.L., Dougherty, P.M., Kress, L.W., and King, J.S. (1998) Leaf area and above- and belowground growth responses of loblolly pine to nutrient and water additions. Forest Science 44: 317–328.

Albaugh, T.J., Allen, H.L., and Fox, T.R. (2008) Nutrient use and uptake in *Pinus taeda*. *Tree Physiology* 28: 1083–1098.

Albaugh, T.J., Allen, H.L., Fox, T.R., Carlson, C.A., and Rubilar, R.A. (2009) Opportunities for fertilization of loblolly pine in the Sandhills of the southeastern United States. *Southern Journal of Applied Forestry* 33: 129–136.

Allen, H.L. (1987) Forest fertilizers: nutrient amendment, stand productivity, and environmental impact. *Journal of Forestry* 85: 37–46.

Allen, H.L., Fox, T.R., and Campbell, R.G. (2005) What is ahead for intensive pine plantation silviculture in the South? *Southern Journal of Applied Forestry* 29: 62–69.

Andersson, F.O., Agren, G.I., and Fuhrer, E. (2000) Sustainable tree biomass production. *Forest Ecology and Management* 132: 51–62.

Aronsson, P.G. and Bergstrom, L.F. (2001) Nitrate leaching from lysimeter-grown short-rotation willow coppice in relation to N-application, irrigation, and soil type. *Biomass and Bioenergy* 21: 155–164.

Astrom, B. and Ramstedt, M. (1994) Stem cankers on Swedish biomass willow caused by *Crtyptodiaporthe salicella* and other fungi. *European Journal of Forest Pathology* 24: 264–276.

Aylott, M.J., Casella, E., Tubby, I., Street, N.R., Smith, P., and Taylor, G. (2008) Yield and spatial supply of bioenergy poplar and willow short-rotation coppice in the UK. *New Phytologist* 178: 358–370.

Baker, J.L. and Mickelson, S.K. (1994) Application technology and best management practices for minimizing herbicide runoff. *Weed Technology* 8: 862–869.

Barbash, J.E., Thelin, G.P., Kolpin, D.W., and Gilliom, R.J. (2001) Major herbicides in ground water: results from the national water-quality assessment. *Journal of Environmental Quality* 30: 831–845.

Birk, E.M. and Vitousek, P.M. (1986) Nitrogen availability and nitrogen use efficiency in loblolly pine stands. *Ecology* 67: 69–79.

Blanco-Canqui H. (2010) Energy crops and their implications on soil and environment. *Agronomy Journal* 102: 403–419.

Borjesson, P. and Berndes, G. (2006) The prospects for shrub willow plantations for wastewater treatment in Sweden. *Biomass and Bioenergy* 30: 428–438.

Bungart, R. and Huttl, R.F. (2004) Growth dynamics and biomass accumulation of 8-year-old hybrid poplar clones in a short-rotation plantation on a clayey-sandy mining substrate with respect to plant nutrition and water budget. *European Journal of Forest Research* 123: 105–115.

Burger, J.A. and Kelting, D.L. (1999) Using soil quality indicators to assess forest stand management. *Forest Ecology and Management* 122: 155–166.

Christersson, L. (2010) Wood production potential in poplar plantations in Sweden. *Biomass and Bioenergy* 34: 1289–1299.

Coyle, D.R., Coleman, M.D., Durant, J.A., and Newman, L.A. (2006a) Survival and growth of 31 *Populus* clones in South Carolina. *Biomass and Bioenergy* 30: 750–758.

Coyle, D.R., Coleman, M.D., Durant, J.A., and Newman, L.A. (2006b) Multiple factors affect pest and pathogen damage on 31 *Populus* clones in South Carolina. *Biomass and Bioenergy* 30: 759–768.

Dalin, P., Kindvall, O., and Bjorkman, C. (2009) Reduced population control of an insect pest in managed willow monocultures. *PloS ONE* 4: e5487.

Dawson, M. (2007) Short rotation coppice willow—best practice guidelines, RENEW Project, Belfast, Northern Ireland, 50p.

Dickmann, D.I. (2006) Silviculture and biology of short-rotation woody crops in temperate regions: then and now. *Biomass and Bioenergy* 30: 696–705.

Dicus, C.A. and Dean, T.J. (2001) Nitrogen and phosphorus use efficiency in stands of loblolly and slash pine, in *Proceedings of the Eleventh Biennial Southern Silvicultural Research Conference*, General Technical Report SRS-48 (ed. K.W. Outcalt), U.S. Department of Agriculture, Forest Service, Southern Research Station,Asheville, NC, pp. 38–41.

Drew, A.P., Zsuffa, L., and Mitchell, C.P. (1987) Terminology relating to woody plant biomass and its production. *Biomass* 12: 79–82.

Evans, J. (1997) Bioenergy plantations-experience and prospects. *Biomass and Bioenergy* 13: 189–191.

Everett, C.J. and Palm-Leis, H. (2009) Availability of residual phosphorus fertilizer for loblolly pine. *Forest Ecology and Management* 258:2207–2213.

Field, J.P., Farrish, K.W., and Carter, E.A. (2003) Soil and nutrient losses following site preparation burning in a harvested loblolly pine site. *Transactions of the American Society of Agricultural Engineers* 46: 1697–1703.

Fisher, G., Prieler, S., and van Velthuizen, H. (2005) Biomass potentials of miscanthus, willow, and poplar: results and policy implications for Eastern Europe, Northern and Central Asia. *Biomass and Bioenergy* 28: 119–132.

Fox, T.R., Allen, H.L., Albaugh, T.J., Rubilar, R., and Carlson, C.A. (2007a) Tree nutrition and forest fertilization of pine plantations in the southern United States. *Southern Journal of Applied Forestry* 31: 5–11.

Fox, T.R., Jokela, E.J., and Allen, H.L. (2007b) The development of pine plantation silviculture in the southern United States. *Journal of Forestry* 2007: 337–347.

Goncalves, J.L. de, M., Stape, J.L., Laclau, J.P., Smethurst, P., and Gava, J.L. (2004) Silvicultural effects on the productivity and woody quality of eucalypt plantations. *Forest Ecology and Management* 193: 45–61.

Grace III, J.M. (2004) Soil erosion following forest operations in the Southern Piedmont of central Alabama. *Journal of Soil and Water Conservation* 59: 160–166.

Gresham, C.A. (2002) Sustainability of intensive loblolly pine plantation management in the South Carolina Coastal Plain, USA. *Forest Ecology and Management* 155: 69–80.

Gustavsson, L. (1997) Energy efficiency and competitiveness of biomass-based energy systems. *Energy* 22: 959–967.

Hansen, E.A. (1991). Poplar woody biomass yields: a look to the future. *Biomass and Bioenergy* 1: 1–7.

Hansen, E.A. (1993) Soil carbon sequestration beneath hybrid poplar plantations in the north central United States. *Biomass and Bioenergy* 5: 431–436.

Haycock, N.E. and Pinay, G. (1993) Groundwater nitrate dynamics in grass and poplar vegetated riparian buffer strips during the winter. *Journal of Environmental Quality* 22: 273–278.

Haywood, J.D. (1988) Loblolly pine response to weed control in central Louisiana. *Weed Technology* 2: 490–494.

Heilman, P. and Norby, R.J. (1998) Nutrient cycling and fertility management in temperate short rotation forest systems. *Biomass and Bioenergy* 14: 361–370.

Herve, C. and Ceulemans, R. (1996) Short-rotation coppiced vs non-coppiced poplar: a study at two different field sites. *Biomass and Bioenergy* 11: 139–150.

Johansson, T. and Karacic, A. (2011) Increment and biomass in hybrid poplar and some practical implications. *Biomass and Bioenergy* 35: 1925–1934.

Johnson, D.W., Todd, D.E., and Tolbert, V.R. (2003) Changes in ecosystem carbon and nitrogen in a loblolly pine plantation over the first 18 years. *Soil Science Society of America Journal* 67: 1594–1601.

Johnston, J.M. and Crossley Jr., D.A. (2002) Forest ecosystem recovery in the southeast US: soil ecology as an essential component of ecosystem management. *Forest Ecology and Management* 155: 187–203.

Jokela, E.J. (2004) Nutrient management of southern pines, in *Slash Pine: Still Growing and Growing! Proceedings of the Slash Pine Symposium*, General Technology Report SRS–76 (eds E.D. Dickens, J.P. Barnett, W.G. Hubbard, and E.J. Jokela), U.S. Department of Agriculture, Forest Service, Southern Research Station, Asheville, NC, pp. 27–35.

Jokela, E.J., Dougherty, P.M., and Martin, T.A. (2004) Production dynamics of intensively managed loblolly pine stands in the southern United States: a synthesis of seven long-term experiments. *Forest Ecology and Management* 192: 117–130.

Jug, A., Makeschin, F., Rehfuess, K.E., and Hofmann-Schielle, C. (1999) Short-rotation plantations of balsam poplars, aspen, and willows on former arable land in the Federal Republic of Germany. III. Soil ecological effects. *Forest Ecology and Management* 121: 85–99.

Kauter, D., Lewandowski, I., and Claupein, W. (2003) Quantity and quality of harvestable biomass from *Populus* short rotation coppice for solid fuel use-a review of the physiological basis and management influences. *Biomass and Bioenergy* 24: 411–427.

Kimmins, J.P. (1997) Predicting sustainability of forest bioenergy production in the face of changing paradigms. *Biomass and Bioenergy* 13: 201–212.

Kinloch, B.B. and Stonecypher, R.W. (1969) Genetic variation in susceptibility to fusiform rust in seedlings from a wild population of loblolly pine. *Phytopathology* 59: 1246–1255.

Kline, K.L. and Coleman, M.D. (2010) Woody energy crops in the southeastern United States: two centuries of practitioner experience. *Biomass and Bioenergy* 34: 1655–1666.

Kopp, R.F., Abrahamson, L.P., White, E.H., Burns, K.F., and Nowak, C.A. (1997) Cutting cycle and spacing effects on biomass production by a shrub willow clone in New York. *Biomass and Bioenergy* 12: 313–319.

Kuzovkina, Y.A. and Quigley, M.F. (2005) Willow beyond wetlands: Uses of *Salix* L. species for environmental projects. *Water, Air, and Soil Pollution* 162: 183–204.

Labrecque, M. and Teodorescu, T.I. (2003) High biomass yield achieved by *Salix* clones in SRIC following two 3-year coppice rotations on abandoned farmland in southern Quebec, Canada. *Biomass and Bioenergy* 25: 135–146.

Labrecque, M. and Teodorescu, T.I. (2005) Field performance and biomass production of 12 willow and poplar clones in short-rotation coppice in southern Quebec (Canada). *Biomass and Bioenergy* 29: 1–9.

Lal, R. (1998) Soil erosion impact on agronomic productivity and environment quality. *Critical Reviews in Plant Sciences* 17: 319–464.

Land Jr., S.B., Roberts, S.D., and Duzan Jr., H.W. (2004) Genetic and spacing effects on loblolly pine plantation development through age 17, in *Proceedings of the 12th Biennial Southern Silvicultural Research Conference*, General Technical Reports SRS–71 (ed. F.K. Connor), U.S. Department of Agriculture, Forest Service, Southern Research Station, Asheville, NC, pp. 413–419.

Laureysens, I., Blust, R., De Temmerman, L., Lemmens, C., and Ceulemans, R. (2004) Clonal variation in heavy metal accumulation and biomass production in a poplar coppice culture: I. Seasonal variation in leaf, wood, and bark concentrations. *Environmental Pollution* 131: 485–494.

Laureysens, I., Deraedt, W., Indeherberge, T., and Ceulemans, R. (2003) Population dynamics in a 6-year old coppice culture of poplar. I. Clonal differences in stool mortality, shoot dynamics, and shoot diameter distribution in relation to biomass production. *Biomass and Bioenergy* 24: 81–95.

Lee, K. and Jose, S. (2005) Nitrate leaching in cottonwood and loblolly pine biomass plantations along a nitrogen fertilization gradient. *Agriculture, Ecosystems and Environment* 105: 615–623.

Li, B., McKeand, S.E., and Allen, H.L. (1991) Genetic variation in nitrogen use efficiency of loblolly pine seedlings. *Forest Science* 37: 613–626.

Logan, T.J. (1993) Agricultural best management practices for water pollution control: current issues. *Agriculture, Ecosystems and Environment* 46: 223–231.

Mann, L. and Tolbert, V. (2000) Soil sustainability in renewable biomass plantings. *AMBIO: A Journal of the Human Environment* 29: 492–498.

Matson, P.A., Hain, F.P., and Mawby, W. (1987) Indices of tree susceptibility to bark beetles vary with silvicultural treatment in a loblolly pine plantation. *Forest Ecology and Management* 22: 107–118.

McLaughlin, R.A., Pope, P.E., and Hansen, E.A. (1985) Nitrogen fertilization and ground cover in a hybrid poplar plantation: effects on nitrate leaching. *Journal of Environmental Quality* 14: 241–245.

Mead, D.J. (2005) Opportunities for improving plantation productivity. How much? How quickly? How realistic? *Biomass and Bioenergy* 28: 249–266.

Mitchell, C.P. (1992) Ecophysiology of short rotation forest crops. *Biomass and Bioenergy* 2: 25–37.

Mola-Yudego, B. (2010) Regional potential yields of short rotation shrub willow plantations on agricultural land in Northern Europe. *Silva Fennica* 44: 63–76.

Munsell, J.F. and Fox, T.R. (2010) An analysis of the feasibility for increasing woody biomass production from pine plantations in the southern United States. *Biomass and Bioenergy* 34: 1631–1642.

Nordman, E.E., Robison, D.J., Abrahamson, L.P., and Volk, T.A. (2005) Relative resistance of willow and poplar biomass production clones across a continuum of herbivorous insect specialization: univariate and multivariate approaches. *Forest Ecology and Management* 217: 307–318.

Park, B.B., Yanai, R.D., Sahm, J.M., Ballard, B.D., and Abrahamson, L.P. (2004) Wood ash effects on soil solution and nutrient budgets in a shrub willow bioenergy plantation. *Water, Air, and Soil Pollution* 159: 209–224.

Park, B.B., Yanai, R.D., Sahm, J.M., Lee, D.K., and Abrahamson, L.P. (2005) Wood ash effects on plant and soil in a shrub willow bioenergy plantation. *Biomass and Bioenergy* 28: 355–365.

Perttu, K.L. and Kowalik, P.J. (1997) *Salix* vegetation filters for purification of waters and soils. *Biomass and Bioenergy* 12: 9–19.

Pimentel, D. and Krummel, J. (1987) Biomass energy and soil erosion: assessment of resource costs. *Biomass* 14: 15–38.

Pritchett, W.L. and Comerford, N.B. (1982) Long-term response to phosphorus fertilization on selected southeastern Coastal Plain soils. *Soil Science Society of America Journal* 46: 640–644.

Quaye, A.K., Volk, T.A., Hafner, S., Leopold, D.J., and Schirmer, C. (2011) Impacts of paper sludge and manure on soil and biomass production of shrub willow. *Biomass and Bioenergy* 35: 2796–2806.

Reijnders, L. (2006) Conditions for the sustainability of biomass based fuel use. *Energy Policy* 34: 863–876.

Royle, D.J. and Ostry, M.E. (1995) Disease and pest control in the bioenergy crops poplar and willow. *Biomass and Bioenergy* 9: 69–79.

Sanchis, J., Kantiani, L., Llorca, M., *et al.* (2012) Determination of glyphosate in groundwater samples using an ultrasensitive immunoassay and confirmation by on-line solid-phase extraction followed by liquid chromatography coupled to tandem mass spectrometry. *Analytical and Bioanalytical Chemistry* 402: 2335–2345.

Sartori, F., Lal, R., Ebinger, M.H., and Eaton, J.A. (2007a) Changes in soil carbon and nutrient pools along a chronosequence of poplar plantations in the Columbia Plateau, Oregon, USA. *Agriculture, Ecosystems and Environment* 122: 325–339.

Sartori, F., Markewitz, D., and Borders, B.E. (2007b) Soil carbon storage and nitrogen and phosphorus availability in loblolly pine plantations over 4 to 16 years of herbicide and fertilizer treatments. *Biogeochemistry* 84: 13–30.

Sebastiani, L., Scebba, F., and Tognetti, R. (2004) Heavy metal accumulation and growth responses in poplar clones Eridano (*Populus deltoides* x *maximowiczii*) and I-214 (*P.* x *euramericana*) exposed to industrial waste. *Environmental and Experimental Botany* 52: 79–88.

Singh, B. (1998) Biomass production and nutrient dynamics in three clones of *Populus deltoides* planted on Indogangetic plains. *Plant and Soil* 203: 15–26.

Smart, L.B., Volk, T.A., Lin, J., *et al.* (2005) Genetic improvement of shrub willow (*Salix spp.*) crops for bioenergy and environmental applications in the United States. *Unasylva* 221 56: 51–55.

Souch, C.A., Martin, P.J., Stephens, W., and Spoor, G. (2004) Effects of soil compaction and mechanical damage at harvest on growth and biomass production of short rotation coppice shrub willow. *Plant and Soil* 263: 173–182.

Spalding, R.F., Exner, M.E., Snow, D.D., Cassada, D.A., Burbach, M.E., and Monson, S.J. (2003) Herbicides in ground water beneath Nebraska's management systems evaluation area. *Journal of Environmental Quality* 32: 92–99.

Tharakan, P.J., Volk, T.A., Lindsey, C.A., Abrahamson, L.P., and White, E.H. (2005) Evaluating the impact of three incentive programs on the economics of cofiring shrub willow biomass with coal in New York State. *Energy Policy* 33: 337–347.

Tilman, D., Cassman, K.G., Matson, P.A., Naylor, R., and Polasky, S. (2002) Agricultural sustainability and intensive production practices. *Nature* 418: 671–677.

Toome, M., Heinsoo, K., Ramstedt, M., and Luik, A. (2009) Rust severity in bioenergy willow plantations treated with additional nutrients. *Forest Pathology* 39: 28–34.

U.S. Department of Energy. (2011) U.S. Billion-Ton Update: Biomass Supply for a Bioenergy and Bioproducts Industry, (Leads R.D. Perlack and B.J. Stokes), ORNL/TM-2011/224, Oak Ridge National Laboratory, Oak Ridge, TN, 227p.

Vitousek, P.M., Andariese, S.W., Matson, P.A., Morris, L., and Sanford, R.L. (1992) Effects of harvest intensity, site preparation, and herbicide use on soil nitrogen transformations in a young loblolly pine plantation. *Forest Ecology and Management* 49: 277–292.

Volk, T.A., Abrahamson, L.P., Nowak, C.A., Smart, L.B., Tharakan, P.J., and White, E.H. (2006) The development of short-rotation shrub willow in the northeastern United States for bioenergy and bioproducts, agroforestry and phytoremediation. *Biomass and Bioenergy* 30: 715–727.

Will, R.E., Markewitz, D., Hendrick, R.L., Meason, D.F., Crocker, T.R., and Borders, B.E. (2006) Nitrogen and phosphorus dynamics for 13-year old loblolly pine stands receiving complete competition control and annual N fertilizer. *Forest Ecology and Management* 227: 155–168.

Zalesny, J.A., Zalesny Jr., R.S., Coyle, D.R., and Hall, R.B. (2007) Growth and biomass of *Populus* irrigated with landfill leachate. *Forest Ecology and Management* 248: 143–152.

Zalesny Jr., R.S., Wiese, A.H., Bauer, E.O., and Riemenschneider, D.E. (2009). *Ex situ* growth and biomass of *Populus* bioenergy crops irrigated and fertilized with landfill leachate. *Biomass and Bioenergy* 33: 62–69.

Chapter 7

Biomass Feedstock Production Impact on Water Resource Availability

K.C. Stone[1], P.G. Hunt[2], K.B. Cantrell[2], and K.S. Ro[2]

[1] *USDA-ARS Coastal Plains Soil, Water, and Plant Research Center, SC, USA*

[2] *USDA-ARS Coastal Plains Soil, Water, and Plant Research Center, Florence, SC, USA*

Introduction

In order to meet increasing biofuel demands agriculture will require greater land and water resources. This will likely necessitate: (1) conversion of existing crop land to grow biofuel crops, (2) changes in other land uses (like forest and pastureland) to grow biofuel crops, and (3) increases in the use of fertilizer and agrochemicals (Uhlenbrook, 2007; USDA-ERS, 2008). Ultimately, all these actions will heighten potential agricultural impacts on natural resources. If local agriculture shifts to biofuel/bioenergy crops that require more than current agricultural water supplies, there is a likelihood of deleterious impacts on limited water resources. To be sustainable, bioenergy production must conserve and protect natural resources, including fresh water.

Fresh water is different from other commodities in that it has no substitutes (Postel *et al.*, 1996). Moreover, only 2.5% of all the water on earth is fresh water. The majority of fresh water, 70%, is stored in polar ice caps and essentially unavailable for human use (UNESCO, 2007). The remaining fresh water, 30%, is held in aquifers, soils, lakes, rivers, and the atmosphere. In 1996, it was estimated that humanity used 54% of the runoff that was geographically and temporally accessible and 26% of the total terrestrial evapotranspiration (Postel *et al.*, 1996). This estimate assumed that freshwater usage for humanity was distributed among many uses including transportation, navigation, industrial consumption, direct human consumption, and food production (Postel *et al.*, 1996). Among these current global uses, water is now being called upon for biofuel production. This reduces availability of an already stretched resource.

Freshwater scarcities have already been reported in many parts of the world (Postel, 2000; Brown, 2003). Further complicating these freshwater scarcities, the world population is expected to increase by an additional two billion people by the year 2030 (United Nations, 1998). Historically, water scarcity has been the subject of lawsuits, conflicts, and wars. Civilizations have risen and fallen because of the availability or lack of water (Sadler and Turner, 1993; Postel, 2001; Montgomery, 2007; Diamond, 2005).

Biofuel Crop Sustainability, First Edition. Edited by Bharat P. Singh.

Water availability has not only directly impacted humans and civilizations, but it has impacted the environment. In many areas of the world, freshwater extraction for agriculture, industry, or cities places at risk the health of aquatic ecosystems and the lives those ecosystems support (Postel, 2000). These ecosystems may be further at risk as bioenergy crop production grows and the demand for fresh water increases. Unfortunately, even today fresh water from many aquifers and river systems is being overutilized to meet societal demands (Brown, 2001; Falkenmark and Lannerstad, 2005). It is projected that these water supplies will be further depleted as both the population and associated fuel consumption increase (Postel, 2000; Brown, 2003).

Thus, it would seem that there are some critical underlying problems if bioenergy development was pursued in the United States and across the globe without very careful considerations of the water resource limitations and their critical connections to ecosystem integrity and sustainability of human food.

In this chapter, we review the potential impact bioenergy production will have on water supplies. We assess the following: (1) climate and weather impact on water supplies for biomass production, (2) water use for major bioenergy crop production, and (3) potential alternatives to improve water supplies for bioenergy.

Climate and Weather Impact on Water Supply

Climate Change

Climate change is likely to impact agriculture and food security across the world (Slingo *et al.*, 2005). Climatic variability such as that from El Niño has already had large impacts on crop production. Slingo *et al.* (2005) reported that in future climatic change scenarios, critical temperature thresholds for food crops will be exceeded with increasing frequency. Long *et al.* (2005) concluded that major agronomic crops grown in carbon dioxide enrichment chambers may have significantly overestimated reported yields. Based on their findings, they reported current projections in future global food security are overoptimistic. Meza and Silva (2009) used simulation modeling to analyze maize and wheat production changes with climate change. They estimated that climate change may result in 5–10% reduction in yields of both maize and wheat. Alternative adaptation strategies, such as changing planting dates, could be implemented to help counterbalance the impacts of ensuing warmer and drier environment.

The US Office of Technology Assessment (U.S. Congress, Office of Technology Assessment, 1993) in the report "Preparing for an Uncertain Climate Volume I" discussed the wide-ranging impacts that climate change would have on all sectors of the economy. The report recognized that agriculture would be sensitive to changes in climate and climatic variability. While climatic change impacts may be offset by intensive management over short time frames, agricultural productivity would be at risk with increasing temperature and more frequent droughts. Agriculture's use of scarce water resources for food production during drought periods could become increasingly contentious with urban, industrial, and environmental sectors.

The Western United States is probably the most recognizable area of the country impacted by climate changes. In particular, Western US agriculture is highly dependent on surface runoff for water supplies. Mote (2006) reported that in the Western US river basins, snow was the largest component of water storage. In testimony before the US Congress, Mote (2007) reported that about 70% of annual water flow is from snowmelt and that snow provides roughly a half-year delay in runoff. Water supplies in the Western United States would be highly vulnerable to any climatic changes that influence snowpack. Barnett *et al.* (2005) reported that over one-sixth of the world's population relies on glaciers and snow packs for their water supply, and that the hydrological changes due to climatic change for future water availability are likely to create severe consequences.

Hamlet *et al.* (2007) studied Western US trends in runoff, evapotranspiration, and soil moisture. They found over the last century, runoff had occurred earlier in spring primarily due to increasing midwinter temperatures. These earlier spring runoff events resulted in earlier spring soil moisture recharge. These earlier trends also corresponded with a shift in evapotranspiration from midsummer to late spring and early summer. Combined, these shifts in runoff, evapotranspiration, and soil moisture require adaptations in water management and cropping systems.

Climatic Variability

Agricultural adaptation to changing climatic conditions will depend on how climate change affects the variation of temperature and precipitation (Negri *et al.*, 2005). Negri *et al.* (2005) estimated the effects of climatic variability on US irrigation. They reported that higher temperatures and less rainfall would increase the need for irrigation. Yet, any increase in irrigation to adapt to climate change would be constrained by water availability. Water availability is the primary factor in present irrigation capacity and would likely be much further limited under future climatic change and the increased production of biomass for biofuels.

Kangas and Brown (2007) studied the spatial and temporal characteristics of drought and pluvial events from 1895 to 2003. They observed that the largest annual droughts or pluvial events occurred more frequently in the Central United States. The Western and Eastern United States had a higher percentage of extreme events. They found that of four large pluvial events occurring in the United States during their study period, three occurred during the past 30 years.

In 2008, the major maize producing states of the upper Midwestern US (e.g., Iowa) experienced extreme flooding due to excess rainfalls over an extended period of weeks. This flooding affected early-season planting operations. Previously in 1993, a more widespread area of the Midwest was affected by similar floods. Both events exceeded the historical 100-year return interval.

Additionally, floodwater has the potential for enormous impacts on downstream water quality. The National Research Council (NRC, 2008) reported on the potential impacts of excess nutrient runoff on water quality. They reported that crops with the greatest nutrient inputs would have the greatest potential for impacting water quality. During periods of excess rainfall, there is potential for the flooding of wastewater treatment lagoons in Iowa and their impact on downstream water quality (Simpkins *et al.*, 2002). Not only would flooded soils delay crop production, but excess nutrients in the water could also deteriorate water quality. Strategies would be needed to reduce nutrient losses while maintaining productivity.

Drought and subsequent reduced production could greatly impact the biomass available for bioenergy production. Woodhouse and Overpeck (1998) analyzed Central US drought through reconstructed climatic data for the last 2000 years. They used current land-use practices (increased cultivation of marginal lands and the escalated groundwater usage from the Ogallala Aquifer) along with Global Climatic Model predictions. They found numerous pre-1900 droughts eclipsing those of the 1930s and 1950s. Some droughts prior to the 1600s had longer multidecadal durations and greater spatial extent than those of the twentieth century. A study using global coupled climate models (Meehl and Tebaldi, 2004) showed a distinct geographical pattern for recent heat waves in North America and the models predicted more frequent, more intensive, and longer lasting events in the second half of the twenty-first century. Whether from preindustrial, geophysical, or current hypothesized climate change, the Central United States has been and will continue to be vulnerable to droughts and weather extremes. Furthermore, Seager *et al.* (2009) reported that historic tree-ring records from the Southeastern United States show that the twentieth century has been moist from the perspective of the last millennium and free of long and severe droughts that were abundant in previous centuries. Historically, the tree-ring records show a 21-year-long uninterrupted drought in the mid-sixteenth century, a long period of dry conditions in the early- to mid-nineteenth century and that the southeast was also affected by some of the medieval megadroughts centered in the Western United States.

Like many other areas of the world, the United States has recently had extended droughts affecting various areas of the country. While there are too many weather-related droughts to address individually a few that would have a potential impact on future energy crop production can be highlighted. Izaurralde *et al.* (2005) reported that the temperate and subtropical Southeastern United States had the potential for maximum annual biomass net primary production growth rates. The Southeastern United States has one of the highest renewable water supplies in the United States (Solley *et al.*, 1998; Figure 7.1). However, the region is not immune to

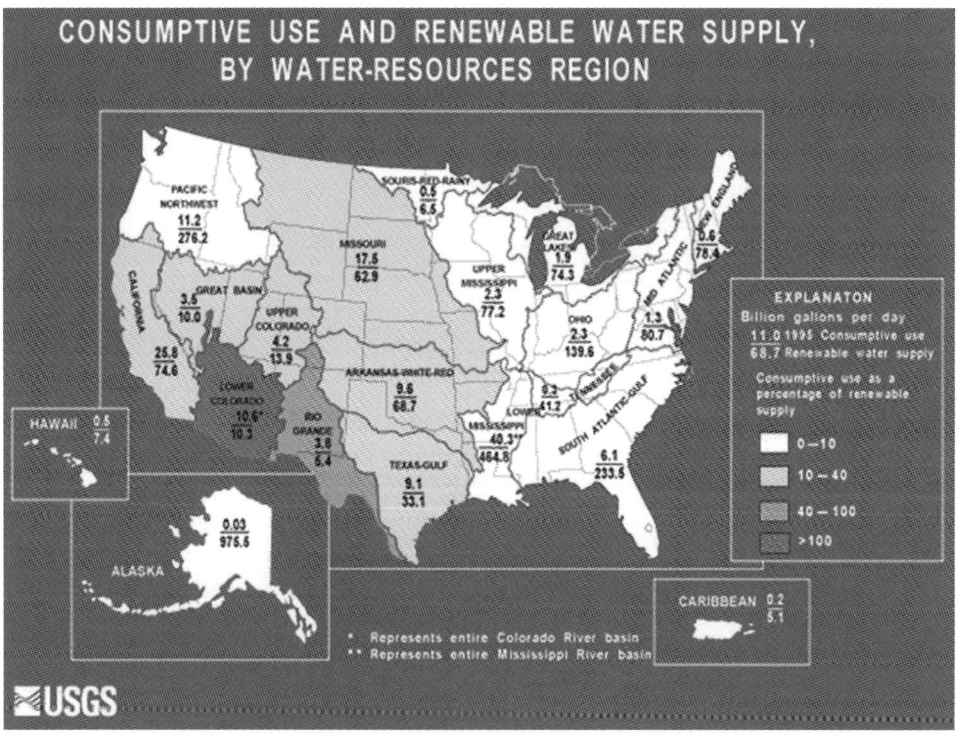

Figure 7.1. Comparisons of average consumptive use and renewable water supply for the 21 water-resources regions of the United States, Puerto Rico, and US Virgin Islands. Adapted from USGS 1995 (http://water.usgs.gov/watuse/misc/consuse-renewable.html). For color detail, please see color plate section.

dramatic climatic extremes. A recent multiyear period with intermittent drought in the Southeastern United States caused serious water shortages in the region. The 2007 Southeastern drought was historic and had record high temperatures (Heim *et al.*, 2008). The drought started in December 2006 and expanded with time, with more than two-thirds of the region in drought from midsummer 2007 through the end of the year. Additionally, unusually warm temperatures and scarce rainfall exacerbated conditions. Many Southeastern states (Kentucky, Tennessee, Alabama, Georgia, South Carolina, and Florida) had a record warm August. The cumulative effects of the drought resulted in the driest year in the 113-year record for North Carolina, second driest for Tennessee, third driest for Alabama, and fourth driest for Georgia. In five southeastern states, the governors declared water conservation measures and drought disasters, or states of emergency. The multiyear drought period also led to the imposition of restrictions on water use and opened up legal conflicts within and between states on the regulation and use of the region's water resources (Seager *et al.*, 2009). Alabama and Florida successfully sued Georgia over withdrawing water from Lake Lanier, the main source of drinking water for the Atlanta metro region (Manuel, 2008). Lake Lanier feeds the Chattahoochee River, which supplies water to towns in Alabama and Florida, and whose flow is key to the survival of a host of endangered species such as freshwater mussels and sturgeon. The three states have feuded since 1989 over how to divide the water, but the drought has worsened the problem as the various parties fight over a much-reduced volume of water.

The water restrictions have also impacted agriculture. To reduce water use during declared droughts, farmers in Southwestern Georgia would be paid not to irrigate crops in order to maintain base stream and river flows exiting the state (USA Today, 2002; GA-DNR, 2008). Similarly, in another major agricultural region of the United States, farmers in Nebraska were paid not to irrigate along the Republican and Platte Rivers. This was also a result of multiyear drought conditions (NE-DNR, 2005; US-Water News Online, 2005; NE-FSA, 2007). Although these reductions in irrigation in the Southeastern and Midwestern United States are troubling for agricultural production, irrigation reductions are more common in the Western United States. In many Western US states, cities have purchased water rights from farmers to meet urban and industrial needs (Brown, 2003). These droughts throughout the United States have highlighted the delicate balance that faces agricultural production in competition with urban, industry, and environmental water uses. The competition for water will only be exacerbated by the energy crop production.

Water Limitation Impacts on Bioenergy

As can be seen in the previous section, climatic variability resulting from flooding, droughts, and the timing in water availability can have a tremendous impact on both crop and biomass production. To examine the potential impact climatic variability would have on bioenergy derived from biomass, Eaves and Eaves (2007) used historical data to estimate the supply risk of ethanol (as an automotive fuel) relative to imported petroleum. They compared historical maize production data (1960–2005) with oil imports to determine the relative reliability of ethanol as an automotive motor fuel. Their analysis fitted distributions to both annual maize yields and yearly oil imports. They found through analyzing the distributions that variations of oil imports were less than half those of annual maize yields concluding that maize production was more volatile than oil imports. Most of this increased volatility of maize and ethanol production was attributed to their dependency on weather. Based on their historical analysis, they surmised that displacing gasoline with ethanol would be exchanging geopolitical risk with yield risks.

All climate change predictions point toward increased variability in both temperature and rainfall extremes. If these extremes were incorporated into the Eaves and Eaves (2007) model, increased variability in grain production would be expected. Climate change is predicted to have significant impacts on agricultural production in the future. Many of these changes have been researched relative to their impact on food productivity (Slingo *et al.*, 2005). Climate change will also impact the productivity of biomass and bioenergy crops. These impacts need to be identified and incorporated into decisions related to bioenergy production. Additionally, Elcock (2010) investigated the potential impact of domestic energy production (fossil and renewable fuels) targets may have on future water demand. He estimated that water consumed for energy production would be expected to increase by nearly 70%, and water consumed for biofuels' (biodiesel and ethanol) production is expected to increase by almost 250%. Most of the increases would be for irrigation in the West North Central region of the United States. Increased water consumption could be significant and contribute to localized unintended impacts on water resources. Future biofuel production locations should be evaluated for their potential impacts on local water resources.

Water Use for Major Bioenergy Crops for Ethanol

Traditional agriculture for food and fiber production is the largest user of fresh water throughout the world. The FAO (2008) estimated that agriculture is using a global average of 70% of all freshwater withdrawals from rivers, lakes, and aquifers. In the United States, it is estimated that agricultural water consumption for irrigation is 80% of the total water consumed (Solley *et al.*, 1998). Currently, the two major crops used for ethanol production are sugarcane and maize in Brazil and the United States, respectively. These crops were analyzed and compared as currently managed to determine their relative water utilization during the production of the biomass feedstocks.

Sugarcane Production in Brazil

Brazil is recognized as the world's second largest producer of ethanol (DOE-EIA, 2007; Trostle, 2008). It began promoting the production of crops for ethanol in the mid-1970s after the first global energy crisis (Rother, 2006). Within 10 years, more than three-quarters of cars made in Brazil were able to run on ethanol. The primary crop that Brazil uses for ethanol production is sugarcane. Brazil is the world's largest producer of sugarcane (FAO, 2008: 420 121 000 metric tonnes). Sugarcane is a tall perennial grass native to warm temperate to tropical regions of the world. It has stout, jointed, fibrous stalks that are rich in sugar and measure 2–6 m tall. Sugarcane's high concentration of sugar, which is readily available to microorganisms, makes it uniquely suitable for ethanol production.

Water use in the production of ethanol can be divided between crop production and ethanol production. The water requirement for sugarcane production is approximately 8–12 mm tonne^{-1} of cane production. The sugarcane growing season is year round, and the annual requirements for sugarcane production are approximately 1500–2500 mm yr^{-1} (Moreira, 2007; Goldemberg *et al.*, 2008). The majority of the sugarcane plantations in Brazil rely on rainfall complemented by partial ferti-irrigation, carried out mainly to manage water wastes. Most plantations limit their production to regions where reasonable rainfall occurs (Moreira, 2007). Therefore, irrigation use in Brazil for agricultural production is generally small. However, due to the increasing demand for ethanol and the high prices paid for it, sugarcane production is expanding to regions where irrigation would be needed to complement rainfall (Goldemberg *et al.*, 2008).

In these cases, Moreira (2007) reported that irrigation can be economically feasible, especially using more efficient application methods such as drip irrigation. In Brazil, traditional surface irrigation accounted for approximately 50% of the total irrigation. This surface water application efficiency is fairly low (~61% on average). New production areas could make use of more efficient irrigation application systems.

Moreira (2007) reported that there is generally sufficient water to supply all foreseeable long-term water requirements of Brazil as a whole, but local water shortages can occur as a result of the occurrence of various water using sectors (competition between industry, agriculture, and urban use). Fortunately, the Pacific Institute (2008) reported that Brazil had the largest annual renewable freshwater supply in the world (8233 $km^3\ yr^{-1}$).

The processing and converting of sugarcane to ethanol requires large amounts of water and is the second major use of water in Brazil. Water is used in four major processes: cane washing, condenser/multijet in evaporation and vacuum, fermentation cooling, and alcohol condenser cooling. In 1997, it was calculated that the water use in processing was approximately 21 $m^3\ tonne^{-1}$ of cane. However, Macedo (2005) reported that most of the water used in the processing is recycled. Improvements in efficiencies in the production processes have reduced the consumption of water from 5.3 $m^3\ tonne^{-1}$ in the 1990s to reported values for 2004 of 1.83 $m^3\ tonne^{-1}$ (Goldemberg *et al.*, 2008).

Maize Production in the United States

In the United States, the current major source of ethanol production is maize. This is expected to increase while the industry develops new methods for producing bioenergy. In 2007, the National Corn Growers Association (NCGA) estimated that 24.7% of the US domestic maize went into ethanol production (see Figure 7.2). This was an 18% increase from only 7% in

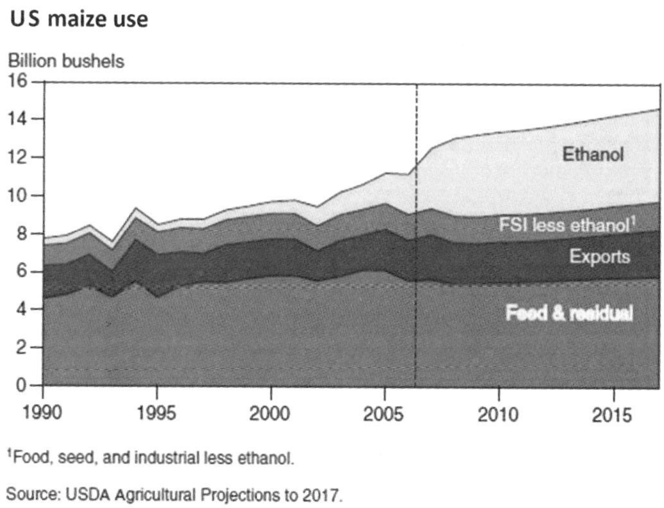

Figure 7.2. Projected U.S. Maize Use from USDA-Economic Research Service (ERS), 2008. For color detail, please see color plate section.

Table 7.1. Top ten maize producing states from the 2002 Census of Agriculture (USDA-NASS, 2004) and 2008 National Agricultural Statistics Service (NASS, USDA-NASS, 2008).

		2002 Census of Agriculture					
Geographic area	2008 NASS Area (ha)	Number of farms	Area (ha)	Number of irrigated land farms	Irrigated land area (ha)	Total area irrigated (%)	Rank in irrigation
United States	34,795,681	348,590	27,611,913	34,278	3,929,446	14	
Iowa	5,382,319	52,806	4,759,666	416	34,909	1	17
Illinois	4,896,696	41,032	4,347,452	913	85,456	2	6
Nebraska	3,561,234	23,889	2,972,301	14,448	1,823,343	61	1
Minnesota	3,116,079	31,782	2,653,152	973	72,219	3	9
Indiana	2,306,708	24,156	2,073,322	767	72,967	4	7
South Dakota	1,922,257	11,446	1,280,907	717	49,869	4	14
Ohio	1,335,463	23,898	1,161,428	26	1,371	0	39
Wisconsin	1,537,805	29,021	1,158,223	501	33,833	3	18
Missouri	1,133,120	15,655	1,083,542	970	99,680	9	5
Kansas	1,558,040	9,552	1,009,358	3,328	545,033	54	2

2001. Although these figures make it appear that a large tonnage of the US maize crop was being diverted into the production of ethanol and away from other uses, the overall impact was offset by increases in maize production and storage drawdown (Trostle, 2008). All other uses of maize have remained approximately unchanged (2001–2007), except for a small decrease in maize exports.

In the United States, the vast majority of maize is produced in the Midwestern states (Table 7.1, USDA-NASS, 2004). Many of the maize producing states (Iowa, Illinois, etc.) have adequate annual rainfall associated with deep rich soils, with adequate water holding capacity to produce maize without supplemental irrigation. However, other maize producing states in the Midwest and High Plains utilize considerable irrigation to produce maize (Table 7.2, e.g., Nebraska, Kansas). Here, the major source of irrigation water in the High Plains region is the

Table 7.2. Top ten irrigated maize producing states from the 2002 Census of Agriculture (USDA-NASS, 2004) and 2008 National Agricultural Statistics Service (NASS, USDA-NASS, 2008).

		2002 Census of Agriculture					
State	2008 NASS Area (ha)	Number of farms	Area (ha)	Number of irrigated land farms	Irrigated land area (ha)	Total area irrigated (%)	Rank in maize production
Nebraska	3,561,234	23,889	2,972,301	14,448	1,823,343	61	3
Kansas	1,558,040	9,552	1,009,358	3,328	545,033	54	10
Texas	930,777	5,102	734,731	1,691	266,355	36	12
Colorado	505,857	1,991	286,597	1,845	256,577	90	16
Missouri	1,133,120	15,655	1,083,542	970	99,680	9	9
Illinois	4,896,696	41,032	4,347,452	913	85,456	2	2
Indiana	2,306,708	24,156	2,073,322	767	72,967	4	5
Michigan	971,246	13,613	812,213	857	72,949	9	11
Minnesota	3,116,079	31,782	2,653,152	973	72,219	3	4
California	271,139	592	68,130	592	68,065	100	29

Ogallala Aquifer. In Nebraska, it is estimated that 95% of the total groundwater withdrawals for irrigation is from the Ogallala Aquifer (Maupin and Barber, 2005).

The Ogallala Aquifer underlies approximately 45 Mha in parts of eight states—Colorado, Kansas, Nebraska, New Mexico, Oklahoma, South Dakota, Texas, and Wyoming. The water from the aquifer was initially tapped around 1900. The Ogallala Aquifer is also the major regional source of water for municipal and industrial users. Starting with the Dust Bowl in the 1930s, the occurrence of repeated droughts along with the widespread installation of irrigation systems, water levels in most regions of the aquifer have declined dramatically. The water stored in the aquifer is generally referred to as geologic water because it is generally thought that rainfall takes hundreds or thousands of years to reach low-permeability areas in the aquifer which impede downward water flows to the water table (Andrews *et al.*, 1999). This slow recharge coupled with large water consumption has resulted in declining water tables over most of the aquifer.

The High Plains Aquifer states are some of the top maize producing states. Nebraska was the third largest maize producing state (3.6 Mha), with Kansas, Texas, and Colorado also producing over 0.5 Mha each (NCGA and NASS data). USDA-NASS (2004) estimated that 19% of all irrigation in the United States was for maize production. The USDA-NASS (2004) also estimated that the states overlying the High Plains Aquifer (Nebraska, Kansas, Colorado, and Texas) accounted for approximately 90% of irrigated US maize acreage (Figure 7.3).

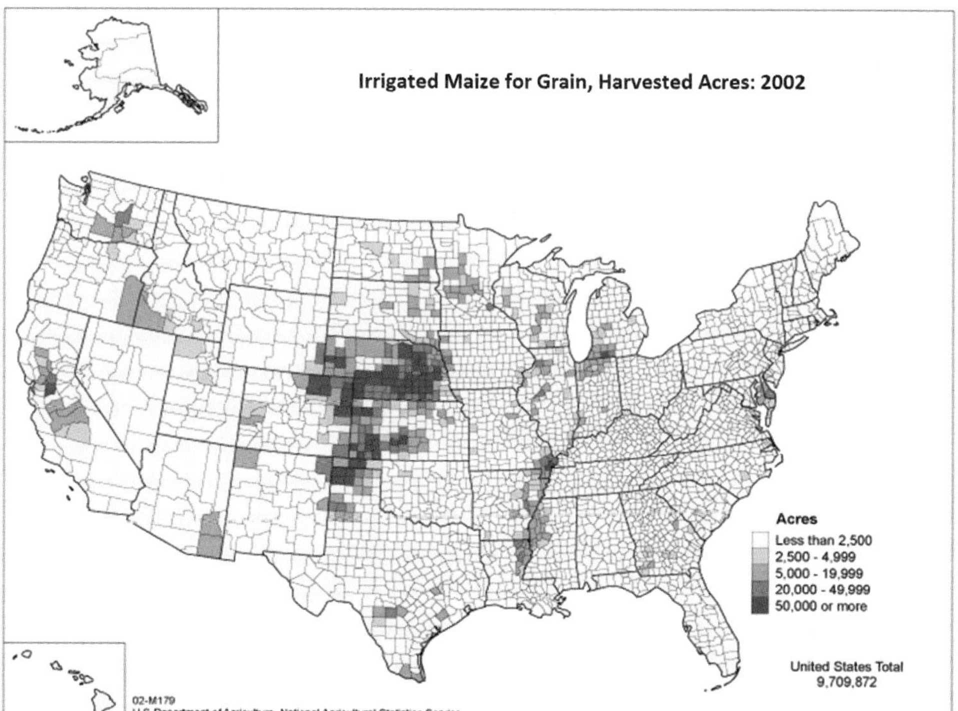

Figure 7.3. US Irrigation maize for grain (http://www.agcensus.usda.gov/Publications/2002/Ag_Atlas_Maps/Crops_and_Plants/index.asp). For color detail, please see color plate section.

Table 7.3. Comparison of water requirements for ethanol production from maize grain, sugarcane, and other potential energy crops.

Crop	Water requirements (m^3 water Mg^{-1} crop)	Biofuel conversion (l fuel Mg^{-1} crop)	Crop water requirement for biofuel (m^3 water Mg^{-1} fuel)	Crop water requirement per unit energy (m^3 water GJ^{-1})
		Ethanol		
World maize (grain)	833	409	2580	97
World sugarcane	154	334	580	22
Nebraska maize (grain)	634	409	1968	74
Maize stover	634	326	2465	92
Maize stover + grain	634	735	1093	41
Switchgrass	525	336	1980	74
Grain sorghum	2672	358	9460	354
Sweet sorghum	175	238	931	35
		Biodiesel		
Soybean	1818	211	9791	259
Canola	1798	415	4923	130.2

Source: World Maize and sugarcane estimates from Postel (1998); Nebraska maize estimates from Nebraska Maize Board (2008); Soybean and grain sorghum (FAO, 1991); Sweet sorghum (Bennett and Anex, 2008; Mastrorilli *et al.*, 1999); Canola (Bauder, 2009); and Switchgrass (Robins *et al.*, 2009; Wright, 2007).

In Nebraska, irrigated maize averages 70% of the total maize acreage (Nebraska Maize Board, 2008, and Table 7.2). In 2007, the Nebraska Maize Board (2008) reported average maize yields of 10 Mg ha^{-1} (160 bu ac^{-1}). The mean 2007 maize yield from Nebraska of 10 Mg ha^{-1} would require approximately 635 mm (~25 in.) of water. Combining the yield with the water requirements for the production would result in approximately 635 m^3 water Mg^{-1} grain. If this is combined with a maize to ethanol conversion rate of 25.9 l ethanol Mg^{-1} grain (2.5 gal ethanol bu^{-1} maize), it would result in a ratio of 1968 m^3 water Mg^{-1} (or 1553 m^3 water/m^3) ethanol. In most of the nonirrigated maize producing areas, the water would be from rainfall and moisture stored in the soil. In irrigated maize production regions such as Nebraska, the NASS reported that the average irrigation for maize production was 365 mm (1.2 ac-ft). This irrigation water requirement would be approximately half of the water needed for maize production. Most of this water in the region would come from the High Plains Aquifer that is already experiencing rapidly declining water levels.

Comparison of Sugarcane and Maize Water Usage

For comparison of maize water use for ethanol production with sugarcane, we used the world average production estimates for sugarcane and maize grain from Postel (1998). We calculated and contrasted the estimates for water requirements for crop production to produce ethanol (Table 7.3). Ethanol conversion from biomass to ethanol was estimated per Mg of maize grain and sugarcane, as 409 and 334 l of ethanol, respectively (James, 2008; US-DOE, 2008). The resulting calculated water requirements were 2580 m^3 water Mg^{-1} (2036 m^3 water/m^3) ethanol for maize grain and 580 m^3 water Mg^{-1} (458 m^3 water/m^3) ethanol for sugarcane. They are also photosynthetically very different. The water requirements to produce maize grain were much higher than water required for sugarcane. The main reason for the greater water requirements for maize grain was that only grain is currently utilized for ethanol production. As new technologies for cellulosic conversion of biomass for ethanol production are discovered, the relative difference between the crop water requirements will be reduced. However, the amount of maize biomass that can be used is limited by soil resource issues such as erosion and nutrient depletion (Doran *et al.*, 1984; Mann *et al.*, 2002; Wilhelm *et al.*, 2004). Again, the large disparity was mainly due to only grain being utilized for ethanol production with maize. Additionally, maize has a much shorter growing season than sugarcane. This shorter growing season with higher water demands and temperature/pollination vulnerability makes maize grain production vulnerable to the following: short-term droughts, lack of supplemental water supplies for irrigation, or excess water from floods. This vulnerability has been previously described with the citation of Eaves and Eaves (2007).

Water Demands for Billion-Ton Vision

Reports by the US Departments of Energy and Agriculture entitled "The Technical Feasibility of a Billion-Ton Annual Supply of Biomass Feedstock for Bioenergy and Bioproducts Industry" (US-DOE, 2005) and "U.S. Billion-Ton Update: Biomass Supply for a Bioenergy and Bioproducts Industry" (US-DOE, 2011) estimated the "potential" biomass within the contiguous United States for bioenergy production. The reports did not directly address the water resource requirements for the Billion-Ton Visions. It assumes that sufficient water resources would be available for increased crop and biomass production. A component of the report focused on the potential for increased grain production for biofuel production.

The US-DOE (2011) study estimated that the US agricultural lands could increase maize production from 27 million dry tonnes in 2012 to 80 million dry tonnes in 2030 at a simulated farmgate price of $40 per dry tonne. These estimates are slightly different from that of a 2005 report as the 2012 update estimated the maize crop availabilities at farmgate prices up to $40, $50, and $60 per dry tonne. Although the 2012 report included the consideration of managing soil carbon during crop residue removal and the irrigation limitation for energy crops especially in western states, the 2012 update still assumed that adequate water resource would be available for maize production to support the Vision. If we assume the same water demands as that for Nebraska maize (634 m^3 H_2O/Mg maize), the Vision currently requires 1.89×10^{10} m^3 (15.3 million ac-ft water), and 5.59×10^{10} m^3 (45 million ac-ft water) by 2030. This is approximately a threefold increase in water demand for production to meet the goals in 2030. Even if geneticists obtained a doubling in water use efficiency to complement the increased yields, the amount of water needed to produce the increased grain would put tremendous stress on current water resources. Alternative cropping and water management approaches must be implemented to meet the Billion-Ton Vision.

Potential Alternatives

Solving our needs for renewable energies while preserving our water resources is an extremely complex problem and will require innovative thinking and adaptation. Researchers throughout the United States and the world are aggressively addressing the issue. In this section, we offer a few examples that could be implemented to address some of the interconnected problems of water and bioenergy and assess their impact.

Crop Production Alternatives and Impact of Cropping Shifts

Karp and Shield (2008) reported on the challenge of producing bioenergy from plants and sustainable yields. They reported that bioenergy from plants, particularly perennial grasses and trees could make substantial contributions to both mitigating climate change and increasing biofuel supplies. The focus of their report was yield traits of key bioenergy crops. They targeted specific traits in these crops for future improvements. From their studies, a common theme was apparent. It was a well-known and very true reality—production of all biomass crops depends greatly on water. Their concluding topic for future work was "*increasing aboveground biomass without increasing water use*." Biomass yields from most of the commonly discussed bioenergy crops (row crops: maize, wheat, etc.; perennial grasses: switchgrass, *Miscanthus*, etc.; and fast-growing trees: poplar, willows, etc.) were all identified as being highly susceptible to shortages of water.

On the other hand, sweet sorghum has long been recognized as a potential sugar crop and more recently for ethanol conversion (Prasad *et al.*, 2007). Sweet sorghum has greater water use efficiency, is more drought tolerant, and requires only 36% the nitrogen fertilizer required by maize. Additionally, sweet sorghum has shown potential for ethanol production due to its rapid growth and early maturity.

Care must be used if major shifts in crop production are implemented without considering the potential changes in water usage. While shifts among crops with similar water usages will have little impacts on water resources, shifts to crops requiring a significant increase in water usage can dramatically impact water supplies. Farley *et al.* (2005) reported that shifting from grasslands to forest could reduce runoff and intensify water shortages. Jackson *et al.* (2005) reported on water impacts of trading water for carbon with biological carbon sequestration. Their study highlighted the potential impacts of introducing tree plantation strategies without considering the full environmental consequences, particularly on water availability. They combined field research with climatic and economic modeling to document substantial losses in stream flows with afforestation. They reported that over the life of the forest plantation, stream flows decreased globally by 53% annually with 13% of streams drying completely for at least 1 year. Powell *et al.* (2005) found a mature forest under varying annual rainfall consumed approximately 85% of annual precipitation. Buytaert *et al.* (2007) reported that converting from natural grasslands to pine plantations resulted in an increase of 40–70% in evapotranspiration.

Reijnders (2006) reported that freshwater resources were not well addressed in previous estimates of biomass for energy potentials. He reported that expanding biomass-for-energy production may substantially exacerbate the world's already scarce water resources for food production. He also pointed out the use of short rotation trees and woody crops consume considerable water and that water availability should be considered as major criteria for site selection. In Mississippi, the unintended impact of large-scale crop conversion was reported by Welch *et al.* (2010). Due to the demand for biofuels and commodity prices, maize acreage increased 288% in the Mississippi Delta along with a concurrent 47% decrease in cotton acreage. This shift to maize production has had implications for both water quantity and quality in the Delta. This includes the loss of cotton seed as a significant accumulator of available soil nitrogen (Hunt *et al.*, 1998). They reported accelerated water-level declines in the Mississippi River Valley alluvial aquifer. Additionally, they estimated through simulation that increased nitrogen fertilizer for maize production will likely result in increased nitrogen export. They concluded that shifting from cotton to maize production may further contribute to the hypoxic conditions in the Gulf of Mexico.

Using Treated Agricultural Effluent for Bioenergy Crop Irrigation

An alternative to utilization of high-quality, fresh ground and surface water for irrigation is to maximize the use of treated and recycled waters for energy crop production. A particularly interesting option would turn liabilities into benefits. For example, the State of North Carolina (NC) is the second largest pork producing state in the United States. Alongside the growth of the pork production is the generation of a large quantity of liquid animal waste. This waste needs to be utilized in an environmentally sustainable manner. This waste from swine production had typically been treated in anaerobic wastewater lagoons. During the 1990s, tropical storms and hurricanes caused many lagoons to fail and spill excess nutrients into surface waters. These lagoon failures along with public outcry led to a search for better treatment methods in 1997 (NC, 1997). During this search, the State of NC along with major pork producers entered into an agreement to investigate new treatment systems for swine wastewater treatment and management. Results from this agreement produced a system that could meet the defined environmental standards (Vanotti *et al.*, 2007). The new system removed solids and significantly reduced the nutrient concentration in effluent waters. This treatment option would essentially convert the on-farm anaerobic lagoons into water storage facilities. These water storage facilities would be easier to manage for water application to crops due to the significantly lower nutrient concentrations. The NC Department of Agriculture estimated that there were approximately 4000 active swine wastewater lagoons on 2500 farms. With an average lagoon size of approximately 1 ha, each would store approximately 23 000 m^3. Statewide, the potential increased water available for potential energy crop production would provide approximately 92 million m^3 of water storage. This quantity of water applied at an average application depth of 150 mm (USDA-NASS, 2004) would provide enough water to irrigate approximately 60 000 ha. This area is potentially double the existing irrigated area in NC and provides an excellent resource for producing biomass energy crops. Stone *et al.* (2008) conducted a study in North Carolina comparing treated effluent with conventional fertilizers for bermudagrass production. They found using treated swine wastewater effluent produced significantly higher bermudagrass hay yields. Cantrell *et al.* (2009) analyzed the biomass samples from the Stone *et al.* (2008) study for the energy content and found with the increased biomass quantity, there was more biomass energy potential from the bermudagrass grown with treated wastewater effluent. Thus, irrigation with treated wastewater provides a means to irrigate future bioenergy crops without burdening local water resources, while at the same time not excessively overloading the crops with nutrients. This utilization of treated wastewater could offset the impacts of utilizing higher quality well and surface water for growing energy crops in regions of the country that utilize similar swine wastewater treatment systems.

Controlled Drainage and Water Table Management

Increased production of energy crops such as maize in the Midwestern states would require additional use of fertilizers to enhance productivity. The NRC (2008) reported that fertilizers applied to increase agriculture yields can result in excess nutrients flowing into waterways via surface runoff and infiltration to groundwater, and will have a significant impact on water quality. Excess nitrogen in the Mississippi River system is known to be a major cause of the oxygen-starved "dead zone" in the Gulf of Mexico. In much of the upper Midwest region, agricultural drainage has been a major contributor to the success of agricultural production. In the United States, Pavelis (1987) estimated that there were over 20 Mha of agricultural drainage in the Midwestern states and that the Southeast and Atlantic regions had more than 9 Mha.

The overall reason for implementing agricultural drainage was to enhance crop production. Drainage systems allowed timely seedbed preparation, planting, harvesting and other field operations while protecting field crops from extended periods of flooded soil conditions.

While there are many positive aspects of land drainage, there are potential adverse aspects as well (Rabalias *et al.*, 1996; Kanwar, 2006; Jaynes and Colvin, 2006; Hunt *et al.*, 2008). Excess nutrients in drainage water can lead to local water quality problems and potentially contribute to hypoxia in larger water bodies, coastal estuaries, and the Gulf of Mexico. Strategies are needed that reduce nutrient loads while maintaining adequate drainage for crop production.

These improved management systems are often referred to as controlled drainage or water table management. These systems utilize structures to control the water levels in agricultural fields, drainage ditches, and even watersheds (Gilliam and Skaggs, 1985; Stone *et al.*, 1992; Evans *et al.*, 1992; Madramootoo *et al.*, 2007). These systems allow timely drawdown of water levels for agricultural operations and prevent excessive nutrient-rich water from being discharged. They can increase the water storage capacity in the soil profile and increase crop water use efficiency (Stampfli and Madramootoo, 2006).

The implementation of controlled drainage systems has been identified as a tool to mitigate the adverse effects of uncontrolled drainage (Thomas *et al.*, 1992; Thomas *et al.*, 1995; Fausey, 2004; Fouss *et al.*, 2004). To address the potential adverse effects of uncontrolled drainage, the Agricultural Drainage Management Systems Task Force (ADMS) was established in 2003 to improve drainage practices and reduce adverse impacts while enhancing crop production and conserving water (http://hostedweb.cfaes.ohio-state.edu/usdasdru/ADMS/ADMSindex.htm). Controlled drainage decreases the peak outflow (Amatya *et al.*, 1998; Tan *et al.*, 1999) from drainage systems and reduces nitrate-nitrogen concentration in drainage outflows (Mejia and Madramootoo, 1998; Elmi *et al.*, 2002; Evans *et al.*, 2007). In a review of several studies, Evans *et al.* (1995) reported nitrogen and phosphorus reductions of 30% and 50% resulting from controlled drainage. Skaggs *et al.* (2010) reported on the effects of controlled drainage on water and nitrogen balances in drained lands. From the literature, they found that controlled drainage reduced drainage volumes and N losses in drainage waters by 17% to over 80%, depending on soil properties, crops, drainage intensities, control strategies, and location. The nitrogen reductions were attributed to denitrification both in the soil profile and in the reduced zones along the seepage paths. Implementation of these systems could improve and mitigate potential water quality problems associated with increased production of potential biofuel crops identified by the National Research Council (NRC, 2008).

Biofuel Generation via Thermochemical Conversion

While ethanol production from maize grain seems to be the current focus of most biofuel production efforts, we have established that if maize grain ethanol is the sole biofuel, then an immense amount of water is needed to supply the Billion-Ton Vision. Thus, endeavors must be made to convert biomass feedstocks beyond maize grain; these feedstocks can include cellulosic biomass as well as agricultural residuals, animal manures, and municipal solid waste (MSW). Even though these feedstocks will eventually be converted biochemically (fermentation), these processes leave a carbon-rich residual that still contains inherent energy; thus, the feedstocks are not broken down to their full energetic potential. Additionally, the biochemical conversion process by nature has a huge water requirement, thereby adding to the sustainability concerns.

Compared to traditional biochemical conversion processes, thermochemical conversion processing of bioenergy crops holds the promise of better feedstock versatility, improved conversion efficiency, greater energy yields, and enormously lower water use. For this type of high-temperature conversion to produce a liquid biofuel, two options stand out—pyrolysis and gasification. Pyrolysis takes advantage of high temperatures and an inert atmosphere to convert organic (carbonaceous) material into one primary product: either a carbonized solid similar to charcoal (bio-char) or a combustible bio-oil. The bio-char is amenable as a feedstock ("green coal") for existing coal combustion and gasification plants.

The bio-oil has potential to be used as a combustible heat source and fungible hydrocarbon fuel. However, the major limitation is the instability of the bio-oil inherent with its acidity, high water content, high oxygen content, and reactive components (e.g., char particulates). Successful stabilization and upgrading of bio-oil would allow for easier assimilation by both existing combustion applications and petroleum refinery infrastructure. Upgrading bio-oils was successfully demonstrated as a jet fuel during a hydroplane test run (http://bioweb.sungrant.org/Technical/Biopower/Technologies/Pyrolysis/Pyrolysis+Oil/Pyrolysis+Oil.htm). This test run, conducted by Honeywell-UOP and Boeing, used a 10% renewable jet fuel containing 2% aromatics from upgraded woody bio-oil. The time to certification for bio-oil fuel substitutes is anticipated to take 5 years (http://bioweb.sungrant.org/Technical/Biopower/Technologies/Pyrolysis/Pyrolysis+Oil/Pyrolysis+Oil.htm). With other commercialization and research and development projects underway, pyrolysis has the potential for farm- and crop-scale implementation (Cantrell *et al.*, 2008).

Gasification is the process that uses gasification media (e.g., steam, air, or oxygen) at high temperatures to convert organic materials into gaseous products. Gasification converts the chemical energy found in the carbon bonds into heat and a combustible gas consisting primarily of CO and H_2, or synthesis gas. This "syngas" product can be purified and used in a variety of ways: heat and power generation, transportation fuels, and chemical intermediates (McKendry, 2002a; Cantrell *et al.*, 2008). For gasification to be effective, the biomass moisture content should be below 10–15% (McKendry, 2002b); a moisture content of about 30% impedes ignition and reduces the heating value of the product gas. Thus, use of this type of bioenergy conversion process eliminates the need for a water input. In fact, recycling and recovering the heat in the product gas as a means to drive away the moisture in the feedstock would be one way farmers and practitioners could recover water without additional energy demands, making the entire gasification process sustainable for bioenergy production.

Conclusions

The expanded production of agricultural crops for bioenergy production has introduced new challenges for management of water. Water is now being called upon for bioenergy production, thereby stretching an already vital resource. Water availability has been widely presumed as a nonlimiting factor in the discussion of bioenergy crop production. However, water is a limited resource. Many parts of the world are experiencing water scarcities complicated by a growing population. Water scarcities are not only impacting humans and agriculture for food production but also the environment.

Bioenergy biomass crop production is highly dependent on water. Thus, weather variability including droughts and floods can greatly impact bioenergy availability. Climate change is impacting water resources throughout the world. In the Western United States, water supplies are highly vulnerable to climatic changes that affect snowpack. Over the last century, runoff from snowmelt in the Western United States is occurring earlier and shifting soil moisture recharge and evapotranspiration earlier in the year. Weather extremes (droughts and flooding) are also affecting other areas of the United States. Flooding and droughts have been occurring with a higher frequency in the Central United States with a higher percentage of extreme events occurring in the Western United States and to some extent the Eastern United States. These climatic extremes highlight the potential vulnerabilities and disruptions in future bioenergy supplies.

Caution should be used in shifting to alternative biomass crops without considering their impact on water resources. Introducing tree plantations and woody crops can significantly increase water use and reduce both runoff and stream flow. Reduced flows may be gradual and initially unnoticed but can be significant over time. Shifting agronomic crops can also impact seasonal water demands and off-site nutrient movement.

Maize and sugarcane are the current major crops used for ethanol production. The water requirements to produce maize grain for ethanol are much higher than to produce sugarcane. The main reason for the greater water requirements for maize grain was that only grain is currently utilized for ethanol production. The water requirements for maize grain production to meet the US-DOE Billion-Ton Vision would increase approximately sixfold to meet the 2030 production goals and may put tremendous stress on current water resources. As new technologies for cellulosic conversion of biomass for ethanol production are demonstrated and improved, the relative difference between the crop water requirements may be reduced. Furthermore, thermochemical conversion utilizing a wider variety of feedstocks for bioenergy may emerge as a more sustainable option.

Alternative water management and technology systems can be implemented to improve water availability and produce bioenergy. New treatment systems for livestock waste offer the potential for utilizing treated effluent to irrigation and grow bioenergy crops. Controlled drainage can increase plant available water and improve water quality. The projected increase in global population and competition for water resources among urban, industrial, economic, and environmental sectors will impact the water available for food and bioenergy production. Consequently, water needs to be incorporated into discussions and decisions related to the implementation and technology for bioenergy. To be sustainable, biomass crop production for bioenergy must conserve and protect natural resources—including fresh water.

References

Amatya, D.M., Gilliam, J.W., Skaggs, R.W., Lebo, M.E., and Campbell, R.G. (1998) Effects of controlled drainage on forest water quality. *Journal of Environmental Quality* 27: 923–935.

Andrews, W.J., Osborn, N.I., and Luckey, R.R. (1999) Rapid recharge of parts of the High Plains Aquifer indicated by a reconnaissance study in Oklahoma. U.S. Geological Survey Fact Sheet 137-00. Available at: http://www.owrb.ok.gov/studies/reports/reports_pdf/high_plains_2.pdf (accessed 19 March 2012).

Barnett, T.P., Pierce, D.W., Hidalgo, H.G., *et al.* (2005) Human-induced changes in the hydrology of the western United States. *Science* 319: 1080–1083.

Bauder, J.W. (2009) The right strategy for irrigating your canola crop. Department of Land Resources and Environmental Sciences, Montana State University, Bozeman, MT. Available at: http://waterquality.montana.edu/docs/irrigation/canolastrategy.shtml (accessed 19 September 2012).

Bennett, A.S. and Anex, R.P. (2008) Farm-gate production costs of sweet sorghum as a bioethanol feedstock. *Transactions of the ASABE* 51: 603–613.

Brown, L.R. (2001) How water scarcity will shape the new century. *Water Science and Technology* 43: 17–22.

Brown, L.R. (2003) *Plan B 2.0: Rescuing a Planet Under Stress and a Civilization in Trouble.* W.W. Norton and Company, New York.

Buytaert, W., Celleri, R., De Bievre, B., and Iniguez, V. (2007) The impact of pine plantations on water yield: A case study from the Ecuadorian Andes. *IAHS-AISH Publication* 317: 225–228.

Cantrell, K.B., Ducey, T.F., Ro, K.S., and Hunt, P.G. (2008) Livestock waste-to-bioenergy generation opportunities. *Bioresource Technology* 99: 7941–7953.

Cantrell, K.B., Stone, K.C., Burns, J.C., Ro, K.S., Vanotti, M.B., and Hunt, P.G. (2009) Bioenergy crop production of swine wastewater treated coastal bermudagrass. *Bioresource Technology* 100: 3285–3292.

Diamond, J. (2005) *Collapse: How Societies Choose to Fail or Succeed*, Viking Press, New York, 592p.

DOE-EIA. (2007) Country analysis briefs: Brazil. Department of Energy, Energy Information Administration, Washington, DC. Available at: http://www.eia.doe.gov/emeu/cabs/Brazil/pdf.pdf (accessed 20 September 2012).

Doran, J.W., Wilhelm, W.W., and Power, J.F. (1984) Crop residue removal and soil productivity with no-till corn, sorghum, and soybean. *Soil Science Society of America Journal* 48: 640–645.

Eaves, J. and Eaves, S. (2007) Renewable corn-ethanol and energy security. *Energy Policy* 35: 5958–5963.

Elcock, D. (2010) Future U.S. water consumption: The role of energy production. *Journal of the American Water Resources Association* 46: 447–460.

Elmi, A.A., Madramootoo, C., Egeh, M., Dodds, G., and Hamel, C. (2002) Water table management as a natural bioremediation technique of nitrate pollution. *Water Quality Research Journal of Canada* 37: 563–576.

Evans, R.O., Bass, K.L., Burchell, M.R., Hinson, R.D., Johnson, R., and Doxey, M. (2007) Management alternatives to enhance water quality and ecological function of channelized streams and drainage canals. *Journal of Soil and Water Conservation* 62: 308–320.

Evans, R.O., Parsons, J.E., Stone, K.C., and Wells, W.B. (1992) Water table management on a watershed scale. *Journal of Soil and Water Conservation* 47: 58–64.

Evans, R.O., Skaggs, R.W., and Gilliam, J.W. (1995) Controlled versus conventional drainage effects on water quality. *Journal of Irrigation and Drainage* 121: 271–276.

Falkenmark, M. and Lannerstad, M. (2005) Consumptive water use to feed humanity – curing a blind spot. *Hydrology and Earth System Sciences* 9: 15–28.

FAO. (1991) *A Manual for the Design and Construction of Water Harvesting Schemes for Plant Production* (eds W. Critchley and K. Siegert), Food and Agricultural Organization of the United Nations, Rome, Italy. Available at: http://www.fao.org/docrep/U3160E/u3160e00.htm (accessed 20 September 2012).

FAO. (2008) Major food and agricultural commodities and producers: Countries by commodity. Available at: http://www.fao.org/es/ess/top/commodity.html?lang = en&item=156&year=2005 (accessed 20 September 2012).

Farley, K.A., Jobbagy, E.G., and Jackson, R.B. (2005) Effects of afforestation on water yield: A global synthesis with implications for policy. *Global Change Biology* 11: 1565–1576.

Fausey, N.R. (2004) Comparison of free drainage, controlled drainage, and subirrigation water management practices in an Ohio lakebed soil. ASABE Annual International Meeting, ASABE, St. Joseph, MI, pp. 3045–3051.

Fouss, J.L., Skaggs, R.W., Fausey, N.R., and Pitts, D.J. (2004) Implementing controlled-drainage technology to reduce nitrate loss in drainage water. Proceedings of the 8th International Drainage Symposium, ASABE, St. Joseph, Michigan, p. 16.

GA-DNR (Georgia Department of Natural Resources, Atlanta, GA) (2008) Available at: http://www.gaepd.org/Files_PDF/news/Flint_River_news_release_2008.pdf (accessed 9 October 2008).

Gilliam, J.W. and Skaggs, R.W. (1985) Use of drainage control to minimize potential detrimental effects of improved drainage systems. Proceedings of Speciality Conference on Development and Management: Aspects of Irrigation and Drainage Systems. ASCE, New York, pp. 352–362.

Goldemberg, J., Coelho, S.T., and Guardabassi, P. (2008) The sustainability of ethanol production from sugarcane. *Energy Policy* 36: 2086–2097.

Hamlet, A.F., Mote, P.W., Clark, M.P., and Lettenmaier, D.P. (2007) Twentieth-century trends in runoff, evapotranspiration, and soil moisture in the western United States. *Journal of Climate* 20: 1468–1486

Heim, R.R., Jr., Gleason, K.L., Kruk, M.C., *et al.* (2008) United States, in *State of the Climate in 2007* (eds D.H. Levinson and J.H. Lawrimore), Bulletin of the American Meteorological Society (special issue) 89: S114–S120.

Hunt, P.G., Bauer, P.J., Camp, C.R., and Matheny, T.A. (1998) Nitrogen accumulation in cotton grown continuously or in rotation with peanut using subsurface microirrigation and GOSSYM/COMAX management. *Crop Science* 38: 410–415.

Hunt, P.G., Matheny, T.A., Ro, K.S., Stone, K.C., and Vanotti, M.B. (2008) Denitrification of agricultural drainage line water via immobilized denitrification sludge. *Journal of Environmental Science and Health Part A* 43: 1077–1084.

Izaurralde, R.C., Thomson, A.M., Rosenberg, N.J., and Brown R.A. (2005) Climate change impacts for the conterminous USA: An integrated assessment part 6. Distribution and productivity of unmanaged ecosystems. *Climatic Change* 69: 107–126.

Jackson, R.B., Jobbágy, E.G., Avissar, R., *et al.* (2005) Atmospheric science: Trading water for carbon with biological carbon sequestration. *Science* 310: 1944–1947.

James, J. (2008) Ethanol from sugar: What are the prospects for U.S. sugar co-ops? U.S. Department of Agricultural, Rural Development, Washington, DC. Available at: http://www.rurdev.usda.gov/rbs/pub/sep06/ethanol.htm (accessed 9 October 2008).

Jaynes, D.B. and Colvin, T.S. (2006) Corn yield and nitrate loss in subsurface drainage from midseason nitrogen fertilizer application. *Agronomy Journal* 98: 479–487.

Kangas, R.S. and Brown, T.J. (2007) Characteristics of US drought and pluvials from a high-resolution spatial dataset. *International Journal of Climatology* 27: 1303–1325.

Kanwar, R.S. (2006) Effects of cropping systems on NO_3-N losses to tile drain. *Journal of the America Water Resources Association* 42: 1493–1501.

Karp, A. and Shield, I. (2008) Bioenergy from plants and the sustainable yield challenge. *New Phytologist* 179: 15–32.

Long, S.P., Ainsworth, E.A., Leakey, A.D.B., and Morgan, P.B. (2005) Global Food Insecurity. Treatment of major food crops with elevated carbon dioxide or ozone under large-scale fully open-air conditions suggests recent models may have overestimated future yields. *Philosophical Transactions of the Royal Society B: Biological Sciences* 360: 2011–2020.

Macedo, I.C. (ed.) (2005) Chapter 5: Impacts on water supply. *Sugarcane's Energy – Twelve Studies on Brazilian Sugarcane Agribusiness and Its Suitability*. Sugar Cane Agroindustry Union, São Paulo, Brazil, pp. 105–118. Available at: http://english.unica.com.br/multimedia/publicacao/ (accessed 18 September 2008).

Madramootoo, C.A., Johnston, W.R., Ayars, J.E., Evans, R.O., and Fausey, N.R. (2007) Agricultural drainage management, quality and disposal issues in North America. *Irrigation and Drainage* 56: S35–S45.

Mann, L., Tolbert, V., and Cushman, J. (2002) Potential environmental effects of corn (*Zea mays* L.) stover removal with emphasis on soil organic matter and erosion. *Agriculture, Ecosystems and Environment* 89: 149–166.

Manuel, J. (2008) Drought in the Southeast: Lessons for water management. *Environmental Health Perspectives* 116(4).

Mastrorilli, M., Katerji, N., and Rana, G. (1999) Productivity and water use efficiency of sweet sorghum as affected by soil water deficit occurring at different vegetative growth stages. *European Journal of Agronomy* 11: 207–215.

Maupin, M.A. and Barber, N.L. (2005) Estimated withdrawals from principal aquifers in the United States, 2000. US Geological Survey Circular 1279. U.S. Geological Survey, Reston, VA, 47p.

McKendry, P. (2002a) Energy production from biomass (Part 1): overview of biomass. *Bioresource Technology* 83: 37–46.

McKendry, P. (2002b) Energy production from biomass (Part 3): gasification technologies. *Bioresource Technology* 83: 55–63.

Meehl, G.A. and Tebaldi, C. (2004) More intense, more frequent, and longer lasting heat waves in the 21st century. *Science* 305: 994–997.

Mejia, M.N. and Madramootoo, C.A. (1998) Improved water quality through water table management in eastern Canada. *Journal of Irrigation and Drainage Engineering* 124: 116–122.

Meza, F.J. and Silva, D. (2009) Dynamic adaptation of maize and wheat production to climate change. *Climate Change* 94: 143–156.

Montgomery, D.R. (2007) *Dirt: The Erosion of Civilizations*, University of California Press, Berkeley, CA. 285p.

Moreira, J.R. (2007) Water use and impacts due ethanol production in Brazil. International Conference on Linkages in Energy and Water Use in Agriculture in Developing Countries, Organized by IWMI and FAO, ICRISAT, India, January 2007. Available at: http://www.iwmi.cgiar.org/EWMA/files/papers/Jose_Moreira.pdf (accessed 18 September 2008).

Mote, P.W. (2006) Climate-driven variability and trends in mountain snowpack in western North America. *Journal of Climate* 19: 6209–6220.

Mote, P.W. (2007) Statement to the U.S. Senate Committee on Energy and Natural Resources, Subcommittee on Water and Power. Western water resources in a changing climate. Hearing 6 June, 2007. Available at: http://energy.senate.gov/public/_files/MoteTestimony.pdf (accessed 18 September 2008).

N.C. (1997) An act to enact the clean water responsibility and environmentally sound policy act, a comprehensive and balanced program to protect water quality, public health, and the environment: Part I. Moratoria on construction or expansion of swine farms. General Assemble of North Carolina, S.L. 1997–458, House Bill 515. Available at: http://www.ncleg.net/EnactedLegislation/SessionLaws/HTML/1997–1998/SL1997–458.html (accessed 29 September 2008).

Nebraska Corn Board. (2008) Nebraska Department of Natural Resources, Lincoln, Nebraska. Available at: http://www.nebraskacorn.org/main-navigation/corn-production-uses/climateirrigation/ (accessed 26 September 2008).

NE-DNR. (2005) Conservation Reserve Enhancement Program Nebraska Platte-Republican Resources Area – Fact Sheet. United States Department of Agriculture, Farm Service Agency. Available at: http://www.cnppid.com/CREP_Rules.htm (accessed 26 September 2008).

NE-FSA. (2007) Fact sheet – Nebraska Central Basins Conservation Reserve Enhancement Program (CREP). United States Department of Agriculture, Farm Service Agency. Available at: Nebraska Farm Service Agency, Lincoln, NE. http://www.fsa.usda.gov/Internet/FSA_File/fsnecrep032307.pdf (accessed 26 September 2008).

Negri, D.H., Gollehon, N.R., and Aillery, M.P. (2005) The effects of climatic variability on us irrigation adoption. *Climatic Change* 69: 299–323.

NRC. (2008) Water implications of biofuels production in the United States. Committee on Water Implications of Biofuels Production in the United States, Water Science and Technology Board, Division on Earth and Life Studies. The National Academies Press: Washington, DC. Available at: http://www.nap.edu/catalog/12039.html (accessed 18 September 2008).

Pacific Institute. (2008) Total renewable freshwater supply, by country (2006 Update). Pacific Institute, Oakland, CA. Available at: http://www.worldwater.org/data20062007/table1.xls (accessed 18 September 2012).

Pavelis, G.A. (1987) Farm drainage in the U.S.- history, status and prospects. U.S. Department of Agriculture, Economic Research Service Miscellaneous, Washington, DC. Publication No. 1455. 170p.

Postel, S.L. (1998) Water for food production: Will there be enough in 2025? *BioScience* 48: 629–637.

Postel, S.L. (2000) Entering an era of water scarcity: The challenges ahead. *Ecological Applications* 10: 941–948.

Postel, S.L. (2001) Growing more food with less water. *Scientific American* 284: 46–49.

Postel, S.L., Daily, G.C., and Ehrlich, P.R. (1996) Human appropriation of renewable fresh water. *Science* 271: 785–788.

Powell, T.L., Starr, G., Clark, K.L., Martin, T.A., and Gholz, H.L. (2005) Ecosystem and understory water and energy exchange for a mature, naturally regenerated pine flatwoods forest in north Florida. *Canadian Journal of Forest Research* 35: 1568–1580.

Prasad, S., Singh, A., Jain, N., and Joshi, H.C. (2007) Ethanol production from sweet sorghum syrup for utilization as automotive fuel in India. *Energy and Fuels* 21: 2415–2420.

Rabalias, N.N., Wiseman, W.J., Turner, R.E., Sen Gupta, B.K., and Dortch, Q. (1996) Nutrient changes in the Mississippi River and system responses on the adjacent continental shelf. *Estuaries* 19: 386–407.

Reijnders, L. (2006) Conditions for the sustainability of biomass based fuel use. *Energy Policy* 34: 863–876.

Robins, J.G., Jensen, K.B., Peel, M.D., and Waldron, B.L. (2009) Establishment of warm-season grasses in summer and damage in winter under supplementary irrigation in a semi-arid environment at high elevation in western United States of America. *Grass and Forage Science* 64: 42–48.

Rohter, L. (2006) With big boost from sugar cane, Brazil is satisfying its fuel needs. The New York Times. 10-Apr-2006, natl. ed. Available at: http://www.nytimes.com/2006/04/10/world/americas/10brazil.html?_r=1&scp=1&sq=&st=nyt&oref=slogin (accessed 10 September 2008).

Sadler, E.J. and Turner, N.C. (1993) Water relationships in a sustainable agricultural system, in *Sustainable Agriculture Systems* (eds J.L. Hatfield and D.L. Karlen), Lewis Press, Boca Raton, FL, pp. 21–46.

Seager, R., Tzanova, A., and Nakamura, J. (2009) Drought in the Southeastern United States: Causes, variability over the last millennium, and the potential for future hydroclimate change. *Journal of Climate* 22: 5021–5045.

Simpkins, W.W., Burkart, M.R., Helmke, M.F., *et al.* (2002) Potential impact of earthen waste storage structures on water resources in Iowa. *Journal of the American Water Resources Association* 38: 759–771.

Skaggs, R.W., Youssef, M.A., Gilliam, J.W., and Evans, R.O. (2010) Effect of controlled drainage on water and nitrogen balances in drained lands. *Transaction of the ASABE* 53: 1843–1850.

Slingo, J.M., Challinor, A.J., Hoskins, B.J., and Wheeler, T.R. (2005) Introduction: Food crops in a changing climate. *Philosophical Transactions of the Royal Society B: Biological Sciences* 360: 1983–1989.

Solley, W.B., Pierce, R.R., and Perlman, H.A. (1998) *Estimated Use of Water in the United States in 1995*. U.S. Geological Survey Circular 1200, Reston, VA, 79p.

Stampfli, N. and Madramootoo, C.A. (2006) Water table management: A technology for achieving crop per drop. *Irrigation and Drainage Systems* 20: 41–55.

Stone, K.C., Hunt, P.G., Millen, J.A., *et al.* (2008) Forage subsurface drip irrigation using treated swine wastewater. *Transactions of the ASABE* 51: 433–440.

Stone, K.C., Sommers, R.C., Williams, G.H., and Hawkins, D.E. (1992) Water table management in the eastern Coastal Plain. *Journal of Soil and Water Conservation* 47: 47–51.

Tan, C.S., Drury, C.F., Ng, H.Y.F., and Gaynor, J.D. (1999) Effect of controlled drainage and subirrigation on subsurface tile drainage nitrate loss and crop yield at the farm scale. *Canadian Water Resources Journal* 24: 177–186.

Thomas, D.L., Hunt, P.G., and Gilliam, J.W. (1992) Water table management for water quality improvement. *Journal of Soil and Water Conservation* 47: 65–70.

Thomas, D.L., Perry, C.D., Evans, R.O., Izuno, F.T., Stone, K.C., and Gilliam, J.W. (1995) Agricultural drainage effects on water quality in southeastern US. *Journal of Irrigation and Drainage Engineering* 121: 277–282.

Trostle, R. (2008) Global agricultural supply and demand: factors contributing to the recent increase in food commodity prices. U. S. Department of Agriculture, Economic Research Service, Washington, DC. Publication WRS-0801, 30p.

U.S. Congress, Office of Technology Assessment. (1993) Preparing for an uncertain climate–Volume I, OTA-O-567, U.S. Government Printing Office, Washington, DC.

Uhlenbrook, S. (2007) Biofuel and water cycle dynamics: what are the related challenges for hydrological processes research? *Hydrological Processes* 21: 3647–3650.

UNESCO (2007) UNESCO water portal weekly update no. 177: Polar Regions. Available at: http://www.unesco.org/water/news/newsletter/177.shtml#know (accessed 10 September 2008). United Nations Educational Scientific and Cultural Organization, New York, USA.

United Nations. (1998) World population prospects: The 2002 revision volume III: Analytical report. Available at: http://www.un.org/esa/population/publications/wpp2002/English.pdf (accessed 18 September 2008) New York, USA.

USA Today (2002) Available at: http://www.usatoday.com/weather/wear.htm (accessed 10 September 2008).

USDA-ERS. (2008) USDA agricultural Projections 2017 (No. OCE-2008-1). U.S. Department of Agriculture, Economic Research Service, Washington, DC.

USDA-NASS. (2004) Farm and Ranch Irrigation Survey (2003). Volume 3, Special Studies Part 1, AC-02-SS-1, of the 2002 Census of Agriculture. U.S. Department of Agriculture, National Agricultural Statistics Service, Washington, DC. Available at: http://www.agcensus.usda.gov/Publications/2002/FRIS/fris03.pdf (accessed 18 September 2012).

USDA-NASS. (2008) Data and Statistics 2008. U.S. Department of Agriculture, National Agricultural Statistics Service, Washington, DC. Available at: http://www.nass.usda.gov/Data_and_Statistics/index.asp (accessed 18 September 2012).

US-DOE. (2005) Biomass as a feedstock for a bioenergy and bioproducts industry: The technical feasibility of a billion-ton annual supply. US Department of Energy, Oak Ridge National Laboratory, Oak Ridge, TN. DOE/GO-102995-2135. Available at: http://www1.eere.energy.gov/biomass/pdfs/final_billionton_vision_report2.pdf (accessed 18 September 2012).

US-DOE. (2008) Biomass Program. Department of Energy, The Office of Energy Efficiency and Renewable Energy, Washington, DC. Available at: http://www1.eere.energy.gov/biomass/biofuels_initiative.html (accessed 18 September 2012).

US-DOE. (2011) *U.S. Billion-Ton Update: Biomass Supply for a Bioenergy and Bioproducts Industry*, R.D. Perlack and B.J. Stokes (Leads). Oak Ridge National Laboratory, Oak Ridge, TN. ORNL/TM-2011/224, 227p.

US-Water News Online. (2005) Available at: http://www.uswaternews.com/archives/arcsupply/5nebrfarm4.html (accessed 10 September 2008).

Vanotti, M.B., Szogi, A.A., Hunt, P.G., Millner, P.D., and Humenik, F.J. (2007) Development of environmentally superior treatment system to replace anaerobic swine lagoons in the USA. *Bioresource Technology* 98: 3184–3194.

Welch, H.L., Green, C.T., Rebich, R.A., Barlow, J.R.B., and Hicks, M.B. (2010) Unintended consequences of biofuels production—The effects of large-scale crop conversion on water quality and quantity: U.S. Geological Survey Open-File Report 2010-1229, 6p.

Wilhelm, W.W., Johnson, J.M.F., Hatfield, J.L., Voorhees, W.B., and Linden, D.R. (2004) Crop and soil productivity response to corn residue removal: a literature review. *Agronomy Journal* 96: 1–17.

Woodhouse, C.A. and Overpeck, J.T. (1998) 2000 years of drought variability in the central United States. *Bulletin of the American Meteorological Society* 79: 2693–2714.

Wright, L. (2007) Historical perspective on how and why switchgrass was selected as a "model" high-potential energy crop. Publication ORNL/TM-2007/109. Environmental Sciences Division, Oak Ridge National Laboratory, Oak Ridge, Tennessee.

Chapter 8

Biofuel Crops and Soil Quality and Erosion

D. Chatskikh[1], A. Ovchinnikova[2], B. Seshadri[3], and N. Bolan[3]

[1] *George Lemaître Centre for Earth and Climate Research, Louvain-la-Neuve, Belgium*
[2] *Cargill Europe BVBA, Mechelen, Belgium*
[3] *Centre for Environmental Risk Assessment and Remediation (CERAR), Mawson Lakes, Australia*

Introduction

Biofuel or energy crop production aims at maximizing the carbon (C) harvest for conversion into fuel. Since soils are involved in the processing chain the question, however, is if this conversion can be done without compromising soil quality. In this chapter we discuss the soil quality aspect of biofuel production. The production of biofuel crops might simultaneously affect a combination of soil properties and stipulating severe human-driven soil quality threats, out of which the decline of soil organic matter (SOM), the increase of erosion risks, and on- and off-site pollution and nutrient losses are the most pronounced. We consider the differences between annual and perennial crops out of the effects of management and land-use change (LUC), including an issue of soil organic carbon (SOC) budget and sustainable removal of crop residues for energy production. Consequently, we discuss soil quality under biofuel crop production as affected by these threats to provide essential soil services. The challenges of the soil quality aspect of sustainable biofuel crop production, which include by-product management, soil remediation potential, and utilization of idle and degraded soils for biofuels, are also covered by this chapter.

Soil Quality Definition and Assessment

Soils are an essential slow renewable resource for food, feed (livestock), and energy production (in both managed and natural systems), and a vital component in all natural cycles. Therefore, the maintenance of soils is crucial for the sustainable development of all human activities and for the agricultural sector in particular. Soil is fundamental to crop production (FAO, 2010), a *sine qua non* element when determining the future of global agriculture (Pretty *et al.*, 2010). Nowadays, the issue of soil quality has become a part of public policy considerations, and in several countries and continents soil legislation is being developed (Hartemink, 2008a).

Biofuel Crop Sustainability, First Edition. Edited by Bharat P. Singh.

In order to classify soil based on resources enabling maximum crop production, the concept of soil quality (Warkentin and Fletcher, 1977) has been introduced; this includes some judgment values in the measurement of soil functioning (Schjønning *et al.*, 2004). Simply put, soil quality is the capacity of soil to perform ecosystem services via dynamic physical processes and chemical pathways (Karlen *et al.*, 1997; Ng Cheong, 2007; Garrigues *et al.*, 2012). Hence, crop productivity is determined by soil quality. The concept of soil quality for growing (biofuel) crops has been reviewed through the years (Larson and Pierce, 1991; Papendick and Parr, 1992; Doran and Safely, 1997; Sojka and Upchurch, 1999; Andrews and Carroll, 2001; Karlen *et al.*, 2003; Sojka *et al.*, 2003; Wienhold *et al.*, 2004; Hartemink, 2008a; Blanco-Canqui, 2010). Modern modifications of the concept of soil quality also consider environmental and climatic constrains (Reeves, 1997), along with sustainable development (Doran, 2002), thus becoming an integral part of up-to-date agroecosystem analysis (Arshad and Martin, 2002).

Globally, there is no single target or optimum quality standard for soil quality assessment. Soil quality might be accessed via soil quality indicators, which can be defined as those soil properties and processes that have the greatest sensitivity to changes in soil function (Andrews *et al.*, 2004). For practical purposes, overall soil quality can be subcategorized into physical, chemical, and biological soil quality, while various indicators are based on measured (physical, chemical, and biological) soil properties (Doran *et al.*, 1994). Consequently, the indicators can be used separately or in various combinations depending on the purpose of the individual assessment. Traditional estimates based on a sample in the field and laboratory analysis involve costly processing of large number of samples (Sanchez *et al.*, 2003); therefore there is a need for newer approaches to estimate soil properties in cost-efficient and nondestructive but also rapid ways (Shepherd and Walsch, 2002).

Soil quality and its assessment is soil- and site-specific and depends on factors including (i) its inherent capabilities—pH, nutrient retention, water holding capacity, ion exchange; (ii) environmental influences, such as climate; (iii) intended land use; and (iv) management goals (Karlen *et al.*, 2006). The above factors may also influence the mobility (leaching and erosion loss) and depletion (removal of residues from the cultivated lands) of nutrients and heavy metals. Hence, the aforementioned complex processes in soils require detailed studies, for example, in the form of life cycle assessment (LCA). Garrigues *et al.* (2012) suggested the use of "minimum data sets" (MDSs) of SOM, soil texture, and bulk density for developing a universal soil quality impact indicator, where existing methods focused only toward SOM (Oberholzer *et al.*, 2012). The LCAs using major soil quality influencing factors will be useful in determining the extent of common soil threats and its impacts on biofuel crop production. This will be detailed under LUCs with example of modeling studies.

Biofuel Crop Production and Soil Quality

Classification of Biofuel Crops

For the purpose of assessing soil quality, biofuel crops can be classified into (i) annual crops (such as maize for bioethanol in Northern America or winter oilseed rape/canola for biodiesel in Europe and Australia) and (ii) perennial crops (sugarcane in South America, oil palm in Asia, emerging warm-season grasses around the world). Although annuals currently dominate, perennials would aid to agricultural diversification in future, offering additional options to change crop production in more environmentally, economically, and energetically rational way (Pimentel *et al.*, 2012).

In general, the magnitude of effects on the production of one or more biofuel crops and parameters used for soil quality assessment largely depends on crop, soil, management, and climate. We have summarized the effects on soil quality reported for some common and potential biofuel crops (Table 8.1).

Some promising genera for fuel wood production, soil quality improvement, and C sequestration and desertification control in arid, semiarid, and dry subhumid areas are *Acacia*, *Albizia*, *Azadirachta*, *Capparis*, *Cassia*, *Casuarina*, *Cupressus*, *Dalbergia*, *Eucalyptus*, *Khaya*, *Leucaena*, *Morus*, *Parkia*, *Pongamia*, *Populus*, *Prosopis*, *Sesbania*, *Tamarix*, *Tectona*, and *Terminalia*.

Meanwhile, ecological literature broadly categorized biofuel crops into (i) high intensity low diversity (HILD) and (ii) low intensity high diversity (LIHD). HILD crops include annually replanted crops (e.g., row crops such as corn) and are characterized by high inputs (e.g., of fertilizer, pesticide, irrigation), while LIHD crops are perennial low-input systems normally composed of a mix of several different plant species. While establishment of HILD agricultural systems is a common strategy for biofuel crop production, biofuels derived from LIHD agricultural systems can potentially provide greater reductions in nutrient losses and less pollution (Tilman *et al.*, 2006). Being by definition a sustainable agricultural crop production system, LIHD cultivation of biofuel feedstock might lead to both diversification of energy production and improvement of soil quality. The impacts of HILD and LIHD crops on soil quality are shown in Figure 8.1. This classification of biofuel crops is in use for larger-scale assessments and projections on biofuel production (Vuichard *et al.*, 2009; Don *et al.*, 2012).

Effects of Biofuel Crops on Soil Quality

Soil quality is determined by a combination of specific characteristics connected to soil type, site management, climate, and previous land use by operating with data at plot or field scale. Considering the complexity of the aforementioned soil quality features, the effects of soil management and land use at different conditions on soil quality are considered in this chapter.

Soil Management

The soil management effects are largely connected to (i) soil compaction, (ii) tillage and crop rotation, and (iii) biomass harvest, including management of crop residues. But before exploring the soil management aspects, the differences in the impacts of annual and perennial crops on soil quality need to be understood:

- Perennial crops in general are more adaptable. Indeed, some biofuel crops do not require tillage operations after the establishment (e.g., perennial grasses for 5–10 years and LIHD for 10–20 years will not be disrupted). Additionally, perennial grasses were found to be tolerant to both saturated and unsaturated soil conditions (Gilker *et al.*, 2002) and able to grow on nutrient-depleted and contaminated soils. On the contrary, annual crops are more sensitive and need regular fertilizer inputs and tillage practices after each harvest.

Table 8.1. Typical management of some of the most important biofuel crops and specific effects on soil quality.

Biofuel crops		Management	Specific effects on soil quality	
			SOC, nutrients removed, soil structure	Soil erosion, land-use change
Annuals	Rapeseed	Only seeds harvested	Improves sustainability as a result of reduced tillage, soil quality might be further improved in rotation (Chatskikh *et al*., 2008)	Stable in good rotation/tillage practice
	Soybean	Only seeds harvested	Low residue input, benefit when grown in rotation (maize/soybean)	Low soil cover creates erosion risks
	Sunflower	Only seeds harvested	Incorporation of surface residues due to extensive tillage reduced SOC (Bowman *et al*., 2000), careful rotations and reduced tillage can help reduce inputs from fertilizer to pesticides	Risks due to incorporation of surface residues, required rotation
	Maize	Kernels only harvested	Removal of corn stover is critical (Karlen *et al*., 2011), combination of tillage and rotation (Munkholm *et al*., 2012)	Erosion risks, environmental pollution
	Sweet sorghum	Annual harvest of whole crop	Low external input and soil biomass removal can reduce nutrient pool and result in N stress. Both tillage and biomass removal caused soil compaction and reduction in SOC (Meki *et al*., 2013)	Continuous tillage can result in soil erosion
	Sugar beets	Tops retained in the field	Compaction due to harvesting of wet soils (Tugrul *et al*., 2012). Probably small SOC decrease (Aubinet *et al*., 2009)	Stable in conservation tillage but affected by climatic changes (Scholz *et al*., 2008)
	Wheat	Annual harvest of whole crop	Sustainable in rotation, increases soil quality risks if crop residues are removed (Blanco-Canqui, 2010)	Soil erosion increased in no-tillage management at >25% residue removal
Perennials	Sugarcane	Annual harvest of whole crop	Manual harvest is beneficial, while mechanical harvest associated with residue burning reduced C input. Use of heavy machinery and large amounts of nutrients removed with the harvest puts high demand on soil (Hartemink, 2008b)	Manual harvest might intensify erosion

Miscanthus	Annual harvest in autumn (wet) to spring (dry)	Low nutrient loss, root system, and high C input improve physical soil quality (Dondini *et al*., 2009)	Potential for soil remediation, planting on degraded lands, and to diversify current cropping systems (contour cropping, intercropping)
Switchgrass	Annual harvest in spring	SOC increased because of belowground C input from root biomass and rhizodeposition (Liebig *et al*., 2008)	No erosion because no tillage requirements for growing grain crops alternatively.
Oil palm	Annual harvest of fruits	Decreasing if cleared and virgin land is drained, creates "carbon debt" (Fargione *et al*., 2008), especially of organic soils. Increasing especially if fruit bunches returned (Abu Bakar *et al*., 2011) and no-burning practice	Increases soil erosion due to clearing. Some practice to reduce the pressure on soil quality (e.g., terrace planting, no burning, intercropping)
Eucalyptus	6–20 years harvest cycle	No net effects on SOC stocks in two rotation cycles (Fialho and Zinn, 2012). Rainfall drop size can influence soil structure in terraced cultivation (Fernandez-Raga *et al*., 2009)	Increase in splash erosion, but relatively close to natural forests
Willow	3 years harvest cycle	Stabilization of SOC in topsoil redistribution of C (Grelle *et al*., 2007; Lockwell *et al*., 2012), depletion of SOC pools in subsoil (Lockwell *et al*., 2012)	Difficult to put back soil into agriculture
Poplar	5–7 years harvest cycle	No significant effects on SOC and total N for both poplar monocropping and agroforestry systems (Mao *et al*., 2011)	Agroforestry practices can decrease erosion effects
Jatropha	Only seeds harvested	Maintained SOC and N stock in soil and improved soil's surface integrity by enhancing water infiltration (Ogunwole *et al*., 2008)	Increased macroaggregate stability ensured no soil erosion (Reubens *et al*., 2011).
Cassava	Annual harvest of roots	Decreased SOC and P contents (Islami *et al*., 2011). Lack of surface cover increased soil crusting (Podwojewski *et al*., 2008)	Retaining residues can reduce soil erosion

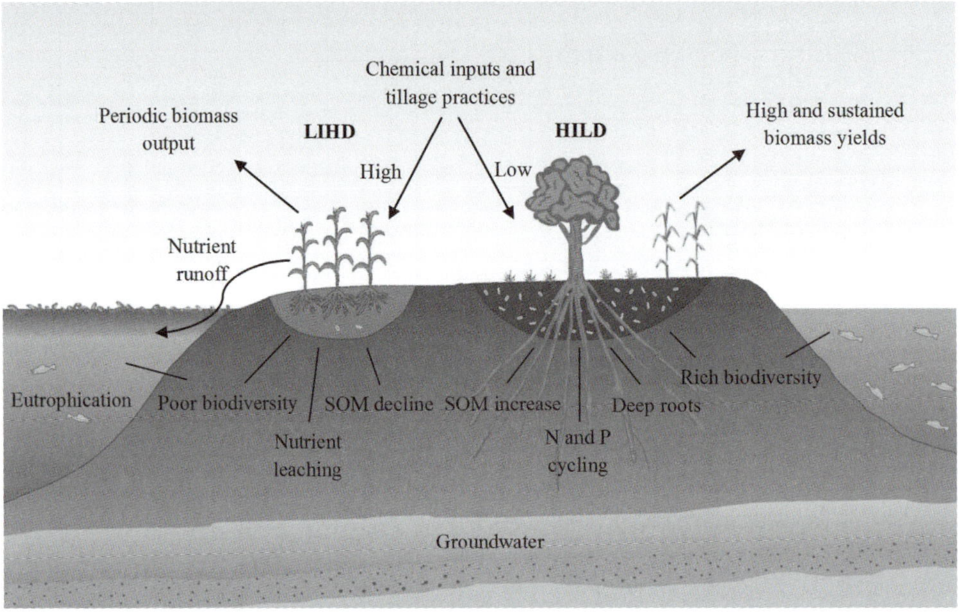

Figure 8.1. The impacts of HILD and LIHD crops on soil quality. For color detail, please see color plate section.

However, rare tillage may, in turn, require intensive row cultivation or pesticide application (e.g., in order to manage weeds in short rotation willow in the establishment year).

- Crop residues provide SOC input and above all help to retain water and nutrients, protecting soils against erosion (Blanco-Canqui, 2010). Conservation agriculture requires retention of crop residues and changes in soil water retention and in some soils might lead to soil compaction (Munkholm *et al.*, 2003). Due to longer roots, perennial crops are able to (i) provide SOC input at different soil depth, compared to annual crops, and (ii) bring water from deeper soil layers, and also capable of improving soil drainage and air circulation at deeper soil profile depths (Clark *et al.*, 1998).
- As opposed to annual biofuel crops, perennial grasses such as *Miscanthus*, switchgrass, or SRC (short rotation willow or poplar coppice) showed consequent increase in soil quality with time (Bessou *et al.*, 2011 for SOC).

Soil compaction is a process that negatively affects soil quality. Rearranging soil aggregates/particles in a denser way, soil compaction results in reduced soil porosity and as a consequence, decreased water and air permeability and increased bulk density (Schjønning *et al.*, 2004), while root penetration decreases as soil compaction increases. Leading not only to soil degradation but also decreasing crop yields, soil compaction consequently is among several main factors that influence soil quality under (biofuel) crop production.

According to Hamza and Anderson (2005), soil compaction is caused by (i) high soil water content, (ii) mechanized farm operations, or (iii) trampling and harvesting, which may negatively affect physical soil quality. This may be of particular concern for biomass harvesting using heavy machinery when the soil is wet. Removal of crop residues for biofuel production with heavier machines and/or several passes in the field may cause soil compaction akin to with annual crops. Harvesting time varies with climate and also its suitability to farming systems (Johnson *et al.*, 2010), but can be manipulated to a certain extent as in the case of perennial grasses (Jørgensen, 2011). Eliminating or postponing autumn tillage operations for annual crops (Hansen and Djurhuus, 1997; Ulén *et al.*, 2010; Myrbeck *et al.*, 2012) can also help reduce soil compaction.

However, biofuel crops may also improve the quality of compacted soils, while controlled traffic might partly solve the problem of compaction. For instance, the roots of some perennial biofuel grasses can penetrate deep into compacted and clayed soil layers, improving water recharge and nutrient uptake, while high amount of biomass produced by some perennial grass species can reduce soil compaction by contributing to the SOM, making soil more resilient against the compactive forces (Thomas *et al.*, 1996).

Extensive conventional tillage (e.g., moldboard ploughing) has a negative effect on overall soil structure, aggregate stability, and SOM, reducing surface crop residues and causing soil compaction and soil erosion in the long run. On the contrary, conservation tillage (reduced tillage, no-till) is able to improve soil physical properties primarily under annual cropping systems and also increase SOC; however, these effects depend on crop residue management. Good tillage practice, especially in combination with proper crop rotation, significantly improves soil quality (Govaerts *et al.*, 2006; Chatskikh *et al.*, 2008; Munkholm *et al.*, 2012).

Combination of tillage and rotation is among the fundamental factors influencing soil quality, and thus the sustainability of cropping systems (Munkholm *et al.*, 2012). Monocropping tends to reduce the production costs in general, but has disadvantages to soil quality (e.g., in sugarcane and oil palm plantations) and yield potential (allelopathy, other unexplained "negative effects"). Govaerts *et al.* (2006) have demonstrated that continuous cropping of maize with no-tillage might be sustainable with proper residue management.

Karlen *et al.* (2006) reported that the lowest values of soil quality index (SQI) were associated with continuously planted maize, while extended rotations (at least 3 years of forage crops) have the highest SQI values, also positively affecting soil quality indicators. Switching to a rotation might also improve yields, and as a consequence total crop biomass (i.e., more roots, more residues). For instance, long-term maize–soybean rotation might provide 5–20% and 10–50% yield increase in irrigated and dryland maize in the US Corn Belt, respectively. On the contrary, Bowman *et al.* (2000) observed that conventionally tilled sunflower (*Helianthus annuus* L.) negatively affected soil quality decreasing by 5 times surface crop residues and by 13% SOC in top soil layer and reducing subsequent winter wheat (*Triticum aestivum* L.) yields by 33%. It is generally suggested that more diverse and extended crop rotations should be followed (Karlen *et al.*, 2006) while incorporating no-till production practices in the rotations (Bowman *et al.*, 2000).

Addition of crop residues and animal manure is vital for maintaining SOM levels and soil fertility (Thomsen and Christensen, 2004). By providing C input they can improve soil nutrient balance and promote soil C sequestration. Adequate manure applications help to stabilize or increase SOC under various soil, management, and climatic conditions (Johnson *et al.*, 2007). Hamelin *et al.* (2012) reported that on manured Danish sandy soils, harvesting of winter wheat crop straw did not cause any reduction in SOC. There are also other practices (e.g., planting of cover crops, leaving mulch, green manures, terrace cropping) to improve soil quality in order to prevent nutrient losses and soil erosion as a result of residue harvest (Blanco-Canqui and Lal, 2009). For example, use of cover crops during the establishment phase of short rotation crops may result over time in SOC increase (Tolbert *et al.*, 2002). Given the need to maintain SOM and nutrients balance, to provide erosion protection, and to secure long-term land productivity while being cost-effective, which part or proportion of residues can be practically and sustainably removed for use as biofuel feedstock still remains to understood (Lal, 2009; Blanco-Canqui, 2010).

In the case of perennial grasses, harvest time is important for the quality as a fuel. For combustion purposes the best quality is obtained from spring-harvested crops, when minerals responsible for fouling and slagging are leached from the crop. At that harvest time, leaves have been lost to soil (Lewandowski and Heinz, 2003) and will supply SOC. However, delaying harvesting until the spring reduced yields by 30–50% compared with autumn yields. Harvesting in autumn to have a higher yield, having less lignified biomass at that time (Hodgson *et al.*, 2010) that can more easily be converted into ethanol, is an important consideration (Jørgensen, 2011). However, the effect on soil quality (SOC) when much fewer leaves are left on the ground must be further investigated.

In general, harvest rates should be limited to those that maintain SOM and do not exacerbate erosion (Johnson *et al.*, 2010). For instance, in case of maize it is advisable to harvest the crop taller leaving some stems. However, a study conducted by Moebius-Clune *et al.* (2008) on the harvesting and tillage practices targeting their long-term effects on maize stover suggested that tillage had more negative effects on soil quality compared to harvesting practice.

The magnitude of benefits of perennial cropping systems depends on the harvest season, soils, and climatic conditions (Hamelin *et al.*, 2012). Hence, the soil quality of agricultural lands should be thoughtfully monitored from the perspective of sustainability by covering the aforementioned factors, because the soil management practices for biomass production and trafficking of farm inputs and outputs could affect the soil quality and consequently (biofuel) crop production.

Agroecosystem or crop models are widely used to study complex interactions of the impact of soil, management, and climate conditions on crop development and also to perform knowledge-based scenario analysis. In order to address the challenges of crop biofuel modeling and data requirement, Nair *et al.* (2012) have reviewed 11 field-scale models (MISCANFOR, MISCANMOD, WIMOVAC for *Miscanthus*; Agro-BGC, ALMANAC, EPIC for switchgrass; APSIM, AUSCANE, CANEGRO for sugarcane; and SECRETS, 3PG for poplar and willow) and two models for regional scale (Agro-IBIS and LPJmL for *Miscanthus* and sugarcane). The ability to simulate dynamically crop growth and soil water and nutrient budgets increases the suitability of these models to assess environmental soil quality associated with biofuel cropping systems (see Figure 8.2 for details).

Land-use Change

There is a growing demand for biofuel crops, and it is therefore important to include associated elements of LUC into assessment of soil quality. The issue of soil quality related to land use has at least three elements: (i) conversion of existing agricultural soils to biofuel crops, (ii) inclusion of newer fields to produce biofuels, and (iii) using of new sites for sustainable energy production, such as idle or degraded lands.

Literature reflects a great debate on LUC impact on biofuel production and reveals that understanding of LUC impact on soil quality is scarce (Fargione *et al.*, 2008; Bessou *et al.*, 2011; Don *et al.*, 2012) and therefore it is only briefly considered here.

Bessou *et al.* (2011) have summarized European studies on conversion of arable fields and grasslands into biofuel crops, such as perennial grasslands (*Miscanthus*, switchgrass) and SRCs (poplar and willow), and showed high variation in soil quality indicators. Converting wastelands into biofuel plantations does not necessarily show any degrading effects associated with LUC, for instance as in *Miscanthus* sp. (Jørgensen, 2011) and *Jatropha curcas* (Firdaus

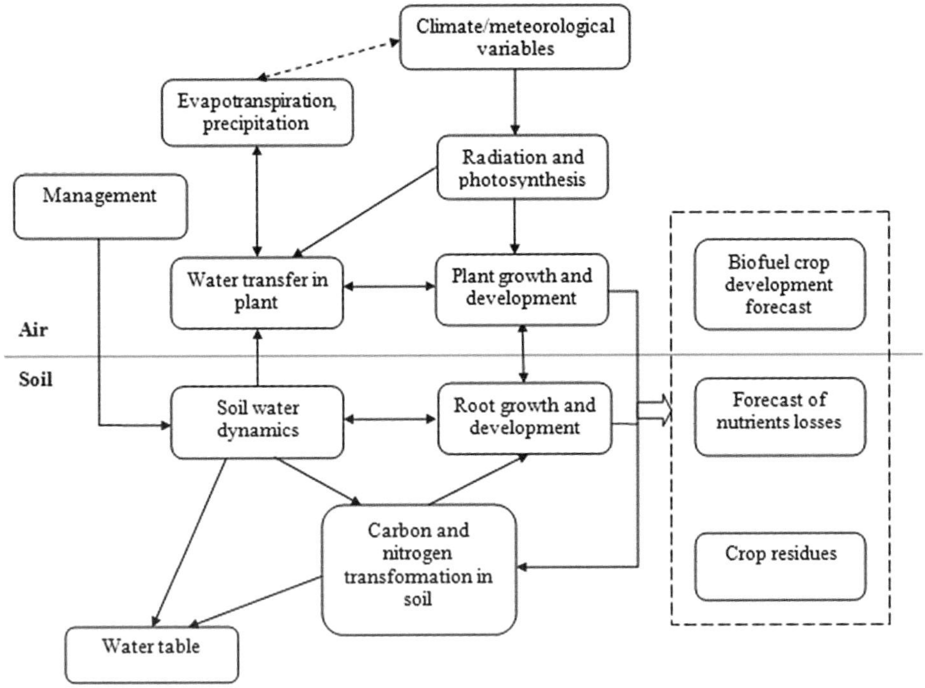

Figure 8.2. Water, C, and N flows in a dynamic biofuel crop model (Poluektov *et al.*, 2002).

and Husni, 2012). On the contrary, in the case of planting oil palm for biofuel in Asia, Fargione *et al.* (2008) has introduced the concept of "carbon debt" in order to demonstrate that land use affects environment more than only in the context of soil environmental quality.

In terms of using idle/degraded lands (e.g., landfill sites and mine sites), the capping material (mainly clay) may have high bulk density and poor nutrient supply. Therefore, biofuel crops that are intended to grow on such sites might require soil amendments to increase plant productivity. For example, Lamb *et al.* (2012) used biosolids for growing giant reed (*Arundo donax*), mustard (*Brassica juncea*), and sunflower (*H. annuus*) in a landfill site and observed increase in biomass production as a result of a rise in nutrient supply and improvements in organic matter content of the soil. However, only LCA of the entire production process of biofuel can determine the overall effects of LUCs on soil quality.

Soil quality, or the impacts on soil quality (e.g., of crop production systems, LUC), is still poorly represented by modeling and in the LCA. LCA (or "cradle-to-grave" analyses) is a common technique which allows us to estimate the environmental impact of biofuel crop throughout its life. An LCA chart for each biofuel crop, which includes the most significant soil-quality-influencing factors (e.g., SOC, nutrient depletion, and tillage practices) and linked to soil variables and processes, can play an effective role in limiting their effects on soil degradation. Several approaches have been developed recently to evaluate the impacts of agricultural activities in LCA (Oberholzer *et al.*, 2006; Milà i Canals *et al.*, 2007), which are, however, still limited with constructive soil quality assessment (Garrigues *et al.*, 2012).

At the same time, it is important to integrate soil quality into model-based assessments to improve credibility of estimations for regional, national, and global scales (Don *et al.*, 2012). Consequently, there is an urgent need to include soil quality into assessment tools to estimate

(i) biofuel production, (ii) residue harvesting from the fields, (iii) land suitability for biofuel production, and (iv) land conversion consequences (Lambin *et al*., 2001; Oberholzer *et al*., 2006; Johnson *et al*., 2010; Garrigues *et al*., 2012; Qin *et al*., 2012).

Models of different complexity are in use to simulate input and output flows of biofuel crop production. For instance, Hamelin *et al.* (2012) has built a life-cycle inventory for analysis of 528 different crop systems (varied in crop types, soil, and tillage) in Denmark in order to address the issue of conversion of food/feed agricultural land into biofuels. They concluded that *Miscanthus* is a promising energy crop, because of the lowest emissions of N compounds, relatively high C turnover efficiency, and increased SOC. At the same time, Thomas *et al.* (2012) used dynamic modeling to assess the environmental impacts of LUC for perennial energy crops on water, C, and N cycling. Nineteen established models were assessed according to suitability criteria out of which ANIMO, CERES-NOE, DNDC, DayCent, ECOSSE, Ecosys, Expert-N, FASSET, and WNMM were identified to be appropriate models (see Thomas *et al*., 2012 for more details).

Biofuel development runs faster than models (Gnansounou and Panichelli, 2008; see Figure 8.3). In addition, overall higher costs limit experimental data availability to provide

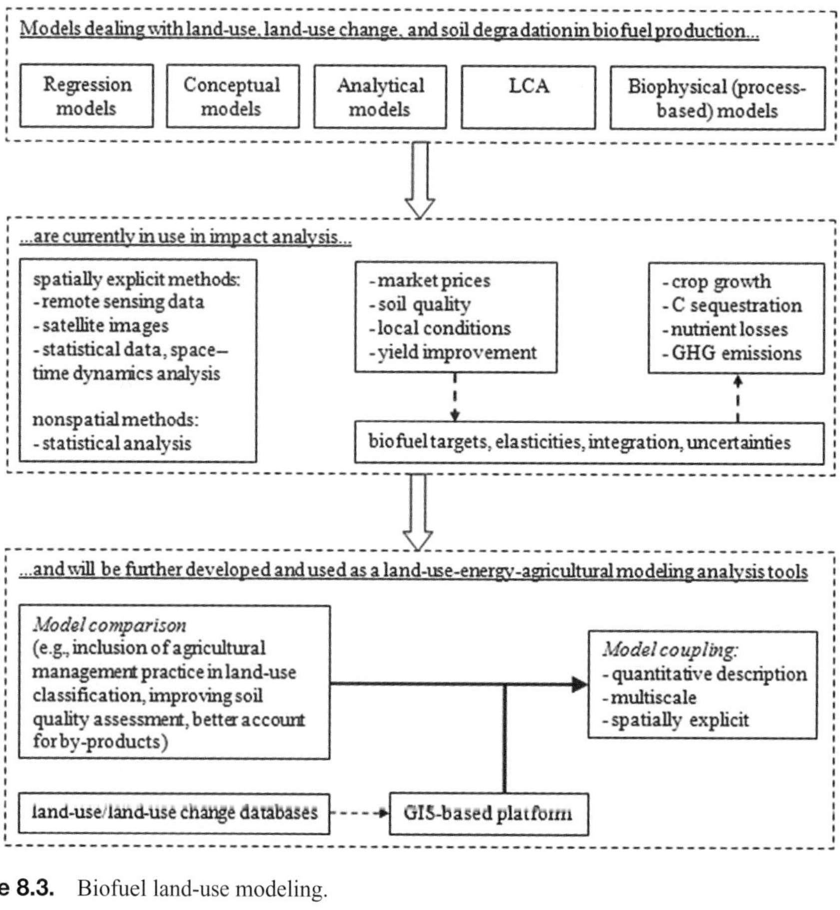

Figure 8.3. Biofuel land-use modeling.

detailed soil quality assessments, which in turn slow down throughout validation of existent biofuel crop models and development of newer modeling approaches. Smith and Searchinger (2012) argued that current LCA models do not take proper account for some important factors (e.g., impact of vehicles and residue burning, N input-driven nitrous oxide emissions). Nevertheless, global increase in the use of modeling and LCA to study biofuel crops will in turn stipulate experimental research in the area.

Effects of LUC due to biofuel crop production can be shaped by a variety of processes operating at different scales. To further illustrate the complexity of LUC effects we have included actual issues on deforestation and emergence of diversified landscapes.

Clearly, deforestation exacerbates risks of soil erosion, soil degradation, and desertification and leads to landscape denudation; therefore, clearing of forests around the world due to biofuel production is at the center of the debate. Thus, Nepstad *et al.* (2008) have concluded that deforestation in Brazil, the biofuel industry leader, is induced by displacement of cattle deeper into the Amazon region due to soybean production, pushed by expansion of land areas under sugarcane. At the same time, Butler (2008) reported that deforestation in Malaysia and Indonesia is primarily caused by expansion of palm oil plantations. However, little research has been carried out to link spatially global deforestation with biofuel production. Searchinger *et al.* (2008) have made an attempt to estimate emissions from indirect LUC using a worldwide agricultural model.

Agricultural fields situate in the landscape, and therefore inclusion of both geomorphic and agroecosystem processes in the landscape-scale vision on biofuel crop production and soil quality is important. Dabney *et al.* (2006) summarized on how in-field and edge-of-field buffers integrated on the landscape scale might improve soil quality and control soil erosion and nutrient leaching and sediment flows, especially if linked to residue management. Furthermore, Wilhelm *et al.* (2012) suggested that diversified landscape could help in addressing a number of interrelated land-use and land management issues (e.g., optimal distribution of productive capacity, harvest of crop residues, mitigation of C and nutrients losses, conservation of energy, improving water management).

To further improve the credibility of scenarios based on crop biofuel modeling, feedback should be included. For instance, intercropping could provide feedstock without primary LUC and therefore might be a good option for biofuel crop production (e.g., intercropping of Jatropha with edible crops or integrating cellulosic crops within managed forests). However, intercropping of oil palm with other crops, particularly on sloping sites can result in serious soil erosion and soil nutrient depletion (Koh *et al.*, 2009). On the other hand, Davis *et al.* (2011) confirmed that all crops are not equal in causing indirect LUC.

To propose reasonable guidelines and environmental regulations for growing biofuel crops in the future LUC effects (especially indirect LUC) should be studied further which opens the door for further analysis and challenges future investigations, including updating of assessment tools (Karlen *et al.*, 2006; Garrigues *et al.*, 2012) and newer modeling approaches.

Soil Quality and Sustainable Biofuel Crop Production

In general, soil quality is determined (i) by the productive capacity of soil on one hand and (ii) by soil interactions with the environment on the other. Complex interaction of soil properties might affect soil quality in an intricate way (Doran *et al.*, 1994). Hence there is always a risk that biofuel-cultivated lands, like with any other crop production system, might result in irreversible soil quality changes associated with soil degradation and soil environmental damage. To secure the soil from further damage while converting soil resources or existing forests and agricultural lands into biofuel production, clear sustainable development targets need to be established.

Sustainable crop production implies that definition of soil quality requires the integration of several major components: The soil must be able (i) to retain the structural integrity, (ii) to secure sustainable productivity by recycling the nutrients, (iii) to provide environmental quality by limiting the loss of nutrients and heavy metals, (iv) to guarantee sustainable soil health[1] management systems, and (v) to sustain the overall ecological health during and after biofuel crop production.

Mueller *et al.* (2010) have suggested the importance of collaboration between soil scientists and plant breeders for enhancing the nutrient capture by plants from soil. It is indeed important to highlight newer areas where research activities might be focused. For example, breeding crops with deeper and bushy roots could provide sustainable yields while improving soil quality in several ways: through overall soil structure, C and N cycles, water and nutrient retention, and biodiversity (Kell, 2011). According to Ragauskas *et al.* (2006) modern genetic tools can be used to manipulate crop productivity to secure sustainable biofuels and biomaterials through (i) photosynthesis, (ii) nutrient metabolism, (iii) stress resistance in crop plants, (iv) energy allocation within valuable plant organs, and (v) growing season soil quality extension.

Common threats to soil quality in some situations might have detrimental effects on productivity and resource-use efficiency (Figure 8.4). That said, energy crops may also sustain or improve soil quality. This can be achieved, for example, by the production of perennial energy crops that ensure a permanent crop cover, high residue cycling, a deep root system, and absence of regular soil tillage (conservation agriculture). Therefore, perennial crops with more in-farm crop diversity will play a significant role in minimizing the soil quality threats and ensure sustainable industrial crop production and environmental quality for future generations (Singh, 2010).

Sustainability Issues of Biofuel Crop Production Systems

Biofuel production is characterized by the wish to maximize the removal of carbon (C) to be converted into energy in a cost-efficient way, although production of sustainable energy source implies understanding of the main threats which might put soil quality at risk under biofuel crops. High biomass requirement for biofuel may necessitate intensified crop production, causing additional threats to soil quality by competing over C necessary for sustaining SOC (Lal, 2005).

While soil quality under biofuel crop production depends on soil type, land-use history, on-site management, and also effects connected to projected climate and environmental change (see Figure 8.1), risks of soil quality threats in agricultural fields under biofuel production might be seen through (i) potential SOM decline, (ii) soil erosion intensification, and also (iii) on- and off-site pollution associated with biofuel crop production (see also Figure 8.4 and Table 8.1).

[1] Soil health presents soil as a finite dynamic living non-/slowly renewable resource and is related to soil function (Karlen *et al.*, 2001; Letey *et al.*, 2003).

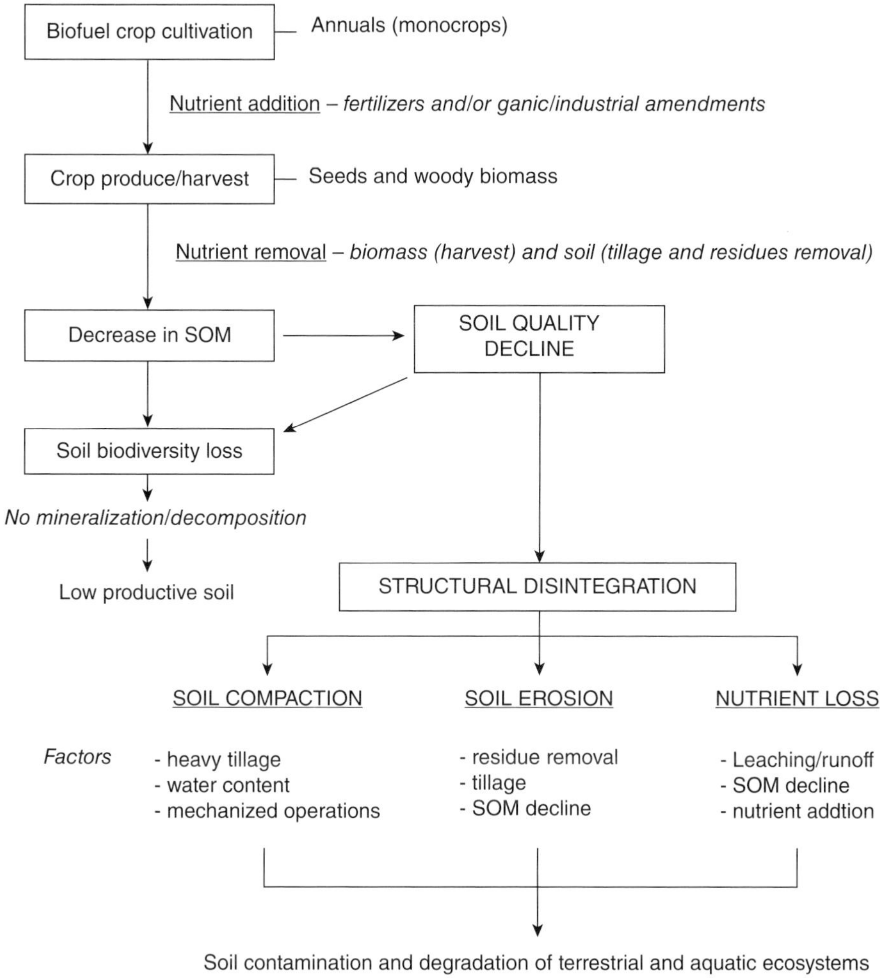

Figure 8.4. The main soil quality threats associated with biofuel crop production.

Soil Organic Matter Decline

SOM is a vital part of soils and it is currently under pressure also due to environmental and climate change (IPCC, 2007). SOM is a major component of global C cycle: by storing approximately one-third of C in biomass, whereas the other two-thirds in soils, terrestrial C pool contains roughly 3 times that of atmospheric CO_2 pool (Lal, 2004). Depending on circumstances, soils therefore may act as a global source or sink for C and nutrients.

SOM is integral for sustainable soil quality, while positive SOC balance is crucial for long-term stability, preservation of soil fertility, nutrient availability, and fluxes of the greenhouse gases (Paustian *et al.*, 1997; Blanco-Canqui and Lal, 2009). In literature the terms SOM and SOC are used interchangeably (it is currently assumed that SOC $= 0.57 \times$ SOM).

The suitability of soil for sustaining (biofuel) crop growth and soil services depends on the quantity and quality of SOC. Thus, increase in SOC not only improves nutrient status, but also provides sustainable soil quality improving microbiological, chemical, and physical soil properties (Bessou *et al.*, 2011). Some recent studies showed benefits of potential biofuel crops (e.g., perennial grasses such as *Miscanthus* and switchgrass) for soil quality. Other research reports, however, have accentuated the risks of lowered SOC stocks in soils under intensive, high-input biofuel cropping systems (mostly annual crops such as maize and soybean) and also due to soil degradation (e.g., oil palm, cassava) caused by LUC.

Bessou *et al.* (2011) have presented the main types of potential biofuel crops from which one can see a positive long-term trend in SOC sequestration. Don *et al.* (2012) have reported such an effect over time. Positive long-term SOC trends were also concluded based on meta-analysis for five biofuel crop rotations by Anderson-Teixeira *et al.* (2009) and also for LUC effects by Nijsen *et al.* (2012) and Qin *et al.* (2012). Anderson-Teixeira *et al.* (2009) summarized that cultivation of temperate-zone perennial grasses (*Miscanthus*, switchgrass, or native mixes) increased SOC in tilled layer annually by 0.1–1 Mg ha^{-1}. In general, higher uncertainties associated with SOC stocks and C sequestration estimations have been reported. For instance, there might be a gap in C sequestration during the first years of biofuel crops establishment (Don *et al.*, 2012).

SOC balance is mediate between C inputs into the soil and C outputs, in the form of harvested biomass or respired carbon dioxide (CO_2) fluxes (Paustian *et al.*, 1997). Consequently, the equilibrium level of SOC depends on the balance between (i) input through plant residues and other biosolids and (ii) output through decomposition, erosion, and leaching (Mulumba and Lal, 2008).

For perennial grasses the C mitigation option has been considered for some time (Clifton-Brown *et al.*, 2007). For instance, Hamelin *et al.* (2012) have compared several biofuel crops and have estimated for *Miscanthus* the lowest emissions of N compounds, the highest amount of CO_2 sequestrated from the atmosphere, a relatively high C turnover efficiency, and also that *Miscanthus* allows us to increase SOC content under Danish soil and climatic conditions.

All energy crops have high biomass, but perennial crops have, in general, higher biomass and more residues than annual crops. In addition, increased aboveground biomass leads to increase in roots and rhizodeposition and consequently provides higher C inputs and lowered C outputs. SOM/SOC models can be used in order to reveal impacts of these complex interactions on SOC turnover and to improve C sequestration estimations. Usually, this includes process-based modeling at plot/field scales based on available data, scenario analysis, and conditional upscaling to the landscape/regional scale. Regional-scale SOC models usually have the form of regression models (Meersmans *et al.*, 2008).

Smith *et al.* (1997) have reported the performance of nine established SOM models in simulation of various land uses (arable, grassland, forestry), soil, management, and climatic conditions. Six models (CANDY, CENTURY, DAISY, DNDC, NCSOIL, RothC) performed significantly better than did three others, while only four models were able to simulate all land uses (CENTURY, NCSOIL, SOMM, RothC). A scheme presented in Figure 8.5 places the SOC modeling into complex assessment of SOM turnover and C sequestration under biofuel crops.

In both previously cultivated and fresh lands, the first casualty will be the disruption of soil macroaggregates, which may directly or indirectly affect SOC. The direct effects are the release of occluded SOC from the disrupted aggregates and indirect effects include reduction in SOM (Six *et al.*, 2002).

Karlen *et al.* (2006) reported that total OC was the most sensitivity indicator of SQI, showing, however, consistent difference between crop rotations and sites. A field history or previous land use might play an important role in C sequestration, but there are a number of

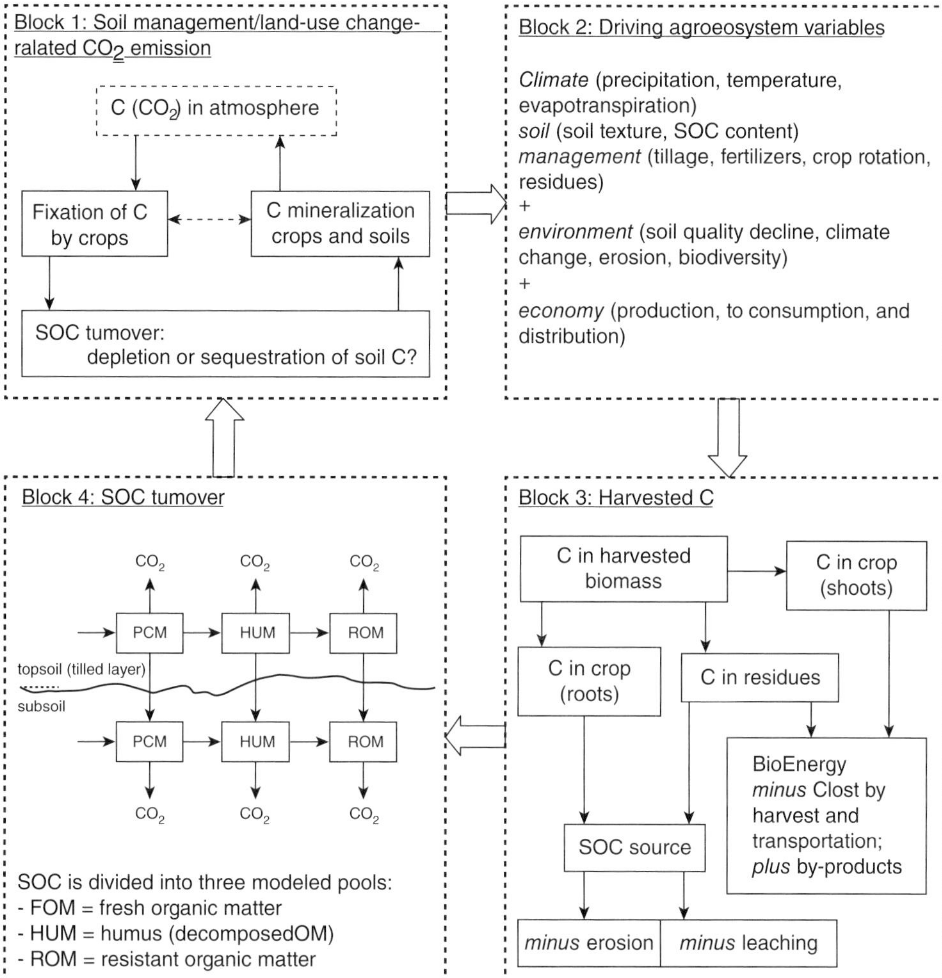

Figure 8.5. SOC modeling for biofuel crop production.

effects on SOC budget which might be generally referred for a field management including crop rotation, soil tillage, and residue management.

Anderson-Teixeira *et al.* (2009) found that clearing of uncultivated land triggered an SOC loss. This may be attributed either (i) to the effects of tillage, which stimulates a release of C from soil, or (ii) to a deficiency of organic inputs to soil. Conservation tillage, that is, reduced or no-tillage in combination with crop residues left in the field, improves physical and chemical soil quality, providing more C input and in general increasing C content in soils previously managed with conventional approaches (Paustian *et al.*, 1997; Lal *et al.*, 1998). Long-term conservation tillage studies demonstrated the ability to maintain or improve soil (SOC) quality in intensive cropping systems such as continuous crops (Reeves, 1997). Taking this into account, perennial crops might seem a better option for C sequestration since soils under these crops are not disrupted for 5–20 years after establishment (see Table 8.1).

Plant residues are a source of annual carbon input and also supply essential plant nutrients (Wilhelm *et al.*, 2004), being, meanwhile, a ground cover to reduce erosion (Johnson *et al.*, 2010). Lal (2005) pointed to short- and long-term impacts of residue removal (for biofuel purposes) on ecosystem degradation, which includes soil, water, and vegetation. Mulumba and Lal (2008) determined optimum mulch rates of 4 Mg ha^{-1} to increase porosity, which was doubled to enhance available water capacity, moisture retention, and aggregate stability.

Crop residues globally have numerous competitive uses, such as energy source, soil amendment, fodder, and constructive material. Crop residues are a good source of feedstock for either direct combustion or production of ethanol using lignocellulosic plant materials (Sanderson, 2006; Graham *et al.*, 2007). More than 50% crop residue removal adversely impacts soil and environmental quality as well as crop yields (Blanco-Canqui, 2010). To make the crop residue option for bioenergy economically sustainable requires removal of significant amounts of crop residues; however, it should not be excessive so as to compromise soil quality, and special harvesting machines (e.g., single pass) should be deployed to avoid the extra pressure on soils that leads to compaction (Lal, 2008; Lal, 2009). Therefore, if there is a risk of soil erosion, SOC losses or water deficit, especially on marginal soils, it is recommended to obtain detailed assessment of the amount of residues that can be safely removed from the field (Blanco-Canqui and Lal, 2009). Most of the studies on crop residues have been short-term and greenhouse experiments; thus, in the absence of data on medium- and long-term consequences of straw residue export on SOC, such assessments are difficult (Saffih-Hdadi and Mary, 2008).

The combination of soil, management, and climatic conditions plays an important role in sustaining SOC budget. To avoid or to reduce the loss of SOM and to solve problems arising from biomass harvest several compensation strategies were presented by Johnson *et al.* (2010). It should be noted here that there is a difference in SOC turnover caused by climate zone (West and Post, 2002); for example, in wet soils with higher temperatures soil processes such as SOC turnover are accelerated, and therefore SOC changes can be detected in shorter time (Smith, 2004). In compensational experimental modeling exercise Chatskikh *et al.* (2009) have demonstrated that partial crop residue removal under no-tillage clearly had a climate-driven nature: in order to sustain SOC content soils in wetter and warmer climates (e.g., of Brazil) required higher annual C inputs compared to cold and dry sites. In turn, benefits of introduction of perennial grass cropping systems for C sequestration in such climates might be detected in a shorter time period as compared to drier and colder sites.

In general, the meta-analysis has showed that annual crops with well-established management in rotation and perennial crops (grasses) have some advantages for soil (SOM) quality (Anderson-Teixeira *et al.*, 2009; Bessou *et al.*, 2011; Don *et al.*, 2012). However, experimental data used in these analyses are scarce. The results were affected by soil, management, and climate; therefore, it is difficult to provide any global analysis on soil quality apart from meta-analysis. Consequently, to follow the evolution of SOC stock precisely over time is difficult due to its small changes and lack of data on soil management (Goidts and van Wesemael, 2007) on the one hand and unclear effects of LUC (Don *et al.*, 2012) on the other.

SOC is converted into carbon dioxide also with LUC to cropping (Davidson and Ackerman, 1993). For instance, Bell *et al.* (2011) reported that the United Kingdom's soils are a sink of C, with the gains from LUC offsetting those due to climate change. Davis *et al.* (2011) have listed the elements that are critical for calculating a complete LUC-related flux of C:

(a) the type of vegetation change (e.g., deforestation/afforestation, logging, cropland and pasture establishment and abandonment, shifting cultivation),
(b) the amount of C vegetation and soils before and after the change,

(c) the method used for the clearing event,
(d) the fate of biomass following clearing,
(e) regrowth of vegetation following clearing, and
(f) SOC dynamics following clearing.

Nijsen *et al.* (2012) and Qin *et al.* (2012) have recently upscaled the effects of SOM changes under biofuel crops to estimate the potential effects on degraded soils. In both studies, they found a high level of uncertainty. If former marginal soil reaches C equilibrium in 50 years an annual soil C sequestration rate of 0.62–0.70 Mg C ha^{-1} (switchgrass, *Miscanthus*) might be foreseen (Qin *et al.*, 2012) or 0.50 Mg C ha^{-1} (*Miscanthus* spp.) in soils of low productivity (Heaton *et al.*, 2004). Coleman *et al.* (2004) have reported a few cases with greater SOC gains when the SRC were grown in marginal lands compared to more fertile agricultural soils.

To further help the modelers formalize our knowledge in the form of algorithms, more experimental data should be available, for instance, on the profile C stratification (Lal, 2009). Subsoil C pool is more stable and also should be carefully accounted for in C sequestration (Jobbagy and Jackson, 2000). Perennial crops have deep and extensive root systems, and increasing the amount of belowground biomass can be a desirable goal (Kell, 2011), a route to sequester more C. However, profile distribution of soil C is less studied (Blanco-Canqui, 2010). On the other hand, inclusion of landscape-scale (geomorphic) processes into long-term SOC analysis might be an essential step forward to deal with the uncertainties connected to spatial variability of SOC stocks (Van Oost *et al.*, 2005; Chatskikh *et al.*, 2009; Wells *et al.*, 2012).

Even though use of biofuel crops to sequester soil C is an intriguing perspective, it is only underpinned by very scarce empirical data and averaged modeling projections, and therefore calls for further research in this area (Blanco-Canqui, 2010). The critical cause–effect relationship between soil management and soil quality, however, can only be established through long-term experiments (Lal and Stewart, 1995; Wilson *et al.*, 2009), while an integrative experimental modeling approach is needed to improve soil (SOC) quality assessment.

Soil Erosion Intensification

Soil erosion is the most important threat to the soil quality of agricultural fields (Boardman and Poesen, 2006). Soil erosion, which is a physical translocation of soil material through space, depending on its driving nature, might be classified as water, wind, or tillage erosion and might occur at the same field.

Biofuel crops may both mitigate and accelerate soil quality threats of soil erosion through (i) crop residue management, that is, harvesting of agricultural residues, incorporation; (ii) effects associated with LUC, that is, converting of existing agricultural lands into biofuel cropping systems, use of marginal lands, land clearing; and (iii) effects associated with land management (except crop residues), for example, converting into rotation/monocropping, intercropping, tillage, irrigation (Karlen *et al.*, 2006; Blanco-Canqui and Lal, 2009; Blanco-Canqui, 2010). Soil erosion occupies an important position among the crucial factors for soil quality decline under biofuel crop production (see Figure 8.4).

Soil erosion risk in the production of biofuel crops can be estimated based on the unique combination of soil, topography, climate, and land use (De Fries, 2012) and also might depend on the particular crop. Annual biofuel crops such as maize and soybean have relatively high erosion rates compared to small grain cereals (wheat, barley, oat, rye, triticale) food crops. But cultivation on slopes of cassava (*Manihot esculenta* Crantz), a starch crop that may be used for producing bioethanol, generally causes more erosion on an annual basis than other crops grown under the same circumstances (Howeler, 2001).

Soil erosion is a threat that creates marginal lands. For instance, erosion, aridity, and field steepness are the primary contributors to the poor quality of the marginal land of the semihumid regions of the Loess Plateau of China, for which Liu *et al.* (2012) proposed to convert traditional agriculture into *Miscanthus* production to prevent further erosion and land degradation. On the other hand, if expansion of maize and soybean production were to occur on marginal land, such as highly erosive post-CRP (the Conservation Reserve Program) lands, the risks of erosion will increase dramatically.

The absence of permanent vegetation on the soil may lead to the loss of nutrients and heavy metals to the soil and water environment, causing pollution. Land-use systems with perennial crops are in general less affected by soil erosion compared to annual cropping systems because the soil is covered throughout the year (Hartemink, 2008b). Bolan *et al.* (2011) listed the major positive effects of vegetation cover on soil toward water and wind erosion as follows:

- interception of the direct impact of rainfall drops and wind;
- decreasing the velocity of runoff, and hence the cutting action of water and its capacity to entrain soil and sediment;
- root-induced compaction and increases in soil strength, aggregation, and porosity;
- enhancement of vegetation-induced biological activities and their influence on soil aggregation and porosity;
- transpiration of water, leading to the subsequent drying out of the soil;
- insulation of the soil against temperature variation, which can result in cracking or "frost heave."

In areas of erosion-risk farming, perennial grasses and SRCs might bring less harm to the soil quality compared with the annual industrial crops or the removal of crop residues for energy production (Blanco-Canqui, 2010). An appropriate biofuel crop (e.g., perennial grasses) might effectively protect slopes against concentrated flow and gully formation (Dabney *et al.*, 1995, 2004; Dosskey *et al.*, 2002; Blanco-Canqui *et al.*, 2004), which are the main attributes of water erosion. The long and wide root systems of perennials might help to protect against concentrated flow and also to retain soil particles from being moved downslope. Gyssels *et al.* (2002) have reported that plant roots control soil erosion, which can be reduced by 50% in the early plant development stages. They have also suggested using double sowing as an erosion control method for concentrated flow areas. On the other hand, intensively managed annual biofuel crops might stimulate erosion rates, especially monocrops and irrigated crops. For example, erosion rates for industrial annual crops (soybean, maize) are up to 2.5 times higher as compared to food crops (small grain cereals, such as wheat, barley, rye) on moderately eroded soils, and these erosion rates at least double if the crops are grown on highly eroded soils. Removal of crop residues may further add to soil erosion and overall soil quality (Blanco-Canqui and Lal, 2009).

Agricultural or crop residues are able to maintain water, SOC, and nutrients in soil and also provide physical protection to the surface of soils exposed to erosion. According to a generic relationship of Laflen and Colvin (1981), 30% of soils covered by the residues following planting will reduce soil erosion by 50% (as compared to bare soil). Consequently, use of the residues as a feedstock for cellulosic biofuels will increasingly leave soils vulnerable to erosion. Other soil erosion-protective methods might be applied, for example, planting cover crops, introduction of conservation methods, optimization in tillage direction and intensity for tillage erosion, contour cropping against water- and wind-erosion-driven sediment losses on steeper slopes, to compensate for lower residue amount (Rey, 2004).

Lal *et al.* (2004) have argued that soil degradation and its adverse effects on productivity on convex slopes are as pronounced in tillage erosion as in water erosion, and both forms of erosion accentuate spatial variability in soil quality. There is an urgent need for further research in this area in order to include the underlying geomorphic and agroecosystem processes in the landscape soil quality assessments (Heckrath *et al.*, 2005; Van Oost *et al.*, 2005).

On- and Off-site Pollution

From an ecosystem perspective, a healthy soil does maintain or increase SOC content and does not pollute the environment (FAO, 2010). The most important pollution problems of biofuel crops linked to soil quality are risks of (i) sedimentation, (ii) loss of nutrients (through leaching or runoff), and (iii) air, water, and soil pollution.

The dynamics in the influx of nutrients and their relative output determine their impact on soil quality (Christian *et al.*, 2008; Costello *et al.*, 2009; Börjesson and Tufvesson, 2011). Hence, nutrient budgeting of bioenergy plantations needs to be carried out to quantify the lost nutrients and also to estimate the nutrients required for sustaining the subsequent cropping, thereby ensuring soil quality. In general, perennial biofuel crops are characterized by lower fertilizer and pesticide requirements compared to their annual counterparts, making it a relatively benign crop environmentally.

The loss of nutrients is more pronounced in HILD than LIHD crops and has two pathways: (i) depletion by harvest and (ii) depletion from residues through leaching/runoff, which has serious consequences on soil health and the overall soil quality. The LIHD crops consisting of woody and cellulolytic plants can not only protect the soil from erosive processes but also provide the raw material for nutrient retention, that is, SOM (Johnson, 2008). Maintenance of nutrient balance and SOC is important for the protection of soil against erosion in order to achieve sustainable soil productivity.

Sedimentation and nutrient loss (especially, N and P) are generally caused by the removal of crop residues (Lindstrom, 1986; Pimentel and Lal, 2007). The loss of N and P results in eutrophication and their impacts are based on the type of biofuel crops (see Table 8.2). For example, Börjesson and Tufvesson (2011) report that biofuels based on sugar beet, livestock crops, and willow contribute roughly 2–3 times less than grain crops like wheat toward eutrophication.

Growing HILD crops for biofuel production imparts detrimental effects on the productivity of the soil, and nutrient extraction from soil to plants is specific to certain types of plants. For example, oilseed crops extract more sulfur (S) and nitrogen (N) (McGrath and Zhao, 1996; Abdallah *et al.*, 2010). McGrath and Zhao (1996) found that a tonne of rapeseed production requires approximately 16 kg of S, whereas wheat requires only 2–3 kg of S for the same amount of grain production. Also, S deficiency influences N uptake in some oilseed crops (Fismes *et al.*, 2000). Nutrient use efficiency and nutrient uptake play important roles in cropping systems. For instance, Sainju and Singh (1997) have observed that a maximum of 70% of applied N was used by plants. Also, nutrient depletion occurs in various stages of plant growth. For instance, when a crop reaches 40% of its final dry weight, 60–75% of the total N, P, and K would have been accumulated within the plants. While a portion of N and P are translocated to grains on maturity, K remains in leaves and stems (Malhi *et al.*, 2007).

Therefore, indiscriminate usage of fertilizers can lead to nutrient losses to soil and aquatic environments with limited supply to the bioenergy crops, creating pollution problems and

Table 8.2 Average levels of nutrients removed by bioenergy crops.

		Nutrient removal from biomass						
		N		P		K		
Crop	Biomass yield (Mg ha^{-1})	Tissue level (kg Mg^{-1})	Total removed (kg ha^{-1})	Tissue level (kg Mg^{-1})	Total removed (kg ha^{-1})	Tissue level (kg Mg^{-1})	Total removed (kg ha^{-1})	Reference
Sunflower	55.5	8.5	471.8	0.7	42.1	16.5	915.8	Madejón *et al.* (2003)
Sugarcane	70.8	0.8	56.7	0.3	21.2	1.32	93.5	IFA (1992)
Giant reed	56.4	1.5	90	0.53	30	4.6	260	Nassi o Di Nasso *et al.* (2011)
Hemp	55.7	1.5	83.6	0.42	23.4	2.13	118.6	Prade (2011)
Rapeseed	62.0	7.2	446.5	2.8	173.7	7.7	477.6	IFA (1992)
Maize	61.1	3.1	191	1.45	89	3.84	235	IFA (1992)
SRC poplar	12.5	9.3	116.3	1.2	15	13	162.5	Singh (2008)
SRC willow	12.5	4.3	53.8	0.6	7.5	2.1	26.25	Adler (2007)
Switchgrass	8.6	6.7	58	0.99	8.5	8.9	77	Kering *et al.* (2012)
Miscanthus	21.2	4.2	90	1.9	40	9.9	210	Nassi o Di Nasso *et al.* (2011)
Cassava	35.7	6.7	238.5	0.76	27.1	4.87	174	Howeler (2001)

adding to environmental concerns. Plant growth limiting nutrients such as N and P are also prone to leaching losses, and N in particular is lost through volatilization (Daigger *et al.*, 1976). The loss of other cations such as K, S, Ca, Na, and Mg may lead to structural changes in soil, thereby affecting the overall quality of the soil (Baumgarten *et al.*, 2012).

In practice, biofuel crops might be a source for pollution not only coming from on-site management, but also as a result of LUCs due to biofuel production; well-known examples of such impacts are from soybean in South America or oil palm in Asia. In the United States, the conversion of CRP land to cultivation has been estimated to transfer from land to atmosphere 54–70 Gt (Gt = 10^{12} kg) of C (Amundson, 2001) and consequently, SOC reduced by up to 60% (Guo and Gifford, 2002). Munson *et al.* (2012) linked C uptake and storage to N and stressed its importance toward N pools and fluxes. Soils polluted with biodiesel and their diesel blends showed genotoxic/mutagenic effects (Leme *et al.*, 2012), although, in general, assessments of soil and water biodiesel pollution are rare.

Utilization of Degraded Soils for Biofuel Production

Land degradation is a human-driven threat of decreasing soil quality and productivity, which might be considered and categorized as soil retrogression (loss of soil primarily due to erosion), soil degradation (in-place decline in soil quality), or ecosystem decline (loss of ecosystem services).

Land degradation is a problem largely related to agricultural fields all around the world (Boardman and Poesen, 2006 for Europe). Nutrient-depleted or abandoned (unmanaged) lands constitute an additional category of degraded land. Biofuel crops might be used to restore degraded lands, to improve its soil quality, increasing SOC and soil resistance to erosion or removing contaminants, positively changing the trend (Qin *et al.*, 2012). Gopalakrishan *et al.* (2011) used the term "marginal lands" and proposed a multiple criterion framework to define marginal/degraded lands such as profitability of current land use, soil health indicators (erosion, flooding, drainage, or high slopes), and contamination of water resources. Consequently, some authors claim that the production of energy crops for biofuel in marginal lands could restore previously unproductive or infertile soils (Ma *et al.*, 2000; Coleman *et al.*, 2004; Heaton *et al.*, 2004; Blanco-Canqui, 2010).

Percival *et al.* (2000) found that soils growing warm-season grasses have better soil aggregation compared to row crops; this tendency was attributed to increased biomass return and reduced soil disturbance. Perennial crops can grow in soils with low fertility and limited water supply, because their roots penetrate much deeper than annual crops, securing nutrient uptake and also recharging groundwater (Clark *et al.*, 1998; Gilker *et al.*, 2002). Furthermore, perennial crops (grasses and SRCs) might sequester C and improve soil quality (Coleman *et al.*, 2004; Frank *et al.*, 2004), while C gains might be higher in lands with lower SOC content, such as marginal soils. In fact, Coleman *et al.* (2004) reported that short rotation poplar may even sequester less SOC compared to annual crops and switchgrass when planted on agricultural soils of good fertility.

In the United States, conversion of the CRP lands into perennial grasslands is a preferable option (Munson *et al.*, 2012). Shifting these lands characterized by less intensive management into maize (currently the number 1 bioethanol crop in the United States) will adversely affect the environment (Malcolm and Aillery, 2009), substantially increasing pressure on soils, and therefore soil quality might be affected significantly (e.g., via SOC turnover, nutrient leaching). Williams *et al.* (2010) estimated increased risks under crop production of former CRP; those risks must be taken into account, especially considering intensive continuous cropping, removal of crop residues for biomass fuel production, extensive tillage, and water management. On the other hand, converting post-CRP lands into perennial energy crops might be a useful option for sustainable soil quality (Blanco-Canqui, 2010), which, however, should be better documented.

Vuichard *et al.* (2009) suggest using LIHD grasses or annual biofuel crops in combination with conservation tillage to improve soil quality on large areas. Such an approach might be applied for millions of hectares of former agricultural fields currently unattended (Vuichard *et al.*, 2008, 2009, for the former Soviet Union). Biofuel production might also be an option for sustainable use of organic soils, for example, abandoned or conserved peat soils in Northern Europe (Biemans *et al.*, 2008). Biofuel crops (perennial grasses) might conserve SOC in these soils and prevent nutrient leaching, while controlled traffic will allow improvement in the physical soil properties in a sustainable way. However, Börjesson and Tufvesson (2011) urge that if peat soils are used to grow biofuel crops soil CO_2 emissions could be 10–20 times higher compared to mineral soils.

Degraded lands are not or only limitedly used for agricultural production. Therefore, their use for biofuel crop production may provide a good utilization outlet (Blanco-Canqui, 2010). By using degraded lands the food croplands are not jeopardized for bioenergy, and there are no carbon emissions due to indirect LUC effects (Qin *et al.*, 2012).

Due to its remediation ability, physiology, and lower management costs, biofuel crops such as perennial grasses and SRC might be a win–win option for idle soils, including also former mine spills (Evangelou *et al.*, 2012). From soil quality perspective there is also an important question of how biofuel crops affect biodiversity, reflecting the mix of living organisms in the agricultural landscape. Biemans *et al.* (2008) reported that the number of plant species in a field can be assumed to influence the number of species on a landscape, and therefore HILD crops might be expected to support fewer species than LIHD crops and therefore be less favorable for biodiversity and overall soil quality. Another serious issue is the potential invasiveness of biofuel crops and potential harm to natural habitats and native species (Raghu *et al.*, 2006; Crosti, 2009). Davis *et al.* (2010) stressed the importance of screening bioenergy crops to mitigate invasion risk and suggested a collaborative effort between agronomists and invasion biologists toward the development of crops for biofuel production.

By-product Incorporation to Maintain Soil Quality

Biofuel production chain determines the type and amount of by-products: increase in the production of energy from biomass envisages a steady increase in the production of by-products. Although these by-products were initially considered ag wastes and therefore to be disposed of, some of them returning to the field could positively affect soil quality (Alotaibi and Schoenau, 2012).

Johnson *et al.* (2004) suggested that by-products from bioenergy production should be utilized as a soil amendment to at least limit the magnitude and severity of residue removal impacts. There is an indication that the return of digested residues to farmland may compensate for some soil C losses due to residue removal but further quantification is needed (Thomsen *et al.*, 2013). The study on the impact of these additions on soil C and N cycle is also in preliminary stages (Cayuela *et al.*, 2010).

Cayuela *et al.* (2010) added 10 different by-products to soil and investigated the soil C and N cycling. They found that biofuel by-products containing high amounts of easily decomposable C led to short-term N immobilization in soil, limiting their potential as short-term fertilizers. Similarly, Alburquerque *et al.* (2010) evaluated the C and N mineralization dynamics in a short-term laboratory study for an agricultural soil amended with some digestates produced from anaerobic codigestion of cattle slurry + 4% glycerol and cattle slurry + 6% glycerol. These digestates showed a high organic load and low degree of stability. This fact conditioned the N and C mineralization processes, leading to high CO_2 production and N-immobilization or denitrification. The authors recommended a further stabilization process, such as the exhaustion of the easily degradable organic matter, in order to obtain maximum agricultural and environmental benefits. Laboratory studies also reveal that by-products of cellulosic fermentation containing high lignin concentrations improved soil quality characteristics (Johnson *et al.*, 2004, 2007).

Meunchang *et al.* (2005) found that the sugarcane by-products, filter cake and bagasse, had potential to stabilize SOM and to control N loss. Filter cake seems suitable for agronomic use since it has a neutral pH and low phytotoxicity and nutrient content (Meunchang *et al.*, 2005). Mill by-products are regarded by farmers as products that confer a range of benefits to their soils. Meunchang *et al.* (2006) found sugar mill by-products compost to be an effective soil amendment for promoting the growth of other crops.

Glycerine (glycerol), a coproduct of different stages of biodiesel production, as a soil amendment has implications for C sequestration and emission of greenhouse gases and may therefore be of beneficial use in soil quality improvement. For instance, incorporation of digestates (from glycerol codigestion) into soil provided an important source of SOM and nutrients for plants (Santibáñez *et al.*, 2011). Since glycerine is derived from natural plant materials, it is generally assumed to be nontoxic but studies of its effects on microbial and plant systems are limited. The findings of Schoenau *et al.* (2009) and Soerens (2011) suggest that microbiological, plant, and animal habitats are not in danger at rates lower than 1000 kg ha^{-1}. It has also been found that glycerin-based compositions might be used to improve biocontrol (Siddiqui and Shaukat, 2002) and for control of weeds, phytopathogenic nematodes, and other soilborne plant pests (Rodriguez-Kabana and Walker, 2008). Moreover, there are glycerine-based products used as biofungicide (Chung *et al.*, 2005).

Biochar, a derivative of solid biofuel, can be used as a soil amendment to enhance (biofuel) crop growth due to its potential in improving soil physical and chemical properties, supplying and retaining nutrients (important for crop production on degraded and nutrient-poor soils), and improving soil quality, for example, by removing pollutants (such as nitrates and heavy metals) from soil water (Lehmann and Rondon, 2006). The ability of biochar to improve crop yields depends on its inherent characteristics and also on local soil and climatic conditions (Glaser *et al.*, 2002).

The aforementioned and other biofuel by-products might be of use in ways that allow them to degrade to an extent to maintain biological activity and nutrient cycling but still show some persistence in soil (Cayuela *et al.*, 2010). Alternatively, they might affect soil quality returning via the agricultural production chain. For instance, neutral effect on meat and a high nutritional value might allow us to use glycerine as animal feed (Kerr *et al.*, 2007) and then to apply it in digested form as a manure in agricultural fields to make it more valuable as an organic fertilizer (Alotaibi and Schoenau, 2012).

Biofuel Crops to Remedy Soil Contamination

Many agricultural soils are contaminated due to (i) excessive use of agrochemical products (fertilizers can contribute, e.g., Cd), (ii) wastewater irrigation, (iii) atmospheric deposition, or (iv) current or former application of sludge wastes. Also in urban industrial areas dumped heavy metals can severely pollute soils. Mining areas with mine spoils and spills are sensitive to heavy metal contamination.

Once applied, heavy metals are strongly bound on soil mineral surfaces and, as complexes with SOM, retained within the soil. Because heavy metals are of limited mobility in the soil, surface applications of sewage sludge, for example, result in higher concentrations of these elements in the surface soil. The consequent retention may also pose a risk of soil ingestion by wildlife. In general, solubility and plant availability of heavy metals for plants is higher at low soil pH and their uptake/toxicity increases if the soil pH is allowed to fall (Bolan *et al.*, 2003).

Heavy metals are present in all soils primarily as the result of soil-forming processes (commonly reflect the composition of the parent rock material) and at concentrations that are most often not of concern to the soil environment. Registered excessive levels of heavy metals in soils are usually human driven. Depending on route and dose such levels deplete soil quality and can damage crops, animals, and humans (via metabolic interference, mutagenesis, for example, Khan and Scullion, 2002) after entering the food chains via bioaccumulation. For instance, Abaye *et al.* (2005) demonstrated that heavy metal contamination affects biological soil quality by increasing biomass-specific respiration rates and causes changes in the microbial community while decreasing soil microbial biomass and N mineralization.

The soil quality problem is amplified because unlike organic pollutants, heavy metals do not decay. Therefore, the only alternative to expensive and often environmentally risky soil excavation is extraction, which is most cost-efficiently done by plants (Purakayastha and Chhonkar, 2010). The *in situ* remediation with plants, without disturbing the soil profile, preserves ecosystem functioning. Establishment of permanent crop cover can be particularly effective for (i) phytostabilization of metal-contaminated sites to ensure that soil erosion processes do not spread the contamination (Bolan *et al.*, 2011), (ii) phytocapping of contaminated sites to avoid water infiltration and thereby limit the loss of nutrients and heavy metals (Lamb *et al.*, 2012), and (iii) phytoextraction of contaminated sites with the possibility of subsequent energy utilization (Robinson *et al.*, 2009). However, the long remediation period most often required to reduce heavy metal contents to acceptable levels is the main drawback of method of phytoremediation. When the heavy metals have been taken up by plants the biomass has to be taken off the land; this leads to the idea of connecting phytoextraction with biofuel production. Hence, biofuel crop production has shown promise to be a cheap and environmentally friendly way to deal with heavy metals on contaminated land, even though commercial experience is still limited (Robinson *et al.*, 2009).

High-biomass, high-value, and deep-rooted energy crops have the potential for a sustainable way of pollution control (Evangelou *et al.*, 2012). Hyperaccumulators are used as colonizers in soil to manage nutrient deficiencies and to uptake the mobile heavy metals (Robinson *et al.*, 2009). The common disadvantages of such plants include low biomass production, exotic to most environments, and facilitation of metal into food chain (Bolan *et al.*, 2011).

A number of heavy metal hyper-accumulating plants, for example, *Thlaspi caerulescens*, that may have 100–1000 times higher concentrations than most other plants have been detected (Brooks, 1998). However, they often produce low yields and may not be suitable for energy conversion. On the other hand, dedicated energy crops, despite lower metal concentrations, produce higher yields and are well developed for ease of production and for energy conversion. High-yielding biomass crops can either have a moderately high uptake rate of metals (e.g., willow (Hammer *et al.*, 2003; Jensen *et al.*, 2009)) or discriminate against heavy metal uptake as do many monocots, for example, maize (Meers *et al.*, 2010). The produced biomass can be biologically converted into biogas or ethanol but extraction or concentration of the heavy metals in the fermentation residue is difficult (Meers *et al.*, 2010). On the other hand, when

Table 8.3. Example calculation of the remediation of soils (based on Andersen *et al.*, 2000).

Contaminant	Soil concentration (mg kg^{-1})	Content 0–50 cm soil (kg ha^{-1})	Annual uptake by crop (kg ha^{-1})		Remediation time (yrs)	
			Willow	*Thlaspi*	Willow	*Thlaspi*
Cd	1	7.5	0.15	0.31	25	12
Zn	1000	7500	21	20	179	188
Pb	80	600	0.6	0.038	500	7895

Precondition: Soil heavy metal content exceeds soil quality limits by 100%. The remediation is defined as reduction to the soil quality limit. Willow yield of 15 MgDM ha^{-1} and contents of 10 mg Cd, 1400 mg Zn, and 40 mg Pb per kgDM. *Thlaspi caerulescens*: 2.5 MgDM ha^{-1}, 125 mg Cd, 8000 mg Zn, and 15 mg Pb per kgDM.

biomass is combusted, the heavy metals accumulate mainly in the fly ash fraction (Obernberger *et al.*, 1997), which means that only a small amount needs to be disposed of, and the metals may even be recovered from the ash.

The long remediation period required is illustrated by an example calculation, which indicates that lightly contaminated soils can be remedied for Cd within 12–25 years using either a hyper-accumulating crop or a biomass crop, while Zn and Pb are not removed within a reasonable time scale with any of the species (see Table 8.3).

Extraction time of heavy metals may be decreased by application of chelating agents, such as EDTA or rather more biodegradable chelants, such as NTA or EDDS (Lestan *et al.*, 2008), which can even induce Pb desorption from soil. However, these opportunities must be handled with care as bioavailability is increased as well as the risk of leaching to groundwater when heavy metal desorption from soil is induced. Moreover, there is also evidence that significant downward movement through the soil profile can occur as a result of the complexation of these metals with soluble organics and transport in preferential flow (Richards *et al.*, 1998).

Another important issue is the need to increase tolerance in biofuel crops to be able to grow on contaminated soils. One of the foreseen solutions might be combining heavy metal-tolerant bacteria with plants used to make biofuels. For example, Odhiambo (2011) has reported that inoculating oilseed rape plants with metal-resistant bacteria provided them with sufficient protection and their growth was also enhanced. The survival of the perennial biofuel crops is in general less affected by the toxicity of the contaminated land and the general condition of the soil than annual biofuel crops (Dale *et al.*, 2010).

Energy crops (in particular monocots) can be used for phytostabilization of lightly contaminated sites, which means that the vegetation cover minimizes the risk of soil erosion and reduces the leaching loss of nutrients, while the biomass with only small concentrations of heavy metals is utilized for energy purposes (Witters *et al.*, 2012).

Incorporation of crop residues into heavy-metal contaminated soils improves physical soil quality, may sustain SOM, and improves other soil chemical and biological properties, including decrease of the heavy metal phytoavailability. In his study Ok *et al.* (2011) have demonstrated such effects for Cd and Pd leaving rapeseed residues in rice–rapeseed double cropping system on paddy soil contaminated with heavy metals. These effects may also be obtained by the production of perennial energy crops which will additionally increase SOM content.

The tolerance by perennial grasses and SRCs of nonheavy metal-contaminated soils leads to other implications for soil remediation connected to biofuel crops, for instance, in trinitrotoluene-contaminated soils (Krishnan *et al.*, 2000; Chekol *et al.*, 2002), and soils contaminated with atrazine and metolachlor (Zhao *et al.*, 2003). Euliss *et al.* (2008) found that perennial grasses have shorter remediation periods and are more effective with petroleum hydrocarbon-contaminated sediments compared to the SRCs (poplar and willow). Gaskin and Bentham (2010) grew native Australian grass species (e.g., *Cymbopogon ambiguus*, *Brachiaria decumbens*, and *Microlaena stipoides*) in mine site soil, experimentally contaminated with known ratio of oil mixture. They found that the grass species increased the soil microbial activity specific to petroleum hydrocarbon degradation, even without any nutrient supplementation.

Conclusions

Growing demand for biofuel (bioenergy) crop production seeks to maximize conversion of harvested C into fuel stipulating active use of very slowly renewable soil resource. Meanwhile, there is a risk that biofuels generated from certain (annual) crops might add to soil quality threats such as SOM decline, soil erosion, nutrient depletion, and soil pollution and also through associated direct and indirect LUC.

In general, perennial biofuel crops are superior to the annual crops in terms of sustainability of overall soil quality. However, results of these analyses are scarce while the effects of biofuel crop production on soil quality are largely site specific and depend on crop, climate, and environment on one hand and on soil management and LUC on the other. Above all, the question of partial removal of crop residues in relation to soil quality requires further quantification.

Soil quality (e.g., of crop production systems, LUC) is still poorly represented in the LCA, a common tool to estimate environmental impacts of biofuel production. An LCA chart for each biofuel crop consisting of the most significant soil quality influencing factors (e.g., SOC, nutrient depletion, and tillage practices) linked to soil variables and processes can play an effective role in limiting their negative effects on soil degradation and ecosystem services. Models of different complexity are in use to simulate input and output flows of biofuel cropping systems. However, experimental data for providing detailed soil quality assessments are still very limited consequently slowing down the development and validation of biofuel crop models.

There are also other challenges in biofuel crop production in relation to soil quality, such as potentials of reuse of biofuel by-products, adaptability and high remediation ability of some biofuel crops (perennials), and, as a consequence, the potential for growing of biofuel crops on degraded soils. For instance, there are some studies suggesting the use of ag wastes as soil amendments or organic fertilizers. Moreover, biorefinery vision (conversion of nonrenewable soil resource to renewable sources of energy and biomaterial without soil quality degradation) suggests changing a traditional way to look at the wastes management. Different potential biofuel crops were found to be able to secure crop production on degraded lands in different climate and soil conditions and to improve soil quality (perennial grasses and short rotation coppices); this option might be further explored while combined with the ability of some biofuel crops to grow on contaminated soils. Most importantly, placing on marginal/idle soils, biofuel crops will not compete with crops grown for food.

Due to the overall complexity of biofuel crop production systems, better integration of LCA and dynamic modeling with detailed experimental data to estimate soil quality indicators is needed. Therefore, in general, soil quality and soil erosion monitoring for biofuel crop production, soil management, and associated land use must be improved, while performance of biofuel crops must be carefully examined.

Acknowledgments

The authors are indebted to Uffe Jørgensen (Aarhus University, Denmark) for valuable comments and additional input. Dmitri Chatskikh was financially supported by the European Union under the Seventh Framework Programme (FP7-PEOPLE-2009-IEF, project GAEMASS no. 255042). Anna Ovchinnikova worked on the chapter during her sabbatical fellowship. The authors would also like to thank CRC CARE for providing funding (no. 2-3-09-07/08; 4.2.1.11/12) to undertake research on landfill site remediation; parts of the chapter were derived from these projects.

References

Abaye, D.A., Lawlor, K., Hirsch, P.R., and Brookes, P.C. (2005) Changes in the microbial community of an arable soil caused by long-term metal contamination. *European Journal of Soil Science* 56: 93–102.

Abdallah, M., Dubousset, L., Meuriot, F., Etienne, P., Avice, J., and Ourry, A. (2010) Effect of mineral sulphur availability on nitrogen and sulphur uptake and remobilization during the vegetative growth of *Brassica napus* L. *Journal of Experimental Botany* 61: 2635–2646.

Abu Bakar, R., Darus, S., Kulaseharan, S., and Jamaluddin, N. (2011) Effects of ten year application of empty fruit bunches in an oil palm plantation on soil chemical properties. *Nutrient Cycling in Agroecosystems* 89: 341–349.

Adler, A. (2007) Accumulation of elements in Salix and other species used in vegetation filters with focus on wood fuel quality. PhD thesis submitted to Swedish University of Agricultural Sciences, Uppsala, Sweden.

Alburquerque, J.A., De la Fuente, C., Carrasco, L., Cegarra, J., and Bernal, M.P. (2010) Carbon and nitrogen mineralization in soil amended with digestates from anaerobic codigestion processes. Proceedings of the 14th Ramiran International Conference "Treatment and Use of Organic Residues in Agriculture: Challenges and Opportunities Towards Sustainable Management", 12–15 September 2010, Lisbon, Portugal. Technical University of Lisbon, Instituto Superior de Agronomia, Lisbon.

Alotaibi, K.D. and Schoenau, J.J. (2012) Biofuel production byproducts as soil amendments. *Sustainable Agriculture Reviews* 9: 67–91.

Amundson, R. (2001) The carbon budget in soils. *Annual Review of Earth and Planetary Sciences* 29: 535–562.

Andersen, L., Holm, P.E., Lehmann, N.K.J., Petersen, L.S., Jørgensen, U., and Mortensen, J. (2000) Phytooprensning af metaller. Report 536 of the Danish EPA. Available at. www2.mst.dk/Udgiv/publikationer/2000/87-7944-097-5/pdf/87-7944-098-3.pdf (accessed 14 September 2012).

Anderson-Teixeira, K.J., Davis, S.C., Masters, M.D., and DeLucia, E.H. (2009) Changes in soil organic carbon under biofuel crops. *GCB Bioenergy* 1: 75–96.

Andrews, S.S. and Carroll, C.R. (2001) Designing a soil quality assessment tool for sustainable agroecosystem management. *Ecological Applications* 11: 1573–1585.

Andrews, S.S., Karlen, D.L., and Cambardella, C.A. (2004) The soil management assessment framework: a quantitative soil quality evaluation method. *Soil Science Society of America Journal* 68: 1945–1962.

Arshad, M.A. and Martin, S. (2002) Identifying critical limits for soil quality indicators in agro-ecosystems. *Agriculture, Ecosystems and Environment* 88: 153–160.

Aubinet, M., Moureaux, C., Bodson, B., *et al.* (2009) Carbon sequestration by a crop over a 4-year sugar beet/winter wheat/seed potato/winter wheat rotation cycle. *Agricultural and Forest Meteorology* 149: 407–418.

Baumgarten, W., Neugebauer, Th., Fuchs, E., and Horn, R. (2012) Structural stability of marshland soils of the riparian zone of the Tidal Elbe River. *Soil and Tillage Research*. doi:10.1016/j.still.2012.06.002.

Bell, M.J., Worrall, F., Smith, P., *et al.* (2011) UK land-use change and its impact on SOC: 1925–2007. *Global Biogeochemical Cycles* 25: GB4015

Bessou, C., Ferchaud, F., Gabrielle, B., and Mary, B. (2011) Biofuels, greenhouse gases and climate change. A review. *Agronomy for Sustainable Development* 31: 1–79.

Biemans, M., Waarts, Y., Nieto, A., Goba, V., Jones-Walters, L., and Zöckler, C. (2008) *Impacts of Biofuel Production on Biodiversity in Europe*, ECNC-European Centre for Nature Conservation, Tilburg, The Netherlands.

Blanco-Canqui, H. (2010) Energy crops and their implications on soil and environment. *Agronomy Journal* 102: 403–419.

Blanco-Canqui, H., Gantzer, C.J., Anderson, S.H., Alberts, E.E., and Thompson, A.L. (2004) Grass barriers and vegetative filters strip effectiveness in reducing runoff, sediment, and nutrient loss. *Soil Science Society of America Journal* 68: 1670–1678.

Blanco-Canqui, H. and Lal, R. (2009) Corn stover removal for expanded uses reduces soil fertility and structural stability. *Soil Science Society of America Journal* 73: 418–426.

Boardman, J. and Poesen, J. (2006) *Soil Erosion in Europe*, John Wiley & Sons, Ltd.

Bolan, N.S., Adriano, D.C., and Curtin, D. (2003) Soil acidification and liming interactions with nutrient and heavy metal transformation and bioavailability. *Advances in Agronomy* 78: 216–272.

Bolan, N.S., Park, J.H., Robinson, B., Naidu, R., and Huh, K.Y. (2011) Chapter four – phytostabilization: a green approach to contaminant containment. *Advances in Agronomy* 112: 145–204.

Börjesson, P. and Tufvesson, L.M. (2011) Agricultural crop-based biofuels-resource efficiency and environmental performance including direct land use changes. *Journal of Cleaner Production* 19, 108–120.

Bowman, R.A., Nielsen, D.C., Vigil, M.F., and Aiken, R.M. (2000) Effects of sunflower on soil quality indicators and subsequent wheat yield. *Soil Science* 165: 516–522.

Brooks, R.R. (ed.) (1998) *Plants That Hyperaccumulate Heavy Metals. Their Role in Phytoremediation, Microbiology, Archaeology, Mineral Exploration and Phytomining*, CAB International, Wallingford, UK.

Butler, R. (2008) U.S. biofuels policy drives deforestation in Indonesia, the Amazon. Available at: http://news.mongabay.com/2008/0117-biofuels.html (accessed 12 August 2012).

Cayuela, M.L., Oenema, O., Kuikman, P.J., Bakker, R.R., and Van Groenigen, J.W. (2010) Bioenergy by-products as soil amendments? Implications for carbon sequestration and greenhouse gas emissions. *GCB Bioenergy* 2: 201–213.

Chatskikh, D., Hansen, S., Olesen, J.E., and Petersen, B.M. (2009) A simplified modelling approach for quantifying tillage effects on soil carbon stocks. *European Journal of Soil Science* 60: 924–934.

Chatskikh, D., Olesen, J.E., Hansen, E.M., Elsgaard, L., and Petersen, B.M. (2008) Effects of reduced tillage on net greenhouse gas fluxes from loamy sand soil under winter crops in Denmark. *Agriculture, Ecosystems and Environment* 128: 117–126.

Chekol, T., Vough, L.R., and Chaney, R.L. (2002) Plant-soil-contaminant specificity affects phytoremediation of organic contaminants. *International Journal of Phytoremediation* 4: 17–26.

Christian, D., Riche, A., and Yates, N. (2008) Growth, yield and mineral content of Miscanthus × giganteus grown as a biofuel for 14 successive harvests. *Industrial Crops and Products* 28: 320–327.

Chung, W., Huangb, J., and Huang, H. (2005) Formulation of a soil biofungicide for control of damping-off of Chinese cabbage (*Brassica chinensis*) caused by *Rhizoctonia solani*. *Biological Control* 32: 287–294.

Clark, R., Alberts, E., Zobel, R., *et al.* (1998) Eastern gamagrass (*Tripsacum dactyloides*) root penetration into and chemical properties of claypan soils. *Plant and Soil* 200: 33–45.

Clifton-Brown, J.C., Breuer, J., and Jones, M.B. (2007) Carbon mitigation by the energy crop, *Miscanthus*. *Global Change Biology* 13: 2296–2307.

Coleman, M.D., Isebrands, J.G., Tolsted, D.N., and Tolbert, V.R. (2004) Comparing soil carbon of short rotation poplar plantations with agricultural crops and woodlots in north central United States. *Environmental Management* 33: 229–308.

Costello, C., Griffin, W.M., Landis, A.E., and Matthews, H.S. (2009) Impact of biofuel crop production on the formation of hypoxia in the Gulf of Mexico. *Environmental Science and Technology* 43: 7985–7991.

Crosti, R. (2009) Invasiveness of biofuel crops and potential harm to natural habitats and native species. Final report T-PVS/Inf(2009)6 for Council of Europe, Strasbourg, France.

Dabney, S.M., Meyer, L.D, Harmon, W.C., Alonso, C.V., and Foster, G.R. (1995) Depositional patterns of sediment trapped by grass hedges. *Transactions of the ASAE* 38: 1719–1729.

Dabney, S.M., Moore, M.T., and Locke, M.A. (2006) Integrated management of in-field, edge-of-field, and after-field buffers. *Journal of American Water Resources Association* 42: 15–24.

Dabney, S.M., Shields, F.D., Temple, D.M., and Langendoen, E.J. (2004) Erosion processes in gullies modified by establishing grass hedges. *Transactions of the ASAE* 47: 1561–1571.

Daigger, L., Sander, D., and Peterson, G. (1976) Nitrogen content of winter wheat during growth and maturation. *Agronomy Journal* 68: 815–818.

Dale, V.H., Kline, K.L., Wiens, J., and Fargione, J. (2010) Biofuels: implications for land use and biodiversity. Ecological Society of America special report. Available at: http://www.esa.org/biofuelsreports (accessed 13 September 2012).

Davidson, E.A. and Ackerman, I.L. (1993) Changes in soil carbon inventories following cultivation of previously untilled soils. *Biogeochemistry* 20: 161–193.

Davis, A., Cousens, R., Hill, J., Mack, R., Simberloff, D., and Raghu, S. (2010) Screening bioenergy feedstock crops to mitigate invasion risk. *Frontiers in Ecology and the Environment* 8: 533–539.

Davis, S.C., House, J.I., Diaz-Chavez, R.A., Molnar, A., Valin, H., and DeLucia, E.H. (2011) How can land-use modelling tools inform bioenergy policies? *Interface Focus* 1: 212–223.

De Fries, S.C. (2012) Resource use efficiency and environmental performance of biofuel cropping systems. PhD thesis, Wageningen University, Wageningen, The Netherlands.

Don, A., Osborne, B., Hastings, A., *et al.* (2012) Land use change to bioenergy production in Europe: implications for the greenhouse gas balance and soil carbon. *GCB Bioenergy* 4: 372–391.

Dondini, M., Hastings, A., Saiz, G., Jones, M.B., and Smith, P. (2009) The potential of *Miscanthus* to sequester carbon in soils: comparing field measurements in Carlow, Ireland to model predictions. *GCB Bioenergy* 1: 413–425.

Doran, J.W. (2002) Soil health and global sustainability: translating science into practice. *Agriculture, Ecosystem and Environment* 88: 119–127.

Doran, J.W., Coleman, D.C., Bezdicek, D.F., and Stewart, B.A. (1994) Defining soil quality for a sustainable environment. *SSSA Special Publication No. 35*, Soil Science Society of America and American Society of Agronomy, Madison, Wisconsin, USA.

Doran, J.W. and Safley, M. (1997) Defining and assessing soil health and sustainable productivity, in *Biological Indicators of Soil Health* (eds C. Pankhurst, B.M. Doube, and V.V.S.R. Gupta), CAB International, Wallingford, UK, pp. 1–28.

Dosskey, M.G., Helmers, M.J., Eisenhauer, D.E., Franti, T.G., and Hoagland, K.D. (2002) Assessment of concentrated flow through riparian buffers. *Journal of Soil and Water Conservation* 57: 336–343.

Euliss, K., Ho, C.H., Schwab, A.P., Rock, S., and Banks, A.K. (2008) Greenhouse and field assessment of phytoremediation for petroleum contaminants in a riparian zone. *Bioresource Technology* 99: 1961–1971.

Evangelou, M.W.H., Conesa, H.M., Robinson, B.H., and Schulin, R. (2012) Biomass production on trace element–contaminated land: a review. *Environmental Engineering Science* 29: 823–839.

FAO (2010) *Climate-Smart Agriculture: Policies, Practices and Financing for Food Security, Adaptation and Mitigation*, FAO, Rome.

Fargione, J., Hill, J., Tilman, D., Polasky, S., and Hawthorne, P. (2008) Land clearing and the biofuel carbon debt. *Science* 319: 1235–1238.

Fernandez-Raga, M., Marques, R., Fraile, R., Varela, M., Castro, A., and Keizer, J. (2009) The importance of rainfall drop size characteristics for splash erosion on terraced eucalyptus plantations in north-central Portugal. 11th Plinius Conference on Mediterranean Storms, 7–10 September 2009, Barcelona, Spain. Available at: http://meetings.copernicus.org/plinius11, id. Plinius11-153 (accessed 13 September 2012).

Fialho, R. and Zinn, Y. (2012) Changes in soil organic carbon under eucalyptus plantations in Brazil: a comparative analysis. *Land Degradation and Development*. doi:10.1002/ldr.2158.

Firdaus, M.S. and Husni, M.H.A. (2012) Planting *Jatropha curcas* on constrained land: emission and effects from land use change. *The Scientific World Journal Article*. doi:10.1100/2012/405084.

Fismes, J., Vong, P., Guckert, A., and Frossard, E. (2000) Influence of sulfur on apparent N-use efficiency, yield and quality of oilseed rape (*Brassica napus* L.) grown on a calcareous soil. *European Journal of Agronomy* 12: 127–141.

Frank, A.B., Berdahl, J.D., Hanson, J.D., Liebig, M.A., and Johnson, H.A. (2004) Biomass and carbon partitioning in switchgrass. *Crop Science* 44: 1391–1396.

Garrigues, E., Corson, M.S., Angers, D.A., Van der Werf, H.M.G., and Walter, C. (2012) Soil quality in life cycle assessment: towards development of an indicator. *Ecological Indicators* 18: 434–442.

Gaskin, S.E. and Bentham, R.H. (2010) Rhizoremediation of hydrocarbon contaminated soil using Australian native grasses. *Science of the Total Environment* 408: 3683–3688.

Gilker, R.E., Weil, R.R., Krizek, D.T., and Momen, B. (2002) Eastern gamagrass root penetration in adverse subsoil conditions. *Soil Science Society of America Journal* 66: 931–938.

Glaser, B., Lehmann, J., and Zech, W. (2002) Ameliorating physical and chemical properties of highly weathered soils in the tropics with charcoal – a review. *Biology and Fertility of Soils* 35: 219–230.

Gnansounou, E. and Panichelli, L. (2008) Biofuels and land use change. Background document of the Roundtable on Sustainable Biofuels (RSB) Expert Workshop "Biofuels and Land-Use Change," 20–21 November 2008, Sao Paulo, Brazil.

Goidts, E. and van Wesemael, B. (2007) Regional assessment of soil organic carbon changes in agriculture in Southern Belgium (1955–2005). *Geoderma* 141: 341–354.

Gopalakrishan, G., Cristina Negri, M., and Snyder, S.W. (2011) A novel framework to classify marginal land for sustainable biomass feedstock production. *Journal of Environmental Quality* 40: 1593–1600.

Govaerts, B., Sayre, K., and Deckers, J. (2006) A minimum data set for soil quality assessment of wheat and maize cropping in the highlands of Mexico. *Soil and Tillage Research* 87: 163–174.

Graham, R.L., Perlack, R., Wright, L., Sheenan, J., and Nelson, R. (2007) Current and potential US corn stover supplies. *Agronomy Journal* 99: 1–11.

Grelle, A., Aronsson, P., Weslien, P., Klemedtsson, L., and Lindroth, A. (2007) Large carbon-sink potential by Kyoto forests in Sweden – a case study on willow plantations. *Tellus Series B-Chemical and Physical Meteorology* 59: 910–918.

Guo, L. and Gifford, R. (2002) Soil carbon stocks and land use change: a meta analysis. *Global Change Biology* 8: 345–360.

Gyssels, G., Poesen, J., Nachtergaele, J., and Govers, G. (2002) The impact of sowing density of small grains on rill and ephemeral gully erosion in concentrated flow zones. *Soil and Tillage Research* 64: 189–201.

Hamelin, L., Jørgensen, U., Petersen, B.M., Olesen, J.E, and Wenzel, H. (2012) Modelling the carbon and nitrogen balances of direct land use changes from energy crops in Denmark: a consequential life cycle inventory. *GCB Bioenergy*. doi:10.1111/j.1757-1707.2012.01174.x.

Hammer, D., Kayser, A, and Keller, C. (2003) Phytoextraction of Cd and Zn with Salix viminalis in field trials. *Soil Use and Management* 19: 187–192.

Hamza, M. and Anderson, W. (2005) Soil compaction in cropping systems: a review of the nature, causes and possible solutions. *Soil and Tillage Research* 82: 121–145.

Hansen, E.M. and Djurhuus, J. (1997) Nitrate leaching as influenced by soil tillage and catch crop. *Soil and Tillage Research* 41: 203–219.

Hartemink, A.E. (2008a) Soils are back on the global agenda. *Soil Use and Management* 24: 327–330.

Hartemink, A.E. (2008b) Sugarcane for bioethanol: soil and environmental issues. *Advances in Agronomy* 99: 125–182.

Heaton, E.A., Long, S.P., Voigt, T.B., Jones, M.B., and Clifton-Brown, J. (2004) *Miscanthus* for renewable energy generation: European Union experience and projections for Illinois. *Mitigation and Adaptation Strategies for Global Change* 9: 433–451.

Heckrath, G., Djuhuusm, J., Quine, T.A., Van Oost, K., Govers, G., and Zhang, Y. (2005) Tillage erosion and its effect on soil properties and crop yield in Denmark. *Journal of Environmental Quality* 34: 312–324.

Hodgson, E.M., Lister, S.J., Bridgwater, A,V., Clifton-Brown, J., and Donnison, I.S. (2010) Genotypic and environmentally derived variation in the cell wall composition of *Miscanthus* in relation to its use as a biomass feedstock. *Biomass and Bioenergy* 34: 652–660.

Howeler, R.H. (2001) Cassava mineral nutrition and fertilization, in *Cassava: Biology, Production and Utilization* (eds R.J. Hillocks, M.J. Thresh, and A. Bellotti), CABI Publishing, Wallingford, UK.

IFA (1992) *World Fertilizer Use Manual*, International Fertilizer Industry Association, Paris.

IPCC (2007) Climate change 2007: The physical science basis. Contribution of Working Group I to the Fourth Assessment Report of the Intergovernmental Panel on Climate Change. Cambridge University Press, Cambridge, UK.

Islami, T., Guritno, B., and Utomo, W.H. (2011) Performance of cassava (*Manihot esculenta* Crantz) based cropping systems and associated soil quality changes in the degraded tropical uplands of East Java, Indonesia. *Journal of Tropical Agriculture* 49: 31–39.

Jensen, J.K., Holm, P.E., Nejrup, J., Larsen, M.B., and Borggaard, O.K. (2009) The potential of willow for remediation of heavy metal polluted calcareous urban soils. *Environmental Pollution* 157: 931–937.

Jobbagy, E.G. and Jachson, R.B. (2000) The vertical distribution of soil organic carbon and its relation to climate and vegetation. *Ecological Applications* 10: 423–436.

Johnson, J.M.F., Franzluebbers, A.J., Weyers, S.L., and Reicosky, D.C. (2007) Agricultural opportunities to mitigate greenhouse gas emissions. *Environmental Pollution* 150: 107–124.

Johnson, J.M.F., Karlen, D.L., and Andrews, S.S. (2010) Conservation considerations for sustainable bioenergy feedstock production: if, what, where, and how much? *Journal of Soil and Water Conservation* 65: 88–91.

Johnson, J.M.F., Reicosky, D., Sharratt, B., Lindstrom, M., Voorhees, W., and Carpenter-Boggs, L. (2004) Characterization of soil amended with the by-product of corn stover fermentation. *Soil Science Society of America Journal* 68: 139–147.

Johnson, K.D. (2008) Modeling spatial and temporal patterns of soil organic carbon in two montane landscapes: the northern hardwoods, Vermont and the tabonuco forest, Puerto Rico. PhD thesis, University of Pennsylvania.

Jørgensen, U. (2011) Benefits versus risks of growing biofuel crops: the case of Miscanthus. *Current Opinion in Environmental Sustainability* 3: 24–30.

Karlen, D.L., Andrews, S.S., and Doran, J.W. (2001) Soil quality: current concepts and applications. *Advances in Agronomy* 74: 1–40.

Karlen, D.L., Doran, J.W., Weinhold, B.J., and Andrews, S.S. (2003) Soil quality: humankind's foundation for survival. *Journal of Soil and Water Conservation* 58: 171–179.

Karlen, D.L., Hurley, E.G., Andrews, S.S., *et al.* (2006) Crop rotation effects on soil quality at three northern corn/soybean belt locations. *Agronomy Journal* 98: 484–495.

Karlen, D.L., Mausbach, M.J., Doran, J.W., Cline, R.G., Harris, R.F., and Schuman, G.E. (1997) Soil quality: a concept, definition, and framework for evaluation. *Soil Science Society of America Journal* 61: 4–10.

Karlen, D.L., Varvel, G.E., Johnson, J.M.F., *et al.* (2011) Monitoring soil quality to assess the sustainability of harvesting corn stover. *Agronomy Journal* 103: 288–295.

Kell, D.B. (2011) Breeding crop plants with deep roots: their role in sustainable carbon, nutrient and water sequestration. *Annals of Botany* 108: 407–418

Kering, M.K., Butler, T.J., Biermacher, J.T., and Guretzky, J.A. (2012) Biomass yield and nutrient removal rates of perennial grasses under nitrogen fertilization. *BioEnergy Research* 5: 61–70.

Kerr, B.J., Dozier, W.A., III, and Bregendahl, K. (2007) Nutritional value of glycerin for nonruminants. Proceedings of Minnesota Nutrition Conference, September 18, 2007, Minneapolis, MN

Khan, M. and Scullion, J. (2002) Effects of metal (Cd, Cu, Ni, Pb or Zn) enrichment of sewage sludge on soil micro-organisms and their activities. *Applied Soil Ecology* 20: 145–155.

Koh, L.P., Levang, P., and Chazoul, J. (2009) Designer landscapes for sustainable biofuels. *Trends in Ecology and Evolution* 24: 431–438.

Krishnan, G., Horst, G.L., and Shea, P.J. (2000) Differential tolerance of cool and warm-season grasses to TNT-contaminated soil. *International Journal of Phytoremediation* 2: 369–382.

Laflen, J.M. and Colvin, T.S. (1981) Effect of crop residue on soil loss from continuous row cropping. *Transactions of the ASAE* 24: 605–609.

Lal, R. (2004) Soil carbon sequestration impacts on global climate change and food security. *Science* 304: 1623–1627.

Lal, R. (2005) World crop residues production and implications of its use as a biofuel. *Environment International* 31: 575–584.

Lal, R. (2008) Crop residues as soil amendments and feedstock for bioethanol production. *Waste Management* 28: 747–758.

Lal, R. (2009) Soil quality impacts of residue removal for bioethanol production. *Soil and Tillage Research* 102: 233–241.

Lal, R., Griffin, M., Apt, J., Lave, L., and Morgan, G.M. (2004) Response to comments on "managing soil carbon". *Science* 305: 1567.

Lal, R., Kimble, J.M., Follett, R.F., and Cole, C.V. (1998) *The Potential of U.S. Cropland to Sequester Carbon and Mitigate the Greenhouse Effect*, Ann Arbor Press, Chelsea, MI.

Lal, R. and Stewart, B.A. (1995) *Soil Management: Experimental Basis for Sustainability and Environmental Quality (Advances in Soil Science)*, CRC Press, Boca Raton, FL.

Lamb, D.T., Heading, S., Bolan, N.S., and Naidu, R. (2012) Use of biosolids for phytocapping of landfill soils. *Water, Air, and Soil Pollution* 223: 2695–2705.

Lambin, E.F., Turner, B.L., Geist, H., *et al.* (2001) The causes of land-use and land-cover change: moving beyond the myths. *Global Environmental Change* 11: 261–269.

Larson, W.E. and Pierce, F.J. (1991) Conservation and enhancement of soil quality, in *Evaluation for Sustainable Land Management in the Developing World*, Vol. 2: technical papers (eds J. Dumanski, E. Pushparajah, M. Latham, and R. Myers), IBSRAM, Bangkok.

Lehmann, J. and Rondon, M. (2006) BioChar soil management on highly weathered soils in the humid tropics, in *Biological Approaches to Sustainable Soil Systems*, (eds N. Uphoff, A.S. Ball, E. Fernandes, H. Herren, O. Husson, M. Laing, C. Palm, J. Pretty, P. Sanchez, N. Sanginga, and J. Thies), CRC Press, Taylor and Francis Group, Florida.

Leme, D.M., Grummt, T., Heinze, R., *et al.* (2012) An overview of biodiesel soil pollution: data based on cytotoxicity and genotoxicity assessments. *Journal of Hazardous Materials* 199–200: 343–349.

Lestan, D., Luo, C.L., and Li, X.D. (2008) The use of chelating agents in the remediation of metal-contaminated soils: a review. *Environmental Pollution* 153: 3–13.

Letey, J., Sojka, R.E., Upchurch, D.R., *et al.* (2003) Deficiencies in the soil quality concept and its application. *Journal of Soil and Water Conservation* 58: 180–187.

Lewandowski, L. and Heinz, A. (2003) Delayed harvest of miscanthus—influences on biomass quantity and quality and environmental impacts of energy production. *European Journal of Agronomy* 19: 45–63.

Liebig, M., Schemer, M., Vogel, K., and Mitchell, R. (2008) Soil carbon storage by switchgrass grown for bioenergy. *BioEnergy Research* 1: 215–222.

Lindstrom, M. (1986) Effects of residue harvesting on water runoff, soil erosion and nutrient loss. *Agriculture, Ecosystems and Environment* 16: 103–112.

Liu, W., Yan, J., Li, J., and Sang, T. (2012) Yield potential of *Miscanthus* energy crops in the Loess Plateau of China. *GCB Bioenergy* 4: 545–554.

Lockwell, J., Guidi, W., and Labreque, M. (2012) Soil carbon sequestration potential of willows in short-rotation coppice established on abandoned farm lands. *Plant and Soil.* doi:10.1007/s11104-012-1251-2.

Ma, Z., Wood, C.W., and Bransby, D.I. (2000) Soil management impacts on soil carbon sequestration by switchgrass. *Biomass and Bioenergy* 18: 469–477.

Madejón, P., Murillo, J.M., Marañón, T., Cabrera, F., and Soriano, M.A. (2003) Trace element and nutrient accumulation in sunflower plants two years after the Aznalcóllar mine spill. *The Science of the Total Environment* 307: 239–257.

Malcolm, S. and Aillery, M. (2009) Growing crops for biofuels has spillover effects. *Amber Waves* 7: 10–15.

Malhi, S., Johnston, A., Schoenau, J., Wang, Z., and Vera, C. (2007) Seasonal biomass accumulation and nutrient uptake of canola, mustard, and flax on a black chernozem soil in Saskatchewan. *Journal of Plant Nutrition* 30: 641–658.

Mao, R., Zeng, D.-H., Li, L.-J., and Hu, Y.-L. (2011) Changes in labile soil organic matter fractions following land use change from monocropping to poplar-based agroforestry systems in a semiarid region of Northeast China. *Environmental Monitoring and Assessment.* doi:10.1007/s10661-011-2462-3.

McGrath, S.P. and Zhao, F.J. (1996) Sulphur uptake, yield responses and the interactions between nitrogen and sulphur in winter oilseed rape (Brassica napus). *The Journal of Agricultural Science* 126: 53–62.

Meers, E., Van Slycken, S., Adriaensen, K., *et al.* (2010) The use of bio-energy crops (*Zea mays*) for 'phytoattenuation' of heavy metals on moderately contaminated soils: a field experiment. *Chemosphere* 78: 35–41.

Meersmans, J., De Ridder, F., Canters, F., De Baets, S., and Van Molle, M. (2008) A multiple regression approach to assess the spatial distribution of soil organic carbon (SOC) at the regional scale (Flanders, Belgium). *Geoderma* 143: 1–13.

Meki, M.N., Snider, J.L., Kiniry, J.R., Raper, R.L., and Rocateli, A.C. (2013) Energy sorghum biomass harvest thresholds and tillage effects on soil organic carbon and bulk density. *Industrial Crops and Products* 43: 172–182.

Meunchang, S., Panichsakpatana, S., and Weaver, R.W. (2005) Co-composting of filter cake and bagasse; by-products from a sugar mill. *Bioresource Technology* 96: 437–442.

Meunchang, S., Panichsakpatana, S., and Weaver, R.W. (2006) Tomato growth in soil amended with sugar mill by-products compost. *Plant and Soil* 280: 171–176.

Milà i Canals, L., Romanyà, J., and Cowell, S.J. (2007) Method for assessing impacts on life support functions (LSF) related to the use of 'fertile land' in life cycle assessment (LCA). *Journal of Clean Production* 15: 1426–1440.

Moebius-Clune, B.N., Van Es, H.M., Idowu, O.J., *et al.* (2008) Long-term effects of harvesting maize stover and tillage on soil quality. *Soil Science Society of America Journal* 72: 960–969.

Mueller, K.E., Diefendorf, A.F., Freeman, K.H., and Eissenstat, D.M. (2010) Appraising the roles of nutrient availability, global change, and functional traits during the angiosperm rise to dominance. *Ecology Letters* 13: 1–6.

Mulumba, L.N. and Lal, R. (2008) Mulching effects on selected soil physical properties. *Soil and Tillage Research* 98: 106–111.

Munkholm, L.J., Heck, R.J., and Deen, B. (2012) Long-term rotation and tillage effects on soil structure and crop yield. *Soil and Tillage Research.* http://dx.doi.org/10.1016/j.still.2012.02.007.

Munkholm, L.J., Schjonning, P., Rasmussen, K.J., and Tanderup, K. (2003) Spatial and temporal effects of direct drilling on soil structure in the seedling environment. *Soil and Tillage Research* 71: 163–173.

Munson, S.M., Lauenroth, W.K., and Burke, I.C. (2012) Soil carbon and nitrogen recovery on semiarid Conservation Reserve Program lands. *Journal of Arid Environments* 79: 25–31.

Myrbeck, Å., Stenberg, M., Arvidsson, J., and Rydberg, T. (2012) Effects of autumn tillage of clay soil on mineral N content, spring cereal yield and soil structure over time. *European Journal of Agronomy* 37: 196–104.

Nair, S.S., Kang, S., Zhang, X., *et al.* (2012) Bioenergy crop models: descriptions, data requirements, and future challenges. *GCB Bioenergy*. doi:10.1111/j.1757-1707.2012.01166.x.

Nassi o Di Nasso, N., Roncucci, N., Triana, F., Tozzini, C., and Bonari, E. (2011) Seasonal nutrient dynamics and biomass quality of giant reed (*Arundo donax* L.) and miscanthus (Miscanthus x giganteus Greef et Deuter) as energy crops. *Italian Journal of Agronomy* 6: e24.

Nepstad, D., Stickler, C.M., Filho, B.S., and Merry, F. (2008) Interactions among Amazon land use, forests and climate: prospects for a near-term forest tipping point. *Philosophical Transactions of the Royal Society of London Series B – Biological Sciences* 363: 1737–1746.

Ng Cheong, L.R. (2007) Effects of sugar cane production practices on soil quality in Mauritius. PhD thesis, University of the Free State, Bloemfontein, South Africa.

Nijsen M., Smeets, E., Stehfest, E., and van Vuuren, D.P. (2012) An evaluation of the global potential of bioenergy production on degraded lands. *GCB Bioenergy* 4: 130–147.

Oberholzer, H.-R., Freiermuth Knuchel, R., Weisskopf, P., and Gaillard, G. (2012) A novel method for soil quality in life cycle assessment using several soil indicators. *Agronomy for Sustainable Development* 32: 639–649.

Oberholzer, H.-R., Weisskopf, P., Gaillard, G., Weiss, F., and Freiermuth, R. (2006) Methode zur Beurteilung der Wirkungen landwirtschaftlicher Bewirtschaftung auf die Bodenqualität in Ökobilanzen - SALCA-SQ. Interner Bericht Agroscope FAL Reckenholz, Eidgenössische Forschungsanstalt für Agrarökologie und Landbau.

Obernberger, I., Biedermann, F., Widmann, W., and Riedl, R. (1997) Concentrations of inorganic elements in biomass fuels and recovery in the different ash fractions. *Biomass and Bioenergy* 12: 211–224.

Odhiambo, O. (2011) Developing improved phytoremediation and phytoprotection systems. PhD thesis submitted to Institute of Technology of Carlow, Carlow, Ireland.

Ogunwole, J., Chaudhary, D., Ghosh, A., Daudu, C., Chikara, J., and Patolia, J. (2008) Contribution of Jatropha curcas to soil quality improvement in a degraded Indian entisol. *Acta Agriculturae Scandinavica Section B – Soil and Plant Science* 58: 245–251.

Ok, S.Y., Usman, A.R.A., Lee, S.S., *et al.* (2011) Effects of rapeseed residue on lead and cadmium availability and uptake by rice plants in heavy metal contaminated paddy soil. *Chemosphere* 85: 677–682.

Papendick, R.I. and Parr, J. (1992) Soil quality – the key to a sustainable agriculture. *American Journal of Alternative Agriculture* 7: 2–3.

Paustian, K., Andrén, O., Janzen, H.H., *et al.* (1997) Agricultural soils as a sink to mitigate CO_2 emissions. *Soil Use and Management* 13: 230–244.

Percival, H.J., Parfitt, R.L., and Scott, N.A. (2000) Factors controlling soil carbon levels in New Zealand grasslands is clay content important? *Soil Science Society of America Journal* 64: 1623–1630.

Pimentel, D., Cerasale, D., Stanley, R.C., *et al.* (2012) Annual vs. perennial grain production. *Agriculture, Ecosystems and Environment* 161: 1–9.

Pimentel, D. and Lal, R. (2007) Biofuels and the environment. *Science* 317: 897.

Podwojewski, P., Orange, D., Jouquet, P., *et al.* (2008) Land-use impacts on surface runoff and soil detachment within agricultural sloping lands in Northern Vietnam. *Catena* 74: 109–118.

Poluektov, R.A., Fintushal, S.M., Oparina, I.V., Shatskikh, D.V., Terleev, V., and Zakharova, E.T. (2002) AGROTOOL-A System for crop simulation. *Archives of Agronomy & Soil Science* 48: 609–635.

Prade, T. (2011) Industrial hemp (*Cannabis sativa* L.) – a high-yielding energy crop. PhD thesis submitted to Swedish University of Agricultural Sciences, Alnarp, Sweden.

Pretty, J., Sutherland, W.J., Ashby, J, *et al.* (2010) The top 100 questions of importance to the future of global agriculture. *International Journal of Agricultural Sustainability* 8: 219–236.

Purakayastha, T.J. and Chhonkar, P.K. (2010) Phytoremediation of heavy metal contaminated soils, in *Soil Heavy Metals. Soil Biology* (eds I. Sherameti and A. Varma), Springer, Berlin.

Qin, Z., Zhuang, Q., and Chen, M. (2012) Impacts of land use change due to biofuel crops on carbon balance, bioenergy production, and agricultural yield, in the conterminous United States. *GCB Bioenergy* 4: 277–288.

Ragauskas, A.J., Williams, C.K., Davison, B.H., *et al.* (2006) The path forward for biofuels and biomaterials. *Science* 311: 484–489.

Raghu, S., Anderson, R.C., Daehler, C.C., *et al.* (2006) Adding biofuels to the invasive species fire? *Science* 313: 1742–1743.

Reeves, D.W. (1997) The role of soil organic matter in maintaining soil quality in continuous cropping systems. *Soil and Tillage Research* 43: 131–167.

Reubens, B., Achten, W.M.J., Maes, W., *et al.* (2011) More than biofuel? *Jatropha curcas* root system symmetry and potential for soil erosion control. *Journal of Arid Environments* 75: 201–205.

Rey, F. (2004) Effectiveness of vegetation barriers for marly sediment trapping. *Earth Surface Processes and Landforms* 29: 1161–1169.

Richards, B.K., Steenhuis, T.S., Peverly, J.H., and Mcbride, M.B. (1998) Metal mobility at an old, heavily loaded sludge application site. *Environmental Pollution* 99: 365–377.

Robinson, B.H., Banuelos, G., Conesa, H.M., Evangelou, M.W.H., and Schulin R. (2009) The phytomanagement of trace elements in soil. *Critical Reviews in Plant Sciences* 28: 240–266.

Rodriguez-Kabana, R. and Walker, R.H. (2008) Treated biodiesel glycerin. United States Patent Application 20080214679. Available at: http://www.freepatentsonline.com/20080214679.pdf (accessed 13 September 2012).

Saffih-Hdadi, K. and Mary, B. (2008) Modeling consequences of straw residues export on soil organic carbon. *Soil Biology Biochemistry* 40: 594–607.

Sainju, U.M. and Singh, B.P. (1997) Winter cover crops for sustainable agricultural systems: influence on soil properties, water quality, and crop yields. *HortScience* 32: 21–28.

Sanchez, P.A., Palm, C.A., and Buol, S.W. (2003) Fertility capability soil classification: a tool to help assess soil quality in the tropics. *Geoderma* 114: 157–185.

Sanderson, K. (2006) US biofuels: a field in ferment. *Nature* 444: 673–676.

Santibáñez, C., Varnero, M.T., and Bustamante, M. (2011) Residual glycerol from biodiesel manufacturing, waste or potential source of bioenergy: a review. *Chilean Journal of Agricultural Research* 71: 469–475.

Schjønning, P., Elmholt, S., and Christensen, B.T. (2004) *Managing Soil Quality: Challenges in Modern Agriculture,* CABI Publishing, Wallingford, UK.

Schoenau, J., Qian, P., Alotaibi, K., and Urton, R. (2009) Potential for using biofuel, crop and animal processing by-products as soil amendments to increase fertility: effect of soil amendment with dehydrated alfalfa, dry and wet distillers grains, thin stillage and glycerol on growth of canola, soil properties, and microbial activity. Final report, Department of Soil Science, College of Agriculture, University of Saskatchewan, Saskatoon, SK, Canada.

Scholz, G., Quinton, J.N., and Strauss, P. (2008) Soil erosion from sugar beet in Central Europe in response to climate change induced seasonal precipitation variations. *Catena* 72: 91–105.

Searchinger, T., Heimlich, R., Houghton, R.A., *et al.* (2008) Use of U.S. croplands for biofuels increases greenhouse gases through emissions from land use change. *Science* 319: 1238–1240.

Shepherd, K.D. and Walsh, M.G. (2002) Development of reflectance spectral libraries for characterization of soil properties. *Soil Science Society of America Journal* 66: 988–998.

Siddiqui, I.A. and Shaukat, S.S. (2002) Zinc and glycerol enhance the production of nematicidal compounds in vitro and improve the biocontrol of Meloidogyne javanica in tomato by fluorescent pseudomonads. *Letters in Applied Microbiology* 35: 212–217.

Singh, B. (2008) Nutrient uptake by hybrid poplar in competition with weed species under growth chamber and field conditions using the soil supply and nutrient demand (SSAND) model. PhD thesis submitted to University of Saskatchewan, Saskatoon, SK, Canada.

Singh, B.P. (2010) *Industrial Crops and Uses*, CABI Publishing, Wallingford, UK.

Six, J., Conant, R., Paul, E., and Paustian, K. (2002) Stabilization mechanisms of soil organic matter: implications for C-saturation of soils. *Plant and Soil* 241: 155–176.

Smith, K.A. and Searchinger, T.D. (2012) Crop-based biofuels and associated environmental concerns. *GCB Bioenergy* 4: 479–484.

Smith, P. (2004) How long before a change in soil organic carbon can be detected? *Global Change Biology* 10: 1878–1883.

Smith, P., Powlson, D.S., Smith, J.U., and Elliott, E.T. (1997) Evaluation and comparison of soil organic matter models using datasets from seven long-term experiments. *Geoderma* 81: 1–225.

Soerens, T.S. (2011) Biodiesel Waste Products as Soil Amendments – Evaluation of Microbial, Biological, and Plant Toxicity. MBTC Project 3025 final report. National Technical Information Service, Springfield, Virginia, USA.

Sojka, R.E. and Upchurch, D.R. (1999) Reservations regarding the soil quality concept. *Soil Science Society of America Journal* 63: 1039–1054.

Sojka, R.E., Upchurch, D.R., and Borlaug, N.E. (2003) Quality soil management or soil quality management: performance versus semantics. *Advances in Agronomy* 79: 1–68.

Thomas, A.R.C., Bond, A.J., and Hiscock, K.M. (2012) A multi-criteria based review of models that predict environmental impacts of land use-change for perennial energy crops on water, carbon and nitrogen cycling. *GCB Bioenergy*. doi:10.1111/j.1757-1707.2012.01198.x.

Thomas, G.W., Haszler, G.R., and Blevins, R.L. (1996) The effects of organic matter and tillage on maximum compactability of soils using the proctor test1. *Soil Science* 161: 502–508.

Thomsen, I.K. and Christensen, B.T. (2004) Yields of wheat and soil carbon and nitrogen contents following long-term incorporation of barley straw and ryegrass catch crops. *Soil Use and Management* 20: 432–438.

Thomsen, I.K., Olesen, J.E., Møller, H.B., Sørensen, P., and Christensen, B.T. (2013) Carbon dynamics and retention in soil after anaerobic digestion of dairy cattle feed and faeces. *Soil Biology and Biochemistry* 58: 82–87.

Tilman, D., Hill, J., and Lehman, C. (2006) Carbon-negative biofuels from low-input high-diversity grassland biomass. *Science* 314: 1598–1600.

Tolbert, V.R., Todd, D.E., Jr., Mann, L.K., *et al.* (2002) Changes in soil quality and below-ground carbon storage with conversion of traditional agricultural crop lands to bioenergy crop production. *Environmental Pollution* 116, 97–106.

Tugrul, K.M., Buzluk, S., and Boyacioglu, A. (2012) Seed drilling distance applications in sugar beet cultivation. *African Journal of Agricultural Research* 7: 4110–4117.

Ulén, B., Aronsson, H., Bechmann, M., Krogstad, T., Øygarden, L., and Stenberg, M. (2010) Soil tillage methods to control phosphorus loss and potential side-effects: a Scandinavian review. *Soil Use and Management* 26: 94–107.

Van Oost, K., Govers, G., Heckrath, G., Quine, T.A., Olesen, J.E., and Merckx, R. (2005) Landscape-scale modeling of carbon cycling under the impact of soil redistribution. *Global Biogeochemical Cycles* 19: GP4014.

Vuichard, N., Ciais, P., Belelli, L., Smith, P., and Valentini, R. (2008) Carbon sequestration due to the abandonment of agriculture in the former USSR since 1990. *Global Biogeochemical Cycles* 22: GB4018.

Vuichard, N., Ciais, P., and Wolf, A. (2009) Soil carbon sequestration or biofuel production: new land-use opportunities for mitigating climate over abandoned Soviet farmlands. *Environmental Science and Technology* 43: 8678–8683.

Warkentin, B.P. and Fletcher, H.F. (1977) Soil quality for intensive agriculture. Proceedings of International Seminar on Soil Environment and Fertility Management in Intensive Agriculture, Society, Science, Soil and Manure, National Institute of Agricultural Science, Tokyo.

Wells, T., Hancock, G.R., Dever, C., and Murphy, D. (2012) Prediction of vertical soil organic carbon profiles using soil properties and environmental tracer data at an untilled site. *Geoderma* 170: 337–346.

West, T.O. and Post, W.M. (2002) Soil organic carbon sequestration rates by tillage and crop rotation: a global data analysis. *Soil Science Society of America Journal* 66: 1930–1946.

Wienhold, B.J., Andrews, S.S., and Karlen, D.L. (2004) Soil quality: a review of the science and experiences in the USA. *Environmental Geochemistry and Health* 26: 89–95.

Wilhelm, W.W., Johnson, J.M.F., Hatfield, J.L., Voorhees, W.B., and Linden, D.R. (2004) Crop and soil productivity response to corn residue removal: a literature review. *Agronomy Journal* 96: 1–17.

Wilhelm, W.W., Richard Hess, J., Douglas, L., *et al.* (2012) Balancing limiting factors and economic drivers for sustainable midwestern US agricultural residue feedstock supplies. *Industrial Biotechnology* 6: 271–287.

Williams, J.R., Llewelyn, R.V., Pendell, D.L., Schlegel, A., and Dumler, T. (2010) A risk analysis of converting Conservation Reserve Program acres to a wheat–sorghum–fallow rotation. *Agronomy Journal* 102: 612–622.

Wilson, G.W.T., Rice, C.W., Rillig, M.C., Springer, A., and Hartnett, D.C. (2009) Soil aggregation and carbon sequestration are tightly correlated with the abundance of arbuscular mycorrhizal fungi: results from long-term field experiments. *Ecology Letters* 12: 452–461.

Witters, N., Mendelsohn, R.O., Van Slycken, S., *et al.* (2012) Phytoremediation, a sustainable remediation technology? Conclusions from a case study. I: energy production and carbon dioxide abatement. *Biomass and Bioenergy* 39: 454–469.

Zhao, S., Arthur, E.-L., and Coats, J.-R. (2003) Influence of microbial inoculation (*Pseudomonas* sp. strain ADP), the enzyme atrazine chlorohydrolase, and vegetation on the degradation of atrazine and metolachlor in soil. *Journal of Agricultural and Food Chemistry* 51: 3043–3048.

Chapter 9

Nutrient Management in Biofuel Crop Production

R. Lemus

Mississippi State University, MS, USA

Introduction

Current energy crops, such as maize, require high fertilization to produce the necessary grains for ethanol production. High oil prices have contributed to price increases for most agricultural crops by raising input costs on the one hand and by boosting demand for agricultural crops used as feedstock in the production of alternative energy sources (biofuels) on the other (FAO, 2011). During 2007 and 2008, fertilizer prices in the United States became extremely volatile. Low fertilizer inventories combined with increased domestic and global demand led to an unprecedented spike in fertilizer prices (Figure 9.1). Costs of some fertilizer products more than doubled. Prices began to moderate by the end of 2008 as the fertilizer supply chain responded and demand in both the United States and global markets declined in response to the high prices. However, fertilizer supply–demand uncertainties continue and form part of the reasons behind the uncommon volatility in grain markets in recent years.

World fertilizer supply is expected to increase by some 27 million Mg representing an annual growth rate of 3% (FAO, 2011). The global fertilizer demand has seen an increase in the last 5 years (Table 9.1). Nitrogen (N) demand is seen rising at a more moderate pace than phosphorus (P) and potassium (K) demand, and there is a need to progressively rethink fertilization and nutrient-use efficiency. It has been estimated that global N supply would grow at an annual rate of 3.7% between 2010 and 2015, compared to growth in demand of 2.3% per year (Heffer and Prud'homme, 2011).

In the last decade, the development of transportation biofuel has been focused primarily on ethanol production from sugar- and starch-based products, but food versus fuel debate and questionable energy and environmental returns of starch biofuels have of late shifted the emphasis toward generating a large and sustainable supply of biomass from herbaceous lignocellulosic crops grown specifically for bioenergy production. Comparative energy budgets (Figure 9.2) for producing ethanol by the current maize-based conversion system that supplies 96% of US ethanol provide lower energy conversion per biomass energy, energy ration, and

Biofuel Crop Sustainability, First Edition. Edited by Bharat P. Singh.

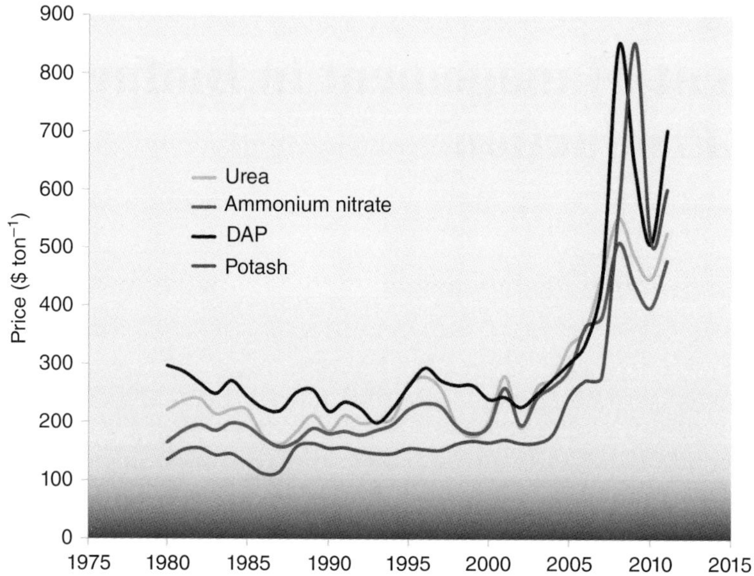

Figure 9.1. Average U S farm prices for selected fertilizers (adapted from USDA-NASS, 2011).

energy in ethanol when compared to switchgrass due to its annual establishment and fertilizer costs (McLaughlin and Walsh, 1998).

Nitrogen fertilization is the largest carbon use in crop production. Innovative bioconversion processes configured to recover key plant nutrients from biomass will allow the recycling of nutrients to soils, thereby closing the nutrient cycles and reducing the energetic and economic costs of fertilization. A system designed to produce ethanol and synthetic fuels from switchgrass indicates that approximately 111 kg ha^{-1} yr^{-1} of N can be recovered (Anex *et al.*, 2007). This is equivalent to 78% of the N fertilizer input required. The proposed system incorporates NH_3 recovery and provides a means of evaluating nutrient cycling potential. The application of this type of technology will allow nutrients that enter the conversion process as part of the lignocellulosic biomass to be recovered and returned to the crop fields instead of being concentrated in the ash (e.g., P and K) or be released as NH_3 during gasification. Cycling captured NH_3 back to production fields has the potential to close an important nutrient cycle by reducing the need for synthetic N fertilizer inputs (NH_3 recycling). This process could be achieved by using common industrial methods for capturing NH_3 gas such as absorption

Table 9.1. Estimated global fertilizer demand (adapted from Heffer and Prud'homme, 2011).

Period	Nitrogen (N)	Phosphate (P_2O_5)	Potash (K_2O)	Total
	Mg nutrients			
2007/2008	100.7	38.5	29	138.2
2008/2009	98.4	33.6	23.4	155.4
2009/2010	101.6	38.3	24	163.9
2010/2011	103.7	40.9	27.5	172.1
2011/2012	105.6	41.7	29	176.4
2015/2016	113	45.2	32.9	191.1

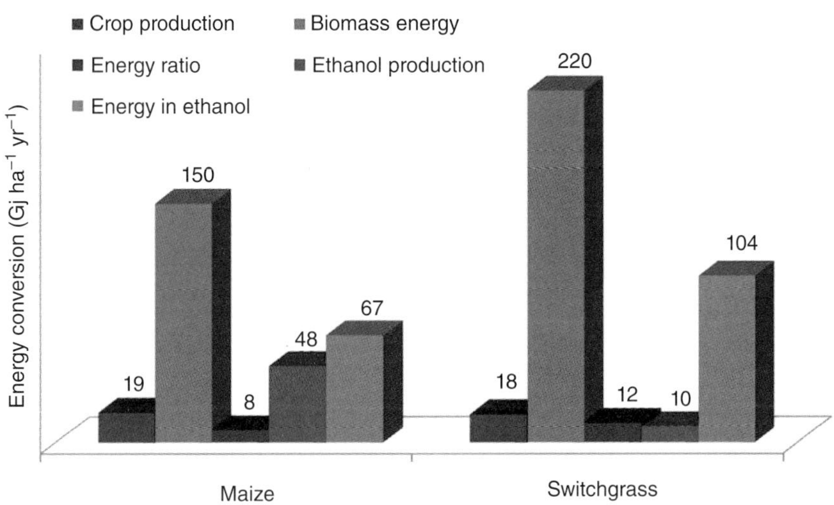

Figure 9.2. Comparative energy flow for ethanol production from maize and switchgrass (adapted from McLaughlin and Walsh, 1998).

and mechanical refrigeration. This type of N recovery and cycling could improve the overall energy balance of ethanol production from lignocellulosic biomass and make biomass production more sustainable. The demand for lignocellulosic biomass from herbaceous crops will also allow the introduction of new crops and cropping systems. The reduction of fertilizer inputs could be enhanced by incorporating legumes in rotations to further reduce fertilizer inputs while improving resource use efficiency and reducing environmental impacts.

In both United States and Europe, there are several candidate perennial herbaceous grasses available, which differ considerably in their productivity potential, chemical and physical properties, environmental demands, and crop management requirements (Lewandowski *et al.*, 2003). These candidates include switchgrass, giant miscanthus, sweet sorghum, energy cane, giant reed (*Arundo* spp.), and reed canarygrass (*Phalaris* spp.).

Fertility Requirement of Bioenergy Crops

The combination of high oil prices and the desire to mitigate environmental issues such as eutrophication, greenhouse gas (GHG) emission, and water pollution is driving the rapid expansion of the biofuel sector. Perennial biomass crops are not fertilized repeatedly like agricultural crops, although specific fertilizer requirements will vary with location and production system. Soil cover provided by biomass crops can increase soil stability and reduce runoff; this in turn can reduce nutrient transport and loss. Studies in switchgrass show that the extensive rooting systems of biomass crops reduce leaching transport of nutrients through the soil column and, thus, can contribute to reducing groundwater contamination from agricultural practices (Ma *et al.*, 2000). Furthermore, due to the recycling of nutrients by their rhizome systems, deciduous perennial grasses have a low demand for nutrient inputs (Christian *et al.*, 1997).

Research suggests that some bioenergy crops can be grown on soils with low fertility and still maintain productivity with low fertilizer inputs (McLaughlin and Walsh, 1998; Lemus, 2004; Lemus *et al.*, 2009; Garten *et al.*, 2011). This is due to the carbon and nitrogen translocation from shoots to belowground parts (roots and crown) at the end of the growing season as a natural nutrient cycling strategy (Lemus *et al.*, 2008). There is evidence that fertilization might be necessary during the first 5 years of establishment (Lemus *et al.*, 2009). By then, switchgrass and giant miscanthus have developed a deep root system (1–3 m) that could provide an adequate nutrient source for sustained growth (Lemus *et al.*, 2008). Other systems that need further exploration to reduce dependence on commercial fertilizer include maintaining a legume component and incorporating manure. However, tropical grasses such as sorghums and energy cane are susceptible to killing frost, thus preventing nutrient translocation to roots and removal of nutrients in harvested biomass.

Perennial grasses have been looked upon as potential energy crops due to many beneficial attributes that include high biomass production and reduced fertilization practices. Some rhizomatous crops include two grasses with a C_3 photosynthetic pathway (reed canarygrass and giant reed) and four greases with a C_4 photosynthetic pathway (miscanthus, switchgrass, energy cane, and sweet sorghum). Nitrogen fertilization is a major issue in the sustainability of energy crops because it is the most energy demanding agronomic practice that accounts for 30–50% of the overall fossil energy used for crop growth (Henlea *et al.*, 2008). Another major factor related to nitrogen fertilization concerns it being a potent GHG responsible for the emission of nitrous oxides (NO_X) to the atmosphere. However, the use of organic fertilizers such as manure applications could help displace the use of commercial nitrogen fertilizer sources, but there is little information on the effects of manure application to lignocellulosic crops and the availability of the material. The rapid expansion in demand for ethanol, the realization that lignocellulosic energy crops must play a major role in attaining mandated goals and the novelty of these energy crops have prompted a growing demand to address establishment, fertilization, energetic efficiency, environmental and economic benefits and trade-offs (Henlea *et al.*, 2008). A well-planned nutrient budget will maintain needed nutrients through recycling and from commercial fertilizers, manure, or other organic sources. Fertilization beyond crop N requirements could be reduced substantially by a shift from yield- to soil-based N management, ideally implemented on a site-specific basis.

Switchgrass (Panicum Virgatum)

Fertility recommendations for switchgrass production have been adapted from grazed forage or hay requirements. Under this management system, nitrogen recommendations are relatively low, ranging from 50 to 100 kg N ha^{-1} (Bredja, 2000). Phosphorus recommendations depend on soil pH, innate P supplying power of the soil, and soil test P (Bredja, 2000). Despite these recommendations, it has been reported that P application is not crucial for biomass production (Muir *et al.*, 2001). Most biomass production has been linked to nitrogen affecting tiller mass, but tiller density showed a very small response (Muir *et al.*, 2001).

Although switchgrass tolerates low fertility soils, optimizing biomass and maintaining quality stands require N fertilizer inputs and proper harvest management (Mitchell *et al.*, 2010). According to the literature, biomass productivity of switchgrass ranges from 9 to 21 Mg ha^{-1} in research trials with an average of 12 Mg ha^{-1} (Mitchell *et al.*, 2010). The fertilizer N requirement is a function of the yield potential of the site, productivity of the cultivar, and time of harvest (Vogel *et al.*, 2002). The optimum N rate generally increases from North to South United States and ranges from 56 kg N ha^{-1} in South Dakota (Mulkey

Table 9.2. A comparison of yield and nutrient removal with three switchgrass cultivars (adapted from McLaughlin *et al.*, 1999).

	Yield		N removal	
	1-cut	2-cut	1-cut	2-cut
Switchgrass	Mg ha^{-1}		kg ha^{-1}	
Alamo	20.2	24.2	52	160
Cave-in-Rock	16.3	22.9	28	157
Kanlow	20.1	22.4	51	152

et al., 2006), 120 kg N ha^{-1} in Nebraska (Vogel *et al.*, 2002), 56 kg N ha^{-1} in Virginia (Lemus *et al.*, 2008) to 224 kg N ha^{-1} in Alabama (Ma *et al.*, 2001). Studies have shown that N removal is significantly reduced by using a one-harvest system at the end of the season instead of a two-harvest system that includes a mid-season harvest (McLaughlin *et al.*, 1999) (Table 9.2). Parrish *et al.* (2003) reported that P, K, and calcium (Ca) concentrations and removal were generally twice as great when biomass was harvested twice versus once per year (Figure 9.3). Similarly, Guretzky *et al.* (2011) indicated that harvesting switchgrass twice per year essentially doubled the amount of N, P, and K removed compared to harvesting once per year. Biomass contained 6.5 and 9.5 g N kg^{-1}, 1 and 1.3 g P kg^{-1}, and 5.4 and 12.2 g K kg^{-1} within the one and two harvests per year systems, respectively. Nitrogen fertilizer application strongly increased concentration of N, but had limited effects on concentrations of P and K. They also observed that P and K removal rates increased by 67–69% on average across harvest systems as N rate increased from 0 to 225 kg ha^{-1}. There is a trade-off to consider when harvesting perennial biomass crops in which a late harvest can affect yields due to leaf loss and an early harvest will present a risk for higher tissue mineral contents, particularly N and K. Allowing the crop to completely senesce and recycle nutrients has several advantages, including improved feedstock quality and reduced fertilizer requirements (Heaton *et al.*, 2009). Although delaying harvest reduces mineral content in biomass feedstock, the crop should be harvested before risk of weather-related losses.

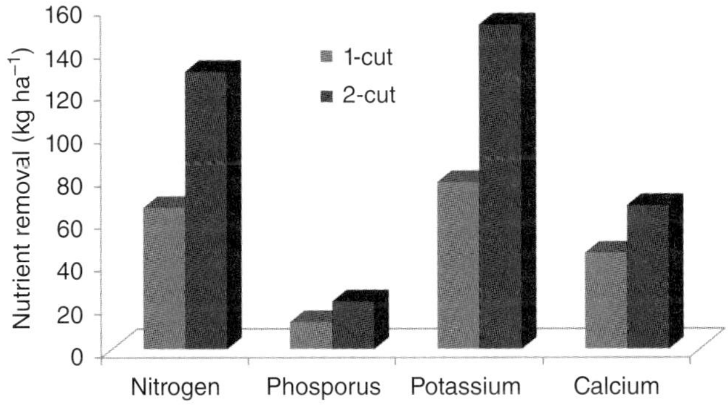

Figure 9.3. Nutrients in Alamo switchgrass sampled in June and November under 1- and 2-cut management. Averages for eight sites from 1997 to 1999 (adapted from Parish *et al.*, 2003).

Switchgrass has relatively low inputs of energy, water, and agrochemicals required per unit of energy produced. There have been several reports that show no N response in switchgrass or only responses to less than 50 kg ha^{-1} yr^{-1} (Turhollow *et al.*, 1988). On the other hand, some locations in the Southern United States can respond to N applications of up to 200 kg ha^{-1} yr^{-1} (Sanderson *et al.*, 1996). This indicates that the effect of N fertilization can be site specific. Despite these findings, Wright (1994) suggested that economically viable yields may require N fertilization rates from 50 to 100 kg N ha^{-1} yr^{-1} but N fertilization was effective only on poorer quality sites with typically little or no effects above 70 kg N ha^{-1} yr^{-1}. On more fertile sites, effects have typically been negative or neutral. Nitrogen fertilizer should be applied in late spring. If the soil tests indicate medium or higher P and K levels, no P and K fertilizer is needed at the time of planting and N should not be applied at planting (Wolf and Fiske, 1995).

One of the advantages of switchgrass as a bioenergy crop is its deciduous nature that allows for the potential for nutrient cycling. Nutrient translocation has been observed in the above- and belowground biomass of different studies. Garten *et al.* (2011) indicated that the application of ammonium nitrate to a 5-year-old stand of Alamo switchgrass at rates of 0, 67 and 202 kg N ha^{-1} changed biomass allocation based on end-of-season root:shoot ratios. They observed that at the end of the season there was more carbon and nitrogen stored in the belowground biomass than the aboveground biomass. Similar trends were documented by Lemus *et al.* (2008). They observed that the application of 0, 90, 180, and 270 kg N ha^{-1} to a well-established Cave-in-Rock switchgrass did not increase yields, but provided residual benefits over a 3-year period. Root N levels at 15 cm depth increased with fertilization, fluctuated seasonally between roots and shoots, and reduced over the course of the study. With two harvests per year, about 100 kg N ha^{-1} yr^{-1} was removed in biomass, even in plots with no N added—suggesting N already present in the soils (at 15 cm depth) contributed to yields; but the soil mineral–N pools were reduced by the end of third year.

Increased root biomass in response to N fertilization would be especially important to maintain soil organic matter (SOM) and sustainable plant yield. Changing root C:N ratios that come about as a result of N fertilization have the potential to alter root decomposition and thereby affect carbon transfer to pools of labile SOM. To accomplish soil carbon sequestration, carbon inputs must exceed carbon losses via decomposition (Garten *et al.*, 2011). Despite these findings, the effect of N fertilization on switchgrass root chemistry and belowground biomass is not well understood. Belowground plant responses to nitrogen fertilization are important to soil carbon cycle processes under bioenergy crops, including soil carbon storage.

Giant Miscanthus (Miscanthus x Giganteus)

Giant miscanthus usually produces higher biomass than switchgrass (Figure 9.4) (Baldwin, 2008). The highest yields are produced on soils with a good water-holding capacity (Eppel-Hotz *et al.*, 1998). Most yield reports for *miscanthus* have been assessed using the species *Miscanthus* × *giganteus* (*Mxg*) after 3–5 years when it becomes fully established and reaches the maximum yield potential (Lewandowski *et al.*, 2003). Yields of 9–23 Mg DM ha^{-1} can be expected with fall harvest (Lewandowski *et al.*, 2000). Late producing and late senescing genotypes have an extended growing season and therefore can form higher yields. On the other hand, their biomass contains higher concentrations of minerals at harvest, especially N, because of the delayed relocation of nutrients from the aboveground shoots to the roots (Lewandowski *et al.*, 2003). Saijonkari-Pahkala (2001) showed an increase in yields with increased N fertilization. The biggest differences occurred between 0 and 50 kg N and

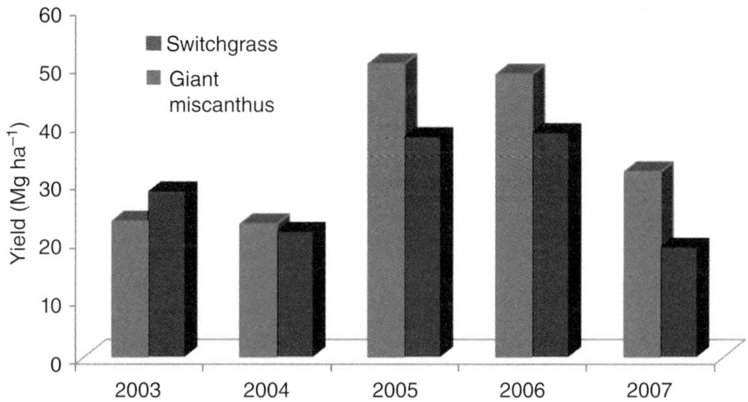

Figure 9.4. Comparison of giant miscanthus and Alamo switchgrass yields from a single harvest in the Southern United States for plots established in 2002 (adapted from Baldwin, 2008).

the lowest between 100 and 150 kg N ha^{-1} yr^{-1}. In the third year of the stand, a rate of 150 kg N ha^{-1} turned out to be unnecessarily high since no yield effects were obtained. The response of *miscanthus* to N fertilization, however, is site specific (Figure 9.5).

Giant miscanthus yield response to N application has been reported with N inputs ranging from 0 to 240 kg N ha^{-1} (Lewandowski *et al.*, 2000). There are reports of N fertilizer increasing biomass yield (Ercoli *et al.*, 1999), but nearly 100 European trials report no relationship between yield and N application (Heaton *et al.*, 2004). At low biomass N concentration (0.4% DM), a 20 Mg ha^{-1} biomass harvest removed 80 kg N ha^{-1} (Christian and Haase, 2001). A 14-year study using three nitrogen fertilizer rates (0, 60, and 120 kg N ha^{-1} yr^{-1}) indicated that N did not influence yield production (average 16 Mg ha^{-1}) (Christian *et al.*, 2008). There was only a 5% difference in total production between the 0 and 120 kg N ha^{-1} application. This supports the hypothesis proposed by several studies (Miyamoto *et al.*, 2004; Davis *et al.*, 2010) that associative N fixation may provide some N for miscanthus. It further suggests that on soils with potential N mineralization from organic matter, N fertilization should be avoided

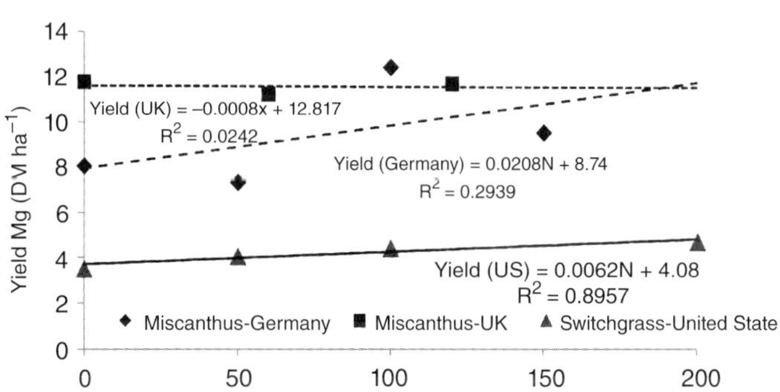

Figure 9.5. Effect on N fertilization on biomass production of different herbaceous energy crops (adapted from Lewandowski and Kicherer, 1997; Lemus *et al.*, 2002; Christian *et al.*, 2008).

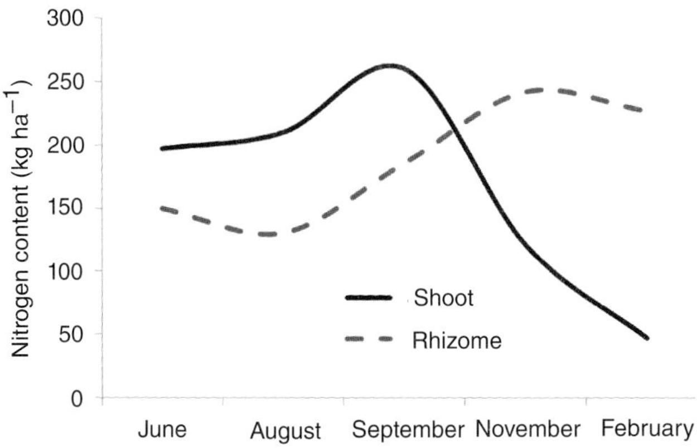

Figure 9.6. Seasonal N content in shoot and rhizome of miscanthus 4 years after planting (adapted from Himken *et al.*, 1997).

or limited to only 50–70 kg N ha^{-1} yr^{-1} (Lewandowski *et al.*, 2003). Over a 10-year period, the amount of P and K removed at harvest in the biomass was 63 kg ha^{-1} and 886 kg ha^{-1}, respectively. Christian *et al.* (2008) recommended an application of 7 kg P ha^{-1} yr^{-1} and 100 kg K ha^{-1} yr^{-1}as replacement rate for minerals removed by harvest to avoid a severe soil nutrient depletion.

Similar to switchgrass, miscanthus has a low requirement for N fertilizer because it efficiently translocates N between aboveground biomass and belowground structures (Himken *et al.*, 1997) (Figure 9.6). The overall nutrient requirements of miscanthus for N, P, and calcium (Ca) are about 2–5, 0.3–1.1, and 0.8–1 mg kg^{-1} of dry matter, respectively (Clifton-Brown and Lewandowski, 2001). Results from these studies indicated that miscanthus N fertilization might be necessary on soils with low N contents and low organic matter content. Applying N at the start of growth may not be appropriate and could result in N losses (leaching and runoff). Occasional applications of organic manure such as poultry litter might be a possible means of ensuring a continuing supply of N and other nutrients if applied following soil test recommendations.

Himken *et al.* (1997) reported that nutrient mobilization from rhizomes to shoots during the growing season was 55 kg N ha^{-1}, 8 kg P ha^{-1}, 39 kg K ha^{-1}, and 11 kg Mg ha^{-1}. These numbers represent approximately 21% N, 36% P, 14% K, and 27% magnesium (Mg) of the maximum nutrient content in the shoots. Total N, P, and Mg pools in the plants reached a maximum in September corresponding to the time of maximum shoot dry weight. Nutrient remobilization from shoots to rhizomes during September through March of the same growing season was 101 kg N, 9 kg P, 81 kg K, and 8 kg Mg ha^{-1} (Figure 9.7). This represented approximately 46% N, 50% P, 30% K, and 27% Mg of the shoot nutrient content observed in September and amounted to net changes of 20, 2, −71, and 4 kg ha^{-1} in N, P, K, and Mg levels, respectively, between September and March. The 35–51% decrease in N, P, and Mg and 14% in K in the shoot can be attributed to the senescent leaves. Reductions in the stems in N, P, K, and Mg were much smaller, equaling 23, −1, −46, and −1 kg ha^{-1}, respectively. The total N, P, and K content in miscanthus root dry matter at different times of the year as estimated by Neukirchen *et al.* (1999) is presented in Figure 9.8.

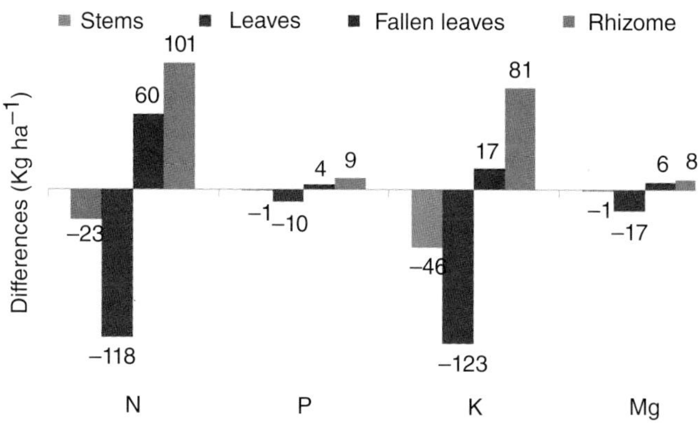

Figure 9.7. Changes in nutrient content (N, P, K, and Mg) in various parts of the plant from September to March (adapted from Himken *et al.*, 1997).

Nutrient requirements of miscanthus are low compared to other crops due to several reasons: (1) high nutrient absorption efficiency through the extensive root system once the crop is established; (2) significant nutrient translocation between rhizomes and aboveground biomass at the end of the season; (3) nutrient cycling through leaf loss in the fall; and (4) possible N contribution from an associative relationship with N-fixing bacteria (Cadoux *et al.*, 2012). Due to the low yield in the establishment phase of the crop, fertilizer application is not recommended during the first 2 years after planting, unless planted on poor soils. From the third year, typically 4.9, 0.45, and 7 g of N, P, and K, respectively, are removed by per kg of dry matter at harvest, and this should be a maximum to be replaced by fertilizers (Cadoux *et al.*, 2012). When comparing the aboveground biomass for N, P, and K concentrations during the growing season, it has been observed that there was 9 times more N than P, 15 times more K than P, and 1.7 times more K than N (Beale and Long, 1997; Himken *et al.*, 1997). This suggests that in situations where the level of K in the soil is high, there could be luxury uptake by giant miscanthus. The P and K content increased during the peak biomass production period

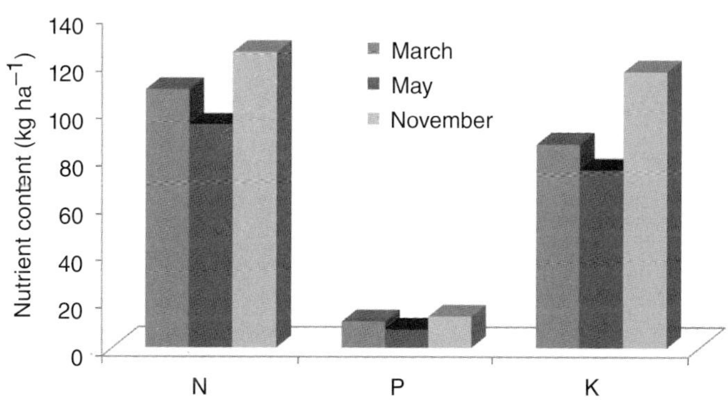

Figure 9.8. Nutrient content of N, P, and K in total root dry matter of miscanthus (adapted from Neukirchen *et al.*, 1999).

and then decreased until the winter harvest. The nutrients removed at harvest are thus low compared to the maximum uptake.

Nutrients may also be translocated within the plant. Using ^{15}N-labeled fertilizer, Christian *et al.* (2006) demonstrated the phenomenon of N translocation from shoots to rhizomes in autumn and from rhizomes to shoots in spring. A proportion of these recycled nutrients should be available for subsequent year's growth, through degradation of the leaf litter and mineralization, and nutrient remobilization from the rhizome to aerial biomass. Nutrient remobilization appears to be greatest for P (18–36% from rhizome), followed by N (9–21%), and then potassium (7–14%). Nutrients removed at harvest are greatest, by proportion, for K (38–57%) and approximately equal for N and P (36–40%). Both studies indicate that about 60% of the N and the P content at the peak of the nutrient uptake are recycled either by translocation or leaf fall. The phenomenon of nutrient cycling from the rhizome to the shoots during growth, then back to the rhizomes during the autumn and winter before harvest is important and must be accounted for in nutrient budget calculations and in order to determine fertilizer requirements (Lemus *et al.*, 2008). Uncertainties in the optimum requirements exist due to the lack of data on N provided by the soil, the actual contribution from sink sources to plant nutrition (rhizomes) and nutrient distribution/partition among aboveground constituents (leaves and shoots).

One of the environmental benefits of giant miscanthus once established is improved groundwater quality compared to row crop production. Christian and Riche (1998) reported that nitrate leaching under newly planted miscanthus with N applications of 0, 60, and 120 kg N ha^{-1} was high in all treatments with values of 154, 187, and 228 kg ha^{-1}, respectively. The large losses in the first year were attributed to previous agricultural management at the site, excessive inputs of N on the fertilized plots, and the spacing of the small plants. During the second year, nitrate losses reduced to 8, 24, and 87 kg ha^{-1} and further declined to 3, 11, and 30 kg ha^{-1}, respectively, under the three N treatments. Third-year leaching losses were closer to those recorded under extensively managed grassland than arable land. These drastic reductions with longevity were attributed to the development of extensive root system that reduced drainage volume and allowed nutrient uptake to be stored as sink in the rhizomes.

Sorghums (Sorghum Vulgare or Sorghum Bicolor)

There have been few studies dedicated to evaluating the nutrient management and nutrient cycling of sweet sorghum as a bioenergy feedstock. Little is known about the N application timing and use efficiency. A study conducted at University of Missouri on sweet sorghum with N application rates of 0, 28, 56, 84, 112, 140, and 168 kg N ha^{-1} indicated a yield response in two soil types (Figure 9.9) (Stevens *et al.*, 2009). Although an exponential yield response, the study indicated that only 67 kg N ha^{-1} was needed to produce optimum sugar yields. Studies in the United States have indicated a response of sorghum types to N fertilization (Table 9.3) (Rooney *et al.*, 2007). Almodares *et al.* (2008) found that a combination of 180 kg urea ha^{-1} and 50 kg potassium sulfate ha^{-1} doubled fresh biomass production. Wiedenfeld (1983) indicated that application of 224 kg N ha^{-1} increase N uptake, but did not further increase yields compared to a 112 kg N ha^{-1} application. Nitrogen uptake efficiency was 33% and 17% for the 112 and 224 kg N ha^{-1} application, respectively. Although sweet sorghum can produce high biomass, the yield is highly affected by harvest management (Figure 9.10) (Ayub *et al.*, 2002). Maximum biomass production and flexibility in harvest are important to extend the harvest season since ratoon capability is genotype dependent (Rooney *et al.*, 2007).

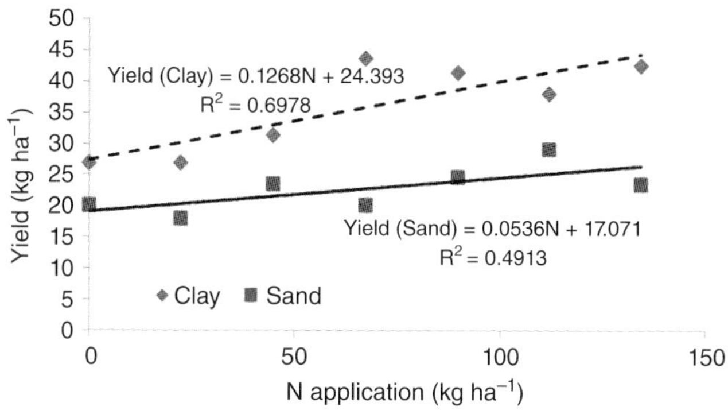

Figure 9.9. Sweet sorghum yields on Sharkey clay and Malden sand in 2008 at the University of Missouri Lee Farm at Portageville and University of Missouri Rhodes Farm at Clarkton, Missouri (adapted from Stevens *et al.*, 2009).

Table 9.3. Biomass yield response of sweet sorghum and forage sorghum to applied nitrogen at the three locations in the United States (adapted from Rooney *et al.*, 2007).

	N rate (kg ha^{-1})	Sweet sorghum	Forage sorghum
Location		Mg ha^{-1}	
Ames, IA	0	11.3	11.3
	140	15.9	14.2
	280	15.2	14.5
Chariton, IA	0	11.6	11.7
	140	16.3	14.1
	280	16.1	16.9
College Station, TX	0	8.6	4.6
	112	17.2	5.9
	224	18.1	7.7

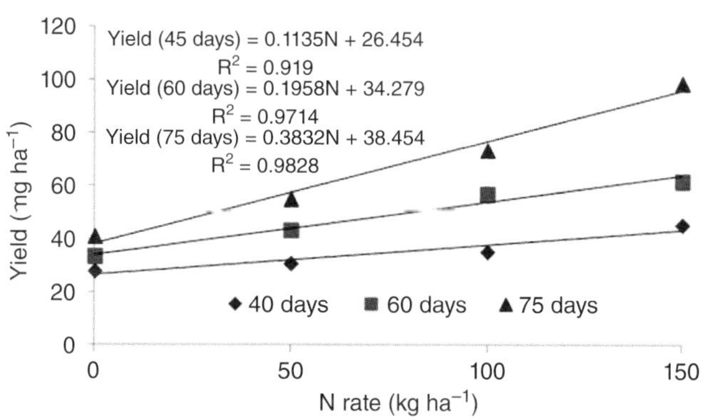

Figure 9.10. Effect of nitrogen levels and harvesting times on green biomass of sorghum (adapted from Ayub *et al.*, 2002).

Han *et al.* (2011) focused on changes in accumulation and partition of N, P, and K in different sweet sorghum varieties (early, mid, and late maturing). They observed that all varieties had a decrease in N, P, and K concentrations in aboveground biomass from elongation to 60 days after anthesis. The decrease was in the order of K > N > P. Conversely, N, P, and K accumulations increased from elongation to anthesis. Comparison of varieties indicated that mid and late maturing varieties had a higher nutrient accumulation ratio (50–82% N, 55–83% P, and 62–88% K) than the early maturing varieties (51–64% N, 40–62% P, and 55–75% K). Early maturing varieties exhibited significant increase in N accumulation compared to late maturing varieties when delaying the harvest period from 40 to 60 days after anthesis while K decreased for all maturing groups (Han *et al.*, 2011). Phosphorus, on the other hand, did not decrease significantly. Nutrient partitioning between straw (stover) and grain indicated that N and P concentrations were significantly lower in the straw, but N and P accumulations were much higher than in the grains due to the amount of biomass. The same principle applied to higher N, P, and K concentrations in the leaves but lower accumulation than in the stems. The increase in nutrient accumulation was found to be due to biomass accumulation with advancement of the crop growth. Although fertilizer requirement of sweet sorghum is much lower than sugarcane, these variations in nutrient efficiency indicate that further genetic improvement might be required to optimize fertilization. This is relevant since the entire aboveground biomass of sweet sorghum could be harvested for ethanol production resulting in high nutrient removal from the soil. There is a need to maintain a long-term soil nutrient balance to reduce nutrient inputs that can affect ethanol production cost.

Giant Reed (Arundo Donax)

Nitrogen fertilization recommendations range from 39 to 56 kg N ha^{-1} yr^{-1} (Bransby *et al.*, 2004). Yields typically increase from the first through the fourth year after establishment (Angelini *et al.*, 2005). In a 15-year field trial, Angelini *et al.* (2009) showed that giant reed could produce an average yield of 35 Mg DM ha^{-1} year^{-1}.

Similar to other perennial herbaceous crops, there is very little response to N application by giant reed; however, before establishing the field, sufficient amount of K and P should be applied if the nutrient status of the soil is poor. Nitrogen applications of 40 and 120 kg N ha^{-1} indicated that highly fertilized plots only yielded 907 kg DM ha^{-1} over that of the lower fertilization rate, an increase that was not significant (Christou *et al.*, 2001). The low fertility requirements are due to the compact masses of rhizomes that develop a tough fibrous root system penetrating deep into the soil. While most of the massive rhizome network is in the upper 5–30 cm, the roots can grow more than 1 m long (Sharma *et al.*, 1998). This system holds the plants in place while allowing more efficient absorption of water and nutrients from the soil. Sharma *et al.* (1998) observed in a fertilization study in which 60 and 120 kg N ha^{-1} were applied, plants in the higher fertilization rate had significant higher N content in the stems 94 days after fertilization and in leaves 30 and 60 days after fertilization. Although not statistically different, the higher N content of the stems and leaves in the 120 K N ha^{-1} remained until the end of the growing season. This indicates that moderate fertilization rates as proposed by Bransby *et al.* (2004) will be favorable for both economic and biofuel quality reasons. Studies conducted by Nassi o Di Nassoa *et al.* (2010) showed that an increase in fertilization rates can reduce ash content. The higher nutrient availability in fertilized crops may lead to a higher translocation rate of nutrients from the aboveground crop to the rhizomes, thus decreasing the ash content of the biomass. Other authors explain the ash content decrease from autumn to winter as a consequence of the nutrient translocation from the aboveground biomass to the rhizomes (Beale and Long, 1997; Himken *et al.*, 1997). The optimum time for fertilization follows initial spring growth, which is driven more from stored energy in the roots than from photosynthesis and external plant nutrition.

Sugarcane/Energy Cane (Saccharum spp.)

Baldwin (2010) reported average energy cane yields of 7.1 tonnes DM ac^{-1}, but that varies by latitude (farther north, lower yields are expected) and the variety, as some are more cold hardy than others. Energy cane produces a harvestable yield in the first full year of growth. Nitrogen recommendations range from 67 to 90 kg N ha^{-1}. Commonly applied nutrients are N, P, and K with N rates of 78 to 100 kg ha^{-1} being applied to the plant cane crop (first growing season) and 100–134 kg ha^{-1} being applied to the subsequent ratoon crops to ensure sustained yields (Mitchell *et al.*, 2011). Heavy, infrequent additions of lime create a cycle of high and low pH that keeps the soil system in a state of flux and has the potential to reduce the best possible crop performance. A lime application of 90 kg ha^{-1} yr^{-1} is recommended to maintain yield potentials, but variable rates might have a greater impact on yields and sugar quality and quantity (Gravois *et al.*, 2011).

Nitrogen is generally the determining factor for sugarcane stalk yield and sugar content (Chapman, 1982). A study conducted in Brazil (Franco *et al.*, 2011) using four N rates (0, 40, 80, and 120 kg ha^{-1} of N as urea) applied at the bottom of the planting furrow indicated that the N in the plant derived from fertilizer (NDDF) contributed up to 40% of the total N in the plant cane at initial stages of development (Figure 9.11). The magnitude of this contribution decreased during stages of maturity to approximately 10% of total N at harvest. In the first ratoon, application of N fertilizer was more effective for crop nutrition, constituting up to 70% of total N in initial stages of development and decreasing through the cycle, reaching approximately 30% at harvest. This contribution is approximately 20% of total plant biomass N, which raises questions regarding the efficiency of N fertilization. Several speculations indicated this low response might be related to biological nitrogen fixation. In a ^{15}N dilution/N balance study, Boddey *et al.* (1991) indicated that certain sugarcane varieties are capable of obtaining large contributions of nitrogen from plant-associated N_2-fixing bacteria. The N_2-fixing, acid-tolerant bacterium *Acetobacter diazotrophicus* was first found to occur within roots, stems, and leaves of sugarcane (Boddey, 1995). It was estimated that up to 60–80% of plant N could be derived from this source, and under good conditions of water

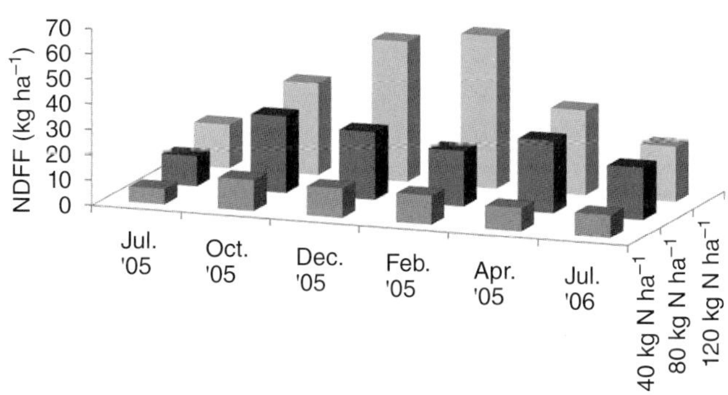

Figure 9.11. Average N plant derived from fertilizer (NDFF) during the plant cane at two ethanol plants as related to N fertilization at planting (adapted from Franco *et al.*, 2011).

and nutrient supply, it may be responsible for N fertilization of these varieties altogether. A study conducted by Muthukumarasamy *et al.* (1999) in India plus mycorrhizal fungi produced a response equivalent to that following a half rate of the recommended nitrogen fertilizer. Subsequently, two species of *Herbaspirillum* also have been found to occur within the interior of all sugarcane tissues (Boddey, 1995). It is not clear how these different microorganisms interact and what synergistic effect they have at different sugarcane crop stages.

Nitrogen fertilization of sugarcane crops is a common practice used to reach sustainable level of productivity, both for plant cane and especially for the ratoon. In Brazil, the amount of N recommended for nonirrigated modern varieties is 50 kg N ha^{-1} yr^{-1} to the plant cane crop, and 100 kg N ha^{-1} yr^{-1} to the first ratoon (Baldani *et al.*, 2002). The response of the planted cane crop to N fertilizer is generally very low or null, although the same does not apply to the first, second, and third ratoons (Baldani *et al.*, 2002). The response varies from country to country and depends on the soil type, region and management, and whether the crop is irrigated or rain-fed. Much higher amounts of nitrogen fertilizer are used in other countries such as Hawaii, United States, India, Mexico, Philippines, and South Africa (Baldani *et al.*, 2002). Courtaillac *et al.* (1998) indicated that a decrease in yield and N absorption by the cane was related to ratoon number. The real utilization coefficient for the fertilizer (RUC%) ranged from 6% to 34%, and a high proportion (30–40%) of fertilizer N was immobilized in the soil after the annual crop cycle. The N absorbed by the cane was essentially derived from the soil but the availability was determined by rainfall at the beginning of (re)growth. To improve fertilizer N efficiency and limit losses, it will be necessary to adapt the amount of N fertilizer applied based on the stage of growth, ratoon age, and crop development.

Reed Canarygrass (Phalaris Arundinacea)

Reed canarygrass responds to N fertilization and to a lesser extent to K and P fertilization (Cherney *et al.*, 2003). For economically optimum forage yield and to enhance crude protein concentration on nonorganic soils, a total of 100–168 kg N ha^{-1} should be applied each year. For best utilization and uniform growth, total N should be divided between at least two applications per year. Reed canarygrass has a superior capacity to persist and remove N when irrigated with municipal and industrial waste effluents (Marten *et al.*, 1979) (Figure 9.12). When grown as a

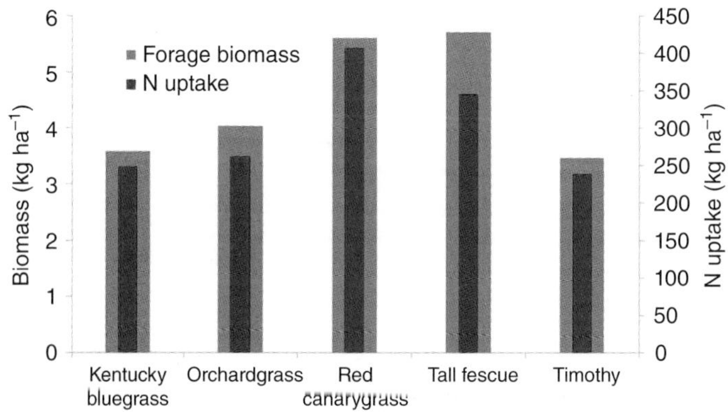

Figure 9.12. Yield comparison and nitrogen uptake of perennial grasses cut three times and treated with municipal wastewater (adapted from Marten *et al.*, 1979).

bioenergy crop, the recommended levels of fertilizer for reed canarygrass differ from those for fodder production. Andersson (2000) recommended a fertilizer application of 40 kg N, 15 kg P, and 50 kg K ha^{-1} during the planting year for crop establishment in Sweden. During the second year, the most economic yield was reached when applying 100, 15, and 80 kg ha^{-1} of N, P, and K, respectively. For the following years, 50, 5, and 20 kg ha^{-1} of N, P, and K, respectively, was found optimal. The decrease in fertilization rates are due to a delayed harvest system for cellulose which allows for translocation of nutrients during the winter (Lewandowski *et al.*, 2003).

The concentration of N in the harvested crop decreased from August to October, but not further to early spring (Katterer *et al.*, 1998). Landstrom *et al.* (1996) indicated that the N concentration was markedly higher in the crop fertilized with 200 kg N ha^{-1} than in the crop fertilized with 100 kg N ha^{-1}. The concentration of K decreased markedly from summer to the following spring and particularly during the winter. Higher nitrogen fertilization, but the same level of K (100 kg ha^{-1}), gave slightly higher concentration of K in the crop harvested in August. The spring harvest resulted in about half of the supplied N and P and one-third of the supplied K being removed with the crop.

Carbon Sequestration Potential of Bioenergy Crops

Carbon sequestration is the uptake and storage of atmospheric carbon in soils and vegetation. Loss of SOM can alter many important aspects of soil quality. Soil moisture holding capacity, soil density and aeration, and soil nutrient availability and conservation are among the essential properties controlled in part by SOM (Buckman and Brady, 1960). Ecosystem productivity as well as organic matter decomposition is strongly determined by environmental conditions such as weather, soil type, and past land management directly affect the carbon sequestering potential (Verardo and Dokken, 2000). This variability is reflected in the decomposition rate as a function of soil temperature and moisture (Vleeshouwers and Verhagen, 2002). High decomposition rates occur in regions where high temperatures in summer coincide with moist conditions whereas low decomposition rates occur in areas with low temperatures and wet conditions.

Globally, about 68 billion Mg of soil per year is removed by wind and water erosion (Wachs and Thibault, 2009). The rate of loss of SOM through annual row cropping in the United States has been estimated at 2.7 million Mg yr^{-1} (CAST, 1992). Herbaceous bioenergy crops may have greater soil carbon storage potentials than conventional row crops such as maize due to their perennial nature and greater root biomass (Lee *et al.*, 2007; Ma *et al.*, 2001). The large active pools of roots are a major source of this carbon due to rhizosphere deposition and root turnover along with an active population of soil microorganisms and invertebrates (Barnes and Taylor, 1985). It has been estimated that the root system and the microbial population may add up to 7 Mg C ha^{-1} yr^{-1} to the soil carbon pool and to the function of soil as a retentive nutrient cycling reservoir (Barnes and Taylor, 1985). Studies by the U.S. National Resources Conservation Service (NRCS) to examine changes in SOM during 5 years of perennial grass production on CRP lands indicate that perennial grasses added 1.1 Mg ha^{-1} yr^{-1} of carbon to the upper 300 cm of midwestern soils (Gebhart *et al.*, 1994). Calculations of net carbon gain for switchgrass versus maize shows large differences between the two crops in the capacity of the ethanol produced to offset the CO_2 emissions of the gasoline (Figure 9.13) (McLaughlin and Walsh, 1998).

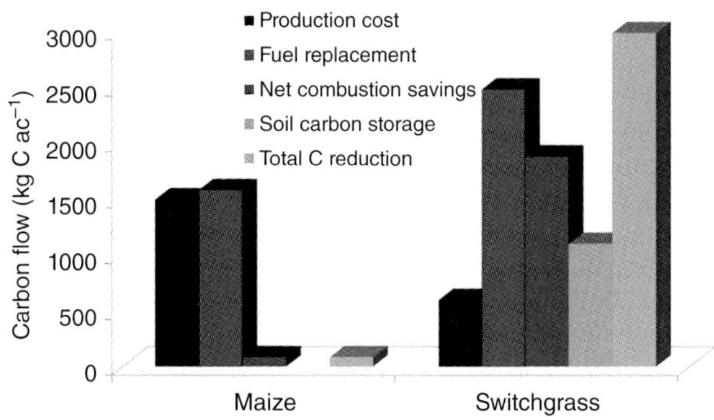

Figure 9.13. Comparative carbon flow in producing ethanol from switchgrass and maize (adapted from McLaughlin and Walsh, 1998).

In general, low crop yields, high soil carbon contents, and high SOM decomposition rates enhance the loss of carbon from agricultural soils. Agriculture is attributed with approximately 7% of total US GHG emissions (USEPA, 2010). Proper management of agricultural systems can reduce direct emissions and offset emissions from other entities by sequestering C in the soil (Follet, 2001; West and Marland, 2002; Lal *et al.*, 2003). Perennial species may contribute to agronomic and environmental sustainability, since they limit soil tillage and reduce soil erosion, enhance soil fertility, and reduce CO_2 emission. In addition, they have low nutritional requirements, due to nutrient "recycling" in species with a rhizomatous root (Consentino *et al.*, 2008). However, the understanding of the short- and long-term effects of removing both grain and biomass on soil nutrient cycling, physical properties, and biological activity of bioenergy cropping systems is limited. In the near term, carbon can be sequestered in the soil during plant growth, thereby reducing GHG concentrations; in the long term, however, the soil's capacity to store carbon is limited (Sanderson and Alder, 2008).

The ecological advantages of the long periods without tilling are reduced risk of soil erosion and a likely increase in soil carbon content (McLaughlin, 1997). Unlike annual crops, the need for soil tillage in perennial grasses is limited to the year in which the crops are established. Biofuels have a large potential to reduce GHG emissions associated with energy use (Adler *et al.*, 2007). Perennial energy crops also lead to less GHG emissions. Compared to annual crops such as cereals, there is a drop in N_2O_x emissions in perennial crops equal to 245–662 kg CO_2 equivalent ha^{-1} (Freibauer *et al.*, 2004). Values for grassland fluxes of carbon range from -1.4 to 1.9 Mg C ha^{-1} yr^{-1} (Freibauer *et al.*, 2004). More importantly, the perennial energy crops can secure carbon storage (sinks) of 1420 Mg CO_2 equivalent ha^{-1} (Anonymous, 2010). Furthermore, due to the recycling of nutrients by their rhizome systems, perennial grasses have a low demand for nutrient inputs (Lemus and Lal, 2005). Whether agricultural soils are a sink or source of carbon depends on the actual organic matter content in the soil (Vleeshouwers and Verhagen, 2002). Establishing a baseline of carbon contents in the topsoil is crucial when establishing sink or source status of agricultural lands. To establish a baseline, there is a need to look at changes in management practices over time; this means establishing studies with a minimum of 10–15 years of continuous practice (Lal *et al.*, 2003). Evaluation of current and new management practices for carbon sequestration should focus on the input and the output of soil organic carbon (SOC) and the complex fluxes that affect the overall system. Possible

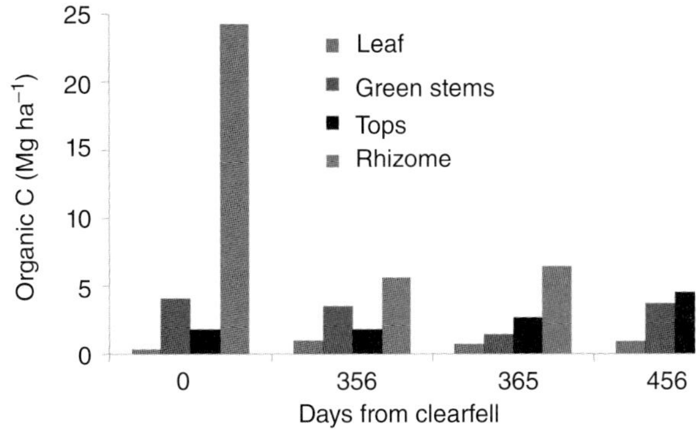

Figure 9.14. Organic carbon sequestered on a 30-year-old stand of *A. donax* under dryland conditions (adapted from Williams *et al.*, 2010).

changes in emissions of N_2O and CH_4 are important when determining the GHG mitigation effect of a given activity, especially when organic amendments and no-till options are involved (Smith et al., 2001a,b). More research is needed to better evaluate the overall GHG effect of C sequestration measures.

Fertilization can impact belowground carbon cycle processes through at least two mechanisms: (1) changes in tissue chemistry that alter root decomposition or otherwise affect the decomposition of SOM produced through root mortality; and (2) changes in plant biomass or carbon allocation that result in increased soil carbon inputs belowground (Garten *et al.*, 2011). Production and turnover of belowground storage organs will add organic matter and carbon to the soil (Zan *et al.*, 2001). Perennials have more extensive root systems present throughout the year, thus providing increased resistance to soil erosion and a more effective means of trapping nutrients and preventing nitrogen loss to drainage water. Accumulation of carbon in giant miscanthus rhizomes and roots was 6.8 to 9.1 Mg C ha^{-1} and represented about 50% of total plant biomass carbon (Amougou *et al.*, 2011). Senescent leaves represented about 1.4 Mg C ha^{-1} yr^{-1}. All residues, especially the roots, had high lignin contents, while the rhizomes also had a high soluble content due to their nutrient storage function. Similar results were reported by Williams *et al.* (2010) in *Arundo donax* (Figure 9.14). Experiments conducted by Ma *et al.* (2000) to determine the effects of row spacing, N rate, switchgrass cultivar, and harvest frequency on carbon sequestration indicated that none of these parameters changed soil organic C in the short-term (2–3 years) after switchgrass establishment. However, after 10 years in switchgrass production, soil organic C storage was 45% and 28% higher at depths of 0–15 cm and 15–30 cm, respectively, compared with fallowed soil in an adjacent area. Interestingly, the C:N ratio did not change for the same layers. This suggests that soil C accumulation, with not only energy crops such as switchgrass but other perennial warm-season crops, might take several years before an increase is quantifiable.

Carbon sequestration is usually measured in terms of the total carbon stored in the soil. The capacity to store carbon depends on several components that include available pools, their recycling potential, the soil type, and the physical location of the carbon in the soil (Freibauer *et al.*, 2004). Soil carbon sequestration is nonlinear due to the conditions described above. Smith *et al.* (1997) indicated with long-term experimentation that the greatest soil carbon increase occurs soon after changes in land use or managements are implemented and the rates of changes decrease as the soil reaches a new equilibrium. This phenomenon is sometimes referred to as sink saturation (Watson *et al.*, 2004). Due to this saturation, soil carbon sequestration does not have unlimited potential to offset CO_2 emissions indicating the yearly benefits will continue for about 20–50 years (Freibauer *et al.*, 2004). The final level at which the soil carbon stabilizes depends on its cation exchange capacity, soil moisture, temperature, the quality of the carbon added to the soil, and the balance between the carbon input to the soil and the carbon lost through land management practices. To support the carbon mitigation policies, regional and local land management strategies data on soil, climate, land cover, land management, and ecosystem productivity will be necessary for accurate assessments. These data are, however, not readily available and quality of the available data varies strongly. There is uncertainty in the estimates of carbon sequestration in soils because the total increase or decrease in SOC is determined by the amount of crop residues left in the field and their turnover time.

Conclusions

Bioenergy crops offer new opportunities to develop markets for co-firing, biofuels, and biomaterials. An advantage of redirecting forage crops as bioenergy crops is that producers are familiar with their management and already have the capacity to grow, harvest, store, and transport them. Perennial crops dedicated to biomass feedstock production seem to be an ideal goal for feedstock because there would be no annual re-establishment costs, annual tillage would be eliminated (reducing inputs, costs, and soil erosion), and a permanent vegetative cover would achieve sustainability of soil conservation and water quality protection.

Ethanol has been one of the prime biofuel options; now the commercial sector is able to convert biomass to petroleum-like products. However, the demand on biofuel from the industry is still placed on biomass. It is expected that dedicated energy crops will be needed to produce the amount of biomass required to meet the projected high demand for ethanol fuel. A suitable energy crop must meet requirements of high yield, resistance to biotic and abiotic stresses, and low-input efficiency (especially fertilizers). Perennial energy crops do not have the annual establishment requirements and associated economic and net energy inputs. Controlling how much, when, and how the bioenergy crop is harvested can help maximize yields, while better managing crop quality, pest control, yields, carbon sequestration, and nutrient loss. The literature indicates that the N application requirement on well-established energy crops is less than needed for growth through the growing season. This is apparently associated with re-translocation of mobile nutrients, such as K, P, and N, and carbohydrates and storage in crowns and root systems later in the growing season as plants approach senescence. Although most of the bioenergy crops tolerate low fertility soils, optimizing biomass and maintaining quality stands requires N fertilizer inputs and proper harvest management to balance nutrient cycling and soil quality (Mitchell *et al.*, 2010).

In addition to water and light, the ideal biomass crop must also efficiently utilize nutrients, especially nitrogen. Nitrogen-use efficiency can be attained at three levels: (1) by maximizing the efficiency of energy transduction into biomass in photosynthesis per unit of nitrogen invested in the photosynthetic process; (2) by maximizing the amount of N and other nutrients translocated to the root system during senescence; and (3) by maximizing capture of nutrients from the soil by establishing a deep and fibrous root system. These properties help to minimize both the quantities of N that need to be applied as fertilizer and the amount lost through leaching.

References

Adler, P.R., Del Grosso, S.J., and Parton, W.J. (2007) Life cycle assessment of net greenhouse gas flux for bioenergy cropping systems. *Ecological Applications* 17: 675–691.

Almodares, A., Taheri, R., Chung, I.M., and Fathi, M. (2008) The effect of nitrogen and potassium fertilizers on growth parameters and carbohydrate contents of sweet sorghum varieties. *Journal of Environmental Biology* 29: 849–852.

Amougou, N., Bertrand, I., Machet, J.M., and Recous, S. (2011) Quality and decomposition in soil of rhizome root and senescent leaf from *Miscanthus x giganteus*, as affected by harvest date and N fertilization. *Plant and Soil* 338: 83–97. doi:10.1007/s11104-010-0443-x.

Andersson, B. (2000) Prospects for breeding and supply of plant material, in *Bioenergy Workshop Report: Renewable Energy and Sustainable Agriculture. The Impact of Perennial Grass Research,* European Commission, Brussels, pp. 50–52.

Anex, P., Lynd, L.R., Laser, M.S., Heggenstaller, A.H., and Liebman, M. (2007) Potential for enhanced nutrient cycling through coupling of agricultural and bioenergy systems. *Crop Science* 47: 1327–1335.

Angelini, L.G., Ceccarini, L., and Bonari, E. (2005) Biomass yield and energy balance of giant reed (*Arundo donax* L.) cropped in central Italy as related to different management practices. *European Journal of Agronomy* 22: 375–389.

Angelini, L.G., Ceccarini, L., Nassi o Di Nasso, N., and Bonari, E. (2009) Comparison of *Arundo donax* L. and *Miscanthus x giganteus* in a long-term field experiment in Central Italy: analysis of productive characteristics and energy balance. *Biomass and Bioenergy* 33: 635–643.

Anonymous. (2010) Perennial energy crops. Green energy: The agricultural contribution. The Danish Ministry of Food, Agriculture and Fisheries.

Ayub, M., Nadeem, M.A., Tanveer, A., and Husnain, A. (2002) Effect of different levels of nitrogen and harvesting times on the growth, yield and quality of sorghum fodder. *Asian Journal of Plant Science* 1: 304–307.

Baldani, J.I., Reis, V.M., Baldani, V.L.D., and Döbereiner, J. (2002) A brief story of nitrogen fixation in sugarcane—reasons for success in Brazil. *Functional Plant Biology* 29: 417–423.

Baldwin, B. (2008) Cultured feedstocks. Annual report. Sustainable Energy Research Center (SERC), Mississippi State University, Mississippi, USA.

Baldwin, B. (2010) Giant miscanthus variety trials. Mississippi State University, Mississippi, USA.

Barnes, R.F. and Taylor, T.H. (1985) Grassland agriculture and ecosystem concepts, in *Forages: The Science of Grassland Agriculture* (eds M.E. Heath, R.F. Barnes, and D.S. Metcalfe), Iowa State University Press, Ames, Iowa, pp. 12–20.

Beale, C.V. and Long, S.P. (1997) Seasonal dynamics of nutrient accumulation and partitioning in the perennial C_4-grasses *Miscanthus x giganteus* and *Spartina cynosuroides*. *Biomass and Bioenergy* 12: 419–428.

Boddey, R.M. (1995) Biological nitrogen fixation in sugarcane: a key to energetically viable biofuel production. *Critical Reviews in Plant Sciences* 14: 263–279. doi:10.1080/07352689509701929.

Boddey, R.M., Urquiaga, S., Reis, V., and Döbereiner, J. (1991) Biological nitrogen fixation associated with sugar cane. *Plant and Soil* 137: 111–117. doi:10.1007/BF02187441.

Bransby, D.I., Gu, H., Duke, J.R., Krishnagopalan, G.A., and Cullinan, H.T. (2004) An update on giant reed and mimosa for energy, fiber, and other uses, in *Annual Meeting of the Association for the Advancement of Industrial Crops/New Uses Council Joint Annual Meeting: Industrial Crops and Uses to Diversify Agriculture*, 19–22 September 2004, Minneapolis, MN. Available at: http://www.aaic.org/04progrm.htm (accessed 21 August 2012).

Bredja, J.J. (2000) Fertilization of native warm-season grasses, in *Native Warm-Season Grasses: Research Trends and Issues*, (eds K.J. Moore and B.E. Anderson), CSSA and ASA, Madison, WI, pp. 177–210. CSSA Special Publication 30.

Buckman, H.O. and Brady, N.C. (1960) *The Nature and Properties of Soils*, Macmillan, New York, 567p.

CAST. (1992) Preparing U.S. agriculture for global climate change. Council for Agricultural Science and Technology, Report No. 119, Ames, Iowa, 93p.

Chapman, L.S. (1982) Estimating sugar yield responses from N, P and K fertilizers in Queensland. Proceedings of Australia Society of Sugar Cane Technologist Conference, pp. 147–153.

Cherney, J.H., Cherney, D.J.R., and Casler, M.D. (2003). Low intensity management of reed canarygrass. *Agronomy Journal* 95: 627–634.

Christian, D.G. and Haase, E. (2001) Agronomy of *Miscanthus*, in *Miscanthus for Energy and Fibre* (eds M. Jones and M. Walsh), Earthscan, London, pp. 21–45.

Christian, D.G., Poulton, P.R., Riche, A.B., Yates, N.E., and Todd, A.D. (2006) The recovery over several seasons of N^{15}-labelled fertilizer applied to *Miscanthus x giganteus* ranging from 1 to 3 years old. *Biomass and Bioenergy* 30: 125–133.

Christian, D.G. and Riche, A.B. (1998) Nitrate leaching losses under *Miscanthus* grass planted on a silty clay loam soil. *Soil Use Management* 14: 131–135.

Christian, D.G., Riche, A.B., and Yates, N.E. (1997) Nutrient requirement and cycling in energy crops, in *Sustainable Agriculture for Food, Energy and Industry* (eds N. El Bassam, R.K. Behl, and B. Prochnow), James and James Ltd, London, pp. 799–804.

Christian, D.G., Riche, A.B., and Yates, N.E. (2008) Growth, yield and mineral content of *Miscanthus x giganteus* grown as a biofuel for 14 successive harvests. *Industrial Crops and Products* 28: 320–327.

Christou, M., Mardikis, M., and Alexopoulou, E. (2001) Research on the effect of irrigation and nitrogen upon growth and yields of Arundo donax L in Greece. *Aspects of Applied Biology 65, Biomass and Energy Crops II*, AAB, Warwick, pp. 47–55.

Clifton-Brown, J.C. and Lewandowski, I. (2001) Screening Miscanthus genotypes in field trials to optimize biomass yield and quality in Southern Germany. *European Journal of Agronomy* 6: 97–110.

Cadoux, S., Riche, A.B., Yates, N.E., and Machet, J.M. (2012) Nutrient requirements of *Miscanthus x giganteus*: conclusions from a review of published studies. *Biomass and Bioenergy* 38: 14–22.

Consentino, S.L., Copani, V., Patanè, C., Mantineo, M. and Agosta, G.M.D. (2008) Agronomic, energetic and environmental aspects of biomass energy crops suitable for Italian environments. *Italian Journal of Agronomy/Rivista di Agronomia* 2: 81–95.

Courtaillac, N., Baran, R., Oliver, R., Casabianca, H., and Ganry, F. (1998) Efficiency of nitrogen fertilizer in the sugarcane-vertical system in Guadeloupe according to growth and ratoon age of the cane. *Nutrient Cycling in Agroecosystems* 52: 9–17.

Davis, S.C., Parton, W.J., Dohleman, F.G., *et al.* (2010) Comparative biogeochemical cycles of bioenergy crops reveal nitrogen fixation and low greenhouse gas emissions in a *Miscanthus x giganteus* agro-ecosystem. *Ecosystems* 13: 144–156.

Eppel-Hotz, A., Jodl, S., Kuhn, W., Marzini, K., and Münzer, W. (1998) Miscanthus: New cultivations and results of research experiments for improving the establishment rate, in

Biomass for Energy and Industry: Proceedings of the 10th European Conference (eds H. Kopetz, T. Weber, T. Palz, P. Chartier, and G.L. Ferrero), Würzburg, Germany, 8–11 June 1998, pp. 780–783.

Ercoli, L., Mariotti, M., Mason, A., and Bonari, E. (1999). Effect of irrigation and nitrogen fertilization on biomass yield and efficiency of energy use in crop production of Miscanthus. *Field Crop Research* 63: 3–11.

Follet, R.F. (2001) Soil management concepts and carbon sequestration in cropland soils. *Soil and Tillage Research* 61: 77–92.

Food and Agriculture Organization (FAO). (2011) Current world fertilizer trends and outlook to 2011/12. FAO.

Franco, H.C.J., Otto, R., Faroni, C.E., Vitti, A.C., de Oliveira, E.C.A., and Trivelin, P.C.O. (2011) Nitrogen in sugarcane derived from fertilizer under Brazilian field conditions. *Field Crop Research* 121: 29–41.

Freibauer, A., Rounsevell, M.D.A., Smith, P., and Verhagen, J. (2004) Carbon sequestration in the agricultural soils of Europe. *Geoderma* 122: 1–23.

Garten, Jr., C.T., Brice, D.J., Castro, H.F., *et al.* (2011) Response of "Alamo" switchgrass tissue chemistry and biomass to nitrogen fertilization in West Tennessee, USA. *Agriculture, Ecosystems and Environment* 140: 289–297.

Gebhart, D.L., Johnson, H.B., Mayeux, H.S., and Polley, H.W. (1994) The CRP increases soil organic carbon. *Journal of Soil and Water Conservation* 49: 48–492.

Gravois, K., LeBlanc, B.D., Sheffield, R., and Nix, K.E. (2011) Sugarcane environmental best management practices. Louisiana State University Publication 2833, Baton Rouge, Louisiana, USA.

Guretzky, J.A., Biermacher J.T., Cook, B.J., Kering, M.K. and Mosali, J. (2011) Switchgrass for forage and bioenergy: harvest and nitrogen rate effects of biomass yields and nutrient composition. *Plant and Soil* 339: 69–81. doi:10.1007/s11104-010-0376-4.

Han, L.P., Steinberger, Y., Zhao, Y.L., and Xie, G.H. (2011). Accumulation and partitioning of nitrogen, phosphorous and potassium in different varieties of sweet sorghum. *Field Crop Research* 120: 230–240.

Heaton, E.A., Dohleman, F.G., and Long, S.P. (2009). Seasonal nitrogen dynamics of *Miscanthus x giganteus* and *Panicum virgatum*. *GCB Bioenergy* 1: 297–307. doi:10.1111/j.1757-1707.2009.01022.x.

Heaton, E., Voigt, T., and Long, S.P. (2004) A quantitative review comparing yields of two candidate C-4 perennial biomass crops in relation to nitrogen, temperature, and water. *Biomass and Bioenergy* 27: 21–30.

Heffer, P. and Prud'homme, M. (2011) Fertilizer Outlook 2011–2012. 79th IFA Annual International Fertilizer Industry Association Conference, 23–25 May, Montreal, Canada.

Henlea, K., Alardb, D., Clitherowc, J., *et al.* (2008) Identifying and managing the conflicts between agriculture and biodiversity conservation in Europe—A review. *Agriculture, Ecosystems and Environment* 124: 60–71.

Himken, M., Lammel, L., Neukirchen, D., CzypionkaKrause, U., and Olfs, H.W. (1997) Cultivation of Miscanthus under west European conditions: seasonal changes in dry matter production, nutrient uptake and remobilization. *Plant and Soil* 189: 117–126.

Katterer, T., Andren, O., and Pettersson, R. (1998) Growth and nitrogen dynamics of reed canarygrass (Phalaris arundinacea L.) subjected to daily fertilization and irrigation in the field. *Field Crops Research* 55: 153–164.

Lal, R., Follett, R.F., and Kimble, J.M. (2003) Achieving soil carbon sequestration in the United States: a challenge to the policy makers. *Soil Science* 168: 827–845.

Landstrom, S., Lomakka, L., and Anderwin, S. (1996) Harvest in spring improves yield and quality of reed canarygrass as a bioenergy crop. *Biomass and Bioenergy* 11: 333–341.

Lee, D.K., Doolittle, J.J., and Owens, V.N. (2007) Soil carbon dioxide fluxes in established switchgrass land managed for biomass production. *Soil Biology and Biochemistry* 39: 178–186.

Lemus, R.W. (2004) Switchgrass as an energy crop: Fertilization, cultivar, and cutting management. Dissertation, Virginia Polytechnic Institute and State University, Blacksburg, Virginia, USA.

Lemus, R., Brummer, E.C., Moore, K.J., Molstad, N.E., Burras, C.L., and Barker, M.F. (2002) Biomass yield and quality of 20 switchgrass populations in southern Iowa, USA. *Biomass and Bioenergy* 23: 433–442.

Lemus, R. and Lal, R. (2005) Bioenergy crops and carbon sequestration. *Critical Reviews in Plant Sciences* 24: 1–21.

Lemus, R., Parrish, D.J., and Abaye, O.A. (2008) Nitrogen-use dynamics in switchgrass grown for biomass. *Bioenergy Research* 1: 153–162. doi:10.1007/s12155-008-9014-x.

Lemus, R., Parrish, D.J., and Wolf, D.D. (2009) Nutrient uptake of 'Alamo' switchgrass used as an energy crop. *Bioenergy Research* 2: 37–50. doi:10.1007/s12155-009-9032-3.

Lewandowski, I., Clifton-Brown, J.C., Scurlock, J.M.O., and Huisman, W. (2000). *Miscanthus*: European experience with a novel energy crop. *Biomass and Bioenergy* 19: 209–227.

Lewandowski, I. and Kicherer, A. (1997). Combustion quality of biomass: practical relevance and experiments to modify the biomass quality of *Miscanthus x giganteus*. *European Journal of Agronomy* 6: 163–177.

Lewandowski, I., Scurlock, J.M.O, Lindvall, E., and Christou, M. (2003) The development and current status of perennial rhizomatous grasses as energy crops in the US and Europe. *Biomass and Bioenergy* 25: 335–362.

Marten, G.C., Clapp, C.E., and Larson, W.E. (1979) Effects of municipal wastewater effluent and cutting management on persistence and yield of eight perennial forages. *Agronomy Journal* 71: 650–658.

Ma, Z., Wood, C.W., and Bransby, D.I. (2000) Soil management impacts on soil carbon sequestration by switchgrass. *Biomass and Bioenergy* 18: 469–477.

Ma, Z., Wood, C.W., and Bransby, D.I. (2001) Impact of row spacing, nitrogen rate, and time on carbon partitioning of switchgrass. *Biomass and Bioenergy* 20: 413–419.

McLaughlin, S.B. (1997) Forage crops as bioenergy fuels: evaluating the status and potential, in *Proceedings of XVIII International Grassland Congress*, Volume 1, Winnipeg, Canada.

McLaughlin, S.B., Bouton, J., Bransby, D., *et al.* (1999) Developing switchgrass as a new crop, in *Perspectives on New Crops and New Uses: Proceedings Fourth National New Crops Symposium, Phoenix, AZ* (ed. J. Janick), ASHS Press, Alexandria, VA, pp. 282–299.

McLaughlin, S.B. and Walsh, M.E. (1998) Evaluating environmental consequences of producing herbaceous crops for bioenergy. *Biomass and Bioenergy* 14: 317–314.

Mitchell, R.B., Vogel, K.P., Schmer, M.R., and Pennington, D. (2010) Switchgrass for biofuel production. Sustainable Ag Energy Community of Practice, eXtension. Available at: http://extension.org/pages/Switchgrass_for_Biofuel_Production (accessed 12 August 2012).

Mitchell, R., Owens, V., Gutterson, N., Richard, E., and Barney, J. (2011) Herbaceous perennials: placement, benefits and incorporation challenges in diversified landscapes, in *Sustainable Alternative Fuel Feedstocks Opportunities, Challenges and Roadmaps for Six U.S. Regions*, (eds R. Braun, D.L. Karlen, and D. Johnson), pp. 84–98.

Miyamoto, T., Kawahara, M., and Minamisawa, K. (2004) Novel endophytic nitrogen-fixing clostridia from the grass *Miscanthus sinensis* as revealed by terminal restriction fragment length polymorphism analysis. *Applied and Environmental Microbiology* 70: 6580–6586.

Muir, J.P., Sanderson, M.A., OCumpaugh, W.R., Jones, R.M., and Reed, R.L. (2001) Biomass production of 'Alamo' switchgrass in response to nitrogen, phosphorous and row spacing. *Agronomy Journal* 93: 896–901.

Mulkey, V.R., Owens, V.N., and Lee, D.K. (2006) Management of switchgrass-dominated Conservation Reserve Program lands for biomass production in South Dakota. *Crop Science* 46: 712–720.

Muthukumarasamy, R., Revathi, G., and Lakshminarasimhan, C. (1999) Diazotrophic associations in sugarcane cultivation in South India. *Tropical Agriculture* 76: 171–178.

Nassi o Di Nassoa, N., Angelini, L.G., and Bonari, E. (2010) Influence of fertilisation and harvest time on fuel quality of giant reed (*Arundo donax* L.) in central Italy. *European Journal of Agronomy* 32: 219–227.

Neukirchen, D., Himken, M., Lammel, J., Czyionka-Krause, U., and Olfs, H.W. (1999). Spatial and temporal distribution of the root system and root nutrient content of an established *Miscanthus* crop. *European Journal of Agronomy* 11: 301–309.

Parish, D.J., Wolf, D.D., Fike, J.H., and Daniels, W.L. (2003) Switchgrass as a biofuels crop for the upper southeast: variety trials and cultural improvements—Final report for 1997 to 2001.ORNL/SUB-03-19XSY163/01. Oak Ridge National Laboratory, Blacksburg, VA.

Rooney, W.L., Blumentjal, J., Bean, B., and Mullet, J.E. (2007) Designing sorghum as dedicated bioenergy feedstock. *Biofuels, Bioproducts and Biorefining* 1: 147–157.

Saijonkari-Pahkala, K. (2001) Non-wood plants as raw material for pulp and paper. *Agricultural and Food Science in Finland* 10: 101.

Sanderson, M.A., and Alder, P.R. (2008). Perennial forages as second generation bioenergy crops. *International Journal of Molecular Science* 9: 768–788. doi:10.3390/ijms9050768.

Sanderson, M.A., Reed, M.L., McLaughlin, S.B., *et al.* (1996) Switchgrass as a sustainable bioenergy crop. *Bioresource Technology* 56: 83–93.

Sharma, K.P., Kushwaha, S.P.S., and Gopal, B. (1998) A comparative study of stand structure and standing crops of two wetland species, Arundo donax and Phragmites karka, and primary production in Arundo donax with observations on the effect of clipping. *Tropical Ecology* 39: 39.

Smith, P., Goulding, K.W., Smith, K.A., *et al.* (2001a) Enhancing the carbon sink in European agricultural soils: including trace gas fluxes in estimates of carbon mitigation potential. *Nutrient Cycling in Agroecosystems* 60: 237–252.

Smith, P., Powlson, D.S., Glendining, M.J., and Smith, J.U. (1997) Potential for carbon sequestration in European soils: preliminary estimates for five scenarios using results from long-term experiments. *Global Change Biology* 3: 67–79.

Smith, P., Smith, J.U., and Powlson, D.S. (2001b) Soil Organic Matter Network (SOMNET): 2001 Model and Experimental Metadata. GCTE Report 7, 2nd edn, GCTE Focus 3, Wallingford, Oxon, 223p.

Stevens, G., Holou, R., Dunn, D., and Wrather, A. (2009) Switchgrass and sweet sorghum fertilization for bioenergy feedstocks. Proceedings of Southern Plant Nutrition Management Conference, 6–7 October, Olive Branch, Mississippi, pp. 38–45.

Turhollow, A.F, Cushman J.H., and Johnston J.W. (1988) Herbaceous energy crops program: annual progress report for FY, Report ORNL-6639, 1990, Oak Ridge National Laboratory, Oak Ridge, Tennessee, 37831–6285, 76p.

United States Department of Agriculture, National Agricultural Statistic Service (USDA-NASS). (2011) Census of Agriculture, USDA, Washington, DC.

United States Environmental Protection Agency (USEPA). (2010) Sources of greenhouse gas emissions: Overview. Available at: http://epa.gov/climatechange/ghgemissions/sources.html (accessed 4 September 2012).

Verardo, D.J. and Dokken, D.J. (2000) *Land Use, Land Use Change, and Forestry*, Cambridge University Press, Cambridge.

Vleeshouwers, L.M. and Verhagen, A. (2002) Carbon emission and sequestration by agricultural land use: a model study for Europe. *Global Change Biology* 8: 519– 530.

Vogel, K.P., Brejda, J.J., Walters, D.T., and Buxton, D.R. (2002) Switchgrass biomass production in the Midwest U.S.A: harvest and nitrogen management. *Agronomy Journal* 94: 413–420.

Wachs, T. and Thibault, M. (2009) Benefits of sustainable land management. WOCAT (World Overview of Conservation Approaches and Technologies), Berne, Switzerland.

Watson, R.T., Noble, I.R., Bolin, B., Ravindranath, N.H., and Freibauer, A. (2004) *Geoderma* 122: 1–23.

West, T.O. and Marland, G. (2002) A synthesis of carbon sequestration, carbon emissions, and net carbon flux in agriculture: comparing tillage practice in the United States. *Agriculture, Ecosystems and Environment* 91: 261–287.

Wiedenfeld, R.P. (1983) Nutrient requirements and use efficiency by sweet sorghum. *Energy in Agriculture* 3: 46–59.

Williams, C., Biswas, T., Chvyl, L., Harris, P., and Dyson, C. (2010) Commercial potential of giant reed for pulp, paper and biofuel production, (eds C. William and T. Biswas), Australian Government, Rural Industries Research and Development Corporation. Publication No. 10/215, Project No. PRJ-00070.

Wolf, D.D. and Fiske, D.A. (1995) Planting and managing switchgrass for forage, wildlife, and conservation. Extension agronomist, Forages. Virginia Polytech Institute and State University Publication No. 418-013. Available at: http://www.caf.wvu.edu/~forage/vpi/418-013/418-013.htm (accessed 18 January 2013).

Wright, L. (1994) Production technology status of woody and herbaceous crops. *Biomass and Bioenergy* 6: 191–209.

Zan, C.S., Fyles, J.W., Girouard, P., and Samson, R.A. (2001) Carbon sequestration in perennial bioenergy, annual corn and uncultivated systems in southern Quebec. *Agriculture, Ecosystems and Environment* 86: 135–144.

Chapter 10
Food, Farming, and Biofuels

J. Popp

University of Debrecen, Debrecen, Hungary

Introduction

In the last 35 years global energy supplies have nearly doubled but the relative contribution from renewables has hardly changed at around 13%. Global energy demand is increasing, as is the environmental damage due to fossil fuel use. Continued reliance on fossil fuels will make it very difficult to reduce emission of greenhouse gases (GHGs) that contribute to global warming. Experts warn that GHG emissions must peak before 2020 to avoid unacceptable risks.

With world population expected to reach 9 billion people before 2050, higher food, feed, and fiber demand will place an increasing pressure on land and water resources, whose availability and productivity in agriculture may themselves be under threat from climate change. The additional impact on food prices of higher demand for crops as energy feedstock is of real concern. Since biomass can substitute for petrochemicals too, higher oil prices will trigger new nonenergy demands on bioresources as well.

Bioenergy currently provides roughly 10% of global supplies and accounts for roughly 80% of the energy derived from renewable sources (IEA Bioenergy, 2009). By far the largest element of the bioenergy sector globally is wood used for cooking and heating in developing countries. The "new" renewables (e.g., solar, wind, biofuel) have been growing fast from a very low base, but their contribution is still marginal of the global renewable supply. Bioenergy has the potential to play an increasing but modest role in ensuring future global energy demand, keeping in perspective the relative significance of the renewables sector in the overall energy mix. Bioenergy was the main source of power and heat prior to the industrial revolution. Since then, economic development has largely relied on fossil fuels. A major impetus for the development of bioenergy has been the search for alternatives to fossil fuels, particularly those used in transportation. The renewed interest in biofuel is driven by a range of considerations, including climate change and the potential economic contribution of the development of the biofuel industry in terms of income and employment.

Biofuel Crop Sustainability, First Edition. Edited by Bharat P. Singh.

The development of biofuels has been one of the most visible and controversial manifestations of the use of biomass for energy. Biofuel policies in the EU, the United States, and Brazil have been particularly important for the development of the industry in these important markets where a variety of measures, including consumption mandates, tax incentives, and import protection to promote the production and use of biofuels, have been used. Despite this, the EU and the United States may run into difficulties in meeting consumption mandates for biofuels. Questions are being raised about the continuation of tax credits and import protection in the EU and the United States; however, US ethanol import tariff on fuel ethanol, federal tax credit of blended ethanol, and tax credit for biodiesel blending expired on 31 December 2011 (Yacobucci, 2012). Furthermore, an ongoing debate about the benefits of reliance on biofuels derived from food crops and concern about the efficacy of current biofuel policies may contribute to the doubts of future policy. Brazil has liberalized its domestic ethanol market and adopted a more market-oriented approach to biofuel policy, but the management of domestic petroleum prices and the interrelationship between the sugar market and ethanol production are important factors affecting domestic consumption and exports.

While biofuel has the potential to be more environmentally friendly in terms of reduced GHG emissions, it may have unintended negative environmental consequences, particularly relating to changes in land use. Characterizing and quantifying the relationship between biofuel production and the environment poses a considerable challenge. Much of the focus has been on the implications of expanded use of biofuels and there are concerns that the accounting of environmental effects remains incomplete. Combination with an improved assessment of the effects of indirect land-use change (iLUC) and an expansion of sustainability criteria to biomass production in general (and not only to biofuels) could help in integrating energy, agricultural, environmental, and international trade policies to develop renewable energy in a sustainable way.

There is substantial uncertainty about the true employment impact of biofuel production in rural areas. Estimated effects are smaller when account is taken of the effect of higher energy costs on economic activity and the diversion of public funds from other uses. Impacts are likely to be stronger in rural areas than at the national level, because of the location of much of the primary bioenergy activity in those areas.

Waste biomass and crops with high energy yields that do not compete for prime cropland are more promising bioenergy feedstocks than food crops. Still, constrained land capacity relative to potential demands means that bioenergy can only be part of the solution. A broader, more integrated approach is needed for energy policy, embracing all renewable energies that reduce GHG emissions without serious side effects. Governments should maximize their efforts to reach global consensus on emission targets and reduction of fossil fuel subsidies.

Risks to Food Security

The combined effect of the Green Revolution has allowed world food production to double in the past 50 years. From 1960 to present, the human population has more than doubled to reach 7 billion people (Figure 10.1). The 7 billion world population is projected to increase by 30% to 9.2 billion by 2050. This increased population density, coupled with changes in dietary habits in developing countries toward high-quality food (e.g., greater consumption of meat and milk products) and the resultant rise in the use of grains for livestock feed, is projected to amplify demand for food by over 40% by 2030 and 70% by 2050, compared with the average

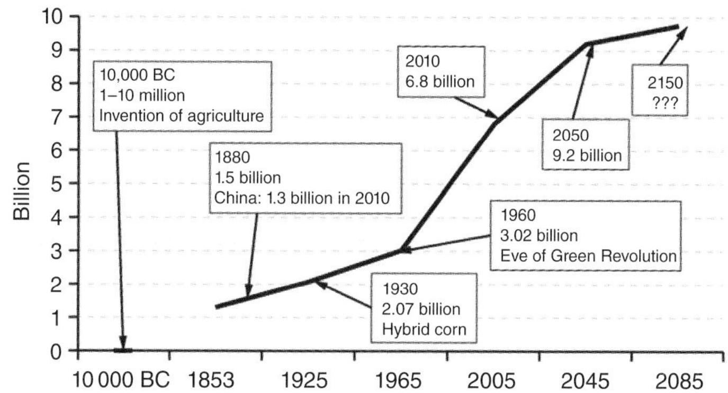

Figure 10.1. World population growth (*Source*: FAO, 2009).

2005–2007 levels (FAO, 2009). At the same time, the increase in arable land between 2005 and 2050 will be just 5% (FAO, 2011a).

Land use for food and feed is typically determined by global diet and agricultural yield improvements. With respect to diet, consumption of meat and dairy products is an important driver for land use since meat and dairy use a lot more basic agricultural production than does the consumption of grain. Livestock products imply an inefficient conversion of calories of the crops used in livestock feeds. On average, 6 kg of plant protein is required to yield 1 kg of meat protein. By 2050 an expanded world population will be consuming two-thirds more animal protein than it does today, bringing new strains to bear on the planet's natural resources. Meat consumption is projected to rise nearly 73% by 2050; dairy consumption will grow 58% over current levels. The surge in livestock production that took place over the last 40 years resulted largely from an increase in the overall number of animals being raised. Meeting projected demand increases in production will need to come from improvements in the efficiency of livestock systems in converting natural resources into food and reducing waste. This will require capital investment and a supporting policy and regulatory environment. Meat consumption in China alone increased from 27 to 60 kg per person per year between 1990 and 2010. Each additional kilogram of meat consumption increase in China results in a need for roughly 4–5 MT of animal feed (FAO, 2011b).

Helping farmers lose less of their crops will be a key factor in promoting food security but even in the poorest countries those rural farmers aspire to more than self-sufficiency. The reduction of current yield losses caused by pests, pathogens, and weeds is a major challenge to agricultural production. Globally, an average of 35% of potential crop yield is lost to pre-harvest pests (Oerke, 2006). In addition to pre-harvest losses transport, pre-processing, storage, processing, packaging, marketing, and plate waste losses are relatively high. If there is going to be enough food at affordable prices for the global population, we may also have to change our food habits and decrease food waste. Food waste in the field pre-processing (broken grains, excessive dehulling), transport (spillage, leakage), storage (insects, bacteria), and processing and packaging (excessive peeling, trimming, and inefficiency) goes up to 10–15% in quantity and 25–50% in value (quality). Marketing (retailing) and plate (by consumers and retailers) waste adds another 5–30% in developed and 2–20% in developing countries to the losses in the food chain (Figure 10.2). Roughly one-third of the edible parts of food produced for human consumption is lost or wasted globally. Food losses in industrialized countries are as high as

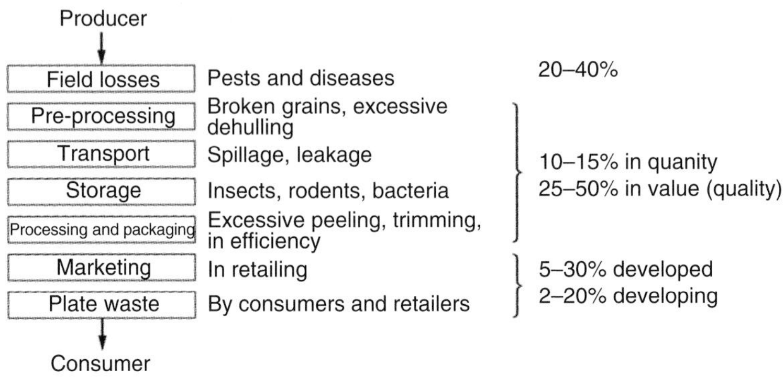

Figure 10.2. Losses along the food chain (*Source*: IWMI, 2007).

in developing countries, but in developing countries more than 40% of the food losses occur at post-harvest and processing levels, while in industrialized countries, more than 40% of the food losses occur at retail and consumer levels (Gustavsson *et al.*, 2011). We can also save water by reducing losses in the food chain.

Competition for Land and Water

LUC is not a new concept but is something that has been taking place since the beginning of civilization and continues to do so. In this context, agriculture has always been an important driver, so far mostly for food and feed production. A growing world population and a changing diet have led to continuously expanding areas of agricultural land, despite parallel increases in yields from existing cropland. In addition, cropland is lost due to erosion through chemical and physical degradation, which further increases the requirement for new agricultural land. On the other hand, the amount of cultivated land is decreasing due to population growth and accelerated urbanization and motorization,[1] changes in lifestyles, falling water tables and diversion of irrigated water toward the cities (Earth Institute, 2005).

The land surface of our planet is equal to 13.4 billion hectares of which 38% is given over to agriculture and 30% to forest (FAO, 2011a). The rest of the total is rounded out through a combination of man-made infrastructure, inland water systems, and land that is unsuited for agriculture and forestry (desert, rocks, etc.). Of the 5 billion hectares of land used for agricultural purposes worldwide, around one-third is suited to annual or permanent crops whereas just over two-thirds are allocated to permanent meadows or pasture. Just 1.6 billion hectares are used for crop production (arable land and land under permanent crops).

Over the last 50 years, land and water management has met rapidly rising demands for food and fiber. In particular, input-intensive, mechanized agriculture and irrigation have contributed to rapid increases in productivity. The world's agricultural production has grown between 2.5 and 3 times over the period while the cultivated area has grown only by 12% as a result of two opposite trends: an increase of 227 Mha in developing countries and a decline of 40 Mha in developed countries. More than 40% of the increase in food production came from

[1] An estimated 40 000 ha of land is needed for basic living space for every 1 million people added and 20 000 ha of land is needed for every 1 million vehicles added.

irrigated areas, which have doubled in area over the same period, accounting for 15% of all arable land. In the same period, the cultivated area of land per person gradually declined from 0.45 to less than 0.23 ha indicating that the largest contribution to increases in agricultural output will most likely come from intensification of production on existing agricultural land (FAO, 2011c).

However, global achievements in production in some regions have been associated with degradation of land and water resources and the deterioration of related ecosystem goods and services. Agriculture also makes use of 70% of all water withdrawn from aquifers, streams, and lakes. Urbanization may double domestic and industrial water use, not to mention climate change and bioenergy production. Without water productivity gains, crop water consumption will double by 2050. The water "bubble" is unsustainable and fragile because 7 billion people at present have to share the same quantity as the 300 million global inhabitants of Roman times. About 80% of water for food production comes directly from rain, but an increasing part is met by irrigation. Both physical water productivity (more crop production per drop water use) and economic water productivity (more production value per drop water use) have to be increased by investing in rainfed agriculture and irrigation. This will require widespread adoption of sustainable land management practices and more efficient use of irrigation water through enhanced flexibility, reliability, and timing of irrigation water delivery. Promoting food trade from water-rich, highly productive areas to water-scarce areas contributes to global water productivity improvement (IWMI, 2007).

Future agricultural production will have to rise faster than population growth largely on existing agricultural land. Improvements will thus have to come from sustainable intensification that makes effective use of land and water resources as well as not causing them harm. Regarding yield improvements, there seems to be a large theoretical potential for yield improvements throughout the world, especially in developing countries, but there are still major uncertainties as to what proportion of this potential can be harvested. The increase in food demand is met to some extent by an increase in agricultural yields. Crop yields would continue to grow, but at a slower rate than in the past. On average, annual growth would be about half that of the historical period: 0.8% per annum from 2005/2007 to 2050, against 1.7% per annum from 1961 to 2007. Nevertheless, agricultural production would still need to increase by 70% by 2050 to cope with a 30% increase in world population. This translates into additional production of 1 billion tonnes of cereals and 200 MT of meat a year by 2050 (compared with production in 2005/2007). In addition to yield growth there will also be a slow expansion of agricultural land. Arable land would expand by 70 Mha (less than 5%), an expansion of about 120 Mha (12%) in developing countries being offset by a decline of 50 Mha (8%) in developed countries. Much of the suitable land not yet in use is concentrated in a few countries in Latin America and Sub-Saharan Africa, not necessarily in Asia (with some 60% of the world's population) where it is most needed, and much is suitable for growing only a few crops, not necessarily those for which the demand is highest (FAO, 2011a).

Large-scale land acquisitions are on the increase in parts of Africa, Asia, and Latin America where land and water resources appear abundant and available. More recent transnational land deals are partly a consequence of the larger changing economic valuation of land and water. Higher agricultural prices generally result in higher land prices because the expected returns to land increase when profits per unit of land increase. Given that the rise of food price has increased competition for land and water resources for agriculture, it is not surprising that farmland prices have risen throughout the world in recent years. Although large-scale land acquisitions remain a small proportion of suitable land in any one country, contrary to widespread perceptions there is very little "empty" land as most remaining suitable land is already used or claimed, often by local people. While they offer opportunities for development, there is a risk that the rural poor could be evicted or lose access to land, water, and other related resources (Braun and Meinzen-Dick, 2009).

Bioenergy may compete with the food sector, either directly if food commodities are used as the energy source, or indirectly if bioenergy crops are cultivated on soil that would otherwise be used for food production. Both effects may impact on food prices and food security if demand for the crops or for land is significantly large. This issue has typically been of concern for the biofuel sector, which uses mainly food crops. Increased biofuel production could also reduce water availability for food production, as more water is diverted to production of biofuel feedstocks (Chakravorty *et al.*, 2009; Hoekstra *et al.*, 2010). Until now, the price increases that this has led to seem to be limited for most crops, and the agricultural sector has responded by increasing production. There are exceptions, though, especially with crops where biofuel demand accounts for a significant share of the total demand (e.g., maize, oilseeds, sugarcane). Besides competition with food and feed, increased use of biomass also has its effects on other sectors. Forest-based industries (pulp and paper, building materials, etc.), for example, will be affected by the increased use of wood for energy conversion, both negatively and positively.

Almost 1 billion people are undernourished. There will always be risks associated with food supply and thus a need to manage these risks. Domestic food supplies are not less risky than, for example, energy imports, but it is sensible to plan for systemic risks (such as nuclear fallout, port strikes). We experience food poverty due to a lack of entitlements, not lack of food (Krugman, 2009). Future food security depends on the development of the political and logistical capacity to make food accessible everywhere, to everyone (FAO, 2011a).

Energy Security

The use of fossil fuels by agriculture has made a significant contribution to feeding the world over the last few decades. The food sector accounts for around 30% of global energy consumption and produces over 20% of global GHG emissions. Around one-third of the food we produce, and the energy that is embedded in it, is lost or wasted (Figure 10.2). The energy embedded in global annual food losses is around 38% of the total final energy consumed by the whole food chain (Gustavsson *et al.*, 2011). Due to high dependence of the global food sector on fossil fuels the volatility of energy markets can have a potentially significant impact on food prices, and this would have serious implications for food security and sustainable development (IPCC, 2011). Rising energy prices may cause spillovers into food markets leading to increasing food insecurity. Furthermore, any increase in the use of fossil fuels to boost production will lead to greater GHG emissions, which the global community has pledged to reduce. The food sector can adapt to future energy supply constraints and to the impacts of climate change with rapid deployment of energy efficiency and renewable energy technologies (FAO, 2011d).

Global primary energy demand is projected to rise from around 12 300 MT oil equivalent (Mtoe) in 2008 to 16 800 Mtoe in 2035—an increase of over 35%. While oil continues to be the dominant fuel in the primary energy mix, its share of the mix drops from 33% in 2008 to 27% in 2035. The average price of imported crude oil will remain high, approaching US$120 per barrel in 2035 (at the rate of US dollar in 2010), that is, more than US$210 per barrel in nominal terms. Natural gas increases from 21% of the global fuel mix in 2008 to 25% in 2035 becoming the second largest fuel in the primary energy mix. The share of primary coal demand declines by 5% from 27% in 2008 to 22% in 2035. The share of nuclear

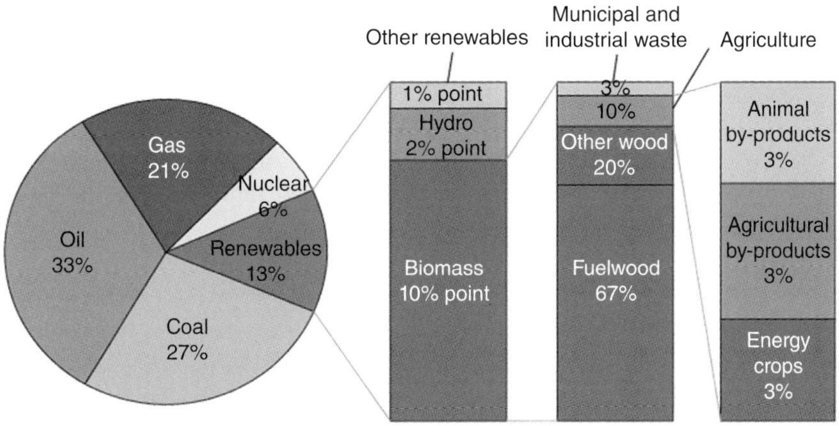

Figure 10.3. World primary energy demand by fuel in 2008 (*Source*: IEA Bioenergy, 2009).

power in global primary energy supply increases from 6% in 2008 to 7% in 2035 (IEA, 2011). Renewables increase from 13% of the mix to 19% over the same period leading to a decreasing share of fossil fuels in the global primary energy consumption from 87% in 2008 to 81% in 2035 (Figures 10.3 and 10.4).

While fossil fuels meet 87% of world energy demand, its subsidies are creating market distortions that encourage wasteful consumption. The International Energy Agency (IEA) estimates governments spent US$409 billion on fossil fuel subsidies in 2010 increasing to US$660 billion (0.7% of global gross domestic product) by 2020 unless action is taken. Eliminating fossil fuel consumption subsidies would slash the growth in energy demand by 4.1% reaching 2020 and contribute to more competitive renewable energy sources. Renewable subsidies of US$66 billion in 2010 (compared with US$409 billion for fossil fuels) need to

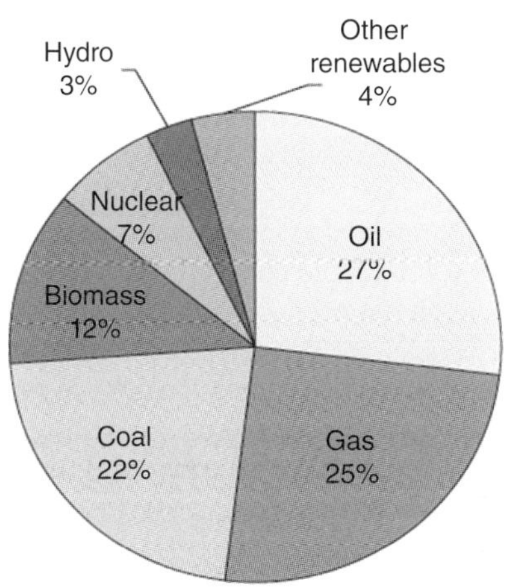

Figure 10.4. World primary energy demand by fuel in 2035 (*Source*: IEA, 2011).

climb to US$250 billion in 2035 as rising deployment outweighs improved competitiveness (IEA, 2011).

On a global basis, it is estimated that renewable energy accounted for 13% of the total 492 exajoules (EJ)[2] of primary energy supply in 2008 (IEA Bioenergy, 2009). The largest contributor to renewable energy with 10% points was biomass. Hydropower represented 2% points, whereas other renewable energy sources accounted for 1% point (Figure 10.3). The contribution of renewable energy to primary energy supply varies substantially by country and region.

Bioenergy Potential of Biomass

Concern about energy security, the threat of climate change, and the need to meet growing energy demand (particularly in the developing world) all pose major challenges to energy decision makers. Advancing the low-carbon technology revolution will involve millions of choices by a myriad of stakeholders. Overall, the global share of biomass has remained stable over the past two decades, but in recent years a sharp decline in share can be observed in China due to a rapid growth of total energy consumption and a steady increase of all types of biomass (for electricity, heat, and biofuels) in the EU.

Biomass is a versatile energy source—it can be stored and converted into practically any form of energy carrier and also into biochemicals and biomaterials from which, once they have been used, the energy content can be recovered to generate electricity, heat, or transport fuels. The worldwide potential of bioenergy is limited because all land is multifunctional and land is also needed for food, feed, timber, and fiber production, as well as for nature conservation and climate protection. In addition, the use of biomass as an industrial feedstock (e.g., plastics) will become increasingly important. Biomass can include land- and water-based vegetation (e.g., trees, algae), as well as other organic wastes. The biomass feedstock can be subdivided into primary, secondary, or tertiary feedstocks. Primary biomass feedstocks are materials harvested or collected directly from forest or agricultural land where they are grown (e.g., grains). Secondary biomass feedstocks are by-products of the processing of primary feedstocks (e.g., maize stover, sawmill residues, black liquor). Tertiary biomass feedstocks are post-consumer residues and wastes (e.g., waste greases, wastewaters, municipal solid waste). At present only a small fraction of biomass is used globally for biofuel production and power generation, but these shares are growing rapidly because of issues like energy security, rising fossil fuel prices and, last but not least, global warming concerns and GHG reduction policies. With demand for energy continuing to rise in absolute terms, the absolute use of biomass will increase even more.

Bioenergy is the single largest source of renewable energy today and has the highest technical potential for expansion amongst renewable energy technologies. In 2008, biomass provided about 10% (50.3 EJ yr^{-1}) of the global primary energy supply (IEA Bioenergy, 2009). More than 80% of the biomass feedstocks are derived from wood (trees, branches, residues) and shrubs. The remaining bioenergy feedstocks came from the agricultural sector (energy crops, residues, and by-products) and from various commercial and post-consumer waste and by-product streams (biomass product recycling and processing or the organic biogenic fraction of municipal solid waste; Figure 10.4).

[2] 1 EJ = 10^{18} J = 23.88 Mtoe.

The majority of biomass (roughly two-thirds) is used inefficiently for traditional domestic cooking, lighting, and space heating in developing countries. More than a third of the world's population depends on this form of energy, which is unhealthy and contributes to the death of 1.5 million people a year from the pollution it causes. The share of the smaller, modern bioenergy use is growing rapidly. High-efficiency modern bioenergy uses more convenient solids, liquids, and gases as secondary energy carriers to generate heat, electricity, combined heat and power, and transport fuels for various sectors. The estimated total primary biomass supply for modern bioenergy is 11.3 EJ yr^{-1} (the secondary energy delivered to end-use consumers is roughly 6.6 EJ yr^{-1}). Additionally, the industry sector, such as the pulp and paper, forestry, and food industries, consumes approximately 7.7 EJ of biomass annually, primarily as a source for industrial process steam (IEA, 2010a).

In developing countries biomass contributes some 22% to the total primary energy mix. The traditional use of biomass is expected to grow with increasing world population, but there is significant scope to improve its efficiency and environmental performance, and thereby help reduce biomass consumption and related impacts. In industrialized countries, the total contribution of modern biomass is on average only about 3% of the total primary energy with large differences among the industrialized countries. Finland and Sweden have shares of around 20%,[3] for example, while for Ireland and the United Kingdom these figures were 1.3% and 1.5%, respectively (IPCC, 2011). In the future biomass could also provide an attractive feedstock for the chemical industry (to produce soap, cosmetics, etc.) and use of biogenic fibers will increase.

The total annual aboveground net primary production (the net amount of carbon assimilated in a time period by vegetation) on the Earth's terrestrial surface is estimated to be about 35 Gt carbon or 1260 EJ yr^{-1} assuming an average carbon content of 50% and 18 GJ $tonne^{-1}$ average heating value (Haberl *et al.*, 2007), which can be compared to the current world primary energy supply of about 500 EJ yr^{-1} (IEA, 2010a). All harvested biomass used for food, fodder, fiber, and forest products, when expressed in equivalent heat content, equals 219 EJ yr^{-1} (2000 data, Krausmann *et al.*, 2008). The global harvest of major crops (cereals, oil crops, sugar crops, roots, tubers, and pulses) corresponds to about 60 EJ yr^{-1} and the global industrial roundwood production corresponds to 15–20 EJ yr^{-1} (FAOSTAT, 2011).

Based on this diverse range of feedstocks, the technical potential for biomass is estimated in the literature to be possibly as high as 1500 EJ yr^{-1} by 2050 (Smeets *et al.*, 2007), although most biomass supply scenarios that take into account sustainability constraints indicate an annual potential of between 200 and 500 EJ yr^{-1} (excluding aquatic biomass owing to its early state of development), representing 40–100% of the current global energy use (IEA Bioenergy, 2009). Forestry and agricultural residues and other organic wastes (including municipal solid waste) would provide between 50 and 150 EJ yr^{-1}, while the remainder would come from energy crops, surplus forest growth, and increased agricultural productivity (Figure 10.5).

Projected world primary energy demand by 2050 is expected to be in the range of 600–1000 EJ yr^{-1} compared to about 500 EJ in 2008. The expert assessment suggests potential deployment levels of bioenergy by 2050 in the range of 100–300 EJ yr^{-1}. However, there are large uncertainties in this potential, such as market and policy conditions, and there is strong dependence on the rate of improvements in the agricultural sector for food, fodder, and fiber production and forest products. The entire current global biomass harvest would be required

[3] Because of the large wood industries (pulp and paper) in both countries there is a large feedstock of black liquor (by-product from paper pulp production) which is used to produce industrial heat.

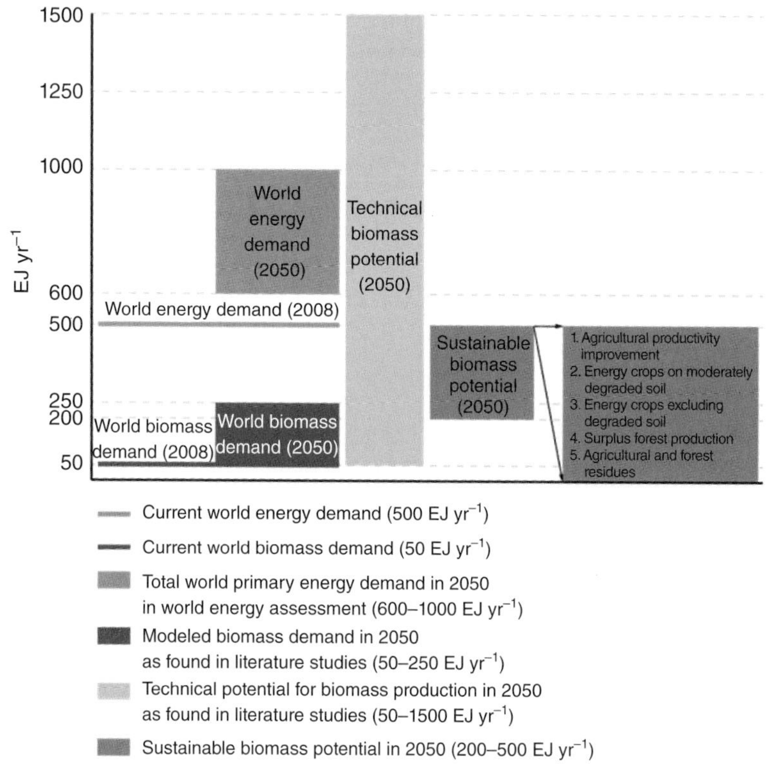

Figure 10.5. Global bioenergy sources (*Source*: IEA Bioenergy, 2009).

to achieve a 200 EJ yr^{-1} deployment level of bioenergy by 2050. Scenarios looking at the penetration of different low-carbon energy sources indicate that future demand for bioenergy could be up to 250 EJ yr^{-1} (Kampman *et al.*, 2010). It is reasonable to assume that biomass could sustainably contribute between a quarter and a third of the future global energy mix.

The transport sector is responsible for about 20% of the world primary energy demand (94 EJ). Transport biofuels are currently the fastest growing bioenergy sector. However, today they represent just 3% of the total road transport fuel consumption and only 5% of the total bioenergy (in energy value). At present only a small fraction of biomass (sugarcane, grain, and vegetable oil crop) is used globally for biofuel production, but these shares are growing rapidly because of issues like energy security, rising fossil fuel prices, and, last but not least, global warming concerns and GHG reduction policies. Liquid transport fuels from biomass represent one of the most important options for the sustainable supply of transport fuels (Kampman *et al.*, 2010).

The projected primary bioenergy demand is 145 EJ (65 EJ for biofuels, 80 EJ mainly for heat and power) in 2050. The total feedstock required to meet the ambitious goals of biofuel production is around 65 EJ of biomass meeting 27% of world demand for transportation fuels by 2050. It is assumed that 50% of the feedstock for advanced biofuels and biomethane will be obtained from wastes and residues (IEA, 2010b). This is a rather conservative estimate, but given the potential constraints regarding collection and transportation of residues and the potentially enormous feedstock demand of commercial advanced biofuel plants, it is not clear if a higher residue share can realistically be mobilized for biofuel production. Advanced biofuels are expected to increase in importance over the next two decades.

The volume of sustainable biomass resources that are economically competitive but do not significantly impact on food production is expected to slowly expand as new feedstock varieties and refining pathways are developed. Availability of land for nonfood crops will be determined by increased yield potential, reducing losses and wastes along the food chain, and lower inputs. However, these volumes will remain limited relative to total energy and transport sector fuel demand. Limited biomass resources will be allocated to the sector (materials, chemicals, energy) that is most able to afford them. This will depend on the price of existing fossil fuel products and the relative cost of converting biomass into substitute final fuels such as bio-derived electricity, ethanol blends, biodiesel and bio-derived jet fuel. It will also depend on factors such as cost of alternative fuel and energy sources, government policies including excise rates, and the emission intensity of each sector.

Competition for land may be limited, as production of feedstocks for advanced biofuels is expected to be grown mainly outside cultivated land, and that some 100 Mha would be sufficient to achieve the target biofuel share in world transport fuels in 2050 (IEA, 2010b). An important step in increasing biofuel production and sustainability is the competitive production of biofuels from (hemi)cellulose. Perennial crops and woody energy crops typically have higher yields than grain and vegetable oil crop used for current biofuels. The extent of grassland and woodland with potential for lignocellulosic feedstocks is about 1.75 billion hectares worldwide. However, much of this grass- and woodland provides food and wood for cooking and heating to local communities or is in use as (extensive) grazing ground for livestock and only some 700 to 800 Mha of this land is suitable for economically viable lignocellulosic feedstock production (Fischer *et al.*, 2009).

The sustainable use of residues and wastes for bioenergy, which do not require any new agricultural land and present limited or zero environmental risks, needs to be encouraged and promoted globally. Several factors may discourage the use of these "lower-risk" resources. Using residues and surplus forest growth, and establishing energy crop plantations on currently unused land, may prove more expensive than creating large-scale energy plantations on arable land. In the case of residues, opportunity costs can occur, and the scattered distribution of residues may render it difficult in some places to recover them (IEA, 2010b). Whatever is actually realized will depend on the cost competitiveness of bioenergy and on future policy frameworks, such as GHG emission reduction targets. The uptake of biomass depends on biomass production costs (US\$4 GJ^{-1} is often regarded as an upper limit if bioenergy is to be widely deployed today in all sectors), logistics, and resource and environmental issues (IPCC, 2011).

Biofuels

World fuels and especially European fuels are moving toward diesel; however, there is more supply of ethanol available than biodiesel. iLUC effects seem to affect more conventional biodiesel than conventional ethanol. Liquid biofuels for transport are generating the most attention and have seen a rapid expansion in production. World fuel ethanol production amounted to 1.8 EJ and biodiesel production increased to 0.6 EJ in 2010. Liquid biofuels make a small but growing contribution to fuel usage worldwide; they covered about 3% (2.4 EJ) of global road transport fuel consumption (in energy value). They accounted for higher shares in some countries (e.g., 4% in the United States) and regions (3% in the EU) and provided a very large

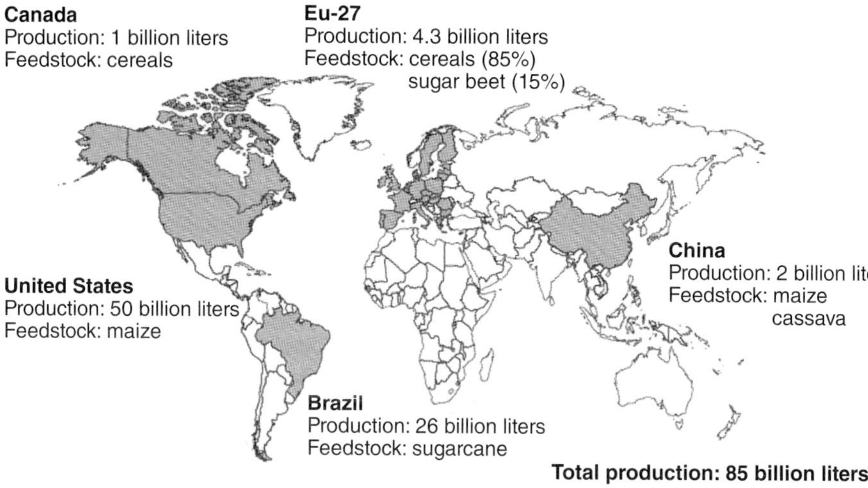

Figure 10.6. World fuel ethanol production, 2010 (*Source*: Licht, 2011; author's calculations).

contribution in Brazil, where ethanol from sugarcane accounted for 41.5% of light-duty transport fuel during 2010. However, biofuels have the potential to meet 27% of the world demand for transportation fuels by 2050 (IEA, 2010b). A considerable share of the required volume will have to come from advanced biofuel technologies that are not yet commercially deployed. Even though liquid biofuels supply only a small share of global energy needs, they still have the potential to have a significant effect on global agriculture and agricultural markets, because of the volume of feedstocks and the relative land areas needed for their production.

Currently, around 80% of the global production of liquid biofuels is in the form of ethanol. In 2010 global fuel ethanol production reached 85 billion liters, global biodiesel production amounted to 16.5 MT, or 18.5 billion liters (Figures 10.6 and 10.7). In 2010, the United States was the world's largest producer of biofuels, followed by Brazil and the European Union. Despite continued increases in production, growth rates for biodiesel slowed again, whereas ethanol production growth picked up new momentum.

The global ethanol industry recovered in 2010 in response to rising oil prices. Some previously bankrupt firms returned to the market, and there were a number of acquisitions as large traditional oil companies entered the industry. The world's top two ethanol producers, the United States and Brazil, accounted for around 90% of the total production, with the remainder accounted for mostly by the EU, China, and Canada (Figure 10.7). The United States is the world's largest bioethanol producer. In 2010, it produced 50 billion liters of ethanol and accounted for nearly 60% of global production. In Brazil fuel ethanol production reached 26 billion liters and in the EU 4.3 billion liters in 2010. China, at 2 billion liters, remained Asia's largest ethanol producer, followed by Thailand and India, which more than doubled its annual production to 0.25 billion liters. Other important producers included Canada, Colombia, Poland, and Spain. Africa represents a tiny share of the world production but saw continued rapid growth in production during 2010 (Licht, 2011).

Global biodiesel production amounted to 16.5 MT (18.5 billion liters) in 2010. Biodiesel production is far less concentrated than ethanol, with the top 10 countries accounting for 75% of the total production in 2010. The European Union remained the center of global biodiesel

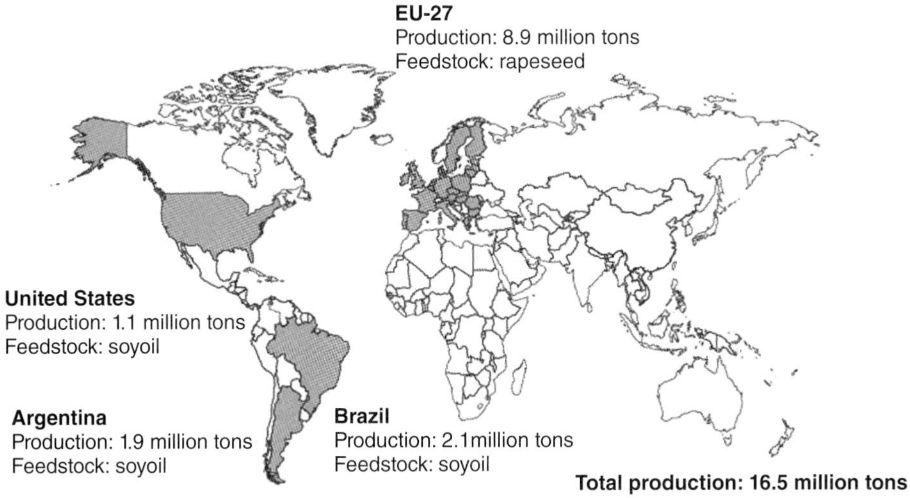

Figure 10.7. World biodiesel production, 2010 (*Source*: Licht, 2011; author's calculations).

production, with 8.9 MT and representing 54% of the total output in 2010. Biodiesel accounted for the vast majority of biofuels consumed in the EU, but growth in the region continued to slow (Licht, 2011). The slowdown of biodiesel output in many countries was due to increased competition with relatively cheap imports from outside the EU (including Canada, Argentina, and increasingly Indonesia). This trend is leading to plant closures from reduced domestic production requirements, an expansion of tariffs on imports, and increases in some blending mandates.

Germany remained the world's top biodiesel producer at 2.3 MT in 2010, followed by Brazil, Argentina, France, and the United States. Consumption in Germany has declined significantly since the elimination of Germany's biodiesel tax credit. The greatest drop in demand has been in pure vegetable oil and B100 (100% unblended biodiesel). In contrast, the use of blended biodiesel has increased during this period due to the national blending quota, and total production rose in 2010. The greatest production increase was seen in Brazil and Argentina, which continued its rapid growth with production, three-quarters of which was exported (and national blending rate has increased from B5 to B7). In the United States, biodiesel production fell more than 40%. Almost 12% of biodiesel production occurred in Asia, with most of this from palm oil in Indonesia and Thailand.

Advanced biofuels are high-energy liquid fuels, usually used for transportation derived from low nutrient input, high-yield crops, agricultural or forestry waste, or other sustainable biomass feedstocks including algae. There is considerable debate on how to classify biofuels. Biofuels are commonly divided into first-, second-, and third-generation biofuels, but the same fuel might be classified differently depending on whether technology maturity, GHG emission balance, or the feedstock is used to guide the distinction. The most transparent way is to follow a definition based on the maturity of a technology and the terms "conventional" and "advanced" for classification (IEA, 2010b). Advanced biofuels are biofuels produced from sustainable feedstock. Sustainability of a feedstock is defined among others by availability of the feedstock, impact on GHG emissions, and impact on biodiversity and land use. The GHG emission balance depends on the feedstock and processes used, and it is important to realize that advanced biofuel performance is not always superior to that of conventional biofuels. Advanced biofuel technologies are conversion technologies which are still in the research and development (R&D), pilot, or demonstration phase.

High capital costs and limited market differentiation (same ethanol molecule/same price) are preventing investment in cellulosic ethanol. It is not attractive to venture capital, banks are not in a hurry either, and strategic partners are balancing against other options. Government policies, incentives, and financial support are necessary to move new technology from lab to commercial deployment. Government can also use "technology push" policies—time-bound subsidies to assist new technologies through the demonstration phase with multiple tools (tax credit, support of FFV sale and blender pumps, higher mandates, low carbon fuel standards, etc.). The biofuel industry must collaborate since new policy support cannot be at the expense of other biofuels, and cellulosic ethanol and grain ethanol need to carry this forward together.

Hydrotreated vegetable oil (HVO) is a drop-in biofuel, that is, has the same molecular structure as hydrocarbons and offers benefits to all stakeholders since it is compatible with existing diesel logistics and existing distribution, engines, and vehicle systems without any modifications. It has outstanding blending properties without blending limits and good storage stability. HVO exceeds minimum requirements of the fuel specification and can be produced from a wide range of vegetable oils and waste animal fats with significant reduction in GHG emissions. The production cost is still much higher than the cost of the conventional biodiesel production. It is available in commercial sale, for example, in Finland.

Algae have been cultivated commercially since the 1950s, mainly for the pharmaceutical industry, but only recently gained attention as a potential source of biomass. Algae oil does have potential as feedstock for biodiesel in future; however, biodiesel produced from algae must also meet given fuel standards. Algae oil is a challenging raw material for biodiesel production if mature conversion technology for reaction and purification is developed. Algae promise a potentially high productivity per hectare, could be grown on nonarable land, can utilize a wide variety of water sources (fresh, brackish, saline, and wastewater), and potentially recycle CO_2 and other nutrient waste streams. However, algae cultivation faces several challenges, related to availability of locations with sufficient sunshine and water, required nutrient inputs, and oil extraction. Traditional oil companies have begun to enter the algae industry. In the future, algae-based biorefinery systems and seaweed production to assimilate dissolved nutrients combined with intensive fish or shrimp culture in integrated multi-trophic aquaculture systems may be a viable option (van Iersel *et al.*, 2009).

The installation of the first commercial-scale advanced biofuel plants is anticipated within the next decade, followed by rapid growth of advanced biofuel production after 2020. Some novel technologies such as algae biofuels and sugar-based hydrocarbons will also need to be developed, but commercialization of these will require more substantial research and development. The installed advanced biofuel capacity today is roughly 200 million liters gasoline equivalent per year and production capacity of another 1.9 billion liters gasoline equivalent is currently under construction (IEA, 2010b).

Ultimately, bioenergy production may increasingly occur in biorefineries where transport biofuels, power, heat, chemicals, and other marketable products could all be co-produced from a mix of biomass feedstocks. Biorefinery is a facility that integrates upstream, midstream, and downstream processing of biomass into a range of products (fuels, power, and chemicals), analogous to today's petroleum refineries, which produce multiple fuels and products from petroleum. Margins on petrol and diesel are very poor. Today about 10% of crude oil is used to make chemicals generating 35% of refinery profits. Why should biorefinering be any different?

The economic competitiveness of the operation is based on the production of high-value, low-volume co-products in addition to comparably low-value biofuels. Biorefineries can potentially make use of a broader variety of biomass feedstocks and allow for a more efficient use of resources than current biofuel production units and reduce competition among different uses of biomass (Jong and Ree, 2009).

Biofuels for Air Travel

The aviation industry supports the efforts to reach a new post-Kyoto deal by ensuring a global commitment to fight climate change effectively, promoting research programs for renewable energy sources such as sustainable biofuels for aviation, and respecting the Chicago Convention (fair treatment of airlines). Aviation is responsible for 2% man-made CO_2 emissions and produces 8% of global gross domestic product (GDP). Over 40 years the focus on innovation has led to 70% reduced aviation fuel consumption and related CO_2 emissions.

The industry is committed to playing its part in reducing global GHG emissions and promotes a global sectoral approach with carbon reduction targets: 1.5% per annum improvement in fuel efficiency until 2020, carbon-neutral growth from 2020, and 50% reduction in net emissions by 2050. Sustainable alternative fuels play an increasing role in achieving these targets. In the United States and around the world, the industry is doing all it can to support sustainable biofuel development and maintain aviation's role in global economic growth. To make that happen, regional supply chains must be developed in addition to supportive government policies that encourage investment in the early stages of this emerging sector. Aviation has a relatively simple supply logistics with about 100 airports requiring over 80% of global jet fuel demand.

Government policies supporting advanced biofuels are essential to ensure that the aviation biofuel industry reaches its full potential and is able to compete against foreign petroleum. Aviation clearly needs a clean energy alternative to fossil fuel. Sustainable biofuels will enable airlines to fly cleaner, foster job growth in a new industry, and can insulate airlines from the volatile price swings of conventional fuel to help make air travel more economical. However, an adequate, affordable, and sustainable supply is needed. Some renewable jet fuel companies could be producing enough renewable fuel to replace 10–20% of the fuel of a typical mid-sized airline in the next 5 years in their current state (Bio Jet Fuel Blog, 2011).

A global issue needs a global solution for a globally operating industry. Holes in the current scheme may lead to unnecessary political and trade issues (retaliation, competition, trade conflicts, etc.). Aviation requires a global scheme developed through the International Civil Aviation Organization (ICAO). The industry supports market-based measures. Aviation biofuels are supposed to compete on equal basis as land transport. They also need to be competitive with current kerosene prices as airlines cannot sustain a premium for biofuels; however, the industry recognizes short-term price challenge. Carbon pricing and rising jet fuel price provide an opportunity. Biodiesel production poses a risk because it is more attractive than aviation biofuel. The industry has made progress in achieving technical fuel approval focusing on second-generation biofuels that avoid negative environmental impacts.

Certification is no longer an issue; technically feasible and certified fuels with no engine or aircraft modifications include maximum 50–50% blend for synthetic paraffinic kerosene (SPK) derived by Fischer–Tropsch process BtL (biomass to liquid) fuels and maximum 50–50% blend hydrotreated renewable jet (HRJ) fuels derived from hydrotreated plant oils. Drop-in fuels are fully compatible and interchangeable with JetA1. Other fuels are also in the pipeline of certification process. In 2011 the American Society for Testing and Materials

(ASTM), an international standards organization, gave the airlines the go-ahead to incorporate biofuels into as much as 50% of the total fuel they use on passenger flights. They certified advanced biofuels as meeting the ASTM international specification for bio-derived aviation fuels, "Hydroprocessed Esters and Fatty Acids" (HEFA) fuel (ICIS Green Chemicals, 2011).

Authoritative figures now call for biofuel resources to be targeted at aviation sector; however, current incentives are misaligned. In the EU Renewable Energy Directive (RED) transposition in several member states only includes incentives for ground transport. Biojet fuel suppliers should qualify for tradable certificates within incentive regimes provided for by national applications of the RED, such as Renewable Transport Fuel Certificate (RTFC) in the United Kingdom. This has been achieved in other EU countries and the United States. Exclusion of aviation prevents a level-playing field with road transport. Unlike aviation other sectors have alternative technologies to liquid fuel like (e.g., electric) and a significant timing advantage versus aircraft in engine technology adoption. In partnership with the Solena Group, British Airways plans to build Europe's first biojet fuel plant that will produce biojet fuel from all types of biomass and residue feedstock that will primarily be sourced from local waste management facilities (British Airways, 2010). The jet fuel plant will help power the fleet of British Airways from 2015 and help the company meet its goal to reduce net carbon emissions by 50% by 2050. Business case is driven around carbon pricing (EU Emissions Trading Scheme (ETS)) and landfill taxes up to US\$100 tonne^{-1} (Counsell, 2012).

The roadmap of the European Commission in 2011 gave a positive signal with clear milestones which targets an annual production of 2 MT of sustainably produced biofuel (4% of EU fuel consumption) for aviation by 2020. The European Commission and Airbus with leading European airlines and European biofuel producers have launched the "Biofuel Flightpath" initiative to try and speed up the commercialization of aviation biofuels in Europe. Priority is now for full-scale production and life-cycle assessment (LCA). Next stage is the biojet take-off, development of new feedstocks and logistics, and improvement of production efficiency. Authorities are in key position to introduce incentives, mandates, and emission trading.

Co-products of Biofuel Production

Maize used to produce ethanol also produces by-products such as distiller's dried grains with solubles (DDGS), maize gluten feed, maize gluten meal, and/or maize oil. Due to lack of production and global trade statistics data related to the co-products of biofuel production are imprecise. The US ethanol industry produced about 36 million metric tonnes of DDGS in the 2010/2011 marketing year. Nearly 25% of the US ethanol feed output is exported to countries around the world to feed livestock and poultry. Bioethanol is used to a far lesser extent in Europe, where about 4 MT of DDGS is produced a year. DDGS is still priced based on the price of maize.

Identifying maize demand for ethanol should take into account that DDGS is used as a high-value animal feed. DDGS and other ethanol feed products significantly reduce the need for maize and soybean meal in animal feed rations. DDGS produced by US ethanol plants have important implications for discussions regarding ethanol's impact on feed grains availability, feed prices, land-use effects, and the GHG impacts of producing maize ethanol (Linwood and Baker, 2011). Over the past several years, DDGS has been one of the most economically competitive sources of energy and protein available on the world feed market.

Feed market impacts of increased maize use for ethanol are smaller than that indicated by the total amount of maize used for ethanol production because of DDGS. By reporting only the gross usage of maize for ethanol, the implication was that all the maize going into ethanol production resulted in fuel ethanol. According to the conventional assumption ethanol producers return a full one-third of the maize processed back to the feeding sector which is the difference between the gross and net volumes of maize used for ethanol. In aggregate, a metric tonne of DDGS can replace, on average, 1.22 metric tonnes of feed consisting of maize and soybean meal in the United States. In fact, the amount of feed (maize and soybean meal) replaced by DDGS represents 38% (on a weight basis) of the maize used in the associated ethanol production process for a given crop year. One of the reasons that 1 tonne of DDGS can replace more than 1 tonne of conventional feed is that its energy and protein content are concentrated. Only the starch portion of the maize kernel is converted to ethanol, while the protein, fat, fiber, and other components are concentrated and passed through the process to DDGS. If coproducts are taken into account the net use of feedstocks declines.

More complicated, but no less important, is the impact of DDGS on LUC and GHG emissions associated with maize ethanol production. Most existing biofuel regulations significantly undervalue the contribution of DDGS when assessing the net GHG impacts of maize ethanol assuming that 1 metric tonne of DDGS replaces only 1 metric tonne of maize, with no substitution of soybean meal. The importance of DDGS is being undervalued by the regulatory agencies requiring a GHG assessment of ethanol. In the future accurate DDGS accounting is of increasing importance.

DDGS can be used to replace some soybean meal as well as maize in livestock and poultry rations. The digestive systems of ruminant animals (cattle, sheep, and goats) are well suited for using DDGS. For cattles, DDGS replaces maize or both maize and soybean meal up to around 40%. Digestive systems of hogs and poultry are less suited for feeding high levels of DDGS, although it can be used at up to 15–20% of their rations. About 90% of DDGS is used in the ruminant sector and only 10% in the hog and poultry sectors.

There is a limited demand for glycerin, the by-product of biodiesel production, for a number of food, beverage, personal care, and oral products, as well as pharmaceutical and other industrial uses. The mandated future levels of biodiesel use in the EU and the United States could create substantial excess supplies of glycerin for these markets. To deal with that potential problem and to create additional markets for the by-product of biodiesel, research in different countries has found that glycerin can be used effectively in livestock rations to replace fossil-based glycerin (Iowa State University, 2012).

Biofuel Production Costs

Estimated biofuel production costs show significant differences depending on factors such as scale of the plant, technology complexity, and feedstock costs. Little detailed data on advanced biofuel production costs are available, because such information is usually confidential and there is as yet no experience from large commercial-scale production plants. Costs vary by world regions, feedstock types, feedstock supply costs for conversion processes, the scale of bioenergy production, and production time during the year.

Conventional biofuels today are not generally competitive with fossil fuels at market prices,[4] although competitiveness varies depending on feedstock costs and oil price. For conventional biofuels today, the main cost factor is feedstock, which accounts for 60–90% of the total production costs. For ethanol production, feedstock costs represent about 60–80% of the total production cost while, for biodiesel from oilseeds, the proportion is 80–90% (author's calculation).

[4]In some cases this can be due to fossil fuel subsidies.

For advanced biofuels the main factor is capital costs, followed by feedstock with 25–40% share in the total production cost. In the long term, reduced feedstock cost volatility will be a vital advantage for advanced biofuels that use lignocellulosic biomass sourced from energy crops, waste, and residues. Handling and transport of biomass from production sites to conversion plants may contribute 20–50% of the total costs of bioenergy production. Factors such as scale increase and technological innovations increase competition and contribute to a decrease in economic and energy costs of supply chains by more than 50%. Densification via pelletization or briquetting is required for transportation distances over 50 km. Making use of co-products such as DDGS, glycerin, bagasse, lignin, or waste heat can reduce biofuel production costs by up to 20% depending on the fuel type and use of co-product[5] (IEA, 2009). Several important bioenergy systems today, most notably sugarcane-based ethanol and heat and power generation from residues and waste biomass, can be deployed competitively. Cultivation of algae and extraction of the oil is currently expensive and production cost estimates for the raw oil vary from US\$0.75 l^{-1} to more than US\$5 l^{-1} excluding costs for conversion to biofuel (USDOE, 2009; Darzins *et al.*, 2010).

Feedstock costs cannot be predicted and are subject to agricultural commodity prices, oil prices, and other factors. Enhancing feedstock flexibility will create access to a broader range of biomass sources with potentially low costs (such as residues) and reduced price volatility. Improving and creating transport infrastructure could further reduce biomass supply costs. The economics of conversion processes need to be further improved for biofuels to be competitive with fossil fuels without subsidies in the longer term (including CO_2 emission pricing). As a first step, fossil fuel subsidies, which are still applied in many countries (IEA, 2010b), should be phased out. Introducing a CO_2 price through a global carbon market will be an important element in fostering the deployment of biofuels and other low-carbon technologies in the longer term and would help considerably to improve their competitiveness.

In 2005/2007 the costs of ethanol production in new plants in Brazil were the lowest in the world at US\$0.20 l^{-1}. As compared to the cost of sugarcane-based ethanol in Brazil, ethanol costs from grains were 50% more in the United States and 100% more in the EU. Transportation and blending and distribution costs can add some US\$0.20 l^{-1} to the retail price. Meanwhile, production costs for ethanol (previously from wheat, but from sweet sorghum and cassava going forward) in China fluctuated between US\$0.28 and US\$0.46 l^{-1}, depending on the price of the feedstock, and sugar-based ethanol production costs in India were around US\$0.44 l^{-1} (IEA, 2009).

The cost of maize is by far the largest expense of producing fuel ethanol. Maize cost is today around 65–75% of the total cost of producing ethanol. Feedstock (sugarcane) purchases account for 60% of the total ethanol production costs in Brazil. Because of the competitive pricing of sugarcane Brazil's ethanol production costs are in general lower (UNICA, 2010).

Average Brazilian ethanol production costs dropped from a high of US\$0.47 l^{-1} in 1996 to average levels of US\$0.21 l^{-1} in 1998–2002 (UNICA 2010). Distillers' costs have since increased due to higher energy, fertilizer, and land prices, and they reached a new high of US\$0.48 l^{-1}in 2008. In 2011 the cost increased further and approached US\$0.60 l^{-1} in 2011. Tight ethanol balance resulted in volatile production margins averaging US\$0.20 l^{-1} ethanol production. In the United States the cost per liter of ethanol was around US\$0.60 fluctuating between US\$0.50 and US\$0.70 in 2011 with a volatile margin of up to US\$0.10 l^{-1}. In the

[5]In Brazil, for instance, sugar mills can sell bioelectricity produced from bagasse; the revenues represent around 15% of the total income of the unit.

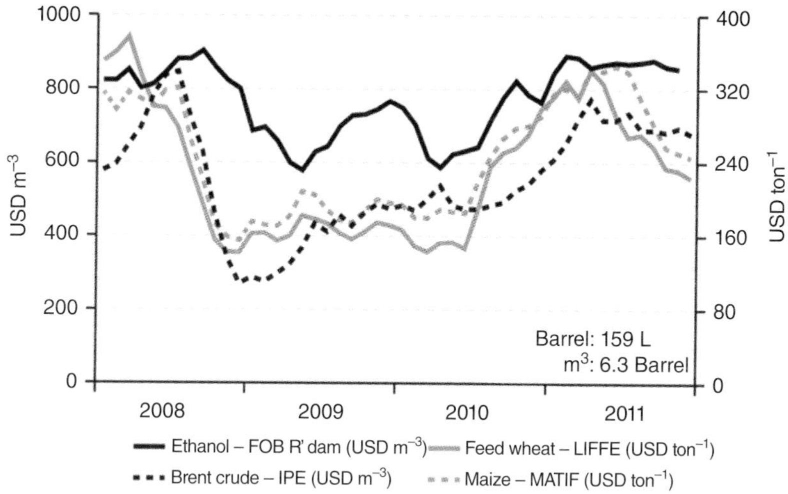

Figure 10.8. Prices of ethanol, crude oil, feed wheat, and maize in the European Union, January 2008 through November 2011 (*Source*: Kingsman, 2012).

EU ethanol production costs reached about US\$0.70 in 2011 with a margin of US\$0.10 l^{-1} (Licht, 2012).

Competitiveness of ethanol production depends on the relative prices of feedstock and fossil fuel (Figure 10.8). If ethanol prices stay strong, the increase in grain and sugar prices will limit ethanol profitability during the coming years. If ethanol prices drop, ethanol profitability may be pushed into the red. So, to absorb the feedstock price increase, ethanol prices will need to remain strong during the coming feedstock marketing year and possibly longer. However, bioethanol may become competitive in the United States and the EU when the oil price approaches US\$100 a barrel, while for Brazil the threshold is much lower. In the long term, technology improvement, higher conversion efficiencies, and better transport logistics could bring costs close to the gasoline reference, if costs for feedstocks remain stable.

The IEA foresees a reduction of one-third in the cost of ethanol by 2030 due to technological improvements and lower costs of feedstock. However, the increasing demand for ethanol due to mandates and targets, the impacts of the fuel versus food debate on its supply, and recent trends of feedstock prices imply that the cost of ethanol may not drop down. Moreover, unless the price of oil is high, production of ethanol may not be competitive without substantial level of subsidies. The cost of cellulosic ethanol, which is still in demonstration stage, is high, typically about 20% higher than conventional ethanol. Pre-treatment and enzymatic treatments are interrelated contributing to about 70% of the total processing costs. Millions of US dollars are spent on R&D by major corporations to optimize pre-treatment. Enzymes are still expensive because they are not tailored to different types of pre-treated biomasses. Enzymes shall be tailored by developing genetically modified ones with high biomass specificity.

Given the speed of technological developments in an emerging field and uncertainty over the long-run costs of feedstock, projections of the future costs of lignocellulosic ethanol differ substantially, but the costs are anticipated to drop to US\$0.50 l^{-1} in the long term. Another path to competitive lignocellulosic ethanol may come from the generation of valuable co-products in biorefinery, which could cut the costs of feedstock. Researchers estimate production costs

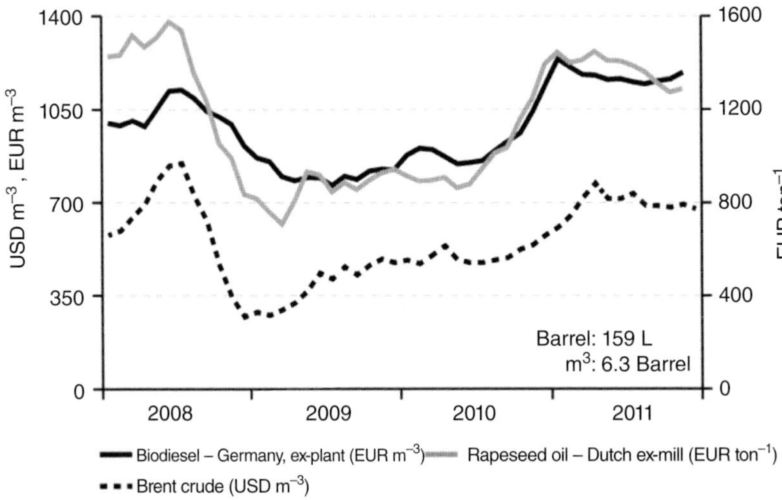

Figure 10.9. Prices of biodiesel, crude oil, and rapeseed oil in the European Union, January 2008 through October 2011 (*Source*: Kingsman, 2012).

of ethanol from the hydrolysis of cellulosic biomass to be US\$0.63 l^{-1} in the following 5–8 years, US\$0.37 l^{-1} in 8–12 years, and US\$0.25 l^{-1} in 13–20 years (IEA, 2010a).

Costs of producing biodiesel vary with the feedstock being used, the plant size, type and design details, when it was built, and how it is managed. By far, the most common feedstock is soybean oil, rapeseed oil, and palm oil. Used cooking oils and animal fats are also feedstocks of biodiesel production. For biodiesel, the feedstock cost has a 75–90% share in the total cost. The higher percentage of total cost reflects higher oilseed prices.

In 2011 the total cost of biodiesel production was US\$1.4 kg^{-1} with a net return overall cost of US\$0.15 kg^{-1} in the United States and US\$1.5 kg^{-1} with a net return overall cost of US\$0.04 kg^{-1} in the EU. The feedstock, soybean oil in the United States and rapeseed oil in the EU, had a share of 85% and 90% in the total production cost, respectively (Iowa State University, 2012; Licht, 2012)

Competitiveness of biodiesel production depends on relative prices of feedstocks and fossil fuels (Figure 10.9). Biodiesel refineries are typically less complex and less expensive to build than those for ethanol production. They also require much less heat and water than ethanol production. The profitability of biodiesel production is extremely variable. Due to the volatile price nature of biodiesel and oilseed oil, its major feedstock, biodiesel profitability can change rapidly from month to month. In addition, price variations of its co-product (glycerin) and its energy source (generally natural gas) add to the variability of biodiesel profits. Most of the cost of producing biodiesel has to be recovered through the price of biodiesel. It means that the main problem comes from low fuel prices.

The co-product of biodiesel production, glycerin, slightly reduces the cost of biodiesel Larger biodiesel plants likely have economics of size that reduce production, marketing, and transportation costs. From a marketing viewpoint, reductions in feedstock costs are very important in making biodiesel competitive with petroleum-based diesel fuel. There is a limited demand for glycerin for a number of food, beverage, personal care, and oral products, as well as pharmaceutical and other industrial uses. Based on demand estimates and estimated growth rates biodiesel production could create substantial excess supplies of glycerin for these markets. Research at Iowa State University and elsewhere has found that glycerin can be used effectively in livestock rations to replace fossil-based glycerin (Iowa State University, 2012). For all biofuels, there is scope for cost reductions that will help to improve competitiveness with fossil fuels and drive commercial deployment: capital costs are expected to come down as a result of scaling up (particularly for advanced biofuels). Co-location with existing biofuel plants, power plants, or other industrial facilities reduces capital costs and can bring further benefits such as more efficient use of by-products. Conversion costs can be brought down through scaling up and technology learning. Further improvement of conversion efficiency (e.g., through more efficient enzymes) and energy efficiency should also help to reduce costs. A wider variety of bio-based products can also be produced at biorefineries to enhance the economics of the overall conversion process.

Financing Advanced Biofuel

Experience of banks with first-generation biofuels shows amateurism, losses, bankruptcies, overcapacity of biodiesel, latent sustainability issues (food/fuel and land), and still a long dependence on policies. Most capacity expansion—and thus financing need—is expected for next-generation biofuels in the longer term (except from sugarcane-based ethanol in Brazil). Ultimately, these biofuels should be produced at lower costs than the current generation but feedstock and technology pose time and money-related barriers since the new supply chains, feedstock, and technology are unproven and investment capital expenditure is very high. The rollout of large-scale next-generation facilities will be a slow process. The key to unlock financing is control or co-operation in the supply chain in addition to lower costs.

General capital constraints make competition for financing from other renewable energy projects (e.g., wind farms) stronger. A strong and clear business case that eliminates or reduces cash flow uncertainties is needed. For example, wind energy often has the advantage of a fixed feed-in-tariff. Pre-requisite for long-term survival is a largely integrated supply chain via contracts, ownership, and agreements. Key success factors of any bioenergy project are logistics and location, price risk management, feedstock supply (easy and assured access), off take (easy and assured contracts), capacity utilization (benchmark is 75%), experienced management, and compliance with sustainability requirements.

The annual value of renewable energy capacity installed will double in real terms to US$395 billion in 2020, rising to US$460 billion in 2030, compared with US$195 billion in 2010—according to analysis company Bloomberg New Energy Finance (BNEF, 2011). Spending on new renewable energy capacity will total US$7 trillion over the next 20 years. The solar and wind sector will continue to expand with a combined share of 70% in total money spent on renewable energy projects but biofuel is projected to reach a share of just 8% or US$510 billion in total spending. Banks are cautious to lend money which means that more sources of capital are needed. Strong competition from other renewable energy projects with lower (perceived) risks (specifically wind) can be experienced. Fuels should be taxed directly proportional to their energy content since competition balances supply and demand. Market prices including CO_2 costs allocate resources most efficiently.

Transport Policies

Passenger vehicle fleet will double to 1.7 billion in 2035. Global car fleet will continue to surge as more and more people in China and other emerging economies buy a car. Alternative technologies, such as hybrid and electric vehicles that use oil more efficiently or not at all, continue to advance but they take time to penetrate markets. Advanced vehicles, which represent 70% of new car sales by 2035, make a big contribution to emissions abatement, underpinned by a dramatic decarbonization of the power sector (IEA, 2011).

Common policies include biofuel subsidies, tax exemptions, or blending mandates. Blending mandates, targets, fuel-tax exemptions, and production subsidies exist in around 50 countries. City and local governments around the world continue to enact policies to reduce GHG emissions and promote renewable energy. In 2010, local governments received official recognition for the first time in international climate negotiations, where they are now designated as "governmental stakeholders." Almost all cities working to promote renewable energy at the local level have established some type of renewable energy or CO_2 emissions reduction target. There are several types of renewable energy-specific targets (IPCC, 2011).

The US renewable fuels standard (RFS) requires fuel distributors to increase the annual volume of biofuels with a specific quota for advanced biofuels blended to 36 billion gallons (136 billion liters) by 2022. The EU is targeting 10% of transport energy from renewables by 2020. The biofuel target refers to road and rail transport but electricity in all transport. Biofuel in aviation and shipping should be included in the biofuel target even if it is not included in the legislation of the member states. Since 1976 the government in Brazil made it mandatory to blend anhydrous ethanol with gasoline, fluctuating between 10 and 25%. In 2011 the mandatory blend of 25% was reduced to 20% (on volumetric basis) due to recurring ethanol supply shortages and high prices that take place between harvest seasons. China targets the equivalent of 13 billion liters of ethanol and 2.3 billion liters of biodiesel per year by 2020 (IPCC, 2011).

To drive development of biofuels that provide considerable emission savings and at the same time are socially and environmentally acceptable, support measures need to be based on the sustainable performance of biofuels. Recent years have also seen increased attention to biofuel sustainability and environmental standards (Licht, 2011). Another approach is to directly link financial support to life-cycle CO_2 emission reductions (calculated with a standard life-cycle analysis methodology agreed on internationally) to support those biofuels that perform best in terms of CO_2 savings. Neither specific advanced biofuel quota nor performance-based support measures on their own seem to be effective to address the higher production costs of advanced biofuels in the short term. Specific transitional measures may thus be needed to support the introduction of the new technologies. Financial incentives, for instance, a tax incentive or perhaps analogous to feed-in tariffs for electricity, could be coupled to the use of co-products such as waste heat to promote efficient use of by-products.

A key requirement for all biofuels to get access to the market will be compliance with international fuel quality standards. This will ensure vehicle and infrastructure compatibility among different regions and promote consumer acceptance for new fuels. End-use infrastructure requirements need also to be addressed to avoid bottlenecks caused by incompatibility with deployed biofuels. The ethanol "blending wall"—the limiting of ethanol in gasoline to 10–15% because of vehicle compatibility constraints—is one example of potential infrastructure bottlenecks that need to be addressed. Evolution of fuel specifications and new fuel grades are taken into account in the development of future vehicles, such as compatibility of vehicles in the fleet with higher biofuel blends or new limits for existing specifications. Backward compatibility of fuel changes is a very difficult issue because it is extremely difficult to cover all the vehicle generations and models combined with reliability risks for the customers and a risk for vehicle manufacturers in meeting legal commitments (CO_2 emissions) and furthermore it is costly. Automotive manufacturers need sufficient protection for the existing fleet at any point in time and a sufficient lead-time and clear fuel specifications for the future. At least 5 years lead-time should enable car industry to adopt new fuel standards. Electric vehicles get much attention and incentives but they still face many barriers. They seem to be viable for light vehicles and short distances.

Introduction of flex-fuel vehicles (FFV) and high-level ethanol blends is a suitable measure to avoid infrastructure incompatibility issues for ethanol, as has been successfully demonstrated in Brazil, the United States, and the EU. Introduced in the market in 2003, flex vehicles became a commercial success in Brazil, reaching a 95% share of all new cars and light vehicle sales today. Most of the cars on the road in the United States can run on blends of up to 15% ethanol, and the use of 10 and 15% ethanol gasoline is mandated in several US states and cities. Well over 90% of US gasoline is blended with ethanol. In the EU member states (Germany and France) the biofuel "blending wall" has been increased up to 10%. Policy measures may be required, such as obligations for retailers to provide high-level biofuel blends (e.g., E85) or tax incentives for FFVs. Ford was the first manufacturer offering in 2001 FFVs in Europe and began to develop market also beyond Sweden. In contrast to Brazil (and Sweden) there are no significant incentives for customers to buy FFVs because the production costs of ethanol exceed gasoline costs. The reason for that is primarily the different feedstock used in Europe versus Brazil. The market of FFVs will remain a niche without substantial and stable net fuel price benefits. The price premium of FFVs ranges between €100 and €300 on consumer price in Germany (IPCC. 2011).

Conventional bioethanol is the most common biofuel and used in blends like E10, E15 (EU/US), E20-25, E100 (Brazil), and E85 (EU/US). Currently approved biofuel blends in the EU are B7, E5, and E10. In blends of ethanol and petrol mainly E5 and/or ethyl-tertiary-butyl-ether (ETBE) are used. High ethanol blends are used in adopted vehicles for niche markets (E85 in FFVs and ED95 for buses). Blend rates of up to 15% of Bio-ETBE is allowed with no material compatibility issues like ethanol. ETBE is not used as a neat fuel but as a gasoline blend component (octane booster). Furthermore, ETBE is more expensive than ethanol and poses a sustainability issue since only 44% of ETBE carbon is of biogenic origin. Ethanol fuel ED95 (95.6% hydrous ethanol and 4.4% fuel additives) is a proven technology and a stable fuel with a long shelf life replacing diesel in heavy transports. In 2010 Sweden opened the first public filling station for ethanol fuel ED95 in Jordbro, south of Stockholm. With currently approved ethanol blends a saturation of the market will occur before the RED target can be fulfilled in 2020. The exact date of blend wall depends on petrol consumption and FFV sales going up from 0.5% of fleet today up to 1% in 2020. However, it takes a long way to go to have E10 across the EU27. Focus on proper implementation of E10 is a precondition for any additional future steps because a high share of noncompatible vehicles in the fleet reduces the acceptance of biofuels (see German E10 introduction). E10+ and B7+ require new standards because new parameters and adaptions are required.

The challenge is the limited backwards fleet capability. In 2008, the German government proposed to increase both biodiesel and bioethanol blend limit from 5% v/v each to 10% v/v. However, after incompatibility concerns regarding B10 biofuel were raised in 2009 by vehicle manufacturers, biodiesel blend limit 7% v/v was agreed upon. The complete vehicle stock in Germany is B7 capable. In 2011 E10 was introduced but only approximately 93% of vehicle stock is E10 capable. Vehicle compatibility lists needed to be issued because protection grade fuel (E5) for 7% of vehicle owners was required. While B7 introduction was easy, E10 introduction is still a nightmare. Customers are unsettled about the right fuel for their car. Furthermore, customers widely boycott E10 leading to much less consumption of E10 than expected. The lesson learned from the experience in Germany is that widespread blending of usual market fuel with biocomponents ideally requires 100% backwards compatibility with vehicle stock. Situations where there is less than 100% backwards compatibility with existing vehicles necessitates the creation of compatibility lists, which is politically a very complicated process. Communication is the key in managing changes. On the other hand blending can lead to new test method demands. Lack of capable test methods during fuel introduction can lead to severe fuel quality issues.

Global Trade in Biomass and Bioenergy

Biomass and biofuel markets have globalized over the last decades but are still immature and face tariffs and nontariff trade barriers leading to radical and mostly undesired shifts in international trade flows. These are aggravated by the fact that international trade flows in bioenergy markets are still relatively low compared to fossil fuel markets. Interactions between national policies, import tariffs, and nontariff trade barriers are also complicated by the cross-links of bioenergy markets to food- or industrial commodities markets. Trade will become increasingly important to promote biofuel production and meet blending mandates, as well as to balance demand and supply fluctuations among different regions.

In general, the most favored nation (MFN) tariffs range from roughly 6 to 50% on an *ad valorem* equivalent basis in the countries of the Organisation for Economic Co-operation and Development (OECD) and up to 186% in the case of India (Steenblik, 2007). The US ethanol import tariff on fuel ethanol (US\$0.54 gal^{-1}) and federal tax credit of US\$0.45 gal^{-1} of blended ethanol and the tax credit for biodiesel blending amounting to US\$1 gal^{-1} expired on 31 December 2011. The new tariff legislation means that fuel ethanol imports now pay the usual 2.5% *ad valorem* import duty (fob value) for undenatured ethanol for nonbeverage purposes under the MFN clause (general rate 20%). Denatured ethanol imports pay 1.9% (20%). Fuel ethanol imports from the Caribbean Basin Initiative (CBI) origin remain completely exempt from all import tariffs. The annual quota always corresponds to 7% vol. of the US fuel ethanol use during the October/September period in the prior season. In the EU the normal MFN trade is subjected to specific import duties of €0.102 l^{-1} on denatured alcohol and €0.192 l^{-1} on undenatured alcohol. Several ethanol exporting countries benefit from EU trade preferences including the Generalized System of Preferences, the Cotonou Agreement, and the Everything But Arms Initiative (Licht, 2009). Biodiesel used to be subject to lower import tariffs than bioethanol, ranging from 0% in Switzerland to 6.5% in the EU and the United States. EU imports of biodiesel are subject to an *ad valorem* duty of 6.5%. No tariff is applied to oilseed in the EU and tariffs on vegetable oils are 0–16%. Imports to the EU of these feedstocks therefore are attractive (for example, imports of palm oil primarily from Malaysia).

However, in the EU import tariffs for biofuels again became an issue, comparable to B99.9 in the years 2007/2009. First, the EU extended the special import duties on US biodiesel imports of B20 and above to lower blends (mainly B19) and to shipments from Canada. The B19 trade volume never reached the extent of B99.9. Second, the EU changed the import duty structures on E90. This blend mainly consists of the US fuel ethanol which has already been subsidized with the subsequently expired blender's credit and additionally paid a lower *ad valorem* duty of 6.5% due to an advantageous custom classification. The product was therefore offered on the EU market at a very competitive price. In autumn 2011, the European Commission reclassified ethanol blends above 70% vol. as denatured alcohol subject to tariff at €0.102 l^{-1} (Licht, 2012).

International trade in biomass and biofuels has become much more important in the last decade but it has decreased over recent years, with roughly 5% of biofuels (fuel ethanol and biodiesel only) traded internationally and one-third of all pellet production for energy use in 2010. Data related to fuel bioethanol trade are imprecise on account of the various potential end uses of ethanol (i.e., fuel, industrial, and beverage use) and also because of the lack of proper codes for biofuels in global trade statistics. Furthermore, nowadays, increasing volumes of ethanol are traded not as pure product but as blends (E90) or ETBE. As an estimate, 3% of global fuel ethanol (2.2 billion liters) production was traded in 2010. International biodiesel trade was around 15% (2.4 MT) of global production in 2010 (Lamers *et al.*, 2011; Licht, 2011).

After several years as a net importer, the United States became a net exporter in 2010, sending a record 1.2 billion liters of fuel ethanol overseas. At the same time Brazil's ethanol exports decreased to 0.8 billion liters. In addition to increased sugar production and exports, in response to higher sugar prices, growing production of ethanol in the United States and the EU contributed to Brazil's loss of global ethanol market share. The world's most important destination for ethanol shipment is the EU, Canada, and the United States. Around 2.4 MT of biodiesel were traded globally in 2010. Argentina is the world's largest biodiesel exporter followed by Indonesia and the United States. The EU remains the leading importer of biodiesel (Licht, 2011).

The development of international markets for bioenergy has become an essential driver to develop available biomass resources and market potential, which are currently underutilized in many world regions. In the short term, trade will include conventional biofuels and feedstocks, but after 2020, lignocellulosic feedstock trade is likely to grow rapidly and supply large advanced biofuel plants in coastal locations. The infrastructure to handle woody resources (both forest residues and plantation-grown wood) already exists in the pulp and paper industry and can be easily used for the biofuel industry. Pelletization, pyrolysis, or torrefaction will become increasingly important since they increase the energy density and thus tradability of lignocellulosic feedstocks (e.g., residues). These intermediate products are relatively homogeneous and thus more suitable for conversion to biofuels. Scale and efficiency improvements will reduce biofuel production costs over time.

Environmental Impact of Land-Use Change for Biofuel

About 84% of current CO_2 emissions are energy-related and about 65% of all GHG emissions can be attributed to energy supply and energy use. All sectors (buildings, transport, industry, and other) will need to reduce dramatically their CO_2 intensity if global CO_2 emissions are to be decreased by 50–85% below 2000 levels by 2050. Energy-related carbon dioxide (CO_2) emissions in 2010 are estimated to have climbed to a record 30.6 gigatonnes (Gt) and concentrations have continued to grow to over 390 parts per million (ppm) CO_2 or 39% above pre-industrial levels. The Cancun Agreements call for limiting global average temperature rises to no more than 2 °C above pre-industrial values. In order to be confident of achieving an equilibrium temperature increase of only 2° to –2.4 °C, atmospheric GHG concentrations would need to be stabilized in the range of 445–490 ppm CO_2 equivalent in the atmosphere. Scientists warn that if the current trend to build high carbon-generating infrastructures continues, the world's carbon budget will be swallowed up by 2017, leaving the planet more vulnerable than ever to the effects of irreversible climate change. The establishment of the required new energy technologies and associated infrastructure will in itself lead to GHG emissions, implying that a portion of the "emission space" allowed within the GHG target will need to be "invested" for energy system transformation (IEA, 2010b).

The transport sector is currently responsible for 23% (10 Gt CO_2 equivalent) of energy-related CO_2 emissions. To achieve the projected target of 50% reduction in energy-related CO_2 emissions by 2050 from 2005 levels sustainably produced biofuel production must provide 27% of the total transport fuel. Reductions in transport emissions contribute considerably to achieving overall targets. India and China show significant increases because of rapidly growing vehicle fleets. Vehicle efficiency improvements account for one-third of emissions reduction in the transport sector; the use of biofuels is the second largest contributor, together with electrification of the fleet accounting for 20% (2.1 Gt CO_2 equivalent) of emissions saving (IEA, 2010b).

Bioenergy's contribution to climate change mitigation needs to reflect a balance between near-term GHG targets and the long-term objective to hold the increase in global temperature below 2 ^oC. Bioenergy has significant potential to mitigate GHGs if resources are sustainably developed and efficient technologies are applied. The impacts and performance of biomass production and use are region- and site-specific. Most current bioenergy systems, including liquid biofuels, result in GHG emission reductions, and advanced biofuels could provide higher GHG mitigation. The GHG balance may be affected by LUCs and corresponding emissions and removals. The possibility of using bioenergy in combination with carbon capture and storage (CCS) is now being actively considered. The idea behind CCS is that capturing the CO_2 emitted during bioenergy generation and injecting it into a long-term geological storage formation could turn "carbon-neutral" emissions into negative emissions (Kraxner *et al.*, 2010). One CCS demonstration project started operation in Illinois in the beginning of 2010 (MGSC, 2010).

The role of bioenergy systems in reducing GHG emissions needs to be evaluated by comparison with the energy systems they replace using LCA methodology. The precise quantification of GHG savings for specific systems is often hampered by lack of reliable data. Furthermore, different methods of quantification lead to variation in estimates of GHG savings. Nonetheless practically all bioenergy systems deliver large GHG savings if they replace fossil-based energy and if the bioenergy production emissions—including those arising due to LUC—are kept low. Currently available values indicate a high GHG mitigation potential of 60–120%,[6] similar to the 70–110% mitigation level of sugarcane ethanol and better than most current biofuels (IEA Bioenergy, 2009). However, these values do not include the impact of LUC[7] that can have considerable negative impact on the life-cycle emissions of advanced biofuels and also negatively impact biodiversity.

To ensure sustainable production of advanced biofuels, it is therefore important to assess and minimize potential indirect LUC caused by the cultivation of dedicated energy crops. This deserves a careful mapping and planning of land use, in order to identify which areas (if any) can be potentially used for bioenergy crops. Brazil is the only emerging country that has initiated the agroecological sugarcane zoning program (ZAE Cana) to direct available land to the production of biofuel feedstock in order to stop deforestation and iLUC. The program constrains the areas in which sugarcane production can be expanded by increasing cattle density, without the need to convert new land to pasture. The program is enforced by limiting access to development funds for sugarcane growers and sugar mill/ethanol plant owners that do not comply with the regulations. The program currently focuses on sugarcane, but it could also be applied to other biofuel feedstocks

[6] An improvement higher than 100% is possible because of the benefits of co-products (notably power and heat).

[7] Two types of LUC exist: Direct LUC occurs when biofuel feedstocks replace native forest, for example; iLUC occurs when biofuel feedstocks replace other crops that are then grown on land with high carbon stocks.

Biomass for energy is only one option for land use among others, and markets for bioenergy feedstocks and agricultural commodities are closely linked. Thus, LUC effects which are "indirect" to bioenergy are "direct" effects of changes in agriculture (food, feed) and forestry (fiber, wood products). They can be dealt with only within an overall framework of sustainable land use and in the context of overall food and fiber policies and respective markets. The direct LUC effects of bioenergy production can, in principle, be controlled through certification systems, wherever biomass is grown. The risks of LUC and resulting emissions can be minimized by focusing on wastes and residues as feedstock; maximizing land-use efficiency by sustainably increasing productivity and intensity and choosing high-yielding feedstocks; using perennial energy crops, particularly on unproductive or low-carbon soils; maximizing the efficiency of feedstock use in the conversion processes; cascade utilization of biomass, that is, linking industrial and subsequent energetic use of biomass; and co-production of energy and food crops.

Changes in land use, principally those associated with deforestation and expansion of agricultural production for food, contribute about 15% of global emissions of GHG. Currently, less than 3% of global agricultural land is used for cultivating biofuel crops and LUC associated with bioenergy represents only around 1% of the total emissions caused by LUC globally most of which are produced by changes in land use for food and fodder production or other reasons (Berndes *et al.*, 2010). iLUCs, however, are more difficult to identify and model explicitly in GHG balances. Most current biofuel production systems have significant reductions in GHG emissions relative to the fossil fuels displaced, if no iLUC effects are considered.

The bioethanol share in total grains demand, that is, maize, wheat, and other coarse grains, in 2010 was 8%, or 143 MT. By adding the feed value of ethanol by-product dried distillers' grains and soluable (DDGS), the net shares decline by one-third to slightly above 5%. The bulk of the worldwide use of grains in alcohol production comprises maize in the United States and China. However, an increase in the offtake of wheat for fuel ethanol can also be observed in Canada and the EU. The fuel ethanol sector, mainly in the United States, accounted for 16% (net 11%) of global maize consumption and 20% of global sugarcane production. The biodiesel share in rapeseed, soybean, and palm oil demand was around 10% of global vegetable oil production. The share of waste biodiesel feedstocks such as animal fat and used cooking oil increased to 15% in total biodiesel output in 2010 (Licht, 2011).

In 2010 about 20 million gross hectares of grains, sugarcane, and cassava for fuel ethanol production and 20 million gross hectares of oilseed feedstock were needed for biodiesel production. A total of 40 million gross hectares was used for biofuels. The proportion of global cropland used for biofuels is currently some 2.5% with wide differences among countries and regions. In the United States some 8% of cropland is dedicated to biofuel production; however, 20% of maize and soybean area is used for biofuel production. In the EU 5–6% of cropland is used for biofuel but 25% of biofuel feedstock or biofuel is imported. In Brazil biofuel just requires 3% (ethanol 1.5%) of all cropland (including pastureland) available in the country even if more than 50% of sugarcane area (20% of global area) is used for ethanol production (author's calculation).

The fuel production processes give rise to by-products which are largely suitable as animal feed. By-products are supposed to be credited with the area of cropland required to produce the amount of feed they substitute. In the cases of grains and oilseeds, DDGS and CGF/CGM (maize gluten feed/meal) and oil cakes (mainly rapeseed and soybean cake/meal) substitute grain and soybean as feed. It means that not all the grains used for ethanol production should be subtracted from the supplies since some 35% is returned to the feed sector in the form of by-products (mainly DDGS) so the land required for feedstock production declines to 15 Mha. In case of biodiesel production 50–60% of rapeseed (rapeseed cake/meal) and 80% of soybean

(soybean meal) are returned to the feed sector and the net land requirement decreases to around 6 Mha. By adding by-products substituted for maize and soybean meal the net hectares needed for fuel ethanol decline to 15 million and for biodiesel to 6 million (author's calculation). By adding by-products substituted for grains and oilseeds the land required for cultivation of feedstocks declines to less than 1.5% of the global crop area (net land requirement).

Based on the land-use efficiencies land use for biofuel production would need to increase from 40 Mha (21 Mha net land requirement by adding by-products substituted for grains and oilseeds) to around 100 Mha in 2050. This corresponds to an increase from 2.5% of total gross arable land today to around 6% in 2050. This expansion would include some cropland, as well as pastures and currently unused land, the latter in particular for production of lignocellulosic biomass (IEA, 2010b; author's calculation).

Social Impact of Biofuel Production

Social impacts associated with large expansions in bioenergy production are very complex and difficult to quantify. The demand for biofuels represents one driver of demand growth in the agricultural and forestry sectors and therefore contributes to global food price increases. Even considering the benefit of increased prices to poor farmers, higher food prices adversely affect poverty levels, food security, and malnourishment of children. On the other hand, biofuels can also provide opportunities for developing countries to make progress in rural development and agricultural growth, especially when this growth is economically sustainable. In addition, expenditures on imported fossil fuels can be reduced. However, whether such benefits end up with rural farmers depends largely on the way production chains are organized and how land use is governed.

In general, bioenergy options have a much larger positive impact on job creation in rural areas than other energy sources. Bioenergy generates more jobs per unit of energy delivered than other energy sources, largely due to feedstock production, especially in developing countries and rural areas (FAO, 2011d). Wage income is a key contribution to the livelihoods of many poor rural dwellers (Ivanic and Martin, 2008). The benefits from bioenergy jobs depend on the relative labor intensity of the feedstock crop compared to the crop that was previously grown on the same land. For example, cultivation of perennial energy crops requires less labor than cereal crop cultivation, and this displacement effect should be taken into account (Thornley *et al.*, 2009).

Worldwide, jobs in renewable energy industries exceeded 3.5 million in 2010. In addition to manufacturing, many of these jobs are in installations, operations, and maintenance, as well as in biofuel feedstocks. Jobs are expected to grow apace with industry and market growth, although increasing automation of manufacturing and economies of scale in installation services may moderate the rate of jobs growth below that of market growth. Estimates of the employment creation potential of bioenergy options differ substantially, but liquid biofuels based on traditional agricultural crops seem to provide the most employment, especially when the biofuel conversion plants are small (Berndes and Hansson, 2007). For ethanol, the number of direct and indirect jobs generated ranges from 45 (maize) to 2200 (sugarcane) jobs PJ^{-1} of ethanol. For biodiesel, the number of direct and indirect jobs generated ranges from 100 (soybean) to 2000 (oil palm) jobs PJ^{-1} of biodiesel (Dias de Moraes, 2007; Clayton *et al.*, 2010). In Europe, Berndes and Hansson (2007) estimate that biomass production for energy can create employment at a magnitude that is significant relative to total agricultural employment (up to 15% in selected countries) but small compared to the total industrial employment in a country.

References

Berndes, G., Bird, N., and Cowie, A. (2010) *Bioenergy, Land-Use Change and Climate Change Mitigation*, IEA, Paris. Available at: http://www.ieabioenergy.com (accessed 10 December 2011).

Berndes, G. and Hansson, J. (2007) Bioenergy expansion in the EU: cost-effective climate change mitigation, employment creation and reduced dependency on imported fuels. *Energy Policy* 35: 5965–5979.

Bio Jet Fuel Blog (2011) Bio jet fuel and renewable diesel: 21st century solutions for the energy markets. Available at: http://biojetfuelblog.com/category/biojet-fuel-airline/ (accessed 28 January 2012).

BNEF (2011) Global *Renewable Energy Market Outlook*, Bloomberg New Energy Finance. Available at: https://www.bnef.com/PressReleases/view/173 (accessed 28 January 2012).

Braun, von J. and Meinzen-Dick, R. (2009) *Land Grabbing by Foreign Investors in Developing Countries: Risks and Opportunities*. Policy Brief 13, International Food Policy Research Institute, Washington, DC.

British Airways (2010) British Airways to build Europe's first biojet fuel plant. Available at: http://www.environmentalleader.com/2010/02/15/british-airways-to-build-europes-first-biojet-fuel-plant/ (accessed 30 January 2012).

Chakravorty, U., Hubert, M-H., and Nostbakken, L. (2009) Fuel versus food. *Annual Review of Resource Economics* 1: 645–663.

Clayton, R., McDougall, G., Perry, M., Doyle, D., Doyle, J., and O'Connor, D. (2010) A study of employment opportunities from biofuel production in APEC economies. APEC#210-RE-01.9, APEC Energy Working Group, Asia-Pacific Economic Cooperation (APEC), Singapore, 82p.

Counsell, J. (2012) Commercial use of aviation biofuels. British Airways, UK. Presented at 2nd Annual Global Biofuels Summit, 26 January 2012, Barcelona, Spain.

Darzins, Al., Pienkos, P., and Edye, L. (2010) Current status and potential for algal biofuels production. Report T39-02. IEA Bioenergy Task 39, 128p.

Dias de Moraes, M.A.F. (2007) Indicadores do mercado de trabalho do sistema agroindustrial da cana-de-acucar do Brasil no periodo 1992–2005. *Estudos Econômicos (São Paulo)* 37: 875–902.

FAO (2009) Proceedings of the expert meeting on how to feed the world in 2050. High-Level Expert Forum on "How to Feed the World in 2050," 12–13 October 2009, FAO, Rome. Available at: http://www.fao.org/wsfs/forum2050/wsfs-background-documents/wsfs-expert-papers/en/ (accessed 9 December 2011).

FAO (2011a) *Looking Ahead in World Food and Agriculture: Perspectives to 2050* (ed. P. Conforti), Agricultural Development Economics Division, Economic and Social Development Department, Food and Agriculture Organization of the United Nations, Paris, 539p. (ISBN 978-92-5-106903-5). Available at: http://www.fao.org/docrep/014/i2280e/i2280e.pdf (accessed 19 December 2011).

FAO (2011b) *World Livestock 2011—Livestock in Food Security*, FAO, Rome.

FAO (2011c) *The State of the World's Land and Water Resources for Food and Agriculture*. Summary report, FAO, Rome. Available at: http://www.fao.org/nr/water/docs/Solaw_ex_summ_web_en.pdf (accessed 22 December 2011).

FAO (2011d) *Energy-Smart Food for People and Climate*. Issue paper, FAO, Rome. Available at: http://www.fao.org/docrep/014/i2454e/i2454e00.pdf (accessed 18 December 2011).

FAOSTAT (2011) *FAOSTAT*, FAO, Rome. Available at: http://www.faostat.fao.org/default.aspx (accessed 28 December 2011).

Fischer, G., Hizsnyik, E., Prieler, S., Shah, M., and van Velthuizen, H. (2009) *Biofuels and Food Security*, The OPEC Fund for International Development (OFID) and International Institute of Applied Systems Analysis (IIASA), Vienna.

Gustavsson, J., Cederberg, C., Sonesson, U., van Otterdijk, R., and Meybeck, A. (2011) *Global Food Losses and Food Wastes—Extent, Causes and Prevention*, FAO, Rome. Available at: http://www.fao.org/fileadmin/user_upload/ags/publications/GFL_web.pdf (accessed 18 December 2011).

Haberl, H., Heinz Erb, K., Krausmann, F., *et al.* (2007) Quantifying and mapping the human appropriation of net primary production in earth's terrestrial ecosystems. *Proceedings of the National Academy of Sciences* 104: 12942–12947.

Hoekstra, A.Y., Bulsink, F., and Booij, M.J. (2010) The water footprint of bio-energy, in *Climate Change and Water: International Perspectives on Mitigation and Adaptation* (eds C.J. Howe, B. Smith, and J. Henderson), American Water Works Association, IWA Publishing, London, pp. 81–95.

ICIS Green Chemicals (2011) Bio-jet fuel momentum soaring high. International Construction Information Society (ICIS). Available at: http://greenchemicalsblog.blogspot.hu/2011_12_01_archive.html (accessed 19 December 2011).

IEA Bioenergy (2009) *A Sustainable and Reliable Energy Source*. Main Report, International Energy Agency, Paris.

IEA (2010a) *Sustainable Production of Second-Generation Biofuels. Potential and Perspectives in Major Economies and Developing Countries*, OECD/IEA, Paris. www.iea.org/papers/2010/second_generation_biofuels.pdf (accessed 19 December 2011).

IEA (2010b) *Energy Technology Perspectives 2010. Scenarios and Strategies to 2050*, OECD/IEA, Paris.

IEA (2011) Are we entering a golden age of gas? Special report. International Energy Agency, Paris. Available at: http://www.worldenergyoutlook.org/media/weowebsite/2011/WEO2011_GoldenAgeofGasReport.pdf (accessed 19 December 2011).

Ivanic, M. and Martin, W. (2008) Food prices and food security. Available at: http://www.voxeu.org/article/food-prices-and-food-security-don-t-blame-liberalisation (accessed 15 January 2012).

Iowa State University (2012) Agriculture Marketing Resource Center, biodiesel profitability. Available at: http://www.extension.iastate.edu/agdm/energy/xls/agmrcbiodieselprofitability (accessed 22 January 2012).

IPCC (2011) *Special Report on Renewable Energy and Climate Change Mitigation*, Intergovernmental Panel on Climate Change, Potsdam. Available at: http://srren.ipcc-wg3.de/report/IPCC_SRREN_Full_Report.pdf (accessed 12 December 2011).

IWMI (2007) *Water for Food, Water for Life: A Comprehensive Assessment of Water Management in Agriculture*, Earthscan, London; International Water Management Institute, Colombo.

Jong, E. and van Ree, R. (2009) Biorefineries: adding value to the sustainable utilisation of biomass. Available at: http://www.ieabioenergy.com/LibItem.aspx?id=6420 (accessed 10 December 2011).

Kampman, B., Bergsma, G., Schepers, B., *et al.* (2010) *BUBE: Better Use of Biomass for Energy*. Background Report to the Position Paper of IEA RETD and IEA Bioenergy, CE Delft/Öko-Institut, Darmstadt.

Kingsman, S.A. (2012) *Kingsman Biofuels Database*, Kingsman SA, Home Grown Cereal Association (H.G.C.A.), Switzerland. Available at: http://www.kingsman.biz/WD150AWP/WD150Awp.exe/CONNECT/Commodities (accessed 29 January 2012).

Krausmann, F., Erb, K.-H., Gingrich, S., Lauk, C., and Haberl, H. (2008) Global patterns of socioeconomic biomass flows in the year 2000: a comprehensive assessment of supply, consumption and constraints. *Ecological Economics* 65: 471–487.

Kraxner, F., Leduc, S., Aoki, K., Kindermann, G., and Obersteiner, M. (2010) Bioenergy use for negative emissions—potentials for carbon capture and storage (BECCS) from a global forest model combined with optimized siting and scaling of bioenergy plants in Europe. Working paper presented at the First International Workshop on Biomass and Carbon Capture and Storage, 14–15 October 2010, University of Orléans, France.

Krugman, P. (2009) Is a new architecture required for financing food and environmental security? Summary of the speech made during the launching event of the Second Forum for the Future of Agriculture, Brussels. Available at: http://www.elo.org (accessed 15 April 2009).

Lamers, P., CarloHamelinck, C., Junginger, M., and Faaij, A. (2011) International bioenergy, trade—a review of past developments in the liquid biofuels market. *Renewable and Sustainable Energy Reviews* 15: 2655–2676.

Licht, F.O. (2009) *World Ethanol and Biofuel Report (Jan.–Dec.)*, Agra Informa, London.

Licht, F.O. (2011) *World Ethanol and Biofuel Report (Jan.–Dec.)*, Agra Informa, London.

Licht, F.O. (2012) *World Ethanol and Biofuel Report, Vol. 10, No. 9, 13 January)*, Agra Informa, London.

Linwood, A. and Baker, A. (2011) *Estimating the Substitution of Distillers' Grains for Corn and Soybean Meal in the U.S. Feed Complex*, USDA/ERS, Washington, DC. Available at: http://www.ers.usda.gov/Publications/FDS/2011/09Sep/FDS11I01/FDS11I01.pdf (accessed 22 January 2012).

MGSC (2010) Illinois Basin—Decatur project moves forward. Groundbreaking project will help determine the future of geologic carbon sequestration (Midwest Geological Sequestration Consortium (MGSC). Available at: http://www.sequestration.org (accessed 11 December 2011).

Oerke, E.C. (2006) Crop losses to pests. *Journal of Agricultural Science* 144: 31–43.

Smeets, E.M.W., Faaij, A.P.C., Iris, M., Lewandowski, I.M., and Turkenburg, W.C. (2007) A bottom-up assessment and review of global bioenergy potentials to 2050. *Energy and Combustion Science* 33: 56–106.

Steenblik, R. (2007) Subsidies: the distorted economics of biofuels. Discussion Paper No. 2007-3. International Transport Forum, Organisation for Economic Co-operation and Development, Geneva, Switzerland, 66p.

The Earth Institute (2005) *The Growing Urbanization of the World*, Columbia University, New York.

Thornley, P., Upham, P., Huang, Y., Rezvani, S., Brammer, J., and Rogers, J. (2009) Integrated assessment of bioelectricity technology options. *Energy Policy* 37: 890–903.

UNICA (2010) Quotes and stats, Brazilian Sugarcane Industry Association. Available at: http://www.portalunica.com.br/portalunica (accessed 11 December 2011).

USDOE (2009) *National Algal Biofuels Technology Roadmap. U.S. Department of Energy Biomass Program*, U.S. Department of Energy, Washington, DC.

van Iersel, S., Gamba, L., Rossi, A., *et al.* (2009) *Algae-Based Biofuels: A Review of Challenges and Opportunities for Developing Countries,* FAO, Rome.

Yacobucci, B.D. (2012) Biofuels incentives: a summary of federal programs, 16p., Congressional Research Service. Available at: www.crs.gov (accessed 10 October 2012).

Chapter 11
Biofuel Crops, Ecosystem Services, and Biodiversity

A. Fieldsend[1] and H.P. Singh[2]

[1]*Research Institute of Agricultural Economics, Budapest, Hungary*
[2]*Fort Valley State University, Fort Valley, GA, USA*

Introduction

An ecosystem is a dynamic complex of plant, animal, and microorganism communities and the nonliving environment interacting as a functional unit. Humans are an integral part of ecosystems. The Millennium Ecosystem Assessment (MA) (2005) is an international synthesis report by over 1000 of the world's leading biological scientists that analyzes the state of the Earth's ecosystems and provides summaries and guidelines for decision makers. The report concludes that human activity is having a significant and escalating impact on the biodiversity of world ecosystems, reducing both their resilience and biocapacity. It refers to natural systems as humanity's "life-support system," providing essential "ecosystem services." The MA deals with the full range of ecosystems, from those relatively undisturbed, such as natural forests, to agricultural land and urban areas.

Ecosystem services are defined as the benefits people obtain from ecosystems (MA, 2003). These include provisioning, regulating and cultural services, which directly affect people, and supporting services needed to maintain the other services. Provisioning services are the products people obtain from ecosystems such as food, fuel, fiber, fresh water, and genetic resources. Regulating services are the benefits people obtain from the regulation of ecosystem processes, including air quality maintenance, climate regulation, erosion control, regulation of human diseases, and water purification. Cultural services are the nonmaterial benefits people obtain from ecosystems through spiritual enrichment, cognitive development, reflection, recreation, and aesthetic experiences. Supporting services are those that are necessary for the production of all other ecosystem services, such as primary production, production of oxygen, and soil formation. Changes in these services affect human well-being through impacts on security, the basic material for a good life, health, and social and cultural relations.

The MA measured 24 ecosystem services and concluded that only four have shown improvement over the last 50 years, 15 were in serious decline, and 5 were in a stable state overall, but under threat in some parts of the world (Table 11.1).

Biofuel Crop Sustainability, First Edition. Edited by Bharat P. Singh.

Table 11.1. Global status of provisioning, regulating, and cultural ecosystem services evaluated in the Millennium Ecosystem Assessment. Status indicates whether the condition of the service globally has been enhanced (if the productive capacity of the service has been increased, for example) or degraded in the recent past. Definitions of "enhanced" and "degraded" are provided in the note below. The fourth category, supporting services, is not included here as they are not used directly by people.

Service	Sub-category	Status	Notes
	Provisioning Services		
Food	Crops	▲	Substantial production increase
	Livestock	▲	Substantial production increase
	Capture fisheries	▼	Declining production due to overharvest
	Aquaculture	▲	Substantial production increase
	Wild foods	▼	Declining production
Fiber	Timber	+/−	Forest loss in some regions, growth in others
	Cotton, hemp, silk	+/−	Declining production of some fibers, growth in others
	Wood fuel	▼	Declining production
Genetic resources		▼	Lost through extinction and crop genetic resource loss
Biochemicals, natural medicines, pharmaceuticals		▼	Lost through extinction, overharvest
Fresh water		▼	Unsustainable use for drinking, industry, and irrigation; amount of hydroenergy unchanged, but dams increase ability to use that energy
	Regulating Services		
Air quality regulation		▼	Decline in ability of atmosphere to cleanse itself
Climate regulation	Global	▲	Net source of carbon sequestration since mid-century
	Regional and local	▼	Preponderance of negative impacts
Water regulation		+/−	Varies depending on ecosystem change and location
Erosion regulation		▼	Increased soil degradation
Water purification and waste treatment		▼	Declining water quality
Disease regulation		+/−	Varies depending on ecosystem change
Pest regulation		▼	Natural control degraded through pesticide use
Pollination		▼[a]	Apparent global decline in abundance of pollinators
Natural hazard regulation		▼	Loss of natural buffers (wetlands, mangroves)
	Cultural Services		
Spiritual and religious values		▼	Rapid decline in sacred groves and species
Aesthetic values		▼	Decline in quantity and quality of natural lands
Recreation and ecotourism		+/−	More areas accessible but many degraded

Adapted from Millennium Ecosystem Assessment (2005).
For provisioning services, enhancement is defined to mean increased production of the service through changes in area over which the service is provided (e.g., spread of agriculture) or increased production per unit area. The production is judged to be degraded if the current use exceeds sustainable levels. For regulating and supporting services, enhancement refers to a change in the service that leads to greater benefits for people (e.g., the service of disease regulation could be improved by eradication of a vector known to transmit a disease to people). Degradation of regulating and supporting services means a reduction in the benefits obtained from the service, either through a change in the service (e.g., mangrove loss reducing the storm protection benefits of an ecosystem) or through human pressures on the service exceeding its limits (e.g., excessive pollution exceeding the capability of ecosystems to maintain water quality). For cultural services, enhancement refers to a change in the ecosystem features that increase the cultural (recreational, aesthetic, spiritual, etc.) benefits provided by the ecosystem.
[a]Indicates low to medium certainty. All other trends are medium to high certainty.

The conceptual framework developed by the MA places human well-being as the central focus for assessment, while recognizing that biodiversity and ecosystems also have intrinsic value and that people take decisions concerning ecosystems based on considerations of well-being as well as intrinsic value (MA, 2003). The status of ecosystem services is affected by "drivers of change." A "driver" can be any factor that changes an aspect of an ecosystem. A direct driver unequivocally influences ecosystem processes and can therefore be identified and measured to differing degrees of accuracy. Direct drivers are primarily physical, chemical, and biological, such as land-use and land-cover change, climate change, air and water pollution, irrigation, fertilizer use, harvesting, the introduction of alien invasive species, and natural drivers such as volcanoes and evolution. An indirect driver operates more diffusely, often by altering one or more direct drivers, and its influence is established by understanding its effect on a direct driver. Examples include demographics, economics (e.g., per capita income, trade and policy), sociopolitical factors (e.g., governance), science and technology, and culture and religion (such as choices about what and how much to consume and what to value).

Estimates suggest that humans are now using or dominating between 39% and 50% of the Earth's terrestrial biological production (Vitousek *et al.*, 1997). As the main activity concerned is agriculture, direct drivers associated with agriculture have a major impact on ecosystem services. Agriculture and ecosystem services are interrelated in at least three ways: firstly, agroecosystems generate beneficial ecosystem services such as soil retention, food production, and aesthetics; secondly, agroecosystems receive beneficial ecosystem services from other ecosystems such as pollination from nonagricultural ecosystems; and thirdly, ecosystem services from nonagricultural systems may be impacted by agricultural practices (Dale and Polasky, 2007).

Within agriculture, biofuel production has become a significant activity. Thus it is reasonable to conclude that clear interrelationships can be observed between biofuel production and ecosystem services that are direct, indirect, or complex. This chapter reviews the evidence in the literature regarding the drivers, impacts, and trade-offs involved in biofuel production with regard to ecosystem services. These impacts can have social, ecological, and economic dimensions. By discussing options that can be developed to enhance the long-term sustainability of biofuels (by mitigating their impact on ecosystem services and biodiversity), the chapter offers ideas for accentuating the value of services of positive value and controlling those of negative value.

The MA Framework as Applied to Biofuels

Using the ecosystem services approach when discussing the trade-offs involved in biofuel production has several advantages. Firstly, the ecosystem services concept explicitly bridges ecosystem impact and human well-being, which are two key components of the biofuel debate. Hence this framework can help to highlight the many values of "natural capital," that is, the natural resource stocks from which resource flows and services (e.g., nutrient cycling, erosion protection) useful for livelihoods are derived (DFID, 1999), as well as the social costs and benefits associated with biofuel expansion. Moreover, the ecosystem services concept has gained popularity in the academic community (Fisher *et al.*, 2009) and, most importantly, has gained understanding and high-level acceptance among policy makers. Hence the MA framework is in a unique position to become a standardized analytical framework for sustainability analysis of biofuels, agreeable and understood by multiple stakeholders.

This section presents how biofuel production can affect provisioning, regulating, and cultural services. It does not address the effect of biofuels on supporting services as the available evidence is still very limited (Stromberg *et al.*, 2010). The evidence presented encompasses the impacts of several feedstocks (e.g., maize, sugarcane, and oil palm) in several areas around the world.

Biofuels and Provisioning Services

The major provisioning services of food, feed, fiber, and fuel are accompanied by a number of others such as pharmaceuticals (Fieldsend, 1996). Biofuel crops currently attract particular attention owing to their current proliferation and the sheer pace and scale of new investment. Under several scenarios (i.e., plausible alternative futures, each an example of what might happen under particular assumptions) considered by Alcamo *et al.* (2005), by 2050 global cereal consumption will increase by a factor of 1.5–1.7 (depending on the scenario), fish consumption (up to the 2020s) by a factor of 1.3–1.4, water withdrawals for households, industry, and other uses by a factor of 1.2–1.8, and biofuel production by a factor of 5.1–11.3.

The term "biofuel" refers to energy derived from recently living plants, animals, or their by-products, thus it is self-evident that biofuels are a source of fuel. For thousands of years humans have used traditional biofuels such as wood, charcoal, and dung to cook, heat, and manufacture goods, and traditional biofuels continue to provide 70–90% of all energy in many African nations and in rural parts of Asia (Kgathi and Zhou, 1995; Ramachandra *et al.*, 2004). In the early twentieth century, the so-called "modern" biofuels began to be used for transport, and now the global development of the biofuel sector is proceeding rapidly, driven by national policy mandates, government subsidies, and profit opportunities for farmers, agribusiness, and energy companies (Milder *et al.*, 2008). Traditional forms of biofuel such as fuelwood still represent a major share of the global energy system, but the MA scenarios assume that they will be gradually replaced by other energy carriers (Alcamo *et al.*, 2005). The substitution of traditional biofuels with modern ones also means that wood, dung, and crop residues that were once burned can now be retained in agroecosystems to reduce erosion and enrich the soil (Milder *et al.*, 2008).

Biofuels can contribute to energy security at the national, local, and community levels. The volatility of world oil prices, uneven global distribution of oil supplies (75% in the Middle East), uncompetitive structures governing the oil supply (i.e., the OPEC cartel), and a heavy dependence on imported fuels leave oil importing countries vulnerable to supply disruption (Dufey, 2006). Recent interruptions in oil supply from Russia to Ukraine because of political disagreements acutely illustrate this vulnerability. Biofuels are often seen as part of a strategy to diversify energy sources to reduce supply risks (de Fraiture *et al.*, 2008). The local use of liquid biofuels across the developing world may increase access to reliable energy (especially through electricity generation), thus spurring rural development and improving quality of life (UN, 2007).

Food crops are the most commonly used feedstocks for "first generation," "modern" biofuel. They can be staple crops (e.g., maize, wheat) or key vegetable oils (e.g., palm oil), while others such as sugarcane and soybeans are important inputs to the food industry. For example, more than 90% of biofuels produced and used in the United States currently come from maize (Groom *et al.*, 2008). Consequently, it has been suggested (Stromberg *et al.*, 2010) that greater biofuel production will compete with food (and feed and fiber) production both directly (e.g., food crops diverted for biofuel production) and indirectly (e.g., competition for land and agricultural labor). Brown (2006) attributes maize price increases in the United States to increased demand owing to new biofuel plants. Other subsequent studies have, however, found that the increases in food prices were primarily due to many other interacting factors: increased demand in emerging economies, soaring energy prices, drought in food-exporting countries, cutoffs in grain exports by major suppliers, market-distorting subsidies, a tumbling US dollar, and speculation in commodity markets. Although ethanol production indeed contributes to higher maize prices, it is not a major factor in world food costs (Dale *et al.*, 2010).

de Fraiture *et al.* (2008) estimate the amount of land devoted to biofuel crop production to be 11–12 Mha, around 1% of the total area under crops. In Brazil, the biggest bioethanol producer, 2.5 Mha (5% of the cropped land) is used for biofuel production, with a production rate of ethanol of 6200 l ha^{-1}, mostly from sugarcane. The United States, the second biggest ethanol producer, allots nearly 4 Mha to biofuel crops (4% of the total cropped area), with yields of roughly 3300 l ha^{-1}, mostly from maize. In Europe, where biodiesel is the main product made from rapeseed, 1 Mha is used, yielding on average 1700 l ha^{-1} of biodiesel. China is now becoming a major player in biofuel production, ranking among the world's top three ethanol producers. In 2002, it produced 3.6 billion liters of bioethanol, of which 76% was derived from maize (China News, AFP, 2006).

Maize is primarily used to feed animals to produce meat and milk. Globally, 65% of all maize is used to feed animals; in the United States, it reaches 75% (de Fraiture *et al.*, 2008). The production of 1 kg of meat can require between 3 and 10 kg of grain, and during the past 40 years, global per capita meat production has increased more than 60%, a trend driven by increasing global per capita incomes (Tilman *et al.*, 2002). Many of the elevated environmental impacts projected for the coming 50 years are tied to increased consumption of livestock products and concomitant greater demand for grains fed to livestock. de Fraiture *et al.* (2008) suggest that the "unfolding global conflict over food," a phrase coined by Brown (2006), may not be between cars and the poor as he envisions, but rather between cars and carnivores.

Freshwater ecosystems provide society with the essential services of water supply for its sustenance, economic activity and recreation, as well as habitat for its freshwater fishery. The production of biomass is a consumptive use of water that may compete directly with food crop production for water as well as land resources (Berndes, 2002) and water scarcity, already a limiting factor in food production in many regions (CA, 2007), may become even more so as a consequence of climate change. For example, agricultural production in the Middle East and North Africa may be at higher risk because climate change decreases the water available for crop irrigation (Alcamo *et al.*, 2005). Biofuel crops such as sugarcane can be water intensive and are often produced under monoculture, leading to increased water scarcity and water pollution. If all national policies and plans for biofuels are successfully implemented, it is estimated that 180 km^3 of additional irrigation water withdrawals annually will be needed (de Fraiture *et al.*, 2008). In addition, processing the harvested crop into biofuel can create high water demands that must be considered in the life-cycle analysis of biofuel production (Dale *et al.*, 2010).

The impact of biofuel production on water supply is, however, likely to differ between regions. From a water perspective it makes a large difference whether biofuel is derived from fully irrigated sugarcane grown in semi-arid areas or rain-fed maize grown in water-abundant regions. In Europe, where rain-fed rapeseed is used, the amount of irrigation for biofuel crops is negligible (de Fraiture *et al.*, 2008). In the United States, where mainly rain-fed maize is used, only 3% of all irrigation withdrawals are devoted to biofuel crop production. In Brazil, where the main biofuel crop (sugarcane) is mostly grown under rain-fed conditions, very little irrigation water is used for ethanol production. By contrast, it is unlikely that fast-growing economies such as China and India will be able to meet future food, feed, and biofuel demand without substantially aggravating already existing water scarcity problems or importing grain (de Fraiture *et al.*, 2008). Irrigation plays a dominant role in China's food production. An estimated 75% of total grain production, 90% of vegetables, and 80% of cotton comes from irrigated areas. Similarly irrigation plays a major role in India's food supply. At present some 63% of cereal production originates from irrigated areas. Close to 85% of the area under sugarcane, the crop currently most used in bioethanol, is irrigated. While both countries already face regional and seasonal water shortages, to meet anticipated biofuel demand China needs to produce 26% more maize and India 16% more sugarcane above the baseline (i.e., for food and feed only) scenario levels.

Biofuels and Regulating Services

Amongst regulating services biofuels are most closely linked with climate regulation. Coupled with effective energy conservation measures increased biofuel use has the potential to slow the rate of global climate change (Groom *et al.*, 2008). Through the efficient use of biofuels in place of fossil fuels (coal, oil, etc.) greenhouse gas (GHG) emissions can be significantly reduced (IEA, 2004) and the increase in concentration of atmospheric carbon dioxide, a major driver of climate change, can consequently be mitigated (Gibbs *et al.*, 2008). For a long time, biofuels were touted as the "green alternative" to petroleum-based gasoline.

Realizing this potential, however, depends on how and where these fuels are produced (Hill *et al.*, 2006; Alder *et al.*, 2007) and some studies indicate that biofuel production generates more GHG than it saves in burning (Pimentel, 2003). The sustainability of any biofuel feedstock depends on good growing practices and sound environmental practices throughout the fuel production life cycle including the production, manufacture, and distribution stages (Gibbs *et al.*, 2008). Groom *et al.* (2008) consider that, in terms of carbon balance, maize-based ethanol is the worst among the alternatives that are available at present. GHG emissions are estimated to be 81–85 kg CO_2 MJ^{-1} compared to 94 kg CO_2 equivalent MJ^{-1} for gasoline. By contrast, results of research on ethanol derived from prairie grasses (Tilman *et al.*, 2006) and from cellulosic woody biomass (Perlack *et al.*, 2005) suggest that these energy crops are at least CO_2 neutral. Biodiesel from soy also generally outperforms maize-derived ethanol (Turner *et al.*, 2007).

An extra dimension to this issue is the question of indirect land-use change caused by expansion in biofuel production areas. Alongside the substitution of biomass fuels for fossil fuels the other primary land-based activity that would help to minimize atmospheric carbon dioxide is carbon storage in the terrestrial biosphere (Huston and Marland, 2003). Natural, undisturbed forests "sequester" carbon (defined by IPCC (2000) as an increase in carbon stocks other than in the atmosphere) and represent a large amount of carbon storage. Similarly, undisturbed soils can accumulate a large amount of carbon over time. Disturbance of these landscapes can result in rapid release of large amounts of carbon that will be recaptured only slowly as the forest regrows or the soil rebuilds (Harmon *et al.*, 1990). According to Fargione *et al.* (2008), if this first-time conversion of natural land is attributed to biofuels, anywhere from 17 to 420 times more carbon dioxide could be released than the annual GHG reductions that these fuels would provide from displacing fossil fuels. Searchinger *et al.* (2008) concluded that indirect land-use change effects associated with an increased use of maize-based ethanol could potentially double GHG emissions in the next 30 years. Depending on the system two to nine times more carbon could be sequestered by conserving or restoring native ecosystems than by producing biofuels on the same land (Righelato and Spracklen, 2007; Fargione *et al.*, 2008).

These conclusions, and the economic modeling approach to land-use change that was used, have been widely disputed (e.g., Kim and Dale, 2011). Several researchers have asserted that adequate land is available for energy crops. For example, assuming no changes in feedstock and conversion efficiency, de Fraiture *et al.* (2008) estimate that biofuels will require 30 million additional hectares of cropped area—too small to lead to major changes in agricultural systems at a global level. Regional difference will occur, however. According to Alcamo *et al.* (2005), in Organization for Economic Cooperation and Development (OECD) countries and former Soviet Union countries the area of agricultural land is likely to decrease because of stabilizing population, saturating per capita food demands, and increasing crop yields due to technological developments. The largest expansion of agricultural land will occur in developing countries because of high population growth and low agricultural investments, leading to slower improvements in crop yield. Agricultural land in Sub-Saharan Africa could increase from approximately 11 to 17 Mha between 2000 and 2050, causing a major shift in ecosystem services from natural grassland and forest to agricultural land (Alcamo *et al.*, 2005).

The "carbon debt controversy" is by no means resolved, and Gibbs *et al.* (2008) provide some quite terrifying estimates of carbon payback times for crop-based biofuel expansion in the tropics. Their results show that even if biofuels are compared to carbon emissions from more "carbon costly" oil sands, the story remains much the same: several decades to centuries are still required to replace lost forest carbon stocks, although converting degraded lands will quickly provide carbon savings. Danielsen *et al.* (2008) calculated that it would take 75–93 years for an oil palm plantation to compensate the carbon lost during the conversion from forest, depending on the forest clearing method used. Building refineries, increasing truck traffic, and changing the economic dynamics of rural areas all constitute land-use changes that will have additional environmental impacts beyond the conversion of land to biofuel feedstock production alone (Dale *et al.*, 2010). It should be noted however that transportation costs tend to concentrate sources of feedstock production close to processing facilities, and this may reduce the environmental footprint of biofuel crops.

Impacts on other regulating services must mostly be attributed to any general expansion of agricultural production rather than directly to biofuel production. Agricultural effects on air pollution include pesticides, odors, smoke, dust, allergenic pollens, and trash (Dale and Polasky, 2007). Oil palm expansion was partially responsible for the devastating 1997–1998 forest fires in Indonesia where satellite imagery showed oil palm companies initially starting the fire to clear land (Dennis *et al.*, 2005).

Regarding water quality, where agricultural practices are intensified, whether due to increased biofuel production or other reasons, the side effects of this intensification (contamination of groundwater, transport of nitrogen to coastal zones) could limit the ecosystem services provided by groundwater, surface waters, and the coastal zone (fishery, recreation). The most common causes of water body impairment are sediments, pathogens, nutrients, metals, dissolved oxygen, and other habitat alteration, and agricultural practices contribute to all (Dale and Polasky, 2007). In particular, increased maize acreage for biofuel production has raised concerns regarding fertilizer and pesticide pollution. Maize is grown with some of the highest fertilizer and pesticide inputs of any major crop in the United States and the highest inputs per hectare of any biofuel crop (NASS, 2007). Similarly, oil palm plantations consume much larger amounts of fertilizers by far than any other crop in Malaysia (FAO, 2004).

The increasingly frequent occurrences of water stress owing to climate change are likely to exacerbate the increase in water pollution loadings to surface and groundwater, and the resulting risk of water contamination and reduced habitat of aquatic ecosystems. The unprecedented increases in return flows resulting from additional water use, especially in Latin America and Sub-Saharan Africa, will lead to large-scale degradation of water quality and interference with the delivery of freshwater services unless these return flows are subjected to wastewater treatment (Alcamo *et al.*, 2005). Further down the production chain, biofuel production has been identified as a major source of water pollution. Waste products and by-products of the industrial processing of sugarcane and palm oil into ethanol and crude palm oil, respectively, can have severe environmental impacts if released without proper treatment. Palm oil mill effluent and vinasse from sugarcane distillation are rich in organic matter and contribute to eutrophication and depletion of dissolved oxygen levels in aquatic systems if left untreated (Donald, 2004; Martinelli and Filoso, 2008).

Thus, in choosing what to grow and where to grow it, available water quantity and crops' effects on water quality emerge as critical limiting factors. The issue of indirect land-use change is relevant here, too. Intact forests can minimize flooding by slowing snowmelt and water discharge, moderate regional climate, and remove and store atmospheric carbon dioxide. Forest and grassland ecosystems can create or regenerate fertile soils, degrade plant litter and animal wastes, and purify water, and this regenerative process is essential for subsistence slash-and-burn farming systems (Nye and Greenland, 1960). The recharge of streams and aquifers by intact ecosystems provides potable water for little more expense than the cost of its extraction. Dale *et al.*, (2010) suggest that there is broad consensus that dedicated bioenergy crops should be grown where irrigation is not required.

The overall capability of soil ecosystems to control soil erosion is viewed as a key regulating ecosystem service. The scenarios of Alcamo *et al.* (2005) indicate that, in Sub-Saharan Africa, the area under high risk of water erosion could approximately double between 1995 and 2050 because of a net increase in precipitation, the widespread replacement of natural vegetation, and the expansion of agriculture onto terrain susceptible to water erosion. Substantial but less extreme increases in the area under risk occur in Latin America and Asia. Mature oil palm plantations in Malaysia have a soil erosion rate of approximately 7.7–14 tonnes ha^{-1} yr^{-1} (Hartemink, 2007) and the problem can be even more serious in the early years before a complete palm canopy has been established. The risk of erosion is magnified by soil tillage and other mechanical disturbances during crop production. On the other hand, it can be minimized by conservation measures such as contour plowing and terracing.

It has been suggested that urban wastes and leftovers or residues from industrial processing and agricultural crops should also be used as biofuel feedstocks (Dale *et al.*, 2010). For example, maize stover includes the stalks, leaves, and empty cobs left from maize production and can be used as feedstocks. However, the environmental implications of residue collection are extremely site specific: soil degradation can occur or increase when residues are collected. There is concern that removal of even half of the stover would exacerbate loss of soil carbon and erosion and would also require additional inputs of fertilizers to replace lost minerals (Lal, 2005). Furthermore, some researchers have voiced concern that birds that regularly use crop residues for food resources may be displaced if these residues are harvested (Dale *et al.*, 2010). By contrast, recent studies of the effects of regulated residue removal on sugarcane cultivation on soil carbon indicate that such practices were sustainable (CGEE, 2008).

Regarding biocontrol services, diverse, small-scale agricultural landscapes with a high proportion of noncrop habitats frequently support a greater abundance of natural enemies and lower pest populations than large-scale, monoculture landscapes with little noncrop habitat (Bianchi *et al.*, 2006). Landis *et al.* (2008) calculated a biocontrol services index (BSI), defined as the proportional decrease in aphid population growth in the presence of natural enemies. They found that BSI declined significantly with increasing proportion of maize in the landscape but BSI was not significantly related to the proportion of soybean. Thus exchanging maize for soybean does not have a neutral effect on biocontrol services. Rather, as maize area increases in the local landscape, biological control services to soybean decline. This analysis suggests that expanded maize in the landscape will reduce biocontrol services and increase reliance on pesticides. By contrast, development of cellulosic ethanol-processing capabilities that can use a variety of feedstocks such as switchgrass, mixed prairie, and woody biomass, create the potential to diversify agricultural landscapes and support multiple ecosystem services (Landis *et al.*, 2008), contributing to flowering plant diversity and supporting a variety of pollinator and natural enemy arthropods.

Biofuels and Cultural Services

Ecosystems provide a range of benefits that are difficult to quantify, and have rarely been priced. Amongst these are cultural services. MA (2003) identifies two paradigms of values associated with ecosystems. One paradigm, known as the utilitarian (anthropocentric) concept, is based on the principle of humans' preference satisfaction (welfare). In this case ecosystems and the services they provide have value to human societies because people derive utility from their use, either directly or indirectly (use values). Within this utilitarian concept of value, people also give value to ecosystem services that they are not currently using (nonuse values). Nonuse values, usually known as "existence value," involve the case where humans ascribe value to knowing that a resource exists even if they never use that resource directly. These often involve the deeply held historical, national, ethical, religious, and spiritual values people ascribe to ecosystems: the values that the MA recognizes as cultural services.

A different, nonutilitarian value paradigm holds that something can have intrinsic value (that is, it can be of value in and for itself) irrespective of its utility for someone else. Thus, for example, villages in India protect "spirit sanctuaries" in relatively natural states, even though a strict cost–benefit calculation might favor their conversion to agriculture. Similarly, many countries have passed laws protecting endangered species based on the view that these species have a right to exist, even if their protection results in net economic costs (MA, 2003).

With regard to biomass crops, making decisions on where, how much, and how the crop is produced requires sensitivity to societal priorities as well as to the relationships among site productivity, the economics of extractive land uses, and biodiversity (Huston and Marland, 2003). Clearly one ecosystem service may be gained at the expense of another, and the demand for ecosystem services is now so great that trade-offs among services have become the rule. A country can increase food or biofuel supply by converting a forest to agriculture, but in so doing it decreases the supply of services that may be of equal or greater intrinsic importance, but of less obvious economic value, such as clean water, timber, ecotourism destinations, or flood regulation and drought control (MA, 2003). Thus, for example, expansion of oil palm plantations comes at the expense of natural habitats in Malaysia (Anon., 2007).

Agriculture, which is arguably the single most dominant influence on the landscape, strongly influences the contribution of landscape to the economic, social, and cultural development of a territory. This contribution, as a specific element of ecosystem services, can be termed "landscape services." The European Landscape Convention defines "landscape" as "an area, as perceived by people, whose character is the result of the action and interaction of natural and/or human factors" (Council of Europe, 2000). Landscape is therefore both a physical entity and a social, cultural, and political representation. There is a potential conflict between the demands of modern farming methods and the aesthetic values of a society that expects agriculture to maintain cherished patterns of land use and distinctive landscape features (ARC, 2010).

In the United Kingdom, the rapidly expanding production area of oilseed rape as a biofuel feedstock has led Blythman (2007) to observe that "the bright yellow tide has upset lovers of traditional country views." The author of this chapter can remark that this is not a new phenomenon. The production area of oilseed rape in the United Kingdom increased from 3857 ha in 1980 to 342 758 ha in 1990 (http://www.ukagriculture.com/crops/oil_seed_rape.cfm) and similar opinions were being expressed even at the beginning of that decade. Milder *et al.* (2008) recommend that, to advance biofuels for sustainable development, investment must shift to include a variety of ecoagriculture-compatible pathways, namely (a) the conservation of native biodiversity and ecosystem services, (b) sustainable production of biofuel feedstocks, and (c) viable livelihoods for local people. Key to this concept is a participatory approach to planning, managing, and supporting agricultural landscapes that engages diverse local and external stakeholders.

The valuation of the cultural benefits provided by agricultural landscapes can be especially difficult (Swinton *et al.*, 2007); similarly, few attempts have been made to quantify the wider economic impacts of environmental land management (and attractive agricultural landscapes) on the local economy (ENRD, 2010), such as via recreation and tourism. A rare example, reported in Winter and Rushbrook (2003), estimated that 3.7 million (79%) of all annual holiday trips to Devon, United Kingdom were motivated by the "conserved landscape," defined as fields, wood, moorland, villages, and coastline. These trips were estimated to generate GBP 749 million, and support a total of 23 900 full time equivalent jobs, of which 16 000 were supported directly by landscape-motivated holiday trips.

Impacts on Biodiversity

Biodiversity is a general term that refers to the variety of living organisms from all sources, including terrestrial, marine, and other aquatic ecosystems, and the ecological complexes of which they are part, found in a particular location. This location might be a study plot, a region or the entire globe. The most common measure of biodiversity is the number of species in an area (Huston and Marland, 2003). Biodiversity and ecosystems are closely related concepts. Diversity is a structural feature of ecosystems and the variability among ecosystems is an element of biodiversity. In addition to the important role of biodiversity in providing ecosystem services, the diversity of living species has intrinsic value independent of any human concern (MA, 2003).

Biodiversity tends to vary with ecosystem productivity (Huston and Marland, 2003). In contrast to the diversity of large animals, which tends to be highest at high productivity levels, the diversity of plants generally declines at high levels of productivity. A widely observed pattern of plant diversity is the unimodal or humpbacked pattern, first described by Grime (1973a, b). The diversity of small animals, in contrast to that of large animals, tends to follow the basic pattern of plant diversity, with a maximum on relatively less productive lands (Huston and Gilbert, 1996). This includes most insects and many birds.

Several organizations have published information regarding global important areas critical to biodiversity preservation. Conservation International has published by-region biodiversity hotspots. These hotspots hold high numbers of endemic species, yet their combined area of remaining habitat covers only 2.3% of the Earth's land surface. The World Wildlife Fund has published "Global 200" ecoregions whose conservation would achieve the goal of saving a broad diversity of the Earth's ecosystems. These ecoregions include areas with exceptional levels of biodiversity in species richness or endemism, or containing unusual ecological or evolutionary phenomena. However, biodiversity remains more of a concept lacking precise definition. All biodiversity initiatives usually incorporate objectives to protect areas of high number of diverse species as well as species endangered or vulnerable. Real problems surface when trying to demarcate areas of high biodiversity importance and limit their accessibility to human exploitation for agriculture and industry. Furthermore, valuable biodiversity ecosystems can be of primary nature as they provide permanent inhabitation to diverse living forms or secondary, such as site of aggregation of migratory birds.

Biodiversity in a land community is dynamic but in equilibrium. However, this equilibrium is disturbed if the existing land-use pattern is altered. MA (2005) lists six main direct drivers of biodiversity loss: habitat loss, overexploitation, invasive alien species, disease, pollution, and climate change, as well as the interactions between them. Biofuel production has been linked to four of these drivers, namely habitat destruction, invasive alien species, pollution, and climate change. The disruption in biodiversity can be at several levels, namely a reduction in the natural variability within a species assembled through the evolutionary process, elimination of specific species from an ecosystem, and replacing the entire spectrum of diverse vegetative and animal habitation of an ecosystem with an alternative and narrow pool of organisms. Practices that change species composition or reduce biodiversity in nonagricultural systems may also diminish goods and services because the ability of ecosystems to provide some services depends both on the number and type of species in an ecosystem (Tilman *et al.*, 2002). Groom *et al.* (2008) express concern that the global loss of habitat for increased food production will be magnified by increasingly large areas being cleared to meet the demand for biofuels, converting biodiverse lands into monocultures.

Furthermore, significantly increasing the scale for replacement of fossil fuel with biomass fuel or substitution of wood products for more energy-intensive alternatives would require relatively frequent harvests on high productivity lands (Huston and Marland, 2003). Depending on economic incentives, this practice may compete with agricultural land uses and/or push more intensive land uses toward the less productive lands that are currently the primary reservoirs for the remaining biodiversity. Without thoughtful planning, conversion of existing natural forest to a biomass production system will have a net negative effect on biodiversity and other ecosystem services. Protection of regrowing forests provides an annual carbon sink and also allows recovery of biodiversity associated with forests. This mutual benefit for biodiversity and carbon sequestration reaches its maximum in relatively unproductive forests, where biodiversity is high and the economics are less favorable for sustainable harvest-and-use systems.

In fact, the extent to which biofuel feedstock production affects biodiversity in a community depends upon the degree of alteration in the land-use pattern, that is, the type of crops that are planted and the previous land use. Biofuel feedstocks belong to several categories with distinct differences in growth, longevity, geographical distribution, and input requirement. Their potential to impact other lives in their ecosystem as well as far-off ecosystems due to common hydrological systems, aerial drift of inputs, and pathogen drift is also variable. Sometimes, substantial off-farm biodiversity loss can occur from reduced water availability, soil erosion, and spread of invasive species (Campbell and Doswald, 2009).

There is also concern regarding the expected increase in the use of genetically modified (GM) cultivars for biofuel feedstocks and their impact on native cultivars and wild relatives. The pollen transfer from GM crops may bring about unwanted or undesired changes to its relatives, and this risk is particularly grave for tree species with great longevity and whose seeds are widely dispersed.

In order to discuss unique biodiversity challenges of different feedstock types, the biofuel crop resources have been separated below into different categories (Table 11.2).

Table 11.2. Biofuel feedstock resource categories based on crop types and production methods.

Row crops	Plantation crops	Grasses	Tree crops	Waste products	Microplant crops
Maize	Palm oil	Miscanthus	Willows	Agricultural waste	Algae
Soybean	Sugarcane	Napiergrass	Poplar	Forestry waste	
Rapeseed camelina	*Jatropha*	Giant reed	Pines	Other organic waste	
Sunflower		Switchgrass			
Sugarbeet					
Wheat					

Row Crops

Row crops are produced on prime agricultural land. The land belonging to this category is limited and for most part already under cultivation. Thus, crop preference on prime land is mostly a matter of prevailing price structure for different crops. For instance, demand for ethanol and the resulting increase in the price of maize has given farmers an incentive to prefer it over other crops. As the crop shift is from one row crop to another, the impact on biodiversity in such situations may be minimal. However, sometimes the search for new prime land can lead to the destruction of natural ecosystems. Marshall (2007) noted the tendency of farmers to take lands enrolled under U.S. Department of Agriculture (USDA) Conservation Reserve Programs to grow maize for ethanol and a similar pattern has also been observed for set-aside land in Europe (Meyerson, 2008). These are matters for concern because the rate of nesting of birds in the set-aside lands in the United States is 10 times higher than in crop land and set-aside lands in Europe provide habitat to farmland birds, mammals, and several endangered species (Campbell and Doswald, 2009). Draining of wetlands to bring them under cultivation of row crops has been a challenge to ecosystems in the United States for decades. Vast land areas in the Prairie Pothole Region, which stretches from Northwest Iowa through Western Minnesota, the Dakotas, Montana and Western Canada, have been lost to farming in past decades (Keeney and Nanninga, 2008). Wetlands are important for the maintenance of the fish population and critical to the migration of migratory birds.

Another significant threat to biodiversity due to the increase in row crops comes from nutrient runoff to the watershed and river systems. Seasonal hypoxia in the Northern Gulf of Mexico is a serious threat to the shrimp, oyster, and fish populations. The maize and soybean production is responsible for 52% N and 25% P load of the Mississippi River, causing oxygen deficiency and threatening aquatic lives (Alexander *et al.*, 2008). The hypoxic zone (dead zone) in the Gulf of Mexico due to nutrient load reached 20 461 km^2 (7900 square miles) in 2007 (Keeney and Nanninga, 2008).

Soybean is increasingly used to produce biodiesel in Brazil and to obtain added production, farmers are resorting to converting Amazon forestland to farmland. Nepstad *et al.* (2008) noted coincidence between the new land brought under soybean cultivation and Amazonian deforestation. The threat to the Amazon is not confined to use of soybean for biodiesel, but is of much larger dimension and emanates from general deforestation to produce row crops for multiple purposes for domestic consumption and export.

Plantation Crops

Generally, new plantations are established either by clearing of forest and other natural ecosystems, converting grassland to plantation, or bringing marginal or degraded land into production. Biodiversity loss occurs when lands with high levels of biodiversity, such as forests, are converted to plantations containing lower levels of biodiversity (Campbell and Doswald, 2009). On the contrary, a gain in biodiversity can be expected by bringing marginal lands under plantation crops.

In Indonesia and Malaysia, extensive tracts of tropical rainforest (including protected areas) have been cleared to create oil palm plantations for biodiesel (Hensen 2005; Dennis and Colfer 2006). Koh and Wilcove (2007) estimate that 55–59% of new oil palm plantations in Malaysia, and 56% in Indonesia resulted from forest clearing, including peatland rainforest. Aside from the primary forest land, oil palm may also infringe upon secondary forests from logged land, which still contain greater diversity than monoculture. In addition, there are diverse agroforestry systems nested within forests that provide continuity for forest lives and oil palm monoculture will destroy the transitional linkages.

Studies until now show that biofuel plantations on productive lands have higher animal biodiversity than do the annual agricultural systems they replace (largely because of the longer harvest intervals and greater physical structure), but lower biodiversity than natural forest stands in the same environments (Cook and Beyea, 2000). The total number of vertebrate species in oil palm plantations was 38% less than in forests (Danielsen *et al.*, 2008). Only 23% of the vertebrates were common to both localities. Venter *et al.* (2009) observed the natural habitat of several endangered mammal species peculiar to these peat forests and biodiversity in general is threatened as forests shrink and oil palm plantations expand. The loss of Sumatran forests to oil palm resulted in the loss of 34 of the 38 medium to large mammals including Sumatran tiger (*Panthera tigris sumatrae*) and clouded leopard (*Neofelis nebulosa*). The void was filled by the wild pig (*Sus scrofa*). Aratrakorn *et al.* (2006) counted fewer bird species and of lower conservation priority in the oil palm plantation than forest. The species loss was not random but concentrated within certain genera; for example, all of the woodpeckers, barbets, and most of the babblers disappeared.

Danielsen *et al.* (2008) found the number of species of invertebrates to be similar in both oil palm plantation and forests, but only 31% of the species were common to both. Bee species in the family Apidae that are the most important forest pollinators were absent from Malaysian oil palm plantations (Liow *et al.*, 2001). The floras of oil palm plantation are also quite different from natural forests. Danielsen *et al.* (2008) reported forest trees, lianas, epiphytic orchids, and indigenous palms present in forest were absent from oil palm plantations. Pteridophytes were common to both communities, but their species compositions were different. The dominant species in the oil palm plantation were common to disturbed lands and early vegetation after burning while forests were rich in epiphytic species growing on large tree branches and shade loving species common to forest understory. Plant diversity within oil palm plantations was also diminished compared to forests due to regular maintenance and replanting.

Tapping the long experience in sugarcane production, after the 1973 world oil crisis Brazil embarked on a national policy of producing sugarcane ethanol for biofuel. Savannah in the Cerrado was cleared to plant sugarcane. The Cerrado is the woodland savannah encompassing most of central Brazil, and is considered the world's most bio-diverse savannah (Sawyer, 2008). It has a large number of unusual mammals, including armadillos, anteaters, tapirs, and maned wolves, as well as thousands of endemic vascular plants adapted to drought and fire. Its destruction would have perilous consequences upon biodiversity. The Brazilian Ministry of Environment has identified 20% of the country's high biological importance areas in the Cerrado (Bustamante *et al.*, 2009). Sawyer (2008) notes 70% of the biodiversity-sensitive areas, however, are suitable for sugarcane production and not protected from encroachment for farming. Ambitious sugarcane plantation schemes are in progress to produce ethanol in the river deltas in Kenya and Tanzania, posing a threat to tropical wetland biodiversity (Campbell and Doswald, 2009).

In the past 15 years, most of the new land for sugarcane planting in Brazil has been incorporated in the southeast state of São Paulo and this has been accompanied by a corresponding decrease in pasture land. Martinelli and Filoso (2008) present data and cite regulatory provisions which point to the arrest of alteration of pristine land to sugarcane planting in Brazil. To start with, less land was needed because of the tremendous improvement in the productivity of sugarcane. New Brazilian laws also restricted sugarcane planting on ecologically sensitive areas. The encroachment of the Amazon by sugarcane growers is not envisioned because its climate is not suitable to sugarcane production. To concentrate sugars, sugarcane needs a period of drought not available in the Amazon. However, an indirect consequence of sugarcane expansion in the Amazon is that depending upon the price and demand equation between sugarcane and soybean, sugarcane has the potential to replace soybean in the Center-West region of the country, and the soybean farmers in turn move north toward the Amazon. Although at present the soybean acreage in this region is 20 times larger than that of sugarcane, adequate safeguard provision for this probable eventuality is prudent.

The current biodiversity threat from sugarcane plantations in Brazil really comes from the destruction of riparian ecosystems, which comprise a narrow belt of land along streams and rivers (Martinelli and Filoso, 2008). The Brazilian law sets a riparian boundary on both margins of streams and rivers, but it has proven inadequate. Silva *et al.* (2007) report that only 500 km^2 (25%) of the riparian forests remain in the seven major agricultural watershed in the state of São Paulo. Riparian ecosystems are complex and contain high levels of biodiversity in both plant and animal species (Naiman *et al.*, 2005). Dotta (2005) observed that 18 species of medium- to large-sized mammals common to larger forested segments were missing in the riparian forest fragments of sugarcane-dominated watersheds, but there was a concomitant increase in the generalist species that have the ability to survive in a wide range of environmental conditions. Restoration of riparian ecosystems in Brazilian sugar plantation therefore deserves attention.

Jatropha is being promoted as a crop that can be grown on degraded land as companion crop on existing pastures (Lovett, 2007) and large-scale *Jatropha* plantations are being created in India, China, Burma, Nicaragua, Africa, the Philippines, and Brazil. *Jatropha* has the potential to be invasive in favorable climates (Invasive Species Compendium Beta, 2011) and eliminate native species. Due to similar concerns, *Jatropha* cultivation was banned in Western Australia in 2006. Buddenhagen *et al.* (2009) observed that 70% of the regionally suitable biofuel crops in Hawaii can become invasive.

Other Biofuel Feedstocks

In recent years, attention has shifted to the production of grasses for biofuel because of their ability to grow on marginal or agriculturally degraded lands. It is expected that grasses will help restore the landscape and provide new habitats for birds and other animal species. Somerville (2007) claims that only 3.2% of terrestrial land area planted with *Miscanthus* will be sufficient to meet world energy needs. However, others (Farrell *et al.*, 2006; Gallagher, 2008; Rubin, 2008) contend that land under agriculture will actually increase as the cellulosic crops do not produce co-products such as animal feed for which crops have to be grown separately. In another scenario, biodiverse prairie pasture lands that are currently part of USDA Conservation Reserve Program may be converted for biofuel to crops such as switchgrass disturbing the current equilibrium (Schmer *et al.*, 2008).

While maize and woody biomass crops are likely to negatively impact biodiversity in North America, perennial biomass crops such as mixed grass prairie have potential to offer important conservation benefits. Grassland birds have experienced radical population declines as grasslands have been converted to annual row crops, but perennial-based biomass feedstocks appear to enhance avian diversity and abundance, providing important habitat both during the breeding and migratory periods. Birds may not only benefit from perennial crops, but provide important pest control services by consuming herbivorous insects. Moreover, by providing important habitat for arthropod predators, perennial-based feedstocks provide enhanced pest control, even to surrounding crops (Robertson 2012).

Rowe *et al.* (2011) show that under current management practices in England, a mixed farming system incorporating mature willow short rotation coppice (SRC) can benefit native farm-scale biodiversity. In particular, the reduced disturbance in willow SRC allows the persistence of perennial plant species, potentially providing a stable refuge and food sources for invertebrates.

Junginger *et al.* (2006) are of the opinion that organic waste and agricultural and forestry residues combined can meet almost one-quarter of the current biofuel need. However, forestry waste is important to the survival of almost 25% of woodland species, and thus the amount to be removed needs proper management (Perlack *et al.*, 2005).

Response Options for Ecologically Sustainable Biofuel Production

Three general principles that should guide investigative efforts to strengthen biofuel practices and policies are as follows: (1) optimize current biofuel crop production practices, (2) develop innovative new approaches to biofuel crop production, and (3) implement policy interventions and related initiatives. Within this framework, we make 18 recommendations to promote sustainably grown, biodiversity-friendly biofuels (Table 11.3).

The following recommendations for enhancing the long-term sustainability of biofuels are amongst the most widely explored in the literature.

Using Degraded Lands

Of all of the potential locations that might be tapped for increased biofuel production, marginal lands, abandoned croplands, and abandoned pasture lands appear to offer significant environmental and economic potential, particularly if the biofuel crops are perennial and if sustainable land management practices are employed. There are approximately 51–67 Mha of this kind of land available in the United States; those areas could produce as much as 321 MT of biomass per year (Dale *et al.*, 2010). Globally, there are about 385–472 Mha of abandoned cropland that could produce between 1.4 and 2.1 billion tons of biomass (Campbell *et al.*, 2008).

However, growing biofuel crops on these marginal lands may require significantly more land area than other regions due to relatively lower yields, and will likely require more energy-intensive management such as fertilizer application or irrigation to remain productive (Gibbs *et al.*, 2008). Campbell *et al.* (2008) estimate that although the energy content of potential biomass grown on 100% of abandoned agriculture lands is less than 10% of primary energy demand for most nations in North America, Europe, and Asia, it represents many times the energy demand in some African nations where grasslands are relatively productive and current energy demand is low.

Table 11.3. Policy recommendations for sustainable ecosystem and stable biodiversity in biofuel production systems.

I. Optimize current biofuel crop production practices
1. Promote research to improve yield and conversion efficiency of biofuel feedstocks to reduce the land needs for biofuel;
2. Develop biofuel crop cultivars for marginal soils and reward farmers for using these soils for production;
3. Integrate biofuel feedstock to the agricultural production systems to avoid environmental damages of monoculture;
4. Use minimum-till, cover crop and integrated pest control methods as low-input means for annual crops;
5. Examine the downstream pollution potential before expanding acreage;
6. Manage plantations as agroforestry systems to reduce the impact of monoculture.

II. Develop innovative new approaches to biofuel crop production
7. Encourage research to design microorganisms that can convert cellulosic biomass to biofuel in one-step low-energy process;
8. Accelerate research on algae for biofuel to take advantage of its high-yield potential and low land requirement;
9. Carefully examine the gene transfer risks and implement precautionary measures to prevent impact on other varieties and wild relatives before large-scale adoption of novel biofuel crops;
10. Promote perennials to establish a stable ecosystem and select species that require low input;
11. Promote the use of native species to preserve the existing ecosystem;
12. Study the full invasive potential before introducing a new feedstock species.

III. Implement policy interventions and related initiatives
13. Devise plans so that local people can fully participate in the biofuel economy and thus share concern for its long-term survival;
14. Legislate to check use of forest or natural savannah for biofuel production;
15. Implement unbiased feedstock certification programs to incentivize co-friendly biofuel production systems;
16. Confine the use of starch, sugar, and oilseed feedstock for biofuel to utilize only the marketplace surplus;
17. Examine the economic and social burdens of biofuel production system by conducting comprehensive life-cycle assessment and devise interventions before entering the production phase;
18. Perform life-cycle assessment of net greenhouse gas flux from biofuel production systems to validate their global climate change advantage.

Choosing to use these previously cleared lands can enhance social and environmental sustainability while reducing pressures on forests. Substantial carbon benefits, especially in terms of payback times, are possible from expanding high-yielding crops, such as sugarcane and oil palm, into already degraded lands (Gibbs *et al.*, 2008). The Brazilian government recently announced that expansion of the crop would be limited to 63.5 Mha. This land could be made available, without the clearing of natural ecosystems, by means of a slight increase in the low stocking density on the estimated 237 Mha of cattle ranching. Many other regions of the world are also well suited to sugarcane production or formerly produced sugarcane on land that has been abandoned (Somerville *et al.*, 2010). Many of the new feedstocks mentioned above could also be efficiently produced on marginal or degraded land (Tilman *et al.*, 2006). Dale *et al.* (2010), however, note that areas containing important habitats for wildlife, which are rich in biodiversity and which were not previously suitable for agriculture, may become at risk of conversion to cropland, and suggest that such areas merit special attention and protection.

Alongside (a) monoculture crops and (b) waste biomass, Tilman *et al.* (2006) suggest that there is potential for a third major source of biofuel biomass, high-diversity mixtures of plants grown with low inputs on agriculturally degraded land. Their data show that biofuels derived from low-input high-diversity (LIHD) mixtures of native grassland perennials can provide more usable energy, greater GHG reductions, and less agrichemical pollution per hectare than can maize grain ethanol or soybean biodiesel. Because LIHD biomass can be produced on abandoned agricultural lands, LIHD biofuels need neither compete for fertile soils with food production nor encourage ecosystem destruction. LIHD biomass could yield 51% more usable energy per hectare from degraded infertile land than does maize grain ethanol from fertile soils. Diverse grasslands for bioethanol have been proposed as a means to restore prairie ecosystems in North America (Palmer and Downing, 2006) and to maintain the financial viability of extensive production systems in parts of Europe. Tilman *et al.* (2006) estimate that around 5×10^8 ha of agriculturally abandoned and degraded land producing biomass at 90 GJ ha^{-1} yr^{-1} could provide, via integrated gasification and combined cycle technology with Fischer–Tropsch hydrocarbon synthesis, about 13% of global petroleum consumption for transportation and 19% of global electricity consumption. Without accounting for ecosystem CO_2 sequestration, this could eliminate 15% of current global CO_2 emissions. LIHD biofuel production should be sustainable with low inputs of agrichemicals.

Improved Plantation Management Practices and "Designer" Landscapes

Because species diversity supports ecosystem health, and in order to maximize tolerance to both biotic and abiotic stresses, it will be desirable to use genetically diverse or species-diverse plantations of energy crops. Perennial polycultures, agroforestry systems, or semi-natural grasslands can maintain or restore native biodiversity. Integrated pest management systems can be applied to oil palm plantations which favor the use of nonchemical pest control methods such as the establishment of beneficial plants (e.g., *Euphorbia heterophylla*) to attract insect predators and parasitoids of oil palm pests such as the wasp *Dolichogenidea metesae* (Corley and Tinker, 2003).

Landscape scale management holds significant potential for reducing the off-site consequences of agriculture. Individual farms, watersheds, and regional planning can take advantage of services provided by adjacent natural, semi-natural, or restored ecosystems. Trees and shrubs planted in buffer strips surrounding cultivated fields decrease soil erosion and can take up nutrients that otherwise would enter surface or groundwaters. Buffer zones along streams, rivers, and lakeshores can decrease nutrient and silt loading from cultivated fields or pastures. Crop pollination can be provided by insects and other animals living in nearby habitats or buffer strips, whereas other organisms from these habitats, such as parasitoids, can provide effective control of many agricultural pests. Buffer strips can also be managed to reduce inputs of weeds and other agricultural pests (Tilman *et al.*, 2002).

For plantation crops such as oil palm, a land-use conversion scheme that requires establishment of agroforestry corridors to natural forests will act as buffer zone to prevent encroachment of residual forest and facilitate movement of wildlife through the landscape. Proper management practices will also help in biodiversity preservation. Koh (2008) suggested setting aside a portion of the plantation to natural forest cover for native butterflies and native birds. He found that such practice increased butterfly species by 1.2–12.9% and bird species by 0.6–53.3%. It also provided practical benefit to the plantation, as birds fed upon herbivorous insects, providing natural pest control. Development of cultivars and management practices to maximize production is another way to help ease land demand.

In line with the above points, Milder *et al.* (2008) suggest that biofuels be produced using surplus or currently wasted resources in the landscape. They construe "resources" broadly to include land, water, nutrients, and labor. What constitutes interstitial space or resource availability varies by context. Growing perennial crops in strips along swales and conventional crop fields and on abandoned or idle cropland, combined with improvements in management and productivity of current cropland–pasture areas, may prove to be one of the easiest initial pathways to expand bioenergy crop production in a manner that is compatible with goals to maintain biodiversity and other ecosystem services, such as water purification and flood protection (Dale *et al.*, 2010).

Develop Alternative Biofuel Feedstocks

Although maize ethanol dominates the economic scene for biofuels in the United States at present, transitioning to other, "second-generation," cellulosic feedstocks based on stems, stalks, or woody components of plants is likely to provide the greatest yields most efficiently with the least environmental repercussions (Dale *et al.*, 2010). Perennial crops such as grasses and fast-growing trees are one such type of feedstock. For example, poplar and willow have been grown successfully with municipal-waste fertilizers and irrigated with municipal or industrial wastewater, thus decreasing waste streams while achieving inputs needed for high yields (Powlson *et al.*, 2005). Results of modeling by Aylott *et al.* (2010) suggest that 7.5 MT of biomass from SRC is realistically available in England. Such production would require 0.8 Mha and could be grown almost entirely on poor quality marginal lands, leading the authors to conclude that SRC energy crops have the potential to play an important role in meeting UK renewable energy targets without compromising environmental sustainability or food production.

Perennial energy crops that are native species, such as switchgrass in many parts of the United States, appear to offer considerable promise for both economic and environmental sustainability, and can be grown with much lower fertilizer inputs than other crops, particularly maize (Kiss *et al.*, 2007, 2009). Their development will allow feedstocks to shift from food crops grown on valuable farmland to nonfood crops grown in low-input systems, on lower value land, in ways that improve habitat, restore soil fertility, and sequester carbon (Milder *et al.*, 2008). Soils benefit when no-till or low-till farming practices are employed, because these practices significantly reduce soil disturbance and erosion (Dale *et al.*, 2010). However, the technology to convert woody biomass into biofuels (i.e., the use of enzymes to ferment straw into lignocellulosic bioethanol) is still in development and not commercial yet, and these feedstocks are not expected to offer a viable, large-scale alternative to contemporary sources until the 2020s or later. Furthermore, due to the high costs of collecting and transporting biomass to processing plants where conversion to liquid fuels occurs, it will be difficult for any currently envisioned dedicated crop to compete economically with the large volumes of sugarcane residues that are already available from ethanol plants in Brazil (Dale *et al.*, 2010).

Because *Jatropha* can grow on marginal land with limited water and its seeds are nonedible, it does not compete directly with food (in terms of land and water resources). Nonedible oils from *Jatropha*, *Pongamia,* and other plants are also well-suited to small-scale production and can be used to power generators, pumps, mills, or diesel vehicles (Milder *et al.*, 2008). The use of water-extensive oil seeds is promising but with existing technologies, biofuel yields from *Jatropha* trees are fairly low (1500 l ha^{-1} biodiesel at most) and processing is relatively expensive (de Fraiture *et al.*, 2008). Again the role of *Jatropha* is likely to remain small until major technology breakthroughs are realized.

Somerville *et al.* (2010) anticipate that relatively water-inefficient C_3 species such as poplar will be grown only where rainfall is abundant and more water-efficient C_4 grasses such as sugarcane, switchgrass, and *Miscanthus* will be grown where rainfall is not in excess. A potential opportunity for production of biomass for fuels from land that is water limited is the use of species with high efficiencies of water use and drought resistance, such as various *Agave* species that use a type of photosynthesis called Crassulacean acid metabolism (CAM) and thrive under arid and semi-arid conditions. Much of the land that has fallen out of agricultural production worldwide is semi-arid and it appears that the amount of land that may be available for cultivation of *Agave* species is vast.

Biofuels produced from waste (wood chips and crop residuals such as straw, maize stover, leaves, and woody biomass) may be more sustainable than any crop-based system (Dale *et al.*, 2010), but the potential importance of crop wastes to maintenance of soil quality must be borne in mind.

Because of their high growth rate, population density, and oil content, algae could be an alternative to oilseed crops for biodiesel and thus relieve the expansion pressure on these crops. The potential biodiesel yield from algae is 90 000 l ha^{-1}, compared to 450 l ha^{-1} from soybean, 1200 l ha^{-1} from rapeseed and 6000 l ha^{-1} from oil palm (Haag, 2007). Sheehan *et al.* (1998) estimate that, even when grown in the least space-efficient manner (in large open ponds), only 200 000 ha would be needed to produce 1 quadrillion BTU from microalgae biodiesel. Furthermore, many of the most promising species are diatoms and green algae that tolerate brackish or salt water and thus can be grown without use of increasingly scarce freshwater resources. Algal biofuel yet to become a market reality because of significant technological, economic, and large-scale production hurdles.

Policy Interventions and Related Initiatives

New incentives and policies for ensuring the sustainability of agriculture and ecosystem services will be crucial if we are to meet the demands of improving yields without compromising environmental integrity or public health (Tilman *et al.*, 2002).

Several voluntary compliance initiatives and regulations have been developed to ensure biofuel is produced in an environmentally, ecologically, and socially responsible manner (Campbell and Doswald, 2009). In 2007, the National Institute of Metrology, Quality, and Technology (INMETRO) instituted a voluntary certification for sugarcane and ethanol production for Brazil that does not encroach upon biodiversity, prevents soil, water, and air pollution, lowers GHG emission, protects workers' rights for equitable wages and safe working conditions, and so on. A voluntary standard scheme can only be effective if it has large-scale grower participation. A common grower's complaint is that standards laid out in the certification schemes are complex and costly to implement. Moreover, as state, national, and international targets drive the biofuel expansion, certifications will need policy level support to be successful. Preferential treatment by the biofuel industry for the certified feedstocks can provide a strong incentive for adherence to certification standards.

Regulations, on the other hand, have been promulgated by governing bodies whose compliance is compulsory. The European Commission has issued the renewable source directive (EU-RES-D) that applies to all biofuel feedstocks whether grown in-country or imported. Switzerland has strict regulatory provisions under which all biofuels from oil palm, maize, and soybean are banned, and there are also rigid guidelines against all other biofuels pertaining to a wide range of sustainability criteria.

Milder *et al.* (2008) express concern that the most influential policies, particularly United States and European biofuel mandates, have not established adequate guidelines for the sustainable production of these fuels. In their view, biofuel demand and investment are growing much faster than the regulatory and governance frameworks needed to guide and support them. Developing countries have legitimately raised the concern that these schemes would be counterproductive if developed countries use them as tools to place trade barriers against biofuel imports to protect their domestic biofuel industries.

Innovative schemes such as Payment for Ecosystem Services (PES) or Reducing Emissions from Deforestation and Degradation (REDD) create financial incentives to divert biofuel feedstock expansion away from forests and onto pre-existing croplands or degraded lands (Stromberg *et al.*, 2010). The establishment and enforcement of protected areas is a legislative tool that remains a top strategic priority for protecting biodiversity. PES and REDD can be used to both protect biodiversity and mitigate other negative effects of biofuels on ecosystem services (e.g., water provision and climate regulation).

Although the simulations of Alcamo *et al.* (2005) raise doubts about the reliability of future ecosystem services, they do not indicate the likelihood that thresholds or breaking points in nature will be reached. The authors recognize, however, that such thresholds are difficult to identify globally, and that data deficiencies (e.g., data on the extent of forests and other land-cover types that can be tracked over time) make it difficult to simulate breaking points with the current generation of global models. They call for efforts to develop global early warning systems for detecting thresholds and breaking points in nature's provision of services to be strengthened. Improved models are needed to anticipate ecosystem collapses so that policies can be developed to avoid or adapt to these collapses.

The quantification and valuation of ecosystem services remains highly uncertain. In addition to the uncertainty about the biophysical production of ecosystem services, there is also uncertainty about the value of services. Dale and Polasky (2007) proposed that a set of ecological indicators for ecosystem services both from and to agriculture should be considered. The Natural Capital Project is developing InVEST, a family of software-based tools for Integrated Valuation of Environmental Services and Tradeoffs (Tallis *et al.*, 2010). By quantifying the values of natural capital, InVEST enables decision makers to quantify the importance of natural capital, to assess the tradeoffs associated with alternative choices, and to integrate conservation and human development. In this way, scientific and economic understanding of natural assets can be integrated into real land use and investment decisions.

The biofuel rush has led some governments and corporations to attempt to appropriate rural lands owned or managed by local people for large-scale feedstock production (Colchester *et al.*, 2006). Formalizing and respecting land tenure rights can help prevent such abuses and enable farmers to earn an equitable share of revenues from biofuel production. In addition to formal land titles, communities' rights to self-determination should be institutionalized through requirements for meaningful local participation and informed decision-making whenever a significant biofuel project is proposed (Milder *et al.*, 2008).

Related to the above, the earlier paradigm of science being developed at the international or perhaps national level and then disseminated to farmers should be replaced by an active exchange of information among scientists and farmers. Scientists in developing countries who understand the ecosystems, human culture, and demands on local agricultural systems must be actively trained, promoted, and brought into the international scientific community (Tilman *et al.*, 2002).

Concluding Remarks

Ecosystems are essential for human well-being through their provisioning, regulating, cultural, and supporting services. While demands for ecosystem services such as food and clean water are growing, human actions are at the same time diminishing the capability of many ecosystems to meet these demands. Progress toward one objective such as increasing food production has often been at the cost of progress toward other objectives, such as conserving biological diversity or improving water quality (MA, 2005). These are important implications for a possible large-scale development of biofuel. Current biofuel production competes for fertile land with food production, can increase pollution from fertilizers and pesticides, and threaten biodiversity when natural lands are converted to cropland. However, while such negative impacts of biofuel production and use have been the focus of much of the public debate, there is evidence to show that the extent of these impacts has often been overstated and, indeed, that biofuel production can also have positive impacts.

At present, the most commonly used biofuel crops are sugar crops and maize (de Fraiture *et al.*, 2008), but choice of an ideal bioenergy crop system will always be location and market specific. The sustainable development of renewable fuel alternatives can offer many benefits but will demand a comprehensive understanding of how our land-use choices affect the ecological systems around us (Dale *et al.*, 2010). Biofuel ecological footprints are a function of many factors. Energy efficiency or net energy balance (energy output:energy input) over the life cycle of the biofuel combined with its fuel yield per hectare will affect the ecological footprint (Groom *et al.*, 2008). The conservation impact of biofuel production depends on: (a) what the system is replacing and (b) whether producers make deliberate efforts to improve sustainability. If the system is replacing another large monocrop or simply changing the mode of processing (as with maize ethanol in the United States), the ecological impact at the site level may be neutral. However, economic pressures may push production into previously forested or otherwise ecologically sensitive areas (Milder *et al.*, 2008).

Several possible response options are described in this chapter. As with all agricultural crops, biofuel feedstocks should be grown using environmentally safe and biodiversity-friendly practices. The choice of production technology (e.g., high-yielding feedstocks) and location (marginal land) can contribute to the sustainability of feedstock production. In other cases, land management measures such as conservation corridors can significantly reduce the negative effects on ecosystems. Sound environmental practices must be applied throughout the fuel-production life cycle. The MA calls for an effective strategy for managing ecosystems (MA, 2003). This will require a mix of interventions including laws, regulations, and enforcement schemes, partnerships and collaborations, the sharing of information and knowledge, and public and private actions. Taken together, these options can help biofuels make a positive contribution both to ecosystem services and to human well-being.

References

Alcamo, J., van Vuuren, D., Ringler, C., *et al.* (2005) Changes in nature's balance sheet: model-based estimates of future worldwide ecosystem services. *Ecology and Society* 10: 19.

Alder, P.R., Grosso, S.J.D., and Parton, W.J. (2007) Life cycle assessment of net greenhouse-gas flux for bioenergy cropping systems. *Ecological Applications* 17: 675–691.

Alexander, R.B., Smith, R.A., Schwarz, G.E., Boyer, E.W., Nolan, J.V., and Brakebill, J.W. (2008) Differences in phosphorus and nitrogen delivery to the Gulf of Mexico from the Mississippi River Basin. *Environmental Science and Technology* 42: 822–830.

Anon. (2007) Numbers in context: are we putting tigers in our tanks? *Conservation* 8: 40–41.

Agricultural and Rural Convention (ARC). (2010) *A Communication from Civil Society to the European Union Institutions on the Future Agricultural and Rural Policy*, Agricultural & Rural Convention, Brussels.

Aratrakorn, S., Thunhikorn, S., and Donald, P.F. (2006) Changes in bird communities following conversion of low land forest to oil palm and rubber plantation in southern Thailand. *Bird Conservation International* 16: 71–82.

Aylott, M.J., Casella, E., Farrall, F., and Taylor, G. (2010) Estimating the supply of biomass from short-rotation coppice in England, given social, economic and environmental constraints to land availability. *Biofuels* 1: 719–727.

Berndes, G. (2002) Bioenergy and water – the implications of large-scale bioenergy production for water use and supply. *Global Environmental Change* 12: 253–271.

Bianchi, F.J.J.A., Booij, C.J.H., and Tscharntke, T. (2006) Sustainable pest regulation in agricultural landscapes: a review on landscape composition, biodiversity, and natural pest control. *Proceedings of the Royal Society of London Series B* 273: 1715–1727.

Blythman, J. (2007) Seeds of discontent. http://www.guardian.co.uk/uk/2007/apr/19/ruralaffairs.lifeandhealth (accessed 25 January 2012).

Brown, L.R. (2006) *Plan B 2.0: Rescuing a Planet Under Stress and a Civilization in Trouble*, Earth Policy Institute, Washington, DC.

Buddenhagen, C.E., Chimera, C., and Clifford, P. (2009) Assessing biofuel crop invasiveness: a case study. *PLoS One* 4: e5261. doi:10.1371/journal.pone.0005261.

Bustamante, M.M.C., Melillo, J., Connor, D.J., *et al.* (2009) What are the final land limits?, in *Biofuels: Environmental Consequences and Interactions with Changing Land Use* (eds R.W. Howarth and S. Bringezu), Proceedings of Scientific Committee on Problems of the Environment (SCOPE) International Biofuels Project Rapid Assessment, 22–25 September 2008, Cornell University, Ithaca, pp. 265–285.

CA (Comprehensive Assessment of water Management in Agriculture). (2007) *Water for Food, Water for Life: A Comprehensive Assessment of Water Management in Agriculture*, Earthscan, London and International Water Management Institute, Colombo.

Campbell, A. and Doswald, N. (2009) *The Impacts of Biofuel Production on Biodiversity: A Review of Current Literature*, UNEP-WCMC, Cambridge.

Campbell, J.E., Lobell, D.B., Genova, R.C., and Field, C.B. (2008) The global potential of bioenergy on abandoned agriculture lands. *Environmental Science and Technology* 42: 5791–5794.

Centro de Gestão e Estudos Estratégicos (CGEE) (2008) *Sugar-cane Based Bioethanol: Energy for Sustainable Development*, Banco Nacional de Desenvolvimento Econômico e Social, Rio de Janeiro.

China News, AFP. (2006) Issue: 21 December 2006. Cited by de Fraiture *et al.*, 2008.

Colchester, M., Jiwan, N., Andiko, Sirait, M., Firdaus, A.Y., Surambo, A., and Pane, H. (2006) *Promised Land: Palm Oil and Land Acquisition in Indonesia – Implications for Local Communities and Indigenous Peoples*, Forest Peoples Programme, Moreton-in-Marsh, United Kingdom and Perkumpulan Sawit Watch, Bogor, Indonesia.

Cook, J. and Beyea, J. (2000) Bioenergy in the United States: progress and possibilities. *Biomass and Bioenergy* 18: 441–455.

Corley, R.H.V. and Tinker, P.B. (2003) *The Oil Palm*, 4th edn. Blackwell Science, Oxford.

Council of Europe. (2000) European Landscape Convention Available online at http://www.conventions.coe.int/Treaty/en/Treaties/Html/176.htm (accessed 25 January 2012).

Dale, V.H., Kline, K.L., Wiens, J., and Fargione, J. (2010) *Biofuels: Implications for Land use and Biodiversity*. Biofuels and Sustainability Reports. Ecological Society of America, Washington, DC.

Dale, V.H. and Polasky, S. (2007) Measures of the effects of agricultural practices on ecosystem services. *Ecological Economics* 64: 286–296.
Danielsen, F., Beukema, H., Burgess, N.D., *et al.* (2008) Biofuel plantations on forested lands: double jeopardy for biodiversity and climate. *Conservation Biology* 23: 348–358.
Dennis, R.A. and Colfer, C.P. (2006) Impacts of land use and fire on the loss and degradation of lowland forest in 1983–2000 in East Kutai District, East Kalimantan, Indonesia. *Singapore Journal of Tropical Geography* 27: 30–48.
Dennis, R.A., Mayer, J., Applegate, G., *et al.* (2005) Fire, people and pixels: linking social science and remote sensing to understand underlying causes and impacts of fires in Indonesia. *Human Ecology* 33: 465–504.
Department for International Development (DFID) (1999) *Sustainable Livelihoods Guidance Sheets*, Department for International Development, London.
Donald, P.F. (2004) Biodiversity impacts of some agricultural commodity production systems. *Conservation Biology* 18: 17–37.
Dotta, G. (2005) Diversidade de mamiferos de médio e grande porte em função da paisagem na sub-bacia do rio Passa-Cinco, São Paulo. Thesis. São Paulo: Universidade de São Paulo.
Dufey, A. (2006) *Biofuels Production, Trade and Sustainable Development: Emerging Issues*, International Institute for Environment and Development, London.
European Network for Rural Development (ENRD). (2010) Thematic Working Group 3: Public Goods and Public Intervention: Final Report.European Network for Rural Development, Brussels.
Food and Agriculture Organization of the United Nations (FAO) (2004) *Fertilizer Use by Crop in Malaysia*, Food and Agriculture Organization of the United Nations, Rome.
Fargione, J., Hill, J., Tilman, D., Polsaky, S., and Hawthorne, P. (2008) Land clearing and the biofuel carbon debt. *Science* 319: 1235–1238.
Farrell, A.E., Plevin, R.J., Turner, B.T., Jones, A.D., O'Hare, M., and Kammen, D.M. (2006) Ethanol can contribute to energy and environmental goals. *Science* 311: 506–508.
Fieldsend, A.F. (1996) Factories in the field. *Green Futures* 2: 20–21.
Fisher, B., Turner, R.K., and Morling, P. (2009) Defining and classifying ecosystem services for decision making. *Ecological Economics* 68: 643–653.
de Fraiture, C., Giordano, M., and Liao, Y. (2008) Biofuels and implications for agricultural water use: blue impacts of green energy. *Water Policy* 10(suppl 1): 67–81.
Gallagher, E. (2008) *The Gallagher Review of the Indirect Effects of Biofuels Production*, Renewable Fuels Agency, St Leonards-on-Sea, UK.
Gibbs, H.K., Johnston, M., Foley, J.A., *et al.* (2008) Carbon payback times for crop-based biofuel expansion in the tropics: the effects of changing yield and technology. *Environmental Research Letters* 3: 1–10.
Grime, J.P. (1973a) Competitive exclusion in herbaceous vegetation. *Nature* 242: 344–347.
Grime, J.P. (1973b) Control of species density in herbaceous vegetation. *Journal of Environmental Management* 1: 151–167.
Groom, M.J., Gray, E.M., and Townsend, P.A. (2008) Biofuels and biodiversity: Principles for creating better policies for biofuel production. *Conservation Biology* 22: 602–609.
Hagg, A.L. (2007) Algae bloom again. *Nature* 447: 520–521.
Harmon, M.E., Ferrell, W.K., and Franklin, J.F. (1990) Effects on carbon storage of conversion of old-growth forests to young forests. *Science* 247: 699–701.
Hartemink, A.E. (2007) Soil erosion: Perennial crop plantations, in *Encyclopedia of Soil Science*, 2nd edn (ed. R. Lal), Taylor & Francis, London, pp. 1613–1617.

Hensen, I.E. (2005) An assessment of changes in biomass carbon stocks in tree crops and forests in Malaysia. *Journal of Tropical Forest Science* 17: 279–296.

Hill, J., Nelson, E., Tilman, D., Polasky, S., and Tiffany, D. (2006) Environmental, economic, and energetic costs and benefits of biodiesel and ethanol biofuels. *Proceedings of the National Academy of Sciences* 103: 11206–11210.

Huston, M.A. and Gilbert, L. (1996) Ecosystem functions of biodiversity: consumers and secondary production, in *Biodiversity and Ecosystem Processes in Tropical Rainforests* (eds G. Orians, R. Dirzo, and J.H. Cushman), SCOPE/Springer, Berlin.

Huston, M.A. and Marland, G. (2003) Carbon management and biodiversity. *Journal of Environmental Management* 67: 77–86.

International Energy Agency (IEA). (2004) *Biofuels for Transport: An International Perspective*, OECD/IEA, Paris.

Intergovernmental Panel on Climate Change (IPCC). (2000) Land use, land–use change, and forestry, in *A special report of the Intergovernmental Panel on Climate Change* (eds R.T. Watson, I.R. Noble, B. Bolin, N.H. Ravindranath, D.J. Verardo, and D.J. Dokken), Cambridge University Press, Cambridge.

Invasive Species Compendium Beta. (2011) *Jatropha gossypiifolia*. http://www.cabi.org./isc/?compid (accessed 19 March 2012).

Junginger, M., Faaij, A., Rosillo-Calle, F., and Wood, J. (2006) The growing role of biofuels: opportunities, challenges, and pitfalls. *International Sugar Journal* 108: 615–629.

Keeney, D. and Nanninga, C. (2008) *Biofuel and Global Diversity*, Institute for Agriculture and Trade Policy, Minneapolis.

Kgathi, D.L. and Zhou, P. (1995) Biofuel use assessments in Africa: implications for greenhouse gas emissions and mitigation strategies. *Environmental Monitoring and Assessment* 38: 253–269.

Kim, S. and Dale, B.E. (2011) Indirect land use change for biofuels: Testing predictions and improving analytical methodologies. *Biomass and Bioenergy* 35: 3235–3240.

Kiss, Z., Fieldsend, A.F., and Wolf, D.D. (2007) Yield of switchgrass (*Panicum virgatum* L.) as influenced by cutting management. *Acta Agronomica Hungarica* 55: 227–233.

Kiss, Z., Fieldsend, A.F., and Wolf, D.D. (2009) Yield and canopy characteristics of switchgrass (*Panicum virgatum* L.) as influenced by cutting management. *Biomass and Bioenergy* 33: 442–448.

Koh, L.P. (2008) Can oil palm plantations be made hospitable for forest butterflies and birds? *Journal of Applied Ecology* 45: 1002–1009.

Koh, L.P. and Wilcove, D.S. (2007) Is oil palm agriculture really destroying tropical biodiversity? *Conservation Letters* 1: 60–64.

Lal, R. (2005) World crop residues production and implications of its use as a biofuel. *Environment International* 31: 575–584.

Landis, D.A., Gardiner, M.M., van der Werf, W., and Swinton, S.M. (2008) Increasing corn for biofuel production reduces biocontrol services in agricultural landscapes. *PNAS* 105: 20552–20557.

Liow, L.H., Sodhi, N.S., and Elmqvist, T. (2001) Bee diversity along a disturbance gradient in tropical lowland forests of South East Asia. *Journal of Applied Ecology* 38: 180–192.

Lovett, J.C. (2007) Biofuels and ecology. *African Journal of Ecology* 45: 117–119.

Marshall, L. (2007) *Thirst for Corn: What 2007 Plantings Could Mean for the Environment*. Policy note, energy: biofuels, no. 2. World Resources Institute, Washington, DC.

Martinelli, L.A. and Filoso, S. (2008) Expansion of sugar cane ethanol expansion in Brazil: environmental and social challenges. *Ecological Applications* 18: 885–898.

Meyerson, L.A. (2008) Biosecurity, biofuels, and biodiversity. *Frontiers in Ecology and the Environment* 6: 291.

Milder, J.C., McNeely, J.A., Shames, S.A., and Scherr, S.J. (2008) Biofuels and ecoagriculture: can bioenergy production enhance landscape–scale ecosystem conservation and rural livelihoods? *International Journal of Agricultural Sustainability* 6: 105–121.

Millennium Ecosystem Assessment (MA). (2003) *Ecosystems and Human Well-being: A Framework for Assessment*, World Resources Institute, Washington, DC.

Millennium Ecosystem Assessment (MA). (2005) *Ecosystems and Human Well-being: Synthesis*, World Resources Institute, Washington, DC.

Naiman, R., Dècamps, N., and McClain, M.E. (2005) *Riparia: Ecology, Conservation, and Management of Streamside Communities*, Elsevier/Academic Press, Burlington, Massachusetts.

NASS (National Agricultural Statistics Service). (2007) *Statistical Highlights of US Agriculture 2006 & 2007: Environmental*, USDA, Washington, DC.

Nepstad, D.C., Stickler, C.M., Soares, B., and Merry, F. (2008) Interactions among Amazon land use, forests and climate: prospects for a near-term forest tipping point. *Philosophical Transactions of the Royal Society B - Biological Sciences* 363: 1737–1746.

Nye, P.H. and Greenland, D.J. (1960) *The Soil under Shifting Cultivation*. Technical Communication No. 51. Commonwealth Agricultural Bureau of Soils, Harpenden.

Palmer, M.W. and Downing, M. (2006) Harvesting our meadows for biofuel? *Science* 312: 1745.

Perlack, R.D., Wright, L.L., Turhollow, A.F., Graham, R.J., Stokes, B.J., and Erhlbach, D.C. (2005) *Biomass as Feedstock for a Bioenergy and Bioproducts Industry: The Technical Feasibility of a Billion-ton Annual Supply*, U.S. Department of Energy and U.S. Department of Agriculture, Oak Ridge, Tennessee.

Pimentel, D. (2003) Ethanol fuels: energy balance, economics and environmental impacts are negative. *Natural Resources Research* 12: 127–134.

Powlson, D.S., Richie, A.B., and Shield, I. (2005) Biofuels and other approaches for decreasing fossil fuel emissions from agriculture. *Annals of Applied Biology* 146: 193–201.

Ramachandra, T.V., Krishna, S.V., and Shruthi, B.V. (2004) Decision support system to assess regional biomass energy potential. *International Journal of Green Energy* 1: 407–428.

Righelato, R. and Spracklen, D.V. (2007) Carbon mitigation by biofuels or by saving and restoring forests? *Science* 317: 902.

Robertson, B.A., Porter, C., and Schemske, D.W. (2012) Cellulosic biomass crops influence the diversity, biomass, and guild structure of terrestrial arthropod communities. *BioEnergy Research* 5: 179–188.

Rowe, R., Hanley, M., Goulson, D., Clarke, D., Doncaster, C.P., and Taylor, G. (2011) Potential benefits of commercial willow Short Rotation Coppice (SRC) for farm-scale plant and invertebrate communities in the agri-environment. *Biomass and Bioenergy* 35: 325–336.

Rubin, E.M. (2008) Genomics of cellulosic biofuels. *Nature* 454: 841–845.

Sawyer, D. (2008) Climate change, biofuels and eco-social impacts in the Brazilian Amazon and Cerrado. *Philosophical Transactions of the Royal Society B – Biological sciences* 363: 1747–1752.

Schmer, M.R., Vogel, K.P., Mitchell, R.B., and Perrin, R.K. (2008) Net energy of cellulosic ethanol from switchgrass. *Proceedings of the National Academy of Sciences* 105: 464–469.

Searchinger, T., Heimlich, R., Houghton, R.A., *et al.* (2008) Use of U.S. croplands for biofuels increases greenhouse gases through emissions from land use change. *Science* 319: 1238–1240.

Sheehan, J., Dunahay, T., Benemann, J., and Roessler, P. (1998) *A Look Back at the U.S. Department of Energy's Aquatic Species Program – Biodiesel from Algae*. Report to the Department of Energy. National Renewable Energy Laboratory, Golden, Colorado.

Silva, A.M., Nalon, M.A., do Nascimento Kronka, F.J., Alvares, C.A., de Camargo, P.B., and Martinelli, L.A. (2007) Historical land-cover/use in different slope and riparian buffer zones in watersheds of the state of São Paulo, Brazil. *Scientia Agricola* 64: 325–335.

Somerville, C. (2007) Biofuels. *Current Biology* 17: R115–R119.

Somerville, C., Youngs, H., Taylor, C., Davis, S.C., and Long, S.P. (2010) Feedstocks for lignocellulosic biofuels. *Science* 329: 790–792.

Stromberg, P.M., Gasparatos, A., Lee, J.S.H., Garcia-Ulloa, J., Koh, L.P., and Takeuchi, K. (2010) Impacts of liquid biofuels on ecosystem services and biodiversity. Policy Report. Yokohama: UNU–IAS.

Swinton, S.M., Lupi, F., Robertson, G.P., and Hamilton, S.K. (2007) Ecosystem services and agriculture: cultivating agricultural ecosystems for diverse benefits. *Ecological Economics* 64: 245–252.

Tallis, H., Yukuan, W., Bin, F., *et al.* (2010) The natural capital project. *Bulletin of the British Ecological Society* 41: 10–13.

Tilman, D., Cassman, K.G., Matson, P.A., Naylor, R., and Polasky, S. (2002) Agricultural sustainability and intensive production practices. *Nature* 418: 671–677.

Tilman, D., Hill, J., and Lehman, C. (2006) Carbon-negative biofuels from low-input high-diversity grassland biomass. *Science* 314: 1598–1600.

Turner, B.T., Plevin, R.J., O'Hare, M., and Farrell, A.E. (2007) *Creating Markets for Green Biofuels*. Research Report UCB–ITSTSRC–RR–2007–1. Institute of Transportation Studies, University of California, Berkeley, California.

United Nations (UN). (2007) *Sustainable Bioenergy: A Framework for Decision Makers*, United Nations, Rome.

Venter, O., Meijaard, E., Possingham, H., *et al.* (2009) Carbon payments as a safeguard for threatened tropical mammals. *Conservation Letters* 2: 123–129.

Vitousek, P.M., Mooney, H.A., Lubchenco, J., and Melillo, J.M. (1997) Human domination of Earth's ecosystems. *Science* 277: 494–499.

Winter, M. and Rushbrook, L. (2003) Literature review of the English rural economy – Final report to Defra.

Chapter 12
Biofuel Crops and Greenhouse Gases

A. Hastings, J. Yeluripati, J. Hillier, and P. Smith
School of Biological Sciences, University of Aberdeen, Aberdeen, UK

Introduction

Climate change is one of the major concerns for humanity in the twenty-first century. Water vapor, carbon dioxide (CO_2), nitrous oxide (N_2O), and methane (CH_4) are the main gases that trap outward going infrared radiation (the greenhouse effect) and keep the planet temperature in the comfort zone for life around 15 °C. Of these greenhouse gases (GHGs), CO_2, N_2O, and CH_4 are generated by anthropogenic activity in sufficient quantities to increase their atmospheric concentrations and thus augment the greenhouse effect. The greenhouse forcing impacts of these gases are classified by their long wave radiation absorption rate and their lifetime in the atmosphere normalized to the impact of CO_2 which is unity, CH_4 is 26, and N_2O is 265 over a 100-year time frame.

These three GHGs are released to the atmosphere during anthropogenic economic, industrial, and land management activities. Carbon dioxide is released by burning fossil fuels, N_2O is released during the production of fertilizers and explosives, and CH_4 is released during the production of fossil fuels and in the processing of waste.

Concern about climate change has led to the need to reduce anthropogenic GHG emissions by replacing the energy obtained from fossil fuels for heating, transportation, and electricity generation by alternative energy sources that emit less GHG such as nuclear, hydro, solar, wind, tidal, and bioenergy. In addition, these energy sources can provide a measure of energy security and economic activity. All alternative energy sources that use natural drivers will cause some carbon emissions, either by embedded carbon in the construction materials such as cement and metals, energy used in the manufacture and operating the equipment, or by land-use change. These GHG emissions need to be quantified and amortized over the life of the equipment and apportioned to the energy produced so that their GHG impact can be compared to those of fossil fuel produced energy. This process is known as life-cycle analysis (LCA).

In theory, burning biomass for energy should be carbon neutral as the carbon absorbed from the atmosphere by the plant by photosynthesis is returned to the atmosphere during the burning

Biofuel Crop Sustainability, First Edition. Edited by Bharat P. Singh.

process, releasing the chemical energy stored in the plant carbon compounds derived from the solar energy. This is true in the case of natural vegetation, where in the life cycle of plants, dead vegetation is burnt rather than just decaying; however, this process speeds up the return of carbon dioxide to the atmosphere and reduces time constant of the carbon cycle of vegetation and soil. However, emissions are saved by the amount that would have been released by the fossil energy replaced. In the case of crops or forest grown or managed specifically for energy there will be an energy input required in this process and hence some GHG emissions.

Any anthropological intervention in the process of growing vegetation whether it be changing the land cover from one vegetation type to another, disturbing the land by tillage, using fertilizers, herbicides, and pesticides, or changing the water supply or drainage leads to changes in the soil's physical and chemical properties. Thus the cultivation of feedstock for bioenergy will create some GHG emissions which need to be compared to emissions from the land use they replace and any change in carbon inventory to estimate the net impact on atmospheric GHG emissions. In addition to the biosphere GHG emissions, the embedded carbon in the machinery and plant manufacture, energy use in cultivation, transport and processing, and the conversion of the feedstock into a useable fuel need to be added to the GHG cost of the bioenergy.

The sustainability of biofuels has to be judged based upon the net energy they contribute to society and the amount of GHG they emit compared to their energy equivalent in fossil fuel. In addition, they also have to be judged by their efficient use of resources such as land and water as these are competed for in the production of food, fiber, and timber or for ecosystem services.

In this chapter we will explore these metrics by the LCA of bioenergy produced from annual and perennial feedstocks and include GHG emission from land-use change. These land-use-change emissions will include the change in carbon stocks in standing vegetation and soils as well as trace gas emissions of CH_4 due to water table changes and N_2O emissions from the use of fertilizers.

Land Cover Change

Annual and perennial crops achieve their maximum growth in one season but slower growing plants, shrubs, and trees exhibit an exponential growth to a maximum implying a slowing of carbon accumulation with each season. This is an important consideration to calculate the frequency of the harvesting of willow in short rotation coppice (SRC) bioenergy production to maximize yield and carbon capture. For example, carbon stored in willow grown for 30 years will be less than that harvested SRC, harvesting every 3 years. However, with a growth time constant of 10 years, the third-year growth will be 75% of the first year. Using this exponential model, it is estimated that the incremental carbon capture with SRC harvests would be 24 units compared to 9 units when leaving the willow to grow to maturity over the same period (as illustrated in Figure 12.1), the incremental harvest being 15 units.

If we only consider the impact of the growth time constants of various trees on the accumulation of carbon in standing vegetation and assume that carbon is assimilated in standing vegetation at a rate that is only related to the growth time constant, the accumulation of carbon from the atmosphere (−ve) in permanent standing vegetation for various species, plus carbon emitted to the atmosphere by burning the same amount of coal to produce the energy represented by the standing vegetation can be calculated. This can be compared to the carbon emitted from both fossil fuels and from producing and burning biomass from a C_4 plant like *Miscanthus* and a C_3 plant such as willow.

Figure 12.2 compares atmospheric C mitigation under different land use. The first curve for fossil fuels considers that one unit of carbon equivalent to that which could be produced

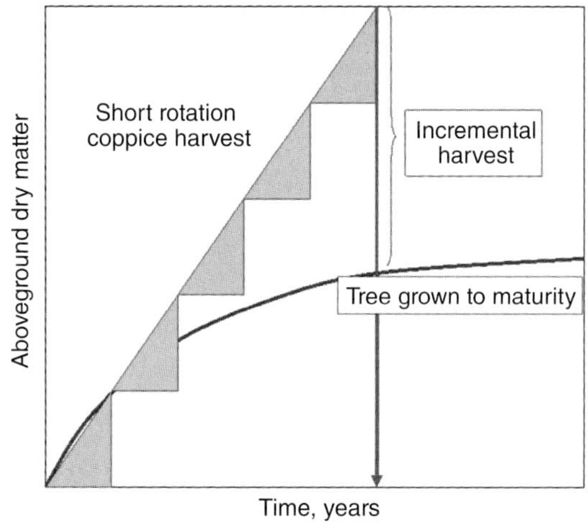

Figure 12.1. Graphical depiction of the carbon in standing vegetation of a tree grown to maturity to incremental harvesting by short rotation coppicing. The mitigated carbon at any time is the difference between the carbon in the tree if left to grow compared to the cumulative harvested carbon at the same time (Adapted from Marland and Schlamadinger, 1995).

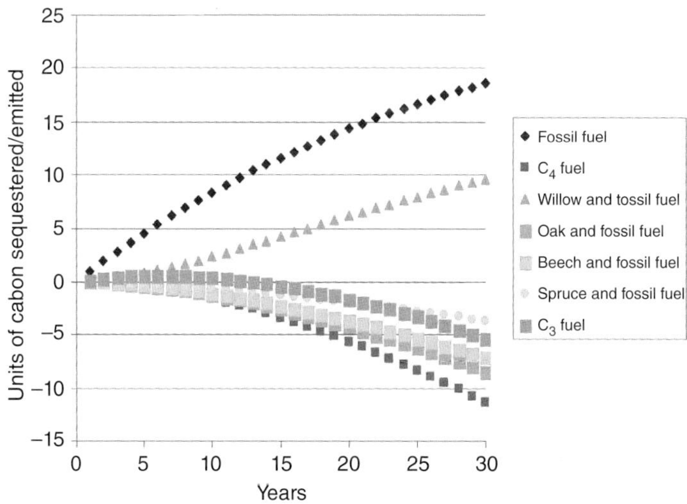

Figure 12.2. Estimation of the best use of land to mitigate atmospheric carbon. Graph of units of carbon sequestered or emitted versus time per hectare of land, considering the rate of absorption of atmospheric CO_2 and the rate of growth of various plants and trees. One unit of carbon represents the amount of carbon that can be fixed by plants or trees on 1 ha of land per year. Curves show the cumulative carbon emitted to or absorbed from the atmosphere. The curve for fossil fuels shows the accumulation of atmospheric carbon from burning the amount of coal that would replace the biomass grown on 1 ha of land. The C_4 and C_3 curves show the impact of using the land for growing *Miscanthus* and SRC willow biomass fuels, respectively, replacing coal. The other curves show the impact of growing various species of trees, willow, oak, beech, and spruce, with different growth rates, and burning the fossil fuel of the same energy value as could be grown on the land as a biomass fuel.

on a unit of land and emitted to the atmosphere each year. By year 30, 30 units would have been emitted to the atmosphere, of which 11.4 would have been absorbed by earth systems and 18.6 would remain in the atmosphere (assuming the 30-year time constant). If the land is used for bioenergy crops such as *Miscanthus* (C_4 crop) then, assuming the carbon cycling is 100% efficient and obtained at no energy and GHG cost, then after 30 years the use of bioenergy would have sequestered 11.4 units of carbon. However, SRC willow (C_3 crop), used as a bioenergy crop which has a 10-year growth time constant, will only sequester 5.4 units of carbon. These fuels are compared to plantations of oak, beech, and spruce, using the same land but at the same time burning the same amount of fossil fuel to replace the bioenergy grown on the same area of land during 30 years. Here we look at the units of carbon stored in the standing vegetation minus the net emissions from burning fossil fuel over the same period. The amount of carbon sequestered is 8.5, 7.1, and 3.7 units, respectively, and depend on the growth time constant of the tree of 150, 100, and 50 years, respectively. Willow left unharvested for the 30 years would result in net emissions of 9.4 units. From this, it can be seen that perennial and annual bioenergy C_4 crops are the most effective at sequestering carbon, trees with growth time constants greater than 50 years being better than C_3 crops bioenergy, such as SRC willow biomass, with a growth time constant of 10 years.

Land as a Limiting Factor

There is a limited availability of suitable land for conversion to grow bioenergy feedstocks caused by competition for other purposes such as the production of food, fiber, wood, and ecosystem services. Thus, it is important that the net energy yield of these feedstocks is maximized. This net energy should be the gross energy of the fuel produced minus the energy cost of production. The starting point for this analysis is finding crops that give the maximum yield of useful products. Crop yields depend on soil conditions, water availability, and climate when the optimum agrometric management is used. Thus, different crops will be more suitable in different areas. As an example, the yields of various crops suitable for production, taken from Eurostat, is shown for European countries and the wide variation in yields can be observed in Table 12.1.

Soil Emissions

Quantitative data and evidence for changes of soil organic carbon (SOC) with agricultural practices have been acquired from soil analysis from long-term, field-scale farm experiments (Smith *et al.*, 2000), some of which have been ongoing for 150 years, especially in Europe and the United States. Initially, these experiments were designed to see how soil fertility could be changed to increase crop yields with various types of fertilizers and farm management. These experiments have recorded changes in SOC from land-use change as some of these sites had been used for arable agriculture for many centuries prior to the start of the crop experiments, as in Rothamsted (Poulton *et al.*, 2003). Others had been natural ecosystems like the Morrow Plot experiments, established on previously C_4 grass prairies (Odell *et al.*, 1984). For the Morrow site, there is a 130-year time series showing the reduction in SOC to a new equilibrium which was reached in ±100 years and which suggests an exponential decay (e-fold) with a time constant of change of around 30 years (Odell *et al.*, 1984). At the same site, there are plots that show an increase in SOC from 1974 onward due to the use of

Table 12.1. Energy feedstock yield for each European country.

Crop yield	Wheat grain[a]	Barley grain[a]	Maize grain[a]	Largest grain yield	Best grain yield[a]	OSR seed[a]	Maize green[a]	Sugar beet yield[b]	Round wood overbark[b]	*Miscanthus* dry matter[c]
Unit	Mg ha^{-1}	Mg ha^{-1}	Mg ha^{-1}		Mg ha^{-1}	Mg ha^{-1}	Mg ha^{-1}	Mg ha^{-1}	Mg ha^{-1}	Mg ha^{-1}
Albania	2.9	2.4	3.7	Maize	3.7		17.9	30.8	4.5	16.5
Austria	5	4.5	9.5	Maize	9.5	2.7	45.7	64.8	8.2	15.5
Belgium	8.2	7.3	11.1	Maize	11.1	3.6	45.9	67.1	8	13.5
Bosnia and Herzegovina	1.7	1.7	2.2	Maize	2.2		20		4.5	17.9
Bulgaria	3	2.7	3.5	Maize	3.5	1.4	10.1	17.9	3.3	2.3
Croatia	3.9	3.1	5.3	Maize	5.3	2.1	24.2	37.4	4.5	17.9
Cyprus	1.8	1.6		Wheat	1.8				1	
Czech Republic	4.7	3.9	6.7	Maize	6.7	2.7	32.5	48.3	8	
Denmark	7.1	5.2		Wheat	7.1	3.1	35.5	56.7	7.3	8
Estonia	2.3	2		Wheat	2.3	1.5	21.3		3.7	
Finland	3.4	3.2		Wheat	3.4	1.7		34.2	3.5	
France	6.9	6.2	8.6	Maize	8.6	3.2	40.3	75.4	6.4	14.8
Germany	7.3	5.9	8.7	Maize	8.7	3.5	43.6	58.2	8.8	13.2
Greece	2.2	2.3	8.9	Maize	8.9		49.5	61.5	1.1	10.4
Hungary	3.9	3.3	5.7	Maize	5.7	1.9	23.9	44.4	5.8	15.6
Iceland										
Ireland	8.7	6.6		Wheat	8.7	3	1	49	6	9.8
Italy	3.3	3.6	9.2	Maize	9.2	1.4	51.8	48.1	3.1	16.4
Latvia	2.9	2		Wheat	2.9	1.7	21.9	35.8	4.6	
Lithuania	3.3	2.4	3.1	Wheat	3.3	1.6	25.2	36.2	5	
Luxembourg	6	5.2	7.9	Maize	7.9	3.3	27		7.8	13.2
Macedonia	2.7	2.5	3.9	Maize	3.9	2.3	22.7	30.1	4.5	16.1
Malta										
Netherlands	8.2	5.9	11.1	Maize	11.2	3.5	44.4	60.5	7	11.9
Norway	4.5	3.6		Wheat	5.4				5.5	
Poland	3.6	3	5.7	Maize	5.7	2.8	41.5	40.9	4.8	
Portugal	1.4	1.4	5.5	Maize	5.5			65.6	6.8	17.3
Romania	2.5	2.3	3.1	Maize	3.1	1.2	15.8	23.4		
Slovakia	3.9	3.2	5	Maize	5	2	21	42.1	7.2	14.5
Slovenia	4.4	3.7	6.9	Maize	6.9	2.4	40.8	44.6	5.9	17.4
Spain	2.7	2.7	9.6	Maize	9.6	1.5	45.5	66.6	2.7	15.3
Sweden	5.9	4.1		Wheat	5.9	2.6		47.8	4	
Switzerland	5.9	6.2	8.9	Maize	8	3.1		71.9	7.7	12
Turkey	2.2	2.4	5.1	Maize	5.1	2.4		42.5	4.5	
United Kingdom	7.7	5.7		Wheat	7.7	3.1	3.1	55	6.9	10.1

[a]Harvest weight with 14% moisture
[b]Harvest weight green.
[c]Dry weight.

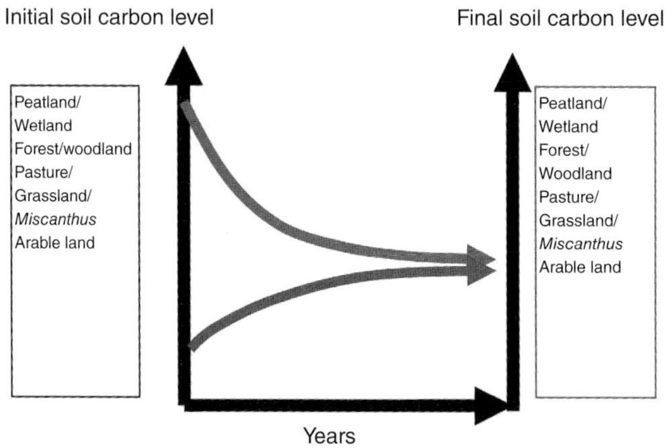

Figure 12.3. Diagrammatic representation of the change in SOC with time for a land-use change. The new equilibrium is reached in ±100 years which implies an e-fold time constant of change of ±30 years.

low tillage farm management on those plots. This increase has a similar time constant for the increase. This change of equilibrium is reversible. In Rothamsted, there had been a 105-year experiment to convert long-term arable land to a wilderness and deciduous forest (Poulton *et al.*, 2003; Blair *et al.*, 2006). Fitting the published data from the Rothamsted sites (Poulton *et al.*, 2003), the Morrow Plots (Odell *et al.*, 1984), and the Sanborn site (Huggins *et al.*, 1998), with an exponential growth (e-fold) curve, the best fit RMSE ($p > 0.001$) is achieved with a time constant of 30, 32, and 32 years, respectively.

Guo and Gifford (2002) reviewed 74 publications and tabulated how the SOC changed with land-use change from a high SOC stock use (peat, forest, woodland, wetlands, or grassland) to a low carbon stock land use (arable land) and vice versa. Changes in agricultural management practices such as less tillage, manure fertilization in lieu of chemical fertilizers, or incorporating straw and crop residue in the soil increase SOC, and increasing tillage and reducing organic additions to the soil reduce SOC (Smith *et al.*, 2000). However, if no such changes are made, then the SOC remains constant. This process is illustrated in Figure 12.3.

Reductions in SOC that result in CO_2 emissions to the atmosphere also release nitrogen compounds that can result in N_2O emissions. However, if the same land management system is used or the same natural ecosystem is maintained, then the soil carbon is in equilibrium and there is no net emission of carbon and the photosynthetic input balances the soil respiration (Smith *et al.*, 2000). So in the case of bioenergy derived from food crops, if there is no land-use or management change and the same intensive cropping system is used to maintain high crop yields, there will be no SOC changes relative to their cultivation for food crops.

The concern regarding using arable land for growing bioenergy crops, either food or second-generation dedicated bioenergy crops such as perennial grasses and SRC woodland, is that the arable land is taken out of food production. As the world's population is growing and there is a general expectation and ambition to increase the consumption of animal protein and a wider variety of vegetables, the area of land required to grow food satisfy this diet will increase. This will occur even if crop yield is increased to maximize the efficiency of land use. So the competition will be between land used for food and land used for energy crops. Mark Twain said, "Buy land as they are not making it anymore"; it is a fixed resource, so it is inevitable that

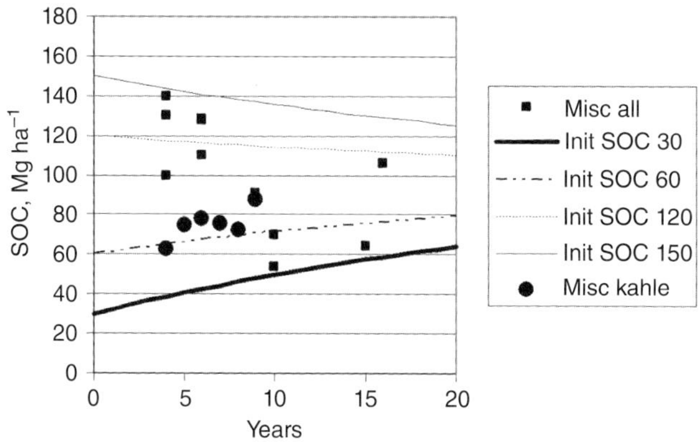

Figure 12.4. Plot of SOC measured in *Miscanthus* trial sites in Germany, Denmark, and Ireland plotted against the length of time under *Miscanthus* (■). Time series from Kahle *et al.* (2001) is plotted (•). In addition four curves representing the exponential change from initial SOCs of 30, 60, 120, and 150 Mg C ha^{-1} to the assumed natural equilibrium level of 100 Mg C ha^{-1} with a time constant of 30 years are plotted.

if more land is used for bioenergy, then more land will be converted from natural grasslands and forest to satisfy both the food and fuel need of the world population. This can lead to the exporting of emissions by one country converting food-producing land to grow energy crops, with a GHG saving, to another that converts natural habitats to arable for food production that is eaten by the country growing the bioenergy feedstock. This is not currently considered in international GHG accounting protocols.

If arable land with a low soil carbon is converted to forest to be used for bioenergy, then the soil carbon will increase—likewise if converted to perennial grasses like *Miscanthus* or to short rotation coppicing of species like willow poplar or eucalyptus. Recent research into soil carbon changes under *Miscanthus* plots has demonstrated this.

As seen in Figure 12.4, the experimental SOC measurements, made on European *Miscanthus* trial sites, fall within the envelope created by the theoretical curves of SOC changes between initial SOC levels of 30 and 150 to equilibrium of ±100 Mg C ha^{-1} with a time constant of ±30 years. The German trial points from Kahle *et al.* (2001) (annotated as "Misc Kahle") fall on the curve that suggests that the SOC is increasing from an initial 60 to an equilibrium level of around 100 Mg C ha^{-1}. The SOC level observed at the long-term C_3 grassland control plot at the same site is 60 Mg C ha^{-1}, which had been C_3 pasture for centuries before the experiment began and was presumed to be at its equilibrium level. To check this, the Kahle *et al.* (2001) time series of SOC measurements made on *Miscanthus* trial and the C_3 control plots were tested (Figure 12.5) using a student t test and linear regression for significant differences or trend versus time.

The linear regression shows an increase in SOC with time on the *Miscanthus* plot $R^2 = 0.57$ and a constant value for C_3 grass. The student t test indicates a significant difference ($p = 0.008$) between the two sets of measurements. The available data, while insufficient to rigorously test statistically, is suggestive of a trend toward a higher equilibrium level of SOC for the *Miscanthus* crop of around 100 Mg C ha^{-1} on this site with a sandy loam soil, and that *Miscanthus* SOC levels are higher than C_3 levels. However, data from Schneckenberger and

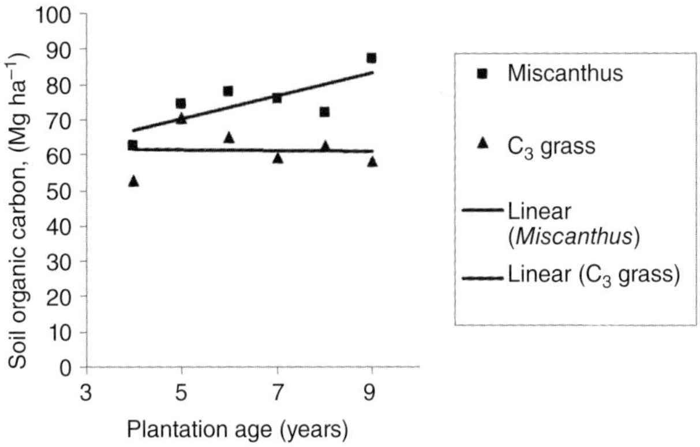

Figure 12.5. Soil organic carbon for both *Miscanthus* trial and the associated C_3 grass controls plotted against time from land-use change from C_3 to *Miscanthus*. Plot shows a constant C_3 SOC value and an increasing *Miscanthus* SOC value.

Kuzyakov (2007) on other sites shows no significant SOC difference between *Miscanthus* and C_3 grasses on both sandy and loamy soils.

More recent work by Dondini *et al.* (2009) on a site in Carlow, Ireland demonstrated a similar equilibrium level for *Miscanthus* plot SOC under temperate climate conditions when experimental results were compared to the model (RothC) predictions (Figure 12.6).

From the available limited published data, it seems that the equilibrium SOC for *Miscanthus* plantations is between 100 and 120 Mg ha^{-1}, similar to natural perennial grasslands, so any conversion from arable land will have a short-term (~30 years) carbon sequestration effect. Work with SRC willow indicates a similar benefit.

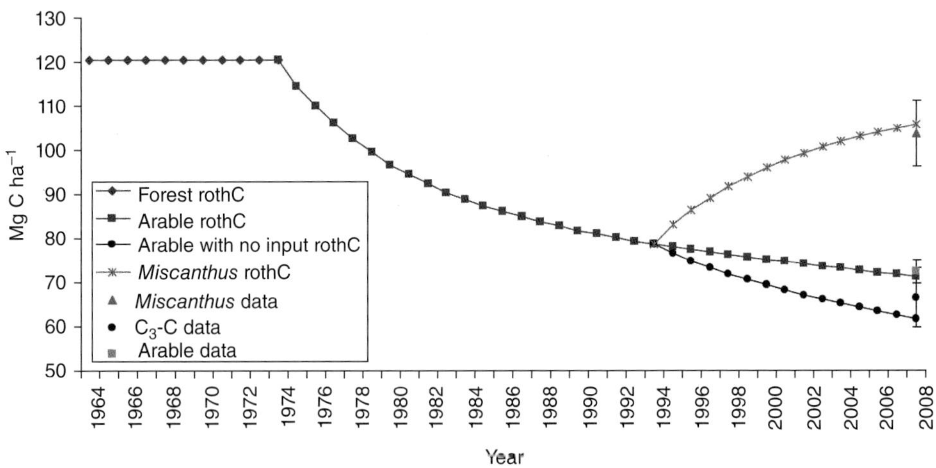

Figure 12.6. Soil C content of different plantations during a period time of 41 years at Carlow, Ireland. Plot shows progression of SOC from forest equilibrium to arable conversion in 1974, conversion to *Miscanthus* in 1995 with the control plot continuing as arable. Differentiation of original carbon from the C_3 arable and forest and the C_4 *Miscanthus* is by C isotope analysis.

Use of Fertilizers

It is important to consider land-use change when evaluating the GHG neutrality of bioenergy systems and include the net carbon and nitrogen budget, but in this LCA we do not consider N_2O emissions for SOC changes. However, when chemical nitrogen fertilizers are used in the crop management, then nitrous oxide (N_2O) emissions due to denitrification are calculated using the default tier 1 IPCC emission factor for N_2O emissions from N fertilizer application (IPCC, 2007). In addition, energy used and N_2O released to the atmosphere during the manufacture of the chemicals is considered (Patzek, 2008). The N_2O emission associated with the manufacture of chemical fertilizers is normally accredited to the chemical industry in GHG emission estimates (IPCC, 2007) and in the LCA (Shapouri *et al.*, 2002), but as fertilizer is only used as a crop nutrient, this is incorrect as the emissions would not occur if the crops were not grown. N_2O is produced primarily from the microbial processes of nitrification and denitrification in the soil. In well-aerated conditions, N_2O emissions from nitrification of ammonium-based fertilizers can be substantial. In wet soils where aeration is restricted, denitrification is generally the source of N_2O. Research has also shown that a number of individual factors are controllers of nitrification, denitrification, and N_2O emissions. Such factors include soil water content, temperature, nitrate or ammonium concentration, available organic carbon and pH. Increases in the amount of N added to the soil generally increases N_2O emissions. The temporal pattern of N_2O emissions following fertilization is generally that of a large efflux of N_2O occurring for a short time (few weeks after fertilization). After this time, emission rates are reduced to fluctuate around a low baseline level independent of the amount of fertilizer applied (Mosier *et al.*, 1982).

Crop Management

The first part of LCA is to determine the energy used and the GHG emitted during the cultivation, farm management, transportation, crop drying, and other conditioning for each bioenergy system feedstock. These costs are divided into those annual fixed costs that relate to a hectare of land each year, including crop establishment costs for perennial *Miscanthus* and SRC willow plantations which are amortized over the life of the plantation, and those that relate to the feedstock yield per hectare, such as fertilizer and transportation costs. This enables the LCA to account for the variable yields in each country.

Food Crops Grown for Energy

Based upon the mean yield of each crop per country, the carbon and energy cost of farming the food crops grown for energy are calculated considering that they are grown on arable land with the same crop management as if grown for food. The SOC is considered to remain constant as current arable land only is considered for energy crops. Soil and the plant respiration are explicitly ignored at this point and the normal inclusion of roots and plant residue is considered to maintain the SOC at a constant level. Here we calculate the emissions of food crops used for bioenergy such as wheat, maize, oilseed rape (OSR), sugar beet, green maize as food crops by using the parameters for farm management from St. Clair *et al.* (2008), who used data for energy costs of mechanical interventions, herbicide, biocide, insecticide, fungicide, and enzyme carbon and energy costs from (2004). Conventional intensive management with conventional mouldboard plough tillage is considered. The energy, carbon, and GHG costs for human work days, seed production, machinery manufacture, and fertilizer manufacture using parameters from Patzek (2008) are also included. This includes the N_2O emission from denitrification using the IPCC default factor of 1% (1 kg N_2O-N per 100 kg of N in fertilizer applied) for arable land, and including the CO_2 emissions and N_2O emission factor from the manufacture of the fertilizer ammonium nitrate of 2%.

All transportation around the farm and transport to the biofuel/energy processing plant is considered. The processing plant is considered to be within 30 km. The energy and GHG costs of farming each crop is broken into a fixed annual cost for the crop and one that is related to the yield. Fertilization is considered as for the intensive chemical fertilization case with only crop residue being left in the field for wheat, OSR, and sugar beet. For grain maize, stover is left on the field and incorporated (Patzek, 2008). For green maize, the anaerobic digester residue is returned to the field as manure, reducing the requirement for chemical fertilizers by 50% (Gerin *et al.*, 2008). The carbon cost of production energy is 21 g C MJ, considering oil is used as the fuel.

SRC Willow

The SRC willow LCA uses overbark roundwood yields (proxy for SRC willow) using SRC crop management to optimize production in terms of fertilization with a harvest every 3 years. The biomass fuel is considered to be used in a co-fired coal power station with a mean transport distance of 200 km. This includes the N_2O emission from denitrification using the IPCC default factor of 1% for arable land and including N_2O emission factor from the manufacture of the fertilizer ammonium nitrate of 2% (Patzek, 2008). Soil carbon emissions are considered based upon the conversion of arable land to SRC plantations using factors from Hillier *et al.* (2009). The plantation life is considered to be 20 years. The carbon cost of production energy is 21 g C MJ, considering oil is used as the fuel.

Miscanthus

The LCA uses modeled yields from MiscanFor (Hastings *et al.*, 2009a,b) to calculate the crop management energy, and the carbon cost is based upon three scenarios: (1) used as a biomass fuel with current crop management at a large-scale coal power station with a mean transportation distance of 200 km; (2) used locally with current crop management as a biomass fuel or other bioenergy system feedstock with a mean transportation distance of 30 km; and (3) used locally with future potential crop management. The carbon cost of production energy is 21 g C MJ, considering oil is used as the fuel. Current management practice is crop establishment by micropropagation into a standard arable crop seedbed, harvested each year following the second/third growing season in the spring after ripening (allowing translocation of nutrients to the rhizome and reduction in moisture content from 50% to about 30%), further swath drying in the field, chipping (large baling is probably best), and transportation to the place of use. Fertilizers are applied to replace nutrients removed at the harvest. The plantation life is considered to be 15 years. This management scenario is considered for both co firing remotely and use locally as a furnace fuel for combined heat and power. As micropropagation is energy intensive, the third scenario considers that this is replaced by rhizome propagation, which is becoming the new standard. This case is only considered for local furnace fuel use. These management costs are summarized in Table 12.2.

Table 12.2. Summary of on-farm energy and carbon costs of production.

Feedstock	Crop management	Transport distance (km)	Energy cost GJ ha^{-1} yr^{-1}	Energy cost GJ Mg^{-1}	Greenhouse gas emissions kg C ha^{-1} yr^{-1}	Greenhouse gas emissions kg C Mg^{-1}[e]
Wheat	Intensive[a]	20	9.97	1.40	219	155
Barley	Intensive[a]	20	9.97	1.40	219	155
Maize	Intensive[a]	20	10.30	1.14	227	125
Sugar beet	Intensive[b]	20	10.88	0.43	239	19
Oil seed rape	Intensive[a]	20	10.40	2.34	229	267
Willow	SRC 3 yrs[c]	200	2.51	1.68	55	75
Willow	SRC 3 yrs[c]	20	2.51	0.57	55	50
Green maize	Intensive[d]	20	9.60	0.10	211	11
Miscanthus	Micropropagation[c]	200	9.07	1.72	199	81
Miscanthus	Micropropagation[c]	20	9.07	0.61	199	57
Miscanthus	Rhizome propagation[c]	20	5.64	0.61	124	57

[a]Harvest weight with 14% moisture.
[b]Harvest weight with 75% moisture and 17% sugar.
[c]Dry weight.
[d]Harvest weight green.
[e]CO_2 equivalent C with N_2O equivalent to 296 × CO_2.

Conversion Processes and Cost

External energy cost of production for each bioenergy system of converting the feedstock to a useful fuel is calculated based upon published conversion costs. The GHG cost of using this external production energy is calculated using the published carbon intensity (CI) of the associated fuel used. For electricity, we use the current (2004) CI for UK grid sourced energy, which is 134 g CO_2 equivalent C MJ^{-1}, and consider the current mix of coal, gas, nuclear, hydro, and wind generation.

Biomass Fuel

Biomass fuels require a dryness of around 15%. In this study, a combination of field drying to 30% and then undercover storage at the generation plant with air drying is considered. The external energy requirement for drying is considered in the fuel conversion cost. The transportation and chopping costs are considered in the farm management component of the LCA. The gross energy at the furnace is calculated considering energy intensity of *Miscanthus* and SRC willow of 18 GJ Mg^{-1} of dry matter and subtracting the latent heat of 30% moisture. The net energy is calculated as the gross energy minus the energy used in the feedstock production.

Ethanol

Ethanol yields from grain were calculated using a conversion factor of 400 l of ethanol per Mg (Patzek, 2008) of dry grain. This includes corrections for losses of 12% of the theoretical thermodynamics and mass balance conversion to ethanol in the fermentation process. For each country, the grain with the highest yield was identified to calculate the maximum possible ethanol yield. Straw, maize stover, and draft are considered by-products with no energy or GHG impact for on-farm use. Ethanol yields from sugar beet were calculated using the conversion factor of 105 l of ethanol per Mg of fresh beet, considering a 17% sugar content and a moisture content of 75%. Koga (2008) uses a similar conversion factor. The crop parameters used are in agreement with the published crop standards from British Sugar (2008) and due to the highly industrialized nature of the sugar beet industry in Europe and the large investment in both breeding and crop to management system to maximize the sugar yield per hectare (Scott and Jaggard, 2000), this represents the quality standard for the EU sugar beet crop. As is standard practice, the beet tops are ploughed in each year. Potatoes were also considered, but as they have a similar geographic distribution to sugar beet (Tuck *et al.*, 2006) but with lower yields, they were not investigated further. The energy cost of the production of bioethanol is 15 GJ kl^{-1}. The carbon cost of production energy is 21 g C MJ^{-1}, considering oil is used as the fuel.

Biodiesel

For first-generation feedstocks for biodiesel, the oil crops grown in Europe that could be used were linseed, sunflower, OSR, hemp, castor, field mustard, or olive; of these, OSR is the most widely gown with the highest yield of oil per hectare in Europe. Olive oil was not considered as a biofuel as Europe does not produce a surplus of this valuable food oil. Sunflower is restricted to Southern Europe at the moment and has a high yield in some climates, so may become more used with climate change (Tuck *et al.*, 2006). The conversion factor for biodiesel yields from OSR is 363.5 l of biodiesel per Mg of rape seed (Karaosmanoğlu *et al.*, 1996; Peterson and Hustrulid, 1998; Rathke *et al.*, 2006). The OSR stem and oil seed cake are considered as by-products with no energy and GHG impact for on-farm use. The energy cost of the production of biodiesel is 6.5 GJ kl^{-1}. The carbon cost of production energy is 21 g C MJ^{-1}, considering oil is used as the fuel.

Biogas

Anaerobic digesters have been used for some time on a large scale at sewage plants to produce biogas for heating and electricity generation. This technology is also used on farms to produce biogas from animal excreta and other farm wastes. This biogas has been used for heating or microgeneration on the farm and the residue incorporated in the land as an organic fertilizer. To increase the organic feedstock to enable farm plants to export energy commercially, green maize is commonly used (Gerin *et al.*, 2008). Green maize use as a feedstock for biogas is analyzed using a conversion factor of 3.87 GJ gas gross energy per tonne of green maize dry matter. The external energy cost of producing this gas is 5% of the gross energy. The carbon cost of production energy is 21 g C MJ^{-1}, considering oil is used as the external fuel. *Miscanthus* is also considered as an anaerobic digester feedstock with the same conversion factors as for green maize, with similar treatment for the residue (Gerin *et al.*, 2008). However, this would entail cutting *Miscanthus* green, before the repartition of the nutrients to the rhizome. This will have an impact on the amount of fertilizer that needs to be added each year to balance the harvest offtake. However, this is offset using the digester residue as a fertilizer and making up the difference with chemical fertilizers.

Cellulosic Ethanol

The production of ethanol from a cellulose feedstock is a process that has been used commercially since the First World War (Harris *et al.*, 1945; Saeman, 1945). The first attempt at commercialization in Germany in 1898 produced 76 l Mg^{-1} wood, based upon a process that hydrolyzed the cellulose to glucose using dilute acid. This process was improved successively during the two world wars and by 1945 it could produce 211 l Mg^{-1}. During the First World War, a single stage dilute sulfuric acid process was developed in the United States, but it only produced 105 l Mg^{-1}. However, the impetus of the Second World War allowed further development to increase this yield to 211 l Mg^{-1}, with the ethanol being used in the production of synthetic rubber. After the war, the plants were closed as they were not profitable in a free market, freed from the constraints and needs of total war.

Global warming and the need to reduce GHG emission, plus the increasing shortage of crude oil, which has driven up energy prices, has helped to create both the political and economic climate in which cellulosic ethanol can be reconsidered as a practical option. Several pilot plants are already in operation and commercial scale production target dates have been set for a number of plants in construction. The technology used in these cellulosic ethanol plants is steam explosion/acid/enzyme extraction of the glucose from the biomass, then fermentation. They currently convert biomass to ethanol at the rate of 280 l Mg^{-1} biomass (Sokhansanj *et al.*, 2002; Pimentel and Patzek, 2005).

This LCA for *Miscanthus* as a feedstock for cellulose ethanol considers the same crop yields and management as for its use as biomass fuel with the processing plant a mean 30 km distant from the source of the crop. Only the current practice crop management case is considered. An ethanol yield, same as corn stover, of 280 l Mg^{-1} (Foutch *et al.*, 1980; Hettenhaus and Wooley, 2000; Sokhansanj *et al.*, 2002) is used in the model as well as an energy cost of cellulose ethanol production of 15 MJ l^{-1} (Pimentel and Patzek, 2005). The carbon cost of production energy is 21 g C MJ^{-1}, considering oil is used as the fuel.

Biorefined Biodiesel

The synthesis of liquid hydrocarbons from coal is an old technology developed by two German researchers Franz Fischer and Hans Tropsch in 1926 (US Patent, 1930). The Fischer–Tropsch process (FTP) entails gasification of the coal by partial oxidization, hydrolyzation by the addition of steam, and then recombination of the resultant CO and H in the presence of catalysts such as the transition metals: cobalt, iron, and ruthenium or nickel to form liquid hydrocarbons. This technique was developed and used by both Japan and Germany during the Second World War but its use decreased thereafter due to the free trade in petroleum products. A few companies continue to use the process commercially, SASOIL in South Africa to overcome the embargo during the apartheid years, and Shell in Bintulu used the process to produce low sulfur diesel from natural gas. The latter was a pilot gas-to-liquid commercial project (Price Waterhouse Coopers, 2003), using the catalytic recombination part of the Fischer–Tropsch process (FTP).

This process can use any feedstock containing carbon and hydrogen and is also used to upgrade tar from tar sands, to produce oil from wood, or biodiesel from livestock or other organic waste. Gasification of waste products such as black liquor and oils from paper production is commonly used for steam, heat, and power generation at paper mills (Vattenfall, 2005). Recently, an FTP has been added to a waste gasification in a plant in Finland to produce

Table 12.3. Summary of feedstock conversion rate, fuel energy intensity costs of conversion.

		Feedstock conversion rate		Fuel energy intensity		Conversion energy cost	
Feedstock.	Biofuel.	Unit	Value	Unit	Value	Unit	Value
Wheat [a]	Ethanol	$l\ Mg^{-1}$	400	$MJ\ kg^{-1}$	35	$MJ\ l^{-1}$	15
Barley [a]	Ethanol	$l\ Mg^{-1}$	400	$MJ\ kg^{-1}$	35	$MJ\ l^{-1}$	15
Maize [a]	Ethanol	$l\ Mg^{-1}$	400	$MJ\ kg^{-1}$	35	$MJ\ l^{-1}$	15
Sugar beet[b]	Ethanol	$l\ Mg^{-1}$	105	$MJ\ kg^{-1}$	35	$MJ\ l^{-1}$	15
Oil seed rape[a]	Biodiesel	$l\ Mg^{-1}$	363.50	$MJ\ kg^{-1}$	43.70	$MJ\ l^{-1}$	6.50
Willow[c]	Biomass	pu	1	$MJ\ kg^{-1}$	18	$MJ\ kg^{-1}$	0.82[e]
Green maize[d]	Biogas	$GJ\ Mg^{-1}$	3.87	$MJ\ kg^{-1}$	55.70	Percentage energy	5[g]
Miscanthus[c]	Biomass	pu	1	$MJ\ kg^{-1}$	18	$MJ\ kg^{-1}$	0.82[e]
Miscanthus[c]	Biogas	$GJ\ Mg^{-1}$	3.87	$MJ\ kg^{-1}$	55.70	Percentage energy	5[g]
Miscanthus[c]	Cellulosic ethanol	$l\ Mg^{-1}$	280	$MJ\ kg^{-1}$	35	$MJ\ l^{-1}$	15
Miscanthus[c]	Biorefined biodiesel	$l\ Mg^{-1}$	199	$MJ\ kg^{-1}$	43.70	$MJ\ l^{-1}$	[f]

[a]Harvest weight with 14% moisture.
[b]Harvest weight with 75% moisture and 17% sugar.
[c]Dry weight assumes 100% conversion, conversion energy cost is latent heat of 30% moisture.
[d]Harvest weight green.
[e]Latent heat of 30% moisture.
[f]Conversion enegy cost included in conversion rate as all energy comes from feedstock.
[g]5% of gas is used on site for heating and pump energy.

biodiesel. In Germany, a pilot commercial plant to produce biodiesel from wood and cellulose-rich crops has been commissioned by CHOREN Industries in Freiberg (Blades *et al.*, 2006) in conjunction with Audi and Volkswagen. The plant has been used with a variety of biological feedstocks including wood, straw, and *Miscanthus*. An LCA was conducted on this plant (Baitz *et al.*, 2004) and reported that if all energy used in the plant was derived from the biomass input, the ratio of conversion of biomass dry matter to biodiesel was $199\ l\ Mg^{-1}$ dry vegetable matter. This gives an effective biodiesel energy output of $5.98\ GJ\ Mg^{-1}$ of dry matter.

A summary of the conversion rates is shown in Table 12.3.

LCA Methodology and Boundary Conditions

A full LCA is required to evaluate the sustainability of a bioenergy system and can be divided into three parts: farming the crop, turning the crop into fuel, and final energy conversion. Part of the energy input and GHG emissions comes from the farm management strategy employed for land preparation, planting and seeding, fertilization, pest and weed control, harvesting, storage, drying, preparing and processing the bioenergy feedstock as well as the final use in energy conversion. Sustainability of bioenergy systems requires that the energy used to produce the fuel is less than the energy harvested and that the GHG cost is less than the equivalent fossil fuel it replaces (Figure 12.7).

Here we present an analysis of the carbon and energy cost for farming the energy crops considering that they are grown on arable land with the same crop management as if grown for

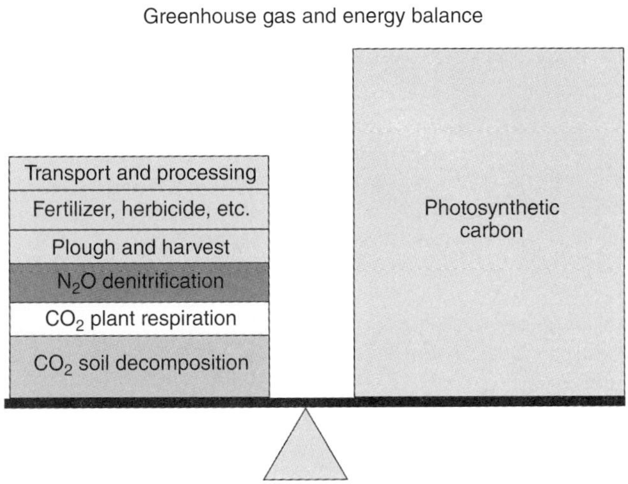

Figure 12.7. Sustainability balance for bioenergy systems.

food. The SOC is considered to remain constant, so soil respiration is explicitly ignored and the plant respiration is implicitly included in the harvest yield. The study is made for wheat, maize, OSR, sugar beet, green maize, and roundwood (proxy for SRC willow) plantations, and a perennial grass *Miscanthus*. We consider the yields for each European country, taken from Eurostat, except for *Miscanthus* which uses modeled results from Hastings *et al.* (2008) which are summarized in Table 12.1.

This comparative study uses the parameters for farm management from St. Clair *et al.* (2008), who used data for energy costs of mechanical interventions from Lai (2004) and herbicide, biocide, insecticide, fungicide, and enzyme carbon and energy costs from Lai (2004). We include the energy, carbon, and GHG costs for human work days, seed production, machinery manufacture, and fertilizer manufacture using parameters from Patzek (2008). This includes the N_2O emission from the manufacture of the fertilizer, ammonium nitrate. All transportation around the farm and transport to the biofuel/energy processing plant is considered. The energy and GHG costs of farming each crop is broken into a fixed annual cost for the crop and one that is related to the yield. Fertilization is considered as for the intensive chemical fertilization case, with only crop residue being left in the field for wheat, OSR, and sugar beet. For grain maize, stover is left on the field and incorporated (Patzek, 2008). For green maize, the anaerobic digester residue is returned to the field as manure, reducing the requirement for chemical fertilizers by 50% (Gerin *et al.*, 2008). The feedstock production costs are summarized in Table 12.2 and the conversion costs are summarized in Table 12.3.

Three metrics are considered for the comparison of "sustainability" of bioenergy systems. The first is the net yield of energy per hectare or land energy intensity (LEI), the second is CI in terms of g C equivalent GHG emissions per MJ of fuel energy, and the final one is the ratio between energy in the fuel and the energy used in production or energy use efficiency (EUE). The resulting metrics for ethanol systems considering feedstocks of wheat, maize, sugar beet, and cellulosic ethanol from *Miscanthus* are listed for each European country in Table 12.4.

Biodiesel systems using pressed oil seed rape, and biorefined *Miscanthus* and SRC willow are presented in Table 12.5.

Table 12.4. Energy use efficiency (EUE), carbon intensity (CI), and net energy yield per hectare for bioethanol with four feedstocks.

	Wheat			Maize			Sugar beet			*Miscanthus* cellulosic		
	EUE E out/E in	CI g C MJ^{-1}	Net yield energy GJ ha^{-1}	EUE E out/E in	CI g C MJ^{-1}	Net yield energy GJ ha^{-1}	EUE E out/E in	CI g C MJ^{-1}	Net yield energy GJ ha^{-1}	EUE E out/E in	CI g C MJ^{-1}	Net yield energy GJ ha^{-1}
Albania	0.87	38	−4	0.95	33	−2	1.05	24.19	3.52	1.23	23.90	20.12
Austria	1	35	0	1.14	29	11	1.14	22.55	19.43	1.22	24.01	18.42
Belgium	1.09	33	6	1.17	29	15	1.14	22.49	20.50	1.20	24.31	14.84
Bosnia and Herzegovina	0.70	44	−7	0.80	38	−5				1.24	23.75	22.66
Bulgaria	0.87	38	−4	0.93	34	−2	0.95	26.44	−2.51	0.75	35.15	−4.98
Croatia	0.94	36	−2	1.03	31	2	1.08	23.64	6.60	1.24	23.76	22.58
Cyprus	0.72	43	−6									
Czech Republic	0.99	35	−1	1.08	30	5	1.11	23.05	11.70			
Denmark	1.07	33	4				1.13	22.76	15.64	1.11	25.85	5.09
Estonia	0.80	40	−5									
Finland	0.91	37	−3				1.06	23.87	5.12			
France	1.06	33	4	1.13	29	9	1.15	22.34	24.41	1.21	24.11	17.13
Germany	1.07	33	5	1.13	29	9	1.13	22.71	16.34	1.20	24.36	14.29
Greece				1.13	29	10	1.13	22.62	17.91	1.16	24.97	9.34
Hungary	0.95	36	−2	1.05	31	3	1.10	23.23	9.91	1.22	24.01	18.54
Iceland												
Ireland	1.10	33	7				1.11	23.03	12.03	1.15	25.15	8.28
Italy	0.90	37	−3	1.14	29	10	1.11	23.06	11.61	1.23	23.91	19.96
Latvia	0.87	38	−4				1.07	23.75	5.88			
Lithuania	0.90	37	−3	0.90	35	−3	1.07	23.72	6.07			
Luxembourg	1.04	34	2	1.11	30	7				1.20	24.36	14.29
Macedonia	0.84	39	−5	0.96	33	−1	1.04	24.25	3.22	1.22	23.95	19.43
Malta												
Netherlands	1.09	33	6	1.17	29	15	1.13	22.65	17.43	1.18	24.61	11.99
Norway	0.98	35	−1									
Poland	0.93	36	−3	1.05	31	3	1.09	23.41	8.27			
Portugal	0.65	46	−7	1.04	31	2	1.14	22.53	19.81	1.23	23.82	21.55
Romania	0.83	39	−5	0.90	35	−3	1.00	25.18	0.06			
Slovakia	0.94	36	−2	1.02	31	1	1.09	23.35	8.81	1.21	24.15	16.60
Slovenia	0.97	35	−1	1.09	30	5	1.10	23.22	9.99	1.23	23.80	21.73
Spain	0.85	39	−5	1.14	29	11	1.14	22.51	20.25	1.22	24.04	18.01
Sweden	1.04	34	2				1.11	23.08	11.46			
Switzerland	1.03	34	2	1.13	29	10	1.15	22.40	22.77	1.18	24.59	12.17
Turkey	0.78	41	−6	1.02	31	1	1.09	23.32	9.01			
United Kingdom	1.08	33	5				1.12	22.81	14.85	1.15	25.06	8.81

Table 12.5. Energy use efficiency (EUE), carbon intensity (CI), and net energy yield per hectare for biodiesel with three feedstocks.

	Oil seed rape pressed			*Miscanthus* biorefined			SRC willow biorefined		
	EUE E out/E in	CI g C MJ^{-1}	Net yield energy GJ ha^{-1}	EUE E out/E in	CI g C MJ^{-1}	Net yield energy GJ ha^{-1}	EUE E out/E in	CI g C MJ^{-1}	Net yield energy GJ ha^{-1}
Albania				7.50	7.94	124.28	7.71	7.16	34.06
Austria	1.86	25.40	19.90	7.28	8.03	116.51	9.93	6.52	64.13
Belgium	2.10	24.04	30.35	6.79	8.25	100.16	9.84	6.54	62.50
Bosnia and Herzegovina				7.79	7.83	135.90	7.71	7.16	34.06
Bulgaria	1.29	30.59	4.94	1.92	16.45	9.63	6.54	7.67	24.31
Croatia	1.65	26.87	13.34	7.78	7.83	135.52	7.71	7.16	34.06
Cyprus							2.82	12.07	5.62
Czech Republic	1.86	25.41	19.86				9.84	6.54	62.50
Denmark	1.96	24.79	23.80	4.99	9.41	55.62	9.52	6.62	56.81
Estonia							6.97	7.46	27.56
Finland	1.45	28.77	8.16				6.76	7.56	25.93
France	1.98	24.66	24.82	7.11	8.10	110.61	9.04	6.74	49.50
Germany	2.07	24.22	28.51	6.70	8.29	97.67	10.17	6.47	69
Greece				5.87	8.75	75.03	3.05	11.50	6.43
Hungary	1.57	27.62	10.98	7.30	8.02	117.08	8.67	6.84	44.62
Iceland									
Ireland	1.96	24.81	23.69	5.66	8.89	70.18	8.80	6.80	46.25
Italy	1.32	30.23	5.49	7.48	7.95	123.55	6.30	7.79	22.68
Latvia	1.46	28.62	8.48				7.79	7.12	34.87
Lithuania	1.43	28.92	7.83				8.11	7.01	38.12
Luxembourg				6.70	8.29	97.67	9.75	6.56	60.88
Macedonia	1.73	26.32	15.42	7.41	7.98	121.12	7.71	7.16	34.06
Malta									
Netherlands	2.07	24.18	28.88	6.34	8.48	87.16	9.37	6.65	54.37
Norway							8.47	6.90	42.18
Poland	1.89	25.23	20.91				7.96	7.07	36.50
Portugal				7.67	7.88	130.82	9.26	6.68	52.75
Romania	1.20	31.98	3.14						
Slovakia	1.60	27.34	11.82	7.04	8.13	108.18	9.47	6.63	56
Slovenia	1.77	26	16.83	7.69	7.87	131.63	8.74	6.82	45.44
Spain	1.36	29.79	6.21	7.23	8.05	114.65	5.80	8.09	19.43
Sweden	1.82	25.65	18.55				7.26	7.33	30
Switzerland	1.98	24.69	24.54	6.37	8.46	87.97	9.71	6.57	60.06
Turkey	1.76	26.04	16.66				7.71	7.16	34.06
United Kingdom	1.98	24.70	24.47	5.77	8.82	72.60	9.31	6.67	53.56

Table 12.6. Energy use efficiency (EUE), carbon intensity (CI), and net yield per hectare for biogas.

	Green maize			*Miscanthus*		
	EUE E out/E in	CI g C MJ^{-1}	Net yield energy GJ ha^{-1}	EUE E out/E in	CI g C MJ^{-1}	Net yield energy GJ ha^{-1}
Albania	4.67	6.68	54.53	5.50	8.65	118.66
Austria	7.68	4.84	153.74	5.38	8.74	111.22
Belgium	7.70	4.83	154.41	5.11	8.96	95.56
Bosnia and Herzegovina				5.65	8.54	129.79
Bulgaria	3.12	9.02	26.64	1.77	17.07	8.84
Croatia	5.61	5.90	76.93	5.65	8.55	129.43
Cyprus						
Czech Republic	6.57	5.32	106.58			
Denmark	6.86	5.18	117.31	4.03	10.11	52.90
Estonia						
Finland						
France	7.28	5	134.45	5.29	8.81	105.57
Germany	7.53	4.89	146.24	5.06	8.99	93.17
Greece	7.94	4.74	167.57	4.58	9.46	71.49
Hungary	5.56	5.93	75.71	5.39	8.73	111.76
Iceland						
Ireland	0.38	59.85	−6.13	4.45	9.59	66.84
Italy	8.08	4.70	175.52	5.49	8.66	117.96
Latvia	5.28	6.14	68.55			
Lithuania	5.74	5.81	80.57			
Luxembourg				5.06	8.99	93.17
Macedonia	5.40	6.05	71.45	5.45	8.69	115.64
Malta						
Netherlands	7.59	4.87	149.30	4.85	9.18	83.10
Norway						
Poland	7.37	4.96	138.67			
Portugal				5.59	8.59	124.93
Romania	4.29	7.10	46.76			
Slovakia	5.29	6.13	68.86	5.25	8.84	103.24
Slovenia	7.32	4.98	136.19	5.60	8.58	125.70
Spain	7.67	4.84	153.18	5.35	8.76	109.44
Sweden						
Switzerland				4.87	9.17	83.88
Turkey						
United Kingdom	1.15	21.01	1.63	4.52	9.52	69.16

A comparison of using green maize and *Miscanthus* for the generation of biogas in anaerobic digesters is shown in Table 12.6.

A comparison of various scenarios for the use of biomass derived from *Miscanthus* and SRC willow is shown in Table 12.7. Local use of *Miscanthus* and SRC willow in furnaces for combined head and power or heat only are compared. For *Miscanthus*, we compare rhizome to vegetative propagation for local use and also for co-firing in a large power station to demonstrate the importance of transportation to the overall GHG emissions. Finally, we show the potential impact of considering the increase on soil carbon under *Miscanthus* on the overall CI of the fuel.

Table 12.7. Energy use effic ency (EUE), carbon intensity (CI), and net energy yield per hectare for *Miscanthus* biomass in four cases: co-firing 200 km, local use at 20 km of crop from vegetative propagation, local use including SOC changes, and local use with rhizome propagation compared to SRC willow.

	Miscanthus co-fire			*Miscanthus* local rhizome			*Miscanthus* local with soil C			*Miscanthus* local			SRC Willow local		
	EUE E out/ E in	CI g C MU^{-1}	Net yield energy GJ ha^{-1}	EUE E out/ E in	CI g C MJ^{-1}	Net yield energy GJ ha^{-1}	EUE E out/ E in	CI g C MJ^{-1}	Net yield energy GJ ha^{-1}	EUE E out/ E in	CI g C MJ^{-1}	Net yield energy GJ ha^{-1}	EUE E out/ E in	CL g C MJ^{-1}	Net yield energy GJ ha^{-1}
Albania	5.83	5	246	10.18	4	268	9.11	1.84	264.24	9.11	3.84	264.24	9.11	3.46	72.25
Austria	5.77	5	231	10.06	4	252	8.96	−0.28	248.32	8.96	3.88	248.32	10.64	3.15	133.72
Belgium	5.61	5	200	9.76	4	218	8.58	−3.05	214.81	8.58	3.99	214.81	10.59	3.16	130.40
Bosnia and Herzegovina	5.92	5	268	10.34	4	291	9.32	0.40	288.07	9.32	3.78	238.07	9.26	3.46	72.25
Bulgaria	279	9	27	4.66	6	33	3.37	−6.11	29.25	3.37	7.95	29.25	8.39	3.70	52.32
Croatia	5.91	5	267	10.34	4	291	9.31	1.23	287.28	9.31	3.78	287.28	9.26	3.46	72.25
Cyprus													4.62	5.83	14.10
Czech Republic													10.59	3.16	130.40
Denmark	4.90	6	115	8.45	4	127	7.03	−7.23	123.52	7.03	4.55	123.52	10.41	3.20	118.77
Estonia													8.72	3.60	58.96
Finland													8.56	3.65	55.64
France	5.72	5	220	9.96	4	240	8.83	−1.19	236.23	8.83	3.91	236.23	10.12	3.26	103.82
Germany	5.58	5	195	9.71	4	213	8.52	−4.62	209.71	8.52	4	209.71	10.77	3.13	143.69
Greece	5.28	6	152	9.14	4	167	7.83	1.19	163.30	7.83	4.23	163.30	4.91	5.56	15.77
Hungary	5.77	5	232	10.07	4	253	8.97	0.46	249.48	8.97	3.88	249.48	9.90	3.30	93.85
Iceland															
Ireland	5.20	6	142	8.99	4	157	7.65	−9.08	153.36	7.65	4.29	153.36	9.98	3.29	97.17
Italy	5.83	5	245	10.17	4	266	9.10	0.83	262.74	9.10	3.84	262.74	8.20	3.76	48.99
Latvia													9.32	3.44	73.91
Lithuania													9.53	3.39	80.56
Luxembourg	5.58	5	195	9.71	4	213	8.52	−2.02	209.71	8.52	4.00	209.71	10.54	3.17	127.08
Macedonia	5.81	5	240	10.13	4	261	9.05	−1.96	257.77	9.05	3.85	257.77	9.26	3.46	72.25
Malta															
Netherlands	5.46	5	175	9.47	4	192	8.23	−8.54	188.16	8.23	4.10	188.16	10.32	3.21	113.79
Norway													9.77	3.33	83.87
Poland													9.43	3.41	77.24
Portugal	5.88	5	258	10.27	4	281	9.23	−4.02	277.66	9.23	3.81	277.66	10.26	3.23	110.47
Romania															
Slovakia	5.69	5	215	9.92	4	235	8.77	0.18	231.25	8.77	3.93	231.25	10.38	3.20	117.11
Slovenia	5.89	5	2.60	10.28	4	2.83	9.24	−1	279.32	9.24	3.80	279.32	9.94	3.30	95.51
Spain	5.75	5	223	10.03	4	243	8.92	−2.95	244.51	8.92	3.89	244.51	7.77	3.91	42.35
Sweden													8.94	3.54	63.95
Switzerland	5.47	5	176	9.49	4	193	8.25	4.09	189.82	8.25	4.09	189.82	10.51	3.17	125.42
Turkey													9.26	3.46	72.25
United Kingdom	5.24	6	147	9.07	4	162	7.75	−4.18	158.33	7.75	4.26	158.33	10.29	3.22	112.13

From the LCA it can be seen that in the European context, the highest energy yield per hectare is achieved by using biomass from perennial grasses or SRC wood. Biomass energy systems also produce the lowest CI and the highest EUE. Local use is important as it minimizes GHG emissions, which are substantially lower than that of coal (33 g C MJ^{-1}) or gas (16 g C MJ^{-1}), even if transported 200 km to a large-scale power station. So, this seems to be the best option to reduce GHG with European-grown bioenergy feedstocks.

European ethanol system produces the lowest energy yield per hectare, EUE's are close to unity, and in all cases with current technology the CI is larger than that of petrol (19) so that there is no net GHG saving. For biodiesel, OSR has the lowest energy yields with a CI higher than diesel. However, the biorefined biodiesel, either from perennial grasses or wood, has a lower CI (9) than diesel (22) and an EUE of 4–9, and so does have a net GHG benefit. Biogas, when using feedstocks from high-yielding countries, gives relatively high energy yields, a lower CI than gas, and a high (4–8) EUE, so in counties with high yields it is a GHG saving option.

GHG Emissions from Indirect Land-use Change

So far in this chapter, we have considered energy crops that could be grown in Europe for use in Europe and looked at the different emissions resulting from the yields achieved and the farming methodologies used in each European country. We have considered the case where agricultural land that is not needed to grow food in Europe is converted to grow bioenergy crops. This is possible as agronometric practices and crop varieties have been optimized in Europe so that sufficient food and fiber can be grown on less land to satisfy the needs of the EU and led to 10% of arable land to be "set aside" to eliminate overproduction. However, the push to increase the use of bioenergy in the EU in order to reduce GHG emissions and to enhance energy security have increased the bioenergy feedstock requirements beyond that which can be produced on the excess land (Don *et al.*, 2012). The EU's ambitious objectives for biofuels have led to an increase in ethanol production from sugarcane in Brazil, palm oil in Africa and Asia, and soya oil production generally. In the United States, the rapid increase in the use of maize to produce ethanol has reduced the amount of maize available for export and this increases the international commodity prices, causing hardship and unrest in several counties such as Mexico in 2008 (Valentine *et al.*, 2011). This pressure on the demand side has led to an increase in land conversion to grow displaced food production and the bioenergy feedstocks. This direct and indirect land-use change increases GHG emissions (Searchinger *et al.*, 2008; Popp *et al.*, 2012). In addition, if subsistence farmland is converted to intensively grow bioenergy feedstock and additional foodstuffs, this will lead to an increase in GHG emissions per hectare (Erb *et al.*, 2008, 2012a,b; Popp *et al.*, 2012). Flynn *et al.* (2012) demonstrated that the increase in GHG due to converting land to grow crops such as rape seed, oil palm, and soya in all continents considering the soil type and original land use was considerable. Thus, indirect land change impacts must be given due consideration while formulating bioenergy policies.

References

Baitz, M., Binder, M., Degen, W., Deimling, S., Krinke, S., and Rudloff, M. (2004) Comparative Life-Cycle assessment for SunDiesel (Choren Process) and Conventional Diesel Fuel. Volkswagen AG and DaimlerChrysler AG. Available at: http://www.pe-international.com/fileadmin/user_upload/images/downloads/Summ_LCA_Choren_en.pdf (accessed 30 June 2012).

Blades, T., Rudloff, M., and Schulze, O. (2006) Sustainable SunFuel from CHOREN's Carbo-V® Process. Presented at ISAF XV, San Diego, September 2005. Available at: http://www.choren.com/ (accessed 28 June 2012).

Blair, N., Faulkner, R.D., Till, A.R., and Poulton, P.R. (2006) Long-term management impacts on soil C, N and physical fertility Part 1: Broadbalk experiment. *Soil and Tillage Research* 91: 30–38.

British Sugar. (2008) UK operations: the sugar beet crop. Available at: http://www.britishsugar.co.uk/ (accessed 30 June 2012).

Don, A., Osborne, B., Hastings, A., *et al.* (2012) Land-use change to bioenergy production in Europe: implications for the greenhouse gas balance and soil carbon. *Global Change Biology Bioenergy* 4: 372–391.

Dondini, M., Hastings, A., Saiz, G., Jones, M., and Smith, P. (2009) The potential of Miscanthus to sequester carbon in soils; comparing field measurements in Carlow, Ireland to model predictions. *Global Change Biology Bioenergy* 1: 413–425.

Erb, K.-H., Gingrich, S., Krausmann, F., and Haberl, H. (2008) Industrialization, fossil fuels, and the transformation of land use. An integrated analysis of carbon flows in Austria 1830–2000. *Journal of Industrial Ecology* 12: 686–703.

Erb, K.-H., Haberl, H., and Plutzar, C. (2012a) Dependency of global primary bioenergy crop potentials in 2050 on food systems, yields, biodiversity conservation and political stability. *Energy Policy* 47: 260–269.

Erb, K.-H., Mayer, A., Krausmann, F., *et al.* (2012b) The interrelations of future global bioenergy potentials, food demand and agricultural technology, in *Socioeconomic and Environmental Impacts of Biofuels: Evidence from Developing Nations*, (eds A. Gasparatos and P. Stromberg), Cambridge University Press, Cambridge, UK, pp. 27–52.

Flynn, H.C., Canals, L.M., Keller, E., *et al.* (2012) Quantifying global greenhouse gas emissions from land use change for crop production. *Global Change Biology* 18: 1622–1635.

Foutch, G.L., Magruder, G.C., and Gaddy, J.L. (1980) Production of ethanol and methane from corn stover. Agricultural energy, Vol. 2. *Biomass Energy Crop Production*. American Society of Agricultural Engineers, St. Joseph, MI, pp. 299–304.

Gerin, P.A., Vliegen, F., and Jossart, J-M (2008) Energy and CO_2 balance of maize and grass as energy crops for anaerobic digestion. *Bioresource Technology* 99: 2620–2627.

Guo, L.B. and Gifford, R.M. (2002) Soil carbon stocks and land use change: a meta analysis. *Global Change Biology* 8: 345–360.

Harris, E.E., Beglinger, E., Hajny, G.J., and Sherrard, E.C. (1945) Hydrolysis of Wood: Treatment with Sulfuric Acid in a stationary digester. *Industrial and Engineering Chemistry* 37: 12–23.

Hastings, A., Clifton-Brown, J.C., Wattenbach, M., Mitchell, C.P., and Smith, P. (2009a) Development of MISCANFOR a new Miscanthus crop growth model: towards more robust yield predictions under different soil and climatic conditions. *Global Change Biology Bioenergy* 1: 154–170.

Hastings, A., Clifton-Brown, J., Wattenbach, M., Stampfl, P.F., Mitchell, C.P., and Smith, P. (2008) Modelling the potential of Miscanthus to meet Europe's energy needs and reduce greenhouse emissions for future climate scenarios. *Agronomy for Sustainable Development* 28: 465–472.

Hastings, A., Clifton-Brown, J.C., Wattenbach, M., Stampfl, P., Mitchell, C.P., and Smith, P. (2009b) Future energy potential of Miscanthus in Europe. *Global Change Biology Bioenergy* 1: 180–196.

Hettenhaus, J.R. and Wooley, R. (2000) Biomass commercialization prospects in the next 2 to 5 years. Biomass Colloquies 2000. National Renewable Energy Laboratory, Golden, Colorado, USA.

Hillier, J., Whittake, C., Dailey, G., *et al.* (2009) Greenhouse gas emissions from four bioenergy crops in England and Wales: Integrating spatial estimates of yield and soil carbon balance in life cycle analyses *Global Change Biology Bioenergy* 1: 267–281.

Huggins, D.R., Buyanovsky, G.A., Wagner, G.H., *et al.* (1998) Soil organic C in the tallgrass prairie-derived region of the corn belt: effects of long term crop management. *Soil and Tillage Research* 47: 219–234.

IPCC. (2007) Fourth Assessment Report (AR4) of Working Group 1: Chapter 2. The Physical science basis. Available at: http://www.ipcc.ch/pub/reports.htm (accessed 1 July 2012).

Kahle, P., Beuch, S., Boelcke, B., Leinweber, P., and Schulten, H.R. (2001) Cropping of Miscanthus in Central Europe, biomass production and influence of nutrients and soil organic matter. *European Journal of Agronomy* 15: 171–184.

Karaosmanoğlu, F., Ciğizoğlu, K.B., Tüter, M., and Ertekin, S. (1996) Investigation of the refining step of biodiesel production. *Energy and Fuels* 10: 890–895.

Koga, N. (2008) An energy balance under a conventional crop rotation system in northern Japan: Perspectives on fuel ethanol production from sugar beet. *Agriculture, Ecosystems and Environment* 125: 101–110.

Lai, R. (2004) Carbon emission from farm operations. *Environment International* 30: 981–990.

Marland, G. and Schlamadinger, B. (1995) Biomass fuels and forest management strategies: how do we calculate the greenhouse-gas emissions benefits? *Energy* 20: 1131–1140.

Mosier, A.R., Hutchinson, G.L., Sabey, B.R., and Baxter, J. (1982) Nitrous oxide emissions from barley plots treated with ammonium nitrate and sewage sludge. *Journal of Environmental Quality* 11: 98–81.

Odell, R.T., Melsted, S.W., and Walker, W.M. (1984) Changes in organic carbon and nitrogen of Morrow Plot soils under different treatments, 1904–1973. *Soil Science* 137: 160–171.

Patzek, T.W. (2008) Thermodynamics of the corn-ethanol biofuel cycle. *Critical Reviews in Plant Science* 23: 519–567.

Peterson C.L. and Hustrulid, T. (1998) Carbon cycle for rapeseed oil biodiesel fuels. *Biomass and Bioenergy* 14: 91–101.

Pimentel, D. and Patzek, T.W. (2005) Ethanol production using corn, switchgrass and wood: Biodiesel production using soybean and sunflower. *Natural Resources Research* 14: 65–76.

Popp, A., Krause, M., Dietrich, J.P., *et al.* (2012) Additional CO_2 emissions from land use change—Forest conservation as a precondition for sustainable production of second generation bioenergy. *Ecological Economics* 74: 64–70.

Poulton, P.R., Pye, E., Hargreaves, P.R., and Jenkinson, D.S. (2003) Accumulation of carbon and nitrogen by old arable land reverting to woodland. *Global Change Biology* 9: 942–955.

Price Waterhouse Coopers. (2003) Shell Middle Distillate Synthesis (SMDS): Update of a life cycle approach to assess the environmental inputs and outputs, and associated environmental impacts, of the production and use of distillates from a complex refinery and SMDA route. Available at: http://www.shell.com.my/products-services/solutions-for-businesses/smds/process-technology.html (accessed 1 July 2012).

Rathke, G.W., Behrens, T., and Diepenbrock, W. (2006). Integrated nitrogen management strategies to improve seed yield, oil content and nitrogen efficiency of winter oil seed rape (*Brassica napus* L.): A review. *Agriculture, Ecosystems and Environment* 117: 80–108.

Saeman, J.F. (1945) Kinetics of wood saccharification: hydrolysis of cellulose and decomposition of sugars in dilute acid at high temperature. *Industrial and Engineering Chemistry* 37: 43–52.

Schneckenberger, K. and Kuzyahov, Y. (2007) Carbon sequestration under Miscanthus in sandy and loamy soils, estimated by natural 13C abundance. *Journal of Plant Nutrition and Soil Science* 170: 538–542.

Scott, R.K. and Jaggard, K.W. (2000) Impact of weather, agronomy and breeding on yields of sugar beet grown in the UK since 1970. *Journal of Agricultural Science, Cambridge* 134: 341–352.

Searchinger, T., Heimlich, R., Houghton, R.A., *et al.* (2008) Use of U.S. croplands for biofuels increases greenhouse gases through emissions from land-use change. *Science* 319: 1238–1240.

Shapouri, J., Duffield, J.A., and Wang, M. (2002) The Energy Balance of Corn Ethanol: An Update. Agricultural Economic Report No. 814, U.S. Department of Agriculture, Economic Research Service, Office of the Chief Economist, Office of Energy Policy and New Uses, Washington, DC.

Smith, P., Powlsen, D.S., Smith, J.U., Falloon, P., and Coleman, K. (2000) Meeting Europe's climate change commitments: quantitative estimates of the potential for carbon mitigation by agriculture. *Global Change Biology* 6: 525–539.

Sokhansanj, S., Turhollow, A., Cushman, J., and Cundiff, J. (2002) Engineering aspects of collecting corn stover for bioenergy. *Biomass and Bioenergy* 23: 347–355.

St. Clair, S., Hillier, J., and Smith, P. (2008) Calculating the pre-harvest greenhouse gas costs of Energy Crops. *Biomass and Bioenergy* 32: 442–452.

Tuck, G., Glendining, M.J., Smith, P., House, J.I., and Wattenbach, M. (2006) The potential distribution of bioenergy crops in Europe under present and future climate. *Biomass and Bioenergy* 30: 183–197.

US patent. (1930) Patent number 1,246,464, 1930.

Valentine, J., Clifton-Brown, J., Hastings, A., Robson, P., Allison, G., and Smith, P. (2011) Food vs. fuel: the use of land for lignocellulosic 'next generation' energy crops that minimise competition with primary food production. *Global Change Biology Bioenergy* 4: 1–19.

Vattenfall. (2005) Life cycle assessment: Vattenfall's electricity in Sweden. Available at: http://www.vattenfall.com/en/file/2005-LifeCycleAssessment_8459810.pdf (accessed 2 July 2012).

Chapter 13

Economics of Biomass Feedstocks and Biofuels

T.A. Maung[1], C. Gustafson[1], B. McCarl[2], D. Ripplinger[1], and D. Saxowsky[1]

[1] *North Dakota State University, Fargo, ND, USA*

[2] *Texas A&M University, College Station, TX, USA*

Introduction

More than a century ago, the United States relied on energy from wood and other biomass resources for most of its energy needs. By the end of the nineteenth century, as the demand for wood outstripped the supply, a shortage was emerging. However, the shortage did not become a crisis as reliance on biomass fuels began to dwindle with a switch to fossil fuels. Increased interest in biofuels arose again in the late 1970s stimulated by rising oil prices during the energy crisis, with biofuels seen as a way to protect against rising fossil fuel prices and the political insecurity of foreign energy supply. Interests in biofuels subsided following the sharp oil price decline in the mid-1980s. Today, with recent increases in oil prices, issues regarding Middle East stability, and concerns for climate change, interest in biofuels is again on the rise.

The primary factor influencing production and consumption of liquid biofuels, such as ethanol and biodiesel in the transportation sector, is crude oil prices since few other substitute fuels are available. However, in the electric power sector, crude oil prices have little influence on the production and consumption of biomass-based electricity because crude oil accounts for only 1% of electricity generation in the United States. Coal, natural gas, and nuclear, respectively, explain 42%, 25%, and 19% of the total US electricity production (EIA, 2012). For biofuels to increase their market share in transportation and electric power sectors, an external factor such as imposing carbon emission pricing through the introduction of a cap and trade permit system would be needed to stimulate interests in producing and using biofuels.

In this chapter, we focus on the economics of liquid biofuel (mainly ethanol) production. First, we look into the background on biofuels including advanced and cellulosic biofuels. Then, the economics of biofuels including advanced and cellulosic feedstock and ethanol production is examined. Finally, biofuel policy uncertainty is discussed and conclusions are drawn.

Biofuel Crop Sustainability, First Edition. Edited by Bharat P. Singh.

Background on Biofuels

Conventional

Renewable Fuel Standard (RFS) and the Energy Independence and Security Act of 2007 (EISA) define biofuels into three classes: conventional, advanced, and cellulosic based on the potential reduction of greenhouse gas (GHG) emissions of 20%, 50%, and 60%, respectively. Today, most maize-based ethanol produced in the United States is qualified as conventional biofuels. Maize is mostly used as livestock feed and is also the single largest source of conventional ethanol production in the country. According to the USDA estimates (USDA, 2011), in 2010, as much as 130 834 million kilograms of maize was used for feed and 136 745 million kilograms was being used to produce conventional ethanol and by-products including distillers' grains, maize gluten feed, maize gluten meal, and maize oil. Because of the limited availability of farmlands and food versus biofuel issues, an increase in the demand of maize-based biofuels may cause not only the price of maize to increase but prices of beef and other food items as well (Gurgel *et al.*, 2007; Tokgoz *et al.*, 2007; Abbott *et al.*, 2011; Harrison, 2009; Pimentel *et al.*, 2009; Gilbert, 2010). High maize prices have induced current and future biofuel producers to seek alternative feedstock sources such as sugar beets, sugarcane, crop and forest residues, and other energy crops.

Advanced

Two crops are uniquely qualified as advanced biofuels: sugar beets and sugarcane. By 2022, advanced biofuel production of 57 billion liters yr^{-1} will be required, creating a niche market opportunity for sugar beets and sugarcane. There are several ongoing efforts to increase sugarcane and beet production for advanced biofuel production. We are involved in one of such endeavors based on sugar beet feedstock. In a consortium configuration, North Dakota State University (NDSU), Green Vision Group (GVG), Inc., and Heartland Renewable Energy (HRE) LLC have been conducting research to commercialize nonfood sugar-based biofuels as advanced biofuels. The long-term goal of this biofuel project is to construct 76 MLY (million liter per year) ethanol plants across North Dakota, improving rural employment across the state. Each 76 MLY ethanol plant will use nonfood energy sugar beets as the primary feedstock. HRE LLC has developed a process to recycle waste materials called "stillage" from ethanol production to produce heat and power at the plant site. Because stillage is spray dried in a patented process and used to generate 75% of the plant's thermal energy needs, ethanol produced from the plants is expected to have GHG emissions that are lower than the advanced biofuel standard. North Dakota and Minnesota account for more than 50% of the total sugar beet production in the United States. Using energy sugar beets to produce advanced biofuels in North Dakota would lower storage, processing, and transportation costs because the region has cool climate, existing beet processing infrastructure, and easy access to the resource. In addition, by using sugar beets land requirements would be minimized because compared to maize, sugar beets can double the ethanol production per unit land. Moreover, growing sugar beets requires less nitrogen, a key contributor to GHG. The process of converting sugar beets to ethanol is simpler and less energy intensive than that of maize to ethanol and thus will result in lowering capital costs and GHG emissions. In order to minimize land competition with existing food crops, North Dakota has potential to expand irrigated sugar beet production.

Sugarcane-based ethanol has a similar production process as sugar beets. Highlands Enviro Fuels LLC (HEF) is proposing to build a 114 MLY advanced biofuel ethanol plant in Highlands County, Florida. The HEF facility will use sugarcane and sweet sorghum as its primary feedstocks. The facility will also use wastes such as sugarcane and sorghum bagasse from the ethanol production process to generate steam and 25 MW yr^{-1} of renewable electricity. HEF has conducted a life-cycle analysis (LCA) of GHG emissions and demonstrated that the planned sugar-based ethanol process will result in 80% lower GHG emissions than the equivalent petroleum-based gasoline. This reduction in GHG emissions will allow its ethanol to qualify as an advanced biofuel.

Cellulosic

Cellulosic feedstock is an attractive alternative for producers to consider due to biofuel price premium in the marketplace, its high potential in offsetting carbon emissions, and lower competition with food uses (Schneider and McCarl, 2003; McCarl *et al.*, 2009). To some extent, the increase in grain-based biofuel production has laid the foundation for a shift to cellulosic biofuel business (Solomon *et al.*, 2007). Several firms have pilot-scale cellulosic ethanol production facilities under construction and testing. However, the transition from pilot scale to full commercialization of cellulosic ethanol will be long and difficult due to financial constraints being imposed on the biofuel industry (Gustafson, 2008). At present, cellulosic biofuel production is on the verge of commercialization. Bacovsky *et al.* (2010) reviewed the status of second-generation biofuel plants globally and discovered more than 20 projects in North America that plan on being operational in the next few years. However, most of the North America plants that are operational today are either demonstration or pilot plants. To the best of our knowledge, currently no cellulosic ethanol plants are commercially operational in the United States. Nevertheless, the RFS requires that by 2022 up to 61 billion liters of cellulosic biofuels must be blended in the nation's fuel supplies. Table 13.1 reports commercial cellulosic ethanol plants in the United States that are planned or under construction.

Table 13.1. Current commercial cellulosic ethanol plants in the United States.

Company	Location	Feedstock	Product	Output (MLY[a])	Status	Start-up year
Abengoa Bioenergy Biomass of Kansas, LLC	Hugoton, Kansas	Maize stover, wheat straw, switchgrass	Ethanol	95	Under construction	2013
BBI BioVentures LLC	Denver, Colorado	Pre-collected feedstocks that require little or no pretreatment	Ethanol	17	Planned	2010
Frontier Renewable Resources	Kincheloe, Michigan	Wood chip	Ethanol; lignin	76	Announced	
POET	Emmetsburg, Iowa	Maize cobs, agricultural residues	Ethanol	95	Planned	2011
Range Fuels, Inc.	Soperton, Georgia	Wood and wood waste	Ethanol; methanol	379	On hold	2010
Coskata	Clewiston, Florida	Sugarcane waste and others	Ethanol	379	Planned	

[a]MLY stands for million liter per year.
Source: IEA Bioenergy Task 39 (2010).

Life-cycle Analysis as Component of Biofuel Economic Value

LCA—as discussed in Wang (1999) or Scientific Applications International Corporation (2006)—is the name of the field that endeavors to look at total GHG emission effects. Looking at the simplest case first (ignoring the market effects), LCA computes the emissions per unit of a crop by totaling

- emissions involved in producing inputs like fertilizer, seed, pesticides, and other items that are used in feedstock production times their use per acre divided by the yield;
- emissions from fossil fuel inputs used in an acre of feedstock production for tilling the land, irrigating, planting, cultivating, harvesting, on farm hauling, storage, drying, and any other farm operations divided by the yield;
- emissions involved with conveying the feedstock to the biorefinery or bioenergy production facility including those arising from transport, off-site storage, and handling normalized to the amount associated with one unit of the feedstock;
- emissions arising from preprocessing and processing one unit of the feedstock at the processing facility including drying, process heat, and so on;
- emissions arising from conveying the energy production to the point of consumption including those in any transport, blending, and storage normalized to the amount of energy crop produced by one unit of the feedstock; and
- credits for the emissions saved by the use of any by-products generated. For example, crediting DDGS (distiller's dried grains with solubles) from one unit of the feedstock at the LCA emissions from the equivalent amount of crops it replaces in livestock feeding.

Generally, the results of the LCA are divided by the emissions from the amount of the fossil fuel replaced by one unit of biofuel to get their relative efficiency. The results of such analyses can be controversial and strongly depend on assumptions. For example, Pimentel and Patzek (2005) provide estimates (in terms of energy, not GHGs) that ethanol production using grain required 29% more fossil energy than the ethanol fuel produced, whereas Shapouri *et al.* (2002) estimate that ethanol yields 34% more energy than it takes to produce it. McCarl and Boadu (2009) provide a set of estimates across a number of biofuel and bioelectricity possibilities (Table 13.2) showing the percent reductions in emissions relative to the replaced fuel.

The data in Table 13.2 show the percentage of direct reduction in CO_2 equivalent emissions when a fossil fuel is replaced. For example, the LCA from maize is 69.5% of the potential emissions savings from replacing the gasoline, so the gain is 30.5%. McCarl and Boadu (2009) draw the following inferences from the data in the above-mentioned table: (1) there are relatively lower offset rates involved with the production of liquid fuels as opposed to electricity; (2) the lowest liquid fuel offsets arise for grain-based ethanol while relatively higher values generally arise from cellulosic ethanol and biodiesel from soybean oil; (3) differential offset rates arise across the different feedstocks and processes; and (4) more GHG emission-intensive transformation processes go into making conventional ethanol along with successively less to make cellulosic ethanol, biodiesel, and electricity. A series of more recent LCA estimates have been done under the EPA rule making relative to the RFS under the EISA (USEPA, 2012a).

LCA as originally conceived limited its attention to the direct production process. However, large programs of feedstock production have market effects on commodity and land markets making it desirable to shift production patterns and add to crop land areas. Consider, for example, the simple graphical exposition in Figure 13.1. The figure uses a three-panel diagram

Table 13.2. Percentage offset of net GHG emissions when raw products replace conventional fossil-fuel-based energy.

Commodity	Crop ethanol	Cellulosic ethanol	Biodiesel	Electricity cofire 10%	Electricity fire 100%
Maize	30.5				
Hard red winter wheat	31.5				
Sorghum	38.4				
Log residue		80		99.1	97.4
Maize residue		74		93.4	87.3
Wheat residue		72.9		95.5	91.1
Lumber milling residue		84.8			
Manure				99.4	96.5
Switchgrass		68.6		94.4	89.5
Hybrid poplar		61.9		94.2	89.1
Willow		67.7		96.7	93.7
Soybean oil			70.2		
Sugarcane	64.8				
Maize oil			55		
Bagasse		90.1		100	100
Lignin				100	100

Source: McCarl and Boadu (2009).

from international trade theory with the panel on the left showing the US market for conventional production of a commodity (consider it), the left giving the rest of the world market for production of the commodity, and the center panel giving the US/world trade market for the commodity involving excess supply curve from the United States and excess demand curve from the rest of the world.

In this diagram the black lines illustrate the pre-ethanol market supply and demand curves and the red line illustrates the market equilibrium with TQ showing the exports from the United States to the rest of the world. SQ_{US} is the US production and SQ_{ROW} is the production of the rest of the world. When the US supply is reduced because some of the commodity is diverted to ethanol production (the blue line in the left panel), the US excess supply shifts up to the blue line in the center panel and raises the equilibrium price (to the dashed horizontal blue

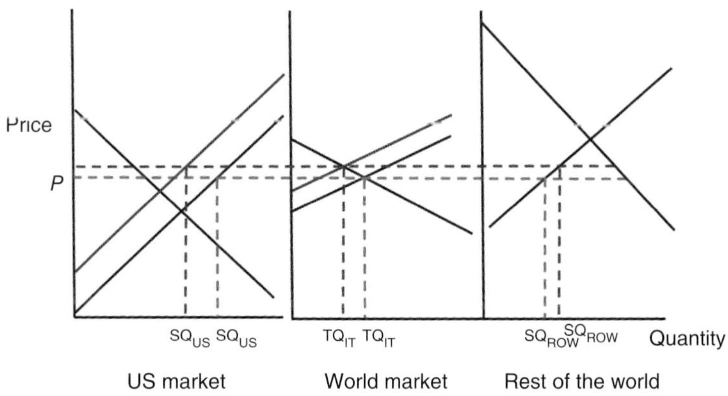

Figure 13.1. Effect of the US production on market price and rest of the world production.

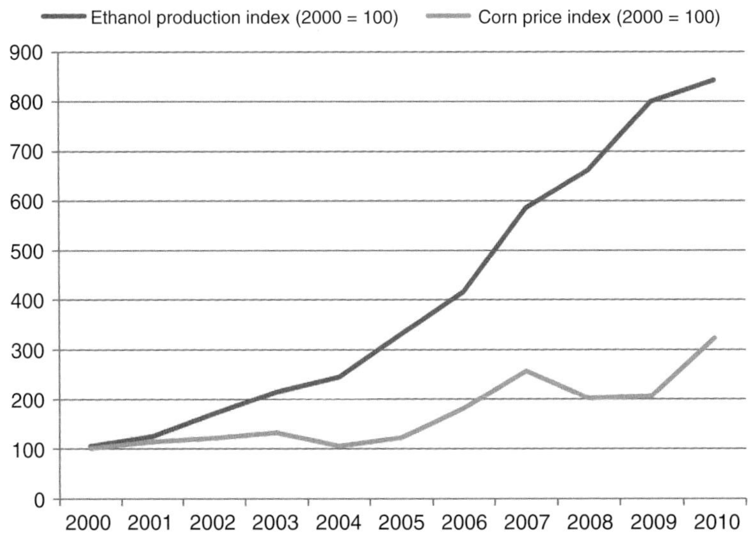

Figure 13.2. Plot of ethanol production and maize price index over time where data are set up so 2000 levels are set to 100.

line) from the original price (dashed horizontal red line). Accompanying this is an increase in production by the rest of the world.

This market interaction shows that a change in market conditions induced by ethanol production stimulates increased market prices and production elsewhere. If additional emissions are associated with this increased production then this change can cause net GHG emission reductions from the biofuel production action to be offset by increased emissions from expanded production in other areas of the world or portions of the economy. This secondary impact has been called leakage of indirect land-use change in the general dialog over this subject (Murray *et al.*, 2004; Fargione *et al.*, 2008; Lee *et al.*, 2007; Searchinger *et al.*, 2008). The subject is anything but academic. Figure 13.2 shows the US history of prices and ethanol production. The figure indicates that the upward movement in ethanol production has been accompanied by an upward movement in price although this is generally a more complex relationship involving biofuel demands, weather, economic growth, exchange rates, stocks, trade, and other factors as discussed in Abbott *et al.* (2011). Today it is common to hear about many forms of such market effects being stimulated by high agricultural commodity prices including

- US forested hectares being deforested and converted to cropland;
- Conservation Reserve Program (CRP) lands not re-enrolling and converting into cropland; or
- expansions of crop hectares in South America, Asia, or Africa at the expense of grasslands and rainforest (Lee *et al.*, 2007; Searchinger *et al.*, 2008).

Key factors in the size of this leakage as discussed in Fargione *et al.* (2008), McCarl (2008), and Murray *et al.* (2004) include the following:

- the amount of diversion of traditionally marketed production—use of residues and waste product feedstocks lower this while use of conventional commodities from current crop lands raises it;

- the land use that replacement or additional production comes from and the embodied emissions—large offsets occur when rainforest, other forest, grassland, or CRP land is involved;
- the supply responsiveness of production and land use when prices rise; and
- the market share of the region producing the biofuel feedstocks.

McCarl (2008) constructed leakage estimates based on a formula by Murray *et al.* (2004) and concluded that international leakage easily offsets nearly 50% of the bioenergy gain when GHG offsets per hectare are equal and an even higher share when hectares with higher emissions are involved. Along this line of thought, Searchinger *et al.* (2008) indicate that when hectares are directly replaced by rainforest reductions, the LCA would be negative. Fargione *et al.* (2008) point out the LCA effect varies with replacement hectares land use and types of feedstocks. Market forces may also cause reduced GHG emissions elsewhere where, for example, commodity price increases for feed may reduce livestock production and feed use with accompanying GHG emission reductions as covered in McCarl (2008). Today, LCA appraisals are evolving toward approaches that consider these market effects moving beyond the attributional LCA as discussed previously to consequential LCA that describes how environmentally relevant flows will change in response to possible decisions and their market consequences (Curran *et al.*, 2005; Finnveden *et al.*, 2009).

LCA and the Market or Regulatory Environment

LCA is a possible factor in both the regulatory environment and the market that will be faced by biofuel producers. Basically, GHG emissions are viewed as an externality where emissions are not fully accounted for in terms of their social costs in the decisions of emitters. In the face of this a variety of regulations and emission pricing mechanisms (taxes and markets) are being considered.

Emissions Pricing Scheme

Under a pricing scheme a tax or an emissions trading market is most commonly discussed. Here emissions would either be taxed (with the possibility of crediting for reductions) or those emitting would need to have permits that they could buy in the market place (as under the European emissions trading scheme). Under either circumstances production of biofuels could involve less tax or less of a need for permits meaning that biofuel production could have enhanced economics. In turn this would raise the willingness of biofuel producers to pay for feedstocks and thus raise the price to producers. LCA would be involved in the development of quantities of emissions relevant to the tax or permit market.

Emissions Regulatory Scheme

Under a regulatory scheme a limit might be imposed on emissions and some adjustment might need to be made for the renewable (biogenic) GHG emissions as opposed to total. Today, the EPA is involved in an effort to account for this and has a formula that is much like the LCA approach that involves changes in site sequestration, losses in conveyance, emissions from processing, and sequestration in products/waste streams (USEPA, 2012b). Net emissions after considering biogenic recycling would be regulated or priced/taxed.

Economics of Biofeedstock and Biofuel Production

Financing is the primary concern for firms entering into new biofuel business venture and offering biofeedstock products to the marketplace. Because of the uncertainty and policy-dependent nature of biofuel market conditions, new start-up firms will incur a significant amount of entrance costs. According to Costello and Finnell (1998), capital and financial markets generally perceive the deployment of new emerging technologies as involving more risk than established technologies. The higher the risk, the higher the rate of return demanded on the investment in these new emerging technologies. Most private commercial banks and other financial institutions are not willing to provide loans to invest in biomass-related businesses due to the perceived high technology and development risks. They constitute failures of capital and financial markets that must be overcome to reach the level of economic potential (Costello and Finnell, 1998; Sathaye and Bouille, 2001). However, public institutions such as USDA and Department of Energy and other local/state public institutions provide funding to conduct research on the desirability of biofuel investments.

Advanced Biofuel Feedstock Production: The Case of Sugar Beet

NDSU is leading a research to examine the viability of commercializing energy sugar beet biofuel production. Initial feasibility study for sugar beet biofuel production funded by the North Dakota Agricultural Products Utilization Commission was completed in 2011. The energy sugar beets used to produce biofuels will be different from conventional sugar beets used to produce food sugar in that seed for energy beets contain genetic impurities which make them unfit for food sugar production. As a consequence of this distinction, an increase in energy sugar beet production would not have any negative economic impact on the conventional sugar beet industry.

Currently, energy beet seeds are being developed by major seed companies. By growing energy beets for biofuel production, North Dakota farmers can generate positive economic impacts to the region's economy through expenditures for production inputs and net returns from production. Table 13.3 shows the economic impacts of dryland and irrigated energy sugar beet production in central North Dakota. The table indicates that the economic impact of farm net returns is \$442 ha^{-1} for dryland energy sugar beets and \$668 ha^{-1} for irrigated energy beets. Due to their higher yield, irrigated energy beets generally have a higher net return than the dryland energy beets.

Total amount of energy sugar beets required each year to produce 76 MLY of fuel ethanol is 652 064 tonnes. Assuming that two-thirds of energy beets are produced on dryland and one-third on irrigated land, the processing plant would need 434 709 tonnes of dryland energy beets and 217 355 tonnes of irrigated energy beets. Total land area required to grow energy beets for a 76 MLY plant is 12 100 ha (9050 dryland hectares plus 3050 irrigated hectares). Table 13.3 suggests that by growing 652 064 tonnes of energy beets in North Dakota, farmers can generate additional gross farm income of about \$6 million as a result of producing 76 MLY energy beet-based fuel ethanol. Similarly, North Dakota's economy can directly benefit from the farmers' expenditures of energy beet production inputs which would be about \$24 million.

Maung and Gustafson (2011a) created a financial pro forma model to analyze the economic viability of producing ethanol using both energy sugar beets and molasses in central North Dakota. Our recent conversation with a molasses marketing firm indicated that there may not be an adequate supply of molasses available for ethanol production. Hence, this chapter looks

Table 13.3. Projected energy sugar beet crop budgets 2012 for East Central North Dakota.

Variable	Dryland	Irrigated
Market yield (tonnes ha^{-1})	48.02[a]	71.26[a]
Market price (\$ $tonne^{-1}$)	46.30	46.30
Market revenue (\$ ha^{-1})	2223.06	3299.30
Direct costs (\$ ha^{-1})		
Seed	430.86	430.86
Fertilizer	203.42	203.42
Crop chemicals	119.18	119.18
Crop insurance	68.58	68.58
Irrigation energy	–	102.54
Fuel and lubrication	157.09	315.06
Repairs	201.54	395.50
Miscellaneous	–	–
Operating interest	9.88	9.88
Total direct costs	1190.55	1645.01
Indirect costs (\$ ha^{-1})		
Miscellaneous overhead	49.42	49.42
Machinery depreciation	148.26	148.26
Machinery investment	86.49	86.49
Irrigation depreciation	–	101.31
Irrigation investment	–	74.13
Land	180.39	308.88
Labor	126.02	217.45
Total indirect costs	590.58	985.95
Total production costs (\$ ha^{-1})	1781.13	2630.96
Net return (\$ ha^{-1})	441.93	668.34

Calculation details and assumptions used are available upon request from the authors.
[a]Based on average energy beet yield trial data in North Dakota.

into economic factors that influence feasibility of producing ethanol using 100% energy sugar beets. Some of the baseline financial assumptions used are described in Table 13.4. As can be seen in the table, to produce 76 MLY of ethanol from a biorefinery, energy sugar beet growers will need to be contracted to supply 652 064 tonnes of energy beets each year. Stillage powder,[1] a renewable feedstock, is used to generate steam to meet 75% of the plant's thermal energy needs. The other 25% of energy needs comes from natural gas. By using the stillage powder, the plant is expected to emit low GHG emissions. Accordingly, the ethanol produced from the plant is expected to qualify as advanced biofuel which is eligible for renewable fuel credits, such as renewable identification numbers (RINs) under the EPA's rules. Presently, NDSU is working with EPA to document an energy beet pathway to secure EPA approval for energy beet ethanol as advanced biofuels.

Financial results generated by our pro forma model are reported in Table 13.5. The table shows that on average the plant is expected to create about \$31.75 million of revenue annually from ethanol sales and about \$9 million of revenue from co-product sales. Total production cost averages about \$37.12 million annually. Net profit for the plant averages

[1]After the ethanol distillation process, the leftover solid residue (also known as stillage) generated from energy beets is converted into syrup through the evaporation process. The syrup is then dried to a powder which can be used to generate steam to meet the plant's thermal energy requirements.

Table 13.4. Baseline assumptions used in the 76 MLY ethanol plant model.

Variable	76 MLY plant
Conversion rate for energy beets (l tonne^{-1}) or (ton gal^{-1})	91 (or) 26.5
Energy beets requirement (tonnes yr^{-1})	6,52,064.51
Electricity requirement (MMkWh yr^{-1})	12
Thermal energy requirement	
Stillage powder (MMBtu yr^{-1})	4,50,000
Natural gas (MMBtu yr^{-1})	1,50,000
Local ethanol transportation (\$ l^{-1}) or (\$ gal^{-1})	0.03 (or) 0.12
Base price for	
Ethanol (\$ l^{-1}) or (\$ gal^{-1})	0.49 (or) 1.84
Yeast (\$ tonne^{-1})	551.16
Ash fertilizer (\$ tonne^{-1})	87.52
Beet pulp (\$ tonne^{-1})	80.67
Energy beets (\$ tonne^{-1} or \$ ton^{-1})	46.30 (or) 42
Electricity (\$ kWh^{-1})	0.06
Natural gas (\$ MMBtu^{-1})	4.5
Engineering and construction cost (million \$)	32.66
Development and start-up cost (million \$)	9.83
Total capital cost (million \$)	42.49

\$3.65 million yr^{-1}. The table also shows that ethanol revenue and energy beet feedstock cost explain about 78% of the total revenue generation and 75% of the total production cost, respectively. These large percentages suggest that changes in ethanol and energy beet prices will significantly impact the net profit of production.

In order to look into the plant's revenue and production cost risks, the analysis simulates the impact of changes in stochastic variables on the net present values (NPVs) of 76 MLY

Table 13.5. Financial results from income statement (11-year average).

Variable	\$ yr^{-1}	Percent of total revenue	Percent of total cost
Sale revenue			
Ethanol	3,17,51,467	77.88	
Yeast	63,24,883	15.51	
Fertilizer	9,51,529	2.33	
Feed	17,40,056	4.27	
Total revenue	4,07,67,935	100	
Production Cost			
Energy beets	2,78,61,448		75.06
Steam	6,71,915		1.81
Chemicals, enzymes, and yeast	5,70,666		1.54
Other utilities	25,85,392		6.96
Direct labor and benefits	6,48,480		1.75
Administrative and operating	22,03,285		5.94
Interest, Income tax, depreciation, and amortization	25,78,826		6.95
Total production cost	3,71,20,013		100
Net revenue	36,47,922		

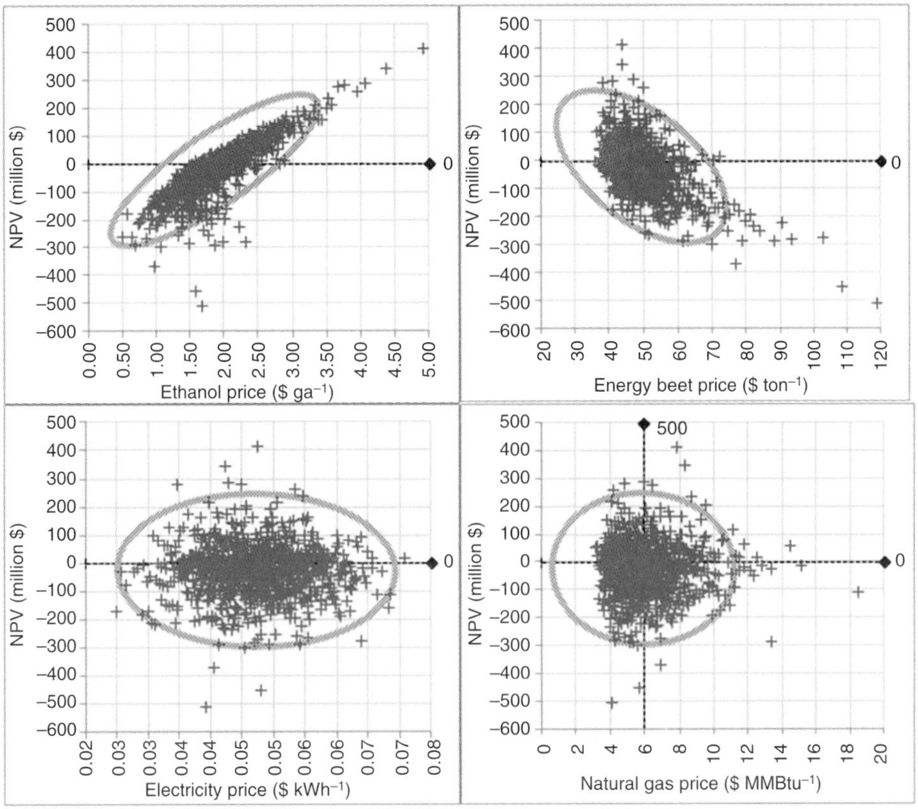

Figure 13.3. Net present value (NPV) versus ethanol, energy beet, electricity, and natural gas prices. For color detail, please see color plate section.

ethanol plant. These stochastic variables include ethanol and co-product prices from the revenue side and electricity and natural gas prices from the production cost side. Using @Risk simulation software package (Palisade Corporation, 2009), the financial model is transformed into a Monte Carlo simulation model. The stochastic price variables are simultaneously simulated 1000 times, each time with different price levels. Other variables are assumed to be constant during simulation. Annual stochastic price data used in simulation are from 2001 to 2010. Distributions of these stochastic prices are determined using the fitting algorithms in @Risk.

Simulation results are shown in Figure 13.3 which suggests that changes in ethanol and energy beet prices will have the most significant impact on the profitability (represented by the NPV) of energy beet ethanol investment. Relationships between the profitability of the investment and other variables including co-product prices (not shown here), and electricity and natural gas prices are shown to be weakly correlated.

Table 13.6 depicts breakeven energy beet prices at different levels of ethanol market prices. The breakeven prices were determined using the NPV values. The table shows that the higher the ethanol prices, the larger the breakeven energy beet prices will be. In a sense, ethanol prices will be the main factor dictating the profitability of energy beet ethanol production.

Table 13.6. Breakeven energy beet prices at various levels of ethanol market prices.

Energy beet price ($ tonne^{-1}) or ($ ton^{-1})	Ethanol price ($ l^{-1}) or ($ gal^{-1})
30.29/27.48	0.32/1.20
36.38/33	0.37/1.40
42.46/38.52	0.42/1.60
49.76/45.14	0.49/1.84
54.62/49.55	0.53/2
60.70/55.07	0.58/2.20
66.79/60.59	0.63/2.40

Cellulosic Biofeedstock Production: The Case of Crop Residues

Farmers may not be willing to supply all of the residue feedstocks because of concerns for soil fertility losses and harvest time limits. For example, in North Dakota due to changing weather, the window for collecting crop residue is short. This reinforces the need to look at multiple crop residues to extend baling time. Maung *et al.* (2012) show that there is a general interest among farmers in exploring an economic opportunity to sell residue feedstocks. However, farmers are unwilling to take any risk in supplying the residues and there is little or no interest on their part in purchasing a baling machine and providing the labor. Most farmers prefer to have an external party bale, store and transport their feedstocks. In order to prevent mold, cellulosic biofuel firms require residue feedstocks be collected dry. This seems to be a challenging issue for farmers since maize stover likely contains a high level of moisture when harvested. Another factor that affects the viability of using biofeedstocks to produce cellulosic biofuels is the harvesting cost.

The feedstock harvesting process involves multiple steps. The process begins with a custom baler that windrows the residues. Then, a large square (or round) baler picks up the feedstocks, compacts, and ties them in the form of large square (or round) bales. After baling is completed, the process of moving the feedstocks will be a multistep process. First, bales are collected from the fields and stacked on a roadside. They are subsequently loaded onto trucks and driven to storage sites for unloading, stacking, and tarping for rain protection. The final step is to load bales onto trucks using a loader and transport them to the final destination for processing and energy conversion.

Total delivered cost which includes on-farm collection cost, farm-to-storage and storage-to-plant transportation costs, and storage cost impacts the feasibility of using biofeedstocks to produce cellulosic biofuels. According to Gustafson *et al.* (2011), on-farm collection cost for maize stover and wheat straw varies from $33 to $61 tonne^{-1} depending on the rate of residue removal and combination of residue ratio. Total transportation cost which includes farm-to-storage and storage-to-plant transportation costs is influenced by the hauling distance and the density of feedstock. The lower the feedstock density around a cellulosic biorefinery plant, the higher the distance traveled to collect the feedstocks and hence larger the transportation cost. Transportation cost can vary from one region to another because of differences in feedstock densities and distances traveled among regions. On average, total transportation cost for maize stover and wheat straw is estimated to be about $22 tonne^{-1}, including loading and unloading costs of $4 tonne^{-1} (Gustafson *et al.*, 2011). Storage cost consists of land preparation and equipment costs such as costs related to handling, hauling, and stacking of bales and tarping

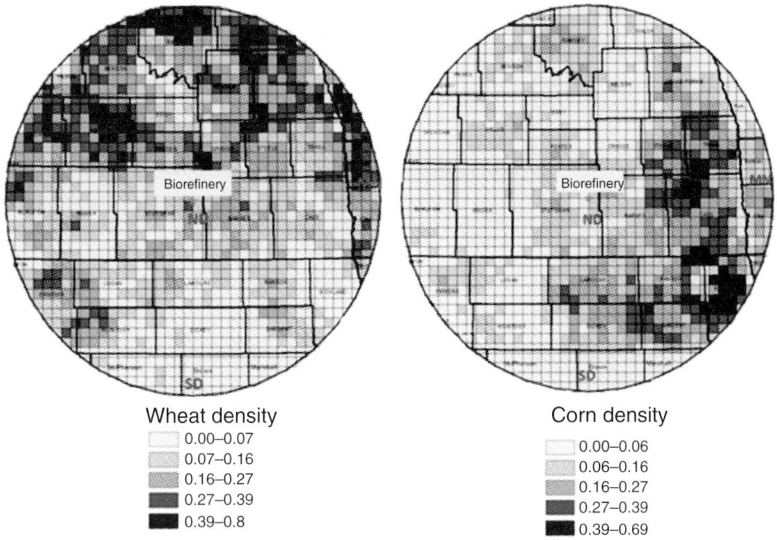

Figure 13.4. Wheat and maize concentration within 100-mile radius of biorefinery in Southeast North Dakota. For color detail, please see color plate section.

for moisture protection. Storage cost is found to be about \$6.3 tonne^{-1} for residue feedstocks like maize stover and wheat straw.

Depending on the residue removal rate, crop density, and distance to the biorefinery, total delivered cost for cellulosic residue feedstocks can vary from \$55 to \$88 tonne^{-1}. Since the density and distance are fixed, one way to decrease total delivered cost is to increase the rate of residue removal. But this rate is dependent on the USDA Natural Resource Conservation Service's (NRCS) environmental, carbon, and sustainability policy decisions. Figure 13.4 defines the geographic area within a 100-mile radius for a planned 76 MLY cellulosic biorefinery located in Southeast North Dakota. The counties in this area are located in three states: North Dakota, South Dakota, and Minnesota. The plan was to supply the biorefinery with 435 449 tonnes of residue feedstocks annually including maize stover and wheat straw. As illustrated in the figure, different residue feedstocks are disparately concentrated in different regions. Relying solely on one particular residue feedstock would likely require traveling longer distances which increases total transportation costs for the plant.

Using linear programming (LP) transportation models and geographic information system (GIS), Gustafson *et al.* (2011) show that employing a combination of maize and wheat residue feedstocks can lower the transportation costs. Costs are lowest with a combination of 50% maize stover and 50% wheat straw. Figure 13.5[2] visually shows that an increase in the residue removal rate would narrow the size of total feedstock supply areas for the biorefinery and hence reduce feedstock production cost. The USDA NRCS future crop residue removal rate policy will be important in determining whether it is economically viable to harvest crop residues as potential feedstocks for energy generation. Another factor (such as timeliness of harvesting discussed below) can have an impact on the amount of residue feedstocks collected.

[2]Results in the figure are developed using a combination of LP transportation model and GIS.

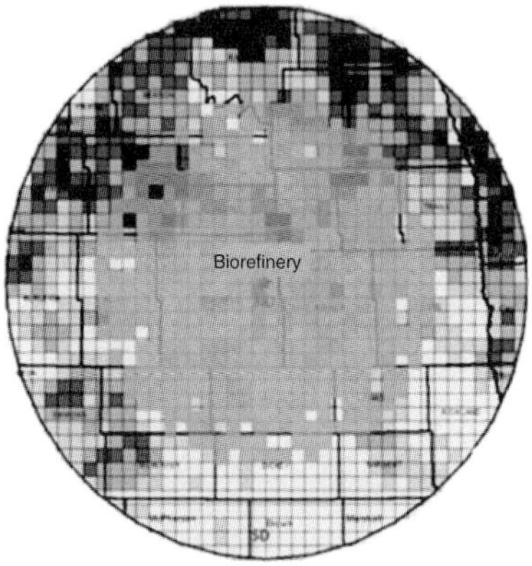

25% removal rate

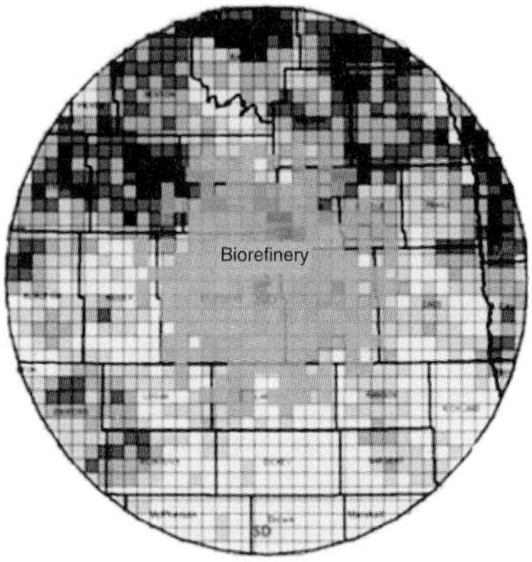

50% removal rate

Figure 13.5. Estimated potential wheat straw supply sites. For color detail, please see color plate section.

Impact of Harvest Field Time on Residue Collection

Because of changing weather conditions such as rainfall and early frost, harvest field time can be limited especially in the northern plains of the United States. Maung and Gustafson (2011b) analyzed net farm returns under changing harvest field time in North Dakota. Randomized fall harvest field time data were incorporated into LP models to create stochastic outcomes which show the impact that variations in the availability of harvest field days can have on net farm profits. The analysis emphasized understanding the impact of harvest field time, feedstock harvest technologies, farm sizes, and tillage practice on profit-maximizing potential of farmers. In order to gain insights into the economic tradeoffs between grain and residue feedstock harvest activities when harvest field time is limited, the analysis considered three harvest options based on the study of Maung and Gustafson (2011b):

1. maize grain-only harvest option;
2. simultaneous maize grain and cob harvest or one-pass option; and
3. separate maize grain and stover harvest or two-pass option.

Maize cobs are desirable as a cellulosic feedstock because they represent about 12% of maize stover remaining on the field, their removal has negligible impact on soil carbon, and they have limited nutrient value to the soil (CVEC, 2009). For example, POET, an ethanol company, has been researching and developing cellulosic ethanol production technology based on maize cobs since 2001. The company plan is to produce 95 MLY of ethanol which will require at least 635 tonnes of maize cobs per day to fuel the plant.

When farmers harvest only maize grain, they are able to complete their harvesting and maximize their profits in a fairly short time as illustrated in Table 13.7. The results in Tables 13.7 and 13.8 are estimated using stochastic LP models described in Maung and Gustafson (2011b). The analysis assumed a maximum of 809 harvestable hectares and four harvest periods. Each harvest period is assumed to last 2 weeks. Harvest field time data were computed based on the data obtained from Crop Progress and Condition Report

Table 13.7. Estimated results for maize grain-only option and simultaneous maize grain and cob harvest option.

		Simultaneous maize grain and cob harvest		
	Maize grain only	No slowdown	25% slowdown	50% slowdown
Net profit ($)	3,83,048.80	4,02,188.80	3,76,588.80	2,63,995.10
Maize grain sold (tonnes)	7,330	7,330	7,330	5,947
Maize cob sold (tonnes)		635	635	515.22
Maize harvested (ha)				
Period 1	282.93	282.93	265.61	176.47
Period 2	227.50	227.50	213.58	141.90
Period 3	271.84	271.84	255.19	169.56
Period 4	27.10	27.10	74.99	168.73
Total maize harvested (ha)	809.37	809.37	809.37	656.66
Harvest field time (d)				
Period 1	10.72	10.72	10.72	10.72
Period 2	8.62	8.62	8.62	8.62
Period 3	10.30	10.30	10.30	10.30
Period 4	10.25	10.25	10.25	10.25
Total harvest field days	39.89	39.89	39.89	39.89

Source: Maung and Gustafson (2011).

(USDA-NASS, 2005–2009). Farmers are assumed to spend 12 hd^{-1} harvesting maize grain and baling cob/stover using the available harvest field time. When maize cob harvest option was added to the grain model, grain harvest speed or capacity was reduced as a result of attaching the cob harvester to the back of combine. According to CVEC (2009), the reduction in grain harvest time can be as high as 50%.

Table 13.8. Estimated results for separate maize grain and stover harvest option.

	Separate grain stover harvest	Shadow price ($)
Net profit ($)	3,77,650.31	
Maize grain sold (tonnes)	7,084.66	
Maize stover sold (tonnes)	587.28	
Maize grain hectares harvested		
Period 1	282.93	
Period 2	227.50	
Period 3	271.84	
Total maize grain hectares harvested	782.27	0
Maize stover harvested (tonnes)		
Period 2	0	−186.30
Period 3	0	−186.30
Period 4	587.28	0
Harvest field time (d)		
Period 1	10.72	1,040.89
Period 2	8.62	1,040.89
Period 3	10.30	1,040.89
Period 4	10.25	60.37
Total harvest field days	39.89	

Source: Maung and Gustafson (2011).

In Table 13.7, the results from three slowdown scenarios are compared. The table shows that farmers can generate an additional revenue of \$19 140 (\$402 188.80 – \$383 048.80) by selling 635 tonnes of maize cobs, if additional cob harvesting does not get in the way of maize grain harvesting and slow down the harvest. This scenario is unrealistic in a real world. Conversely, farmers can incur a loss of \$6460 and \$119 054 if additional cob harvesting reduces the maize grain harvest capacity by 25% and 50%, respectively. Given that maize harvest time is constrained by the limited availability of harvest field days in each period, farmers would not be able to maximize their net returns because of the slowdown.

The table also indicates that about 153 maize hectares are left unharvested if maize combine capacity slows down by 50%; the outcome is a large reduction in net farm profit. All 809 maize hectares are harvested if combine speed reduces by only 25%. However, the net profit of a 25% slowdown scenario is lower (as compared to the net profit of no slowdown scenario). This is because the reduction in combine harvest speed by 25% increases maize grain and cob harvest costs which decrease net profit. Thus, under this grain and cob one-pass harvest option, farmers' harvest activities are not only constrained by the reduction in harvest capacity but also by the limited availability of harvest field days. Because of this reduction in harvest capacity and limited amount of harvest time, farmers' production costs would increase and consequently result in lower net farm revenue compared to the maize grain-only harvest option. Farmers may earn more profit by harvesting maize grain only without venturing into maize cob harvesting business.

Time allocation is the main concern when farmers choose the option of harvesting maize grain and stover separately under the two-pass harvest method. If more labor time is allocated to harvesting stover, less would be available for harvesting maize grain and this impacts farmers' net profits. Table 13.8 reports the results for maize grain and stover harvested separately. The analysis assumes four harvest time periods and the maize grain harvest must be completed within the first three periods. Stover harvesting and baling can take place only when maize grain harvest is completed. Therefore, no stover can be harvested and baled in the first period and no maize grain can be harvested in the final period. All grain and stover harvest activities need to be completed within these four harvest periods. The table indicates that the total hectare of maize grain harvested is 782 ha which fell short of maximum harvestable amount of 809 ha. The incomplete harvest is due in part to the farmer's time being allocated to harvesting and baling stover in the final period.

The table also shows that no stover is harvested in the second and third periods because farmers find it optimal to allocate all of their time to harvest maize grain instead of stover during these two periods. The negative shadow price of \$186.3 implies that if farmers were to attempt to harvest maize stover in either the second or third harvest period, their net profit would decrease by \$205 tonne^{-1} of stover harvested. As can be seen in the table, all stover harvest time is optimally allocated to the final period since no grain can be harvested during this final period. The shadow prices of \$1040.89 (for grain) and \$60.37 (for stover) suggest that it would be much more valuable for farmers to harvest maize grain than stover if an additional harvest field hour is available.

The research suggests that the availability of harvest field time is the main factor dictating farmers' interests in harvesting maize cob/stover. Because of limited harvest field time, a frequent suggestion is for farmers to hire additional labor to operate the farm machinery. However, additional farm labor is difficult to find in rural areas of North Dakota. At present, the state has the lowest unemployment rate in the nation. Moreover, large tractors and new stover collection equipment are highly sophisticated. Any new farm laborer would need extensive training and oversight, again burdening farmers during the busy harvest season. Furthermore, farmers are unwilling to assume risk of holding a large inventory of maize cob/stover, and prefer to have an external party bale, store and transport the goods.

Biomass Pricing and Standards

Biomass pricing and standards can play an important role in accelerating the demand for biomass feedstocks. However, traditional commodity market pricing mechanisms and quality standards are not directly transferable to biomass. For example, No. 2 yellow maize has specific quality standards and market prices that readily exist in established Agricultural Marketing Service (AMS) reporting markets. While sellers and buyers of maize have reliable standard and pricing information for decision making and planning, those wishing to trade agricultural biomass have no unbiased source of information or standards to measure the product. Importantly, the process goal for biomass is to preserve Btus, not nutrient content for animal feed. At the moment, most biomass is contracted and information for decision making is proprietary and difficult to obtain.

Standards play a critical role in the development and operation of markets. Biomass users place value on specific attributes which vary as a result of production, harvest, storage, and transportation conditions and practices. Sellers, in turn, can use knowledge of standards to guide their production, logistics, and marketing practices to maximize profits. The growth of the solid biofuel industry (e.g., wood, pellets) has led to standard development efforts for these goods. These include standards for definitions, testing, specifications, and sustainable supply chains. Much of this activity has occurred in Europe where a larger, more established solid biofuel industry is in place.

Technical Committee 335 of the European Committee for Standardization has developed more than 28 related solid biofuel standards (European Committee for Standardization, 2012). Work to harmonize international solid biofuel standards has been the goal of a technical committee of the International Organization for Standardization. This group currently has eight standards under development including a common set of terminology, definitions, and descriptions; general requirements; and specifications for six distinct fuels: wood pellets, wood briquettes, wood chips, firewood, nonwoody pellets, and nonwoody briquettes. The American Society of Agricultural and Biological Engineers coordinates the US position in this international effort.

The Pellet Fuels Institute has established standards for premium, standard, and utility grade solid biofuel. Mandatory specifications have been created for density, diameter, durability, fines, ash, length, moisture, and chloride content. The Council on Sustainable Biomass Production (CSBP) is developing voluntary sustainability standards for biomass production and conversion (CSBP, 2012). The intent is for the standard to serve as the basis for third-party accreditation. CSBP's provisional standard defines the scope as including dedicated energy crops, crop residue, and native vegetation. Their sustainability criteria include soil quality, biological diversity, water quality and quantity, climate change, socioeconomic well-being, legality, transparency, and continual improvement. The chemical or thermal conversion of cellulosic biomass to biofuels is fundamentally different from the physical processes that are used to mix and increase the density of solid fuels. Consequently the attributes of interest to buyers, and consequently sellers, vary.

The Real Constraint to Financing Biofuel Development—Uncertain Policy

Cellulosic ethanol production technology is on the verge of commercial deployment (Table 13.1), lenders have adequate financial capital to loan, and equity investors are anxious to invest in any new advanced or cellulosic biofuel projects that have merit. Without policy support, the current cellulosic biofuel industry could not have developed as it has to date. On the other hand, policy risk is the core reason which unnerves equity investors and lenders. Not only is one individual policy provision the culprit, several diverse uncertainties have unexpectedly combined to thwart development.

Transitory Biofuel Tax Credit Policy

Federal renewable energy policy has to a degree always been unstable. For example, wind energy tax credits have a history of being transitory and having expired or been set to expire and reinstated five times in the past 15 years (AWEA, 2007). Lapses have resulted in massive declines in annual installation of wind capacity such as the 93% drop from 1999 to 2000 (AWEA, 2011). With a strong political coalition of farmers, rural developers, and environmentalists, biofuel tax credit incentives remained relatively stable until December 31, 2009, when Congress failed to extend the biodiesel tax credit. Similar uncertainty crept into the ethanol tax credit during spring and summer of 2011 when budgetary pressures led to possible and immediate expiration of the Blenders Tax Credit on August 1, 2011. Public debate of the policy created anxiety among investors because they did not know if other biofuel support programs were also under consideration for curtailment. Further, the exact economic impact of terminating the tax credits was unknown.

Research has generally found that farmers and consumers have captured most of the benefit of the tax credit (e.g., consumers required the credit so ethanol was cost competitive with gasoline and farmers reaped benefits under low commodity prices; Moschini *et al.*, 2010). Thus, ethanol producers did not capture most of the benefit which implies they will not be as disadvantaged when it is terminated. On March 13, 2012, the US Senate put an end to the \$0.27 l^{-1} (or \$1.01 gal^{-1}) tax credit for cellulosic ethanol producers and did not extend the \$0.27 l^{-1} tax credit on biodiesel that expires in December 2012.

Renewable Fuel Standard Mandate Waivers

As mentioned, EISA defines three categories of biofuels based on potential reduction of GHG emissions—conventional (20%), advanced (50%), and cellulosic (60%). The legislation also specifies volumetric goals for each biofuel class through 2022 when 136 billion liters yr^{-1} is supposed to be commercially available to the consumers. EPA has been charged with monitoring commercialization and implementation (e.g., rule making). In the past several years, commercialization of cellulosic biofuels has fallen short of RFS goals (USEPA, 2011). EPA lowered volumetric requirements of cellulosic biofuels in 2010 and 2011. The cellulosic industry has been critical of these reductions. They contend that retention of the mandate goals would have led to rising prices for cellulosic biofuels and commercialization would have been stimulated at a faster pace. At this point, uncertainty surrounding the mandate creates anxiety among potential investors. The market effect of the mandate is translated into the value of RINs which vary based on the demand of obligated parties and supplies of various biofuels available in the marketplace (USEPA, 2007). Recently, RIN spot prices for advanced and cellulosic biofuels were, respectively, quoted at \$0.211 and \$0.206 l^{-1} (or \$0.80 and \$0.78 gal^{-1}) (OPIS, 2012). Greater availability of advanced and cellulosic ethanol would lower the trading value of RINs.

Uncertain Biofuel Import Tariffs

When Congress implemented the Blenders Tax Credit, it also placed a tariff on imported biofuels to offset tax subsidies claimed by biofuels entering the country. On June 16, 2011 the US Senate voted down on the Blenders Tax Credit which expired at the end of 2011. Justification of the tariff is also in question with the tax credit coming to an end. Imported biofuel, especially from Brazil, is already quite competitive with other US biofuels.

A unique set of federal policy choices could make importation of Brazilian sugarcane ethanol even more attractive. The biofuel's pathway has already been approved by the EPA. Given the EPA's retention of advanced biofuel volumetric mandates, and elimination of the tariff, Brazilian ethanol produced from sugarcane could fulfill the US advanced biofuel mandate. Analysts believe that the same is true for California which has its own state mandate for biofuel use. At present, efforts are underway in North Dakota to develop an energy-beet-based advanced biofuel industry; importation of additional sugarcane-based advanced biofuels will likely stifle national commercialization.

Carbon Policy

In July 2009, the US Senate failed to pass the American Clean Energy and Security Act of 2009 which would have established a national carbon policy. Recall that the current foundation for biofuel policy in the United States is the degree to which new advanced and cellulosic biofuels reduce GHG emissions. Failure to adopt broader carbon policy initiatives leads the biofuel industry to question the longevity and ongoing support for renewable energy. Moreover, given Congress' failure to pass the act, the Obama administration proceeded to move forward by implementing several provisions unilaterally. Most notable is the EPA's announcement of CO_2 as a threat to public health and the environment (USEPA, 2009). Implementation of such a policy is more unpredictable because public discussion did not signal impending change. Administrative regulations are often promulgated more quickly than legislative actions. This can catch the public off guard and create uncertainty. It is also uncertain which policy initiatives the administration will advance and which will remain stagnant.

Emerging Conflict between USDA Biofuel and Environmental Policy

Gustafson *et al.* (2011) evaluated the feasibility of collecting wheat straw for biofuel production in North Dakota. An interesting constraint that emerged during project evaluation was USDA NRCS policy. One goal of the wheat straw to biofuel study was to understand provisions NRCS might apply to wheat straw removal across varying topographies in the study region. While still mindful of highly erodible land characteristics and the potential for wind/water erosion, NRCS' primary focus now appears to be on carbon. Consequently, the coarseness of straw or amount of stubble remaining after harvest to guard against either wind or water erosion is now secondary considerations. NRCS' emerging policy is to recommend that 907–1361 kg (or 2000–3000 pounds) of organic matter remain in or on the soil after harvest, regardless of crop, for carbon and soil organic matter maintenance. The agency calculates that an average wheat crop produces 1814 kg (or 4000 pounds) of organic matter annually. Therefore, on average about 680 kg (or 1500 pounds) of straw is available for collection annually. However, if low residue crops like soybeans or sugar beets are planted in rotation with wheat and do not leave enough carbon (907–1361 kg), their annual deficit must be made up with residues from rotational crops (e.g., wheat) in succeeding years. Since soybeans are widespread in the region, the amount of wheat straw available for collection is reduced because the straw must replace organic matter removed by the soybean enterprise. In most regions, NRCS would recommend against harvesting wheat straw for biofuel production if soybeans are raised in rotation. This is not just an issue for wheat straw. A recent USDA Agricultural Research Service study also found that recommended stover collection from maize in rotation with soybeans is limited in Iowa (Karlen *et al.*, 2006).

A further restriction is that requirements are imposed by field, unlike the past where only the select areas susceptible to erosion within a field were primary concerns. While most fields have only portions that are at risk for erosion, carbon requirements will likely be imposed field wide. Thus, USDA appears to be sending mixed messages with respect to development of cellulosic biofuel. On the one hand, USDA is publicly supporting the growth of cellulosic biofuels. However, the agency is also implementing new environmental policy that has the potential to significantly constrain gathering of biofuel feedstock materials.

Conclusions

The US ethanol demand has been steadily increasing following passage of RFS and the EISA. Almost all domestic ethanol production utilizes maize grain as a major feedstock. Rising maize prices are encouraging current and potential ethanol producers to seek alternative feedstocks from advanced and cellulosic sources. Advanced and cellulosic biofuels have a promising future because they offer a market premium, they have higher GHG offset rates than grain-based ethanol, and production technology is on the verge of commercialization. This chapter looks into major issues related to LCA GHG emissions of biofuels in terms of concepts and use. Economically, a carbon tax or permit market system could play an important role in stimulating the production and consumption of biofuels in the future.

Energy sugar beet ethanol has the potential to qualify as an advanced biofuel because of its low GHG emission potential. Variations in ethanol and energy beet prices will have a significant impact on the net profit of energy beet ethanol production. Estimated breakeven energy beet prices at various levels of ethanol market prices show that ethanol prices dictate the movement of energy beet prices and play a key role in determining the profitability of energy beet ethanol production.

To reduce feedstock production costs, cellulosic biofuel producers need to rely on multiple cellulosic feedstocks rather than a single one. Diverse feedstocks not only reduce feedstock production costs but also lower feedstock supply risks. For the case of cellulosic crop residue feedstocks, USDA NRCS future crop residue removal rate policy will be critical in determining the viability of harvesting crop residues as potential feedstocks for cellulosic biofuel production. Research indicates that from a farmer's (or feedstock supplier's) perspective, availability of harvest field time is the major factor that dictates the farmer's economic interests in harvesting cellulosic residue feedstocks. Farmers are unwilling to take any risk of holding a large inventory of feedstocks and prefer to have an external party bale, store and transport the goods.

The success of commercializing advanced and cellulosic biofuels hinges on the federal government's policy mandates. Uncertainties in federal carbon policy, environmental stewardship, import tariffs, and variability in biofuel policy implementation have the potential to thwart the development of advanced and cellulosic biofuel investments. The future market penetration of advanced and cellulosic biofuels will be influenced not only by the carbon and other policies, but by financial constraints being imposed internally and externally on the biofuel industry (Gustafson, 2008), as well as by strong competition from conventional biofuels and limits on biofuel blending (Babcock *et al.*, 2011).

References

Abbott, P.C., Hurt, C., and Tyner, W.E. (2011) What's driving food prices in 2011? Farm Foundation Issue Report, July 2011.

American Wind Energy Association (AWEA) (2007) Wind energy and U.S. energy subsidies. Available at: http://www.maine.gov/doc/mfs/windpower/pubs/bill_hopwood/Energy%20Subsidies%20Factsheet_1.2007.pdf (accessed 30 August 2012).

American Wind Energy Association (AWEA) (2011) The reality of U.S. energy incentives. Available at: http://www.awea.org/learnabout/publications/upload/Subsidies-Factsheet-May-2011.pdf (accessed 30 August 2012).

Babcock, B.A., Marette, S., and Treguer, D. (2011) Opportunity for profitable investments in cellulosic biofuels. *Energy Policy* 39: 714–719.

Bacovsky, D., Dallos, M., and Wörgetter, M. (2010) Status of 2nd generation biofuels demonstration facilities. A Report to IEA Bioenergy Task 39, T39–P1b.

Chippewa Valley Ethanol Company (CVEC) (2009) Corn cobs as sustainable biomass for renewable energy, a field-to-facility demonstration and feasibility study. Final Report to the Minnesota Department of Commerce, Office of Energy Security, St. Paul, MN.

Costello, R. and Finnell, J. (1998) Institutional opportunities and constraints to biomass development. *Biomass and Bioenergy* 15: 201–204.

Council on Sustainable Biomass Production (CSBP) (2012) Standard for sustainable production of agricultural biomass. Available at: http://www.csbp.org/Portals/0/Documents/CSBP%20Standard%20For%20Sustainable%20Production%20of%20Agricultural%20Biomass%2006122012_1.pdf (accessed 29 August 2012).

Curran, M.A., Mann, M., and Norris, G. (2005) The international workshop on electricity data for life cycle inventories. *Journal of Cleaner Production* 13: 853–862.

Energy Information Administration (EIA) (2012) Electric power monthly. Available at: http://205.254.135.7/electricity/monthly/epm_table_grapher.cfm?t=epmt_1_1 (accessed 3 September 2012).

European Committee for Standardization (2012) Solid biofuels – published standards. Available at: http://www.cen.eu/cen/sectors/technicalcommitteesworkshops/centechnicalcommittees/pages/Standards.aspx?param=19930&title=Solid%20biofuels (accessed 14 September 2012).

Fargione, J., Hill, J., Tilman, D., Polasky, S., and Hawthorne, P. (2008) Land clearing and the biofuel carbon debt. *Science* 319: 1235–1238.

Finnveden, G., Hauschild, M.Z., Ekvall, T., *et al.* (2009) Recent developments in life cycle assessment. *Journal of Environmental Management* 91: 1–21.

Gilbert, C.L. (2010) How to understand high food prices. *Journal of Agricultural Economics* 61: 398–425.

Gurgel, A., Reilly, J.M., and Paltsev, S. (2007) Potential land use implications of a global biofuels industry. *Journal of Agricultural & Food Industrial Organization* 5, Article 9.

Gustafson, C.R. (2008) Financing growth of cellulosic ethanol. Staff Papers AAE8003, Department of Agribusiness & Applied Economics, North Dakota State University, Fargo, ND.

Gustafson, C.R., Maung, T.A., Saxowsky, D., Nowatzki, J., and Miljkovic, T. (2011) Economics of sourcing cellulosic feedstock for energy production. Selected paper at the Agricultural and Applied Economics Association, 24–26 July 2011, Pittsburgh, PA.

Harrison, R.W. (2009) The food versus fuel debate: implications for consumers. *Journal of Agricultural & Applied Economics* 41: 493–500.

IEA Bioenergy Task 39 (2010) Status of 2nd generation biofuels demonstration facilities. A Report to IEA Bioenergy Task 39. Report T39-P1b. Available at: http://www.task39.org/LinkClick.aspx?fileticket=PBlquceJcEQ%3D&tabid=4426&language=en-US (accessed 2 September 2012).

Karlen, D.L., Birrell, S.J., Radtke, C.J., and Wilhelm, W.W. (2006) Crop residue removal effects on production costs and soil quality. ASA-CSSA-SSSA Annual Meeting Abstracts, ASA-CSSA-SSSA Annual Meeting, 12–16 November 2006, Indianapolis, Indiana.

Lee, H.-C., McCarl, B.A., Schneider, U.A., and Chen, C.C. (2007) Leakage and comparative advantage implications of agricultural participation in greenhouse gas emission mitigation. *Mitigation and Adaptation Strategies for Global Change* 12: 471–494.

Maung, T.A. and Gustafson, C. (2011a) The viability of harvesting corn cobs and stover for biofuel production in North Dakota. Selected paper at the Agricultural and Applied Economics Association, 24–26 July 2011, Pittsburgh, Pennsylvania.

Maung, T.A. and Gustafson, C.R. (2011b) The economic feasibility of sugar beet biofuel production in central North Dakota. *Biomass and Bioenergy* 35: 3737–3747.

Maung, T.A., Gustafson, C.R., Saxowsky, D., Miljkovic, T., and Nowatzki, J. (2012) Market information on sourcing cellulosic feedstock for biofuel production in Northern Plains Region of the United States. *Journal of Agricultural Science and Technology A* 2: 10–23.

McCarl, B.A. (2008) Lifecycle carbon footprint, bioenergy and leakage: empirical investigations. Farm Foundation and USDA Meeting on the Lifecycle Carbon Footprint of Biofuels, January 29, Miami, FL.

McCarl, B.A. and Boadu, F.O. (2009) US bioenergy: potential, policy, and economic/climate change concerns. *Drake Journal of Agricultural Law* 14: 43–73.

McCarl, B.A., Maung, T.A., and Szulczyk, K.T. (2009) Could bioenergy be used to harvest the greenhouse: an economic investigation of bioenergy and climate change? *Handbook of Bioenergy Economics and Policy*, Springer, New York, pp. 195–218.

Moschini, G., Lapan, H., Cui, J., and Cooper, J. (2010) Assessing the welfare effects of US biofuel policies. *AgBioForum* 13: 370–374.

Murray, B.C., McCarl, B.A., and Lee, H.-C. (2004) Estimating leakage from forest carbon sequestration programs. *Land Economics* 80: 109–124.

Oil Price Information Service (OPIS) (2012) Ethanol & Biodiesel Information Service: pricing, news and analysis for buying and supplying ethanol-blended fuel and biodiesel. Available at: http://opisnet.com/news/sample/ebis.pdf (accessed 15 September 2012).

Palisade Corporation (2009) @Risk 4.5 – professional edition software, New York.

Pimentel, D., Marklein, A., Toth, M.A., *et al.* (2009) Food versus biofuels: environmental and economic costs. *Human Ecology* 37: 1–12.

Pimentel, D. and Patzek, T.W. (2005) Ethanol production using corn, switchgrass, and wood; biodiesel production using soybean and sunflower. *Natural Resources Research*, 14(1).

Sathaye, J. and Bouille, D. (2001) Barriers, opportunities and market potential of technologies and practices, in *Climate Change 2001: Mitigation*. Contribution of Working Group III to the Third Assessment Report of IPCC, Cambridge, UK, Cambridge University Press, pp. 345–398.

Schneider, U.A. and McCarl, B.A. (2003) Economic potential of biomass based fuels for greenhouse gas emission mitigation. *Environmental and Resource Economics* 24: 291–312.

Scientific Applications International Corporation (2006) Life cycle assessment: principles and practice. Report produced under contract with the National Risk Management Research Laboratory, Environmental Protection Agency, Contract No. 68-C02-067. EPA/600R-06/060.

Searchinger, T., Heimlich, R., Houghton, R.A., *et al.* (2008) Use of U.S. croplands for biofuels increases greenhouse gases through emissions from land use change. *Science* 319: 1238–1240.

Shapouri, H., Duffield, J.A., and Wang, M.Q. (2002) The energy balance of corn ethanol: an update. Agricultural Economics Report 34075, United States Department of Agriculture, Economic Research Service.

Solomon, B.D., Barnes, J.R., and Halvorsen, K.E. (2007) Grain and cellulosic ethanol: history, economics, and energy policy. *Biomass and Bioenergy* 31: 416–425.

Tokgoz, S., Elobbeid, A., Fabiosa, J., *et al.* (2007) Emerging biofuels: outlook of effects on U.S. grain, oilseed, and livestock markets. CARD Staff Report 07-SR 101, Center for Agriculture and Rural Development, Iowa State University, Ames, IA.

United States Department of Agriculture (USDA) (2011) World agricultural supply and demand estimates. WASDE-499, 12 October 2011. Available at: www.usda.gov/oce/commodity/wasde/latest.pdf (accessed 10 October 2012).

United States Department of Agriculture – National Agricultural Statistics Service (USDA-NASS) (2005–2009) Crop progress and condition report. Available at: http://www.nass.usda.gov/Statistics_by_State/North_Dakota/Publications/Crop_Progress_&_Condition/index.asp (accessed 8 October 2012).

United States Environmental Protection Agency (USEPA) (2007) Regulation of fuels and fuel additives: Renewable Fuel Standard Program. Available at: www.epa.gov/oms/renewablefuels/420r07006chp11.pdf (accessed 15 September 2012).

United States Environmental Protection Agency (USEPA) (2009) EPA: greenhouse gases threaten public health and environment. Available at: http://yosemite.epa.gov/opa/admpress.nsf/7ebdf4d0b217978b852573590040443a/08d11a451131bca585257685005bf252!OpenDocument (accessed 4 October 2012).

United States Environmental Protection Agency (USEPA) (2011) Renewable fuels: regulations and standards. Available at: www.epa.gov/otaq/fuels/renewablefuels/regulations.htm (accessed 4 October 2012).

United States Environmental Protection Agency (USEPA) (2012a) Renewable fuel standard (RFS). Available at: http://www.epa.gov/oms/fuels/renewablefuels/index.htm (accessed 6 October 2012).

United States Environmental Protection Agency (USEPA) (2012b) Accounting framework for biogenic CO_2 emissions from stationary sources. Available at: http://www.epa.gov/climatechange/ghgemissions/biogenic-emissions/study.html (accessed 7 October 2012).

Wang, M. (1999) GREET 1.5. *Transportation Fuel-Cycle Model, Volume 1: Methodology, Development, Use, and Results*. ANL/ESD-39, Volume 1, Centre for Transportation Research, Argonne National Laboratory, Argonne, IL.

Chapter 14
Geospatial Modeling Applications for Biofuel Sustainability Assessment

S.S. Panda

GIS/Environmental Science Gainesville State College, Oakwood, GA, USA

Introduction

Before major decisions of capital investment for locating biorefineries in a region can be made, the answer to the most critical question "is the region suitable climatically to produce sufficient amounts of feedstock?" is essential. Different crops respond differently to a certain climate; thus, the climate prevailing in a region must be optimal to support the growth of a core of high biomass crops to ensure uninterrupted supply of feedstock throughout the year. Sustainable agriculture calls for maximum output while keeping inputs to a minimum by their prudent deployment. Site-specific experiments on a crop performance, in general and at varietal level, and responses to different inputs and the levels of their applications are valuable to making local cropping decisions. However, broad-based inferences can only be derived by pulling together a number of experiments with common denominators. The inference drawing becomes easier if the underlying phenomena can be parameterized mathematically into a model. In this chapter, the use of a geospatial model to determine spatial suitability for the production of individual biofuel crops within a region, precision agriculture of the chosen crops, and analysis of the environmental (soil and hydrologic) impact of production decisions will be reviewed. The application of the model on different crops, such as blueberry, maize, and switchgrass, will be demonstrated. The procedures mentioned hereafter can be replicated for the sustainable production of any biofuel crop to meet future renewable energy needs.

Spatial Suitability Analysis of Biofuel Crops

Major biofuel crops are either row crops like corn, soybean, oilseed rape, sugarcane, ground nut, sunflower, mustard, sesame, rapeseed, sorghum, wild radish, cotton (seed), wheat, and so on; or horticulture crops like coconut (copra), jatropha, palm, castor, and so on; or forest trees like fast growing Chinese tallow tree also known as chicken tree (China native),

Biofuel Crop Sustainability, First Edition. Edited by Bharat P. Singh.

Paulownia (family Paulowniaceae), India native chakunda (*Cassia siamea*) and Nilgiri (*Eucalyptus* hybrid), SRC poplar, SRC willow, and so on. Switchgrass, reed canarygrass, miscanthus (*Miscanthus giganteus*), elephant grass, and other high biomass producing grasses are a good source of biomass for bioenergy production due to their high cellulose content. Forest residues and understory can also be used for biofuel production.

Successful biofuel crop or tree cultivation has specific requirements with respect to weather/climate suitability like chilling hours or sufficient supply of solar radiations in cold climatic locations, temperature, and rainfall; soil characteristics such as pH level, drainage types, soil moisture availability, texture, permeability, and slope; and require defined land use and land cover (LULC) vegetation combination. Land use and other environmental suitability modeling are the main technologies in environmental decision making and can be valuable aids to environmental managers to arrive at informed decisions (Carr and Zwick, 2007; Lance and Panda, 2008). Spatial analysis techniques, such as map algebra and weighted overlay analysis, have the ability to perform land-use, environmental, or crop production location determination suitability analyses (Erwig and Schneider, 2000). Farajzadeh *et al.* (2007) conducted a study to map the suitable spatial locations for saffron cultivation in Iran. They used a geographic information system (GIS) to model different spatial parameters responsible for saffron crop growth and management. The parameters that were used included topography (altitude and slope), climate (precipitation, probability of fall frost, probability of occurrence of 5 °C and lower, probability of occurrence of 10 °C and 22 °C, and growing degree days), hydrology (water quality, accessibility to surface waters, and accessibility to groundwaters), land-use types, and soil. These were then modeled based on assigned weights to each parameter. Bowen and Hollinger (2004) provided insight into the use of GIS and remote sensing to geographically screen potential locations for alternative crops. Few other studies also used geospatial technology to determine the crop suitability on a spatial basis (Hewson and Steininger, 2012; Reshmidevi *et al.*, 2009).

Panda *et al.* (2010b) conducted a study to identify the most suitable locations in Georgia for planting blueberry in order to obtain maximum yields. They used advanced geospatial technology including ArcGIS ModelBuilder to develop a single click automated model to pinpoint these locations. Our crop production spatial suitability model for blueberry is described below. The same procedure can be pursued to identify proper locations for the production of biofuel crops in different parts of the world based upon the suitability factors mentioned above.

Spatial Model Building for Crop Suitability Analysis

Suitable Factor Analysis for a Crop

At the outset, a literature review is conducted to obtain a crop's production suitability factors. In the case of blueberry, five major factors have been reported to have the maximum effect on its production (Krewer and Meyers, 1989; Anderson *et al.*, 1991; Braswell *et al.*, 1991; Williamson, 1994; Williamson and Lyrene, 2004; Krewer and NeSmith, 2006; Braswell *et al.*, 2009). They comprise chilling hours, soil pH, soil drainage, soil permeability, and LULC (Table 14.1).

Hewson and Steininger (2012) have defined climatic parameters for several crops. According to them, optimal min–max temperatures for cassava, oil palm, *Jatropha*, soybean, sugarcane, and switchgrass are 20–29, 11–28, 20–35, 20–33, 24–37, and 17–32 °C, respectively, and optimal min–max annual rainfall for these crops are 100–150, 50–150, 150–300, 60–150, 150–200, and 50–110 cm, respectively. Cassava, oil palm, *Jatropha*, soybean, and sugarcane

Table 14.1. Suitable factors for blueberry production.

Chilling hours (h. <5 °C)	Soil permeability (cm h^{-1})	Soil pH	Soil drainage	Land cover
>600	15–50	≤5 >4.4	Well drained and somewhat excessively drained	Coniferous forest, clear cut/young pine plantation, mixed forest, and cultivated/exposed earth

can grow optimally at elevations of 2000, 1600, 1300, 3000, and 1600 m, respectively. Optimum soil pH requirement for a few biofuel crops like cassava, oil palm, *Jatropha*, soybean, sugarcane, switchgrass, *Eucalyptus globulus*, *Eucalyptus nitens*, *Eucalyptus camaldulensis*, and *Eucalyptus grandis* are 5.5–8, 4.5–6, 5.5–7.5, 5.5–6.5, 5–8, 6–7, 5.5–6.5, 6–6.5, 5–7, and 5.5–6.5, respectively. Medium soil depth is essential for optimum crop growth for cassava and soybean and deep soil depth is required for oil palm, *Jatropha,* sugarcane, switchgrass, *E. globulus*, *E. nitens*, *E. camaldulensis*, and *E. grandis*. Low salinity, very well or well-drained, and high fertility soil is optimum for most of these biofuel crops. Cassava, *Jatropha*, switchgrass, and *E. camaldulensis* can grow optimally in moderately fertile soil. In a field study conducted at farmer fields in Nebraska, and North and South Dakotas, yields of 5.2–11.1 tonnes ha^{-1} of baled switchgrass (Biello, 2012) were obtained. The variations in yield were attributed to rainfall. The Midwestern soil and the forest soil with higher rainfall were found to be favorably suited to switchgrass cultivation. Similar information can be gathered for any biofuel crop of interest.

Data Acquisition in Preparation for Crop Production Area Delineation

Most of the states in the United States have their own geospatial data gateway. A few major federal sources like the USDA-NRCS Geospatial Data Gateway (http://datagateway.nrcs.usda.gov/), the National Map Viewer (http://viewer.nationalmap.gov/viewer/), and the NOAA National Weather Service Database (http://www.nws.noaa.gov/geodata/) can provide most of the geospatial data required for such suitability modeling. Most of the states in United States also have their own weather network to provide climatologic data such as the Automated Environmental Monitoring Network (AEMN, www.Georgiaweather.net) for Georgia. Spatial data can be obtained for other countries from their own database or from the United Nations-Food and Agricultural Organization (UN-FAO) or the United Nations Environment Program (UNEP) database. NASA-sponsored low-resolution Shuttle Radar Topography Mission (SRTM)-based Digital Elevation Model (DEM) is available for all countries of the world. FAO has low-resolution soil maps for the entire world. The UNEP can provide weather data for the world. Below is the example of spatial and nonspatial data collected for the blueberry production suitability analysis along with processing, which can be replicated for any biofuel crop production suitability model development.

For the United States, State Soil Geographic Database (STATSGO) soil data are the low-resolution soil information of any study area that contains the soil pH, drainage, and permeability attributes as spatial factors. The STATSGO soil data for the entire state of Georgia were obtained from the Georgia GIS Data Clearinghouse. The weather data to calculate chilling hours were obtained from the Georgia AEMN (www.Georgiaweather.net) (Hoogenboom,

2012). Total chilling hours were calculated for each weather station in Georgia by summing up the hours of below 5 °C temperatures between October 1 and February 15. Chilling hour data were obtained on a point location basis for all weather stations and were converted to a point shape file using the latitude and longitude information for each station. The attribute table of the vector layer was populated with the total number of chilling hours for the 2004–2005, 2005–2006, 2006–2007, and 2007–2008 winters in four different attribute fields. Another field was added to the attribute table for the calculated average chilling hour data of the 4 years, which was used for raster development of chilling hours. Georgia Land-use Trend Program (GLUT) 2005 (Landsat-derived classification) data were obtained from the Natural Resources Spatial Analysis Laboratory (http://narsal.uga.edu/glut.html) of the College of Agricultural and Environmental Sciences, University of Georgia, Athens, GA. The GLUT 2005 was a 30 m resolution, Landsat satellite imagery, LULC classified map that provided the actual LULC types of the state in 2005.

One common coordinate system, GCS North American Datum 1983, was used for ease of geospatial analysis. The various GIS layers, including both vector (i.e., State County boundary layer, chilling hour point feature layer, and STATSGO Soil layer) and raster, (i.e., GLUT 2005) were projected into the same coordinate system. The following raster layers were developed to use in the crop production suitability model development.

Raster Data Development and Suitability-based Reclassification

LULC raster: The GLUT 2005 LULC map 30 m resolution (cell size of 30 m) raster data were the basis for creating rasters of the other layers that were used in the analysis.

Chilling hour raster: To convert the chilling hour data into a raster file and to obtain chilling hours for locations where there is currently no weather station, surface interpolation was performed using the *Inverse Distance Weighted* (IDW) technique (ArcGIS, ESRI, Redlands, CA). The average chilling hour data (attribute) that were prepared in an earlier subsection were used as the z-value field to create a chilling hour raster for the entire state. The interpolation raster development was conducted with a cell size of 30 m. Generally, interpolation functions generate a raster with an extent of the coordinates of extreme top left and extreme bottom right point features. The entire Georgia boundary was set up as the extent setting in the ArcGIS geoprocessing "Environmental Settings" to get rid of the problem of obtaining chilling hours for missing areas in Georgia. Thus, the interpolated chilling hour raster was generated for the entire state of Georgia.

Soil rasters: The STATSGO soil vector file was converted to three separate raster coverages using the *Feature to Raster* function available with the Conversion Tools toolbox of ArcToolbox (ESRI, Redlands, CA). Permeability, pH, and drainage attributes of a soil layer were used as conversion fields to develop individual raster coverages for soil permeability, soil pH, and soil drainage. The feature to raster conversion was conducted with a cell size of 30 m to conform to the common raster size to be used for geospatial modeling.

The blueberry production suitability criteria were used to reclassify each raster with binary Boolean classes of 0 and 1. Conditions considered suitable for blueberry production were given a value of 1, while the conditions considered unsuitable were given a value of 0. For example, in the raster reclassification process, chilling hours of more than 600 h received a value of 1 and less than 600 h received a value of 0 (Table 14.1). Thus, using the suitable factors presented in Table 14.1, five reclassified rasters, including chilling hours, soil pH, soil permeability, soil drainage, and LULC were produced, which only contained cells with values of 0 and 1.

Spatial Overlay

The five reclassified rasters were overlaid using the map algebra function of ArcGIS and the *Times* tool was used to consolidate them into a single raster. The resultant raster layer contained only values of 0, which denoted unsuitable and 1, which denoted suitable, for each individual cell. Following this operation, any cell that satisfied all five suitable conditions had a value of 1 and was considered to be suitable for blueberry production. It is to be noted that the cell sizes of each raster was 30 m × 30 m. This means that a particular cell of 900 m^2 area was totally suitable for blueberry production. The resultant raster file was overlaid on the Georgia County boundary map to identify the counties that had the largest potential for blueberry production. The counties that contained a significant number of suitable areas were considered as having the potential for blueberry production.

Geospatial Model Development

ModelBuilder is an application that is part of the ArcGIS software and that can be used to create, edit, and manage geospatial models (ESRI, 2012). The procedures that were discussed in the data preparation and geospatial analysis section were included in the ModelBuilder to develop individual models or submodels. In Figure 14.1 a submodel is shown that was developed in ArcGIS ModelBuilder to determine the dominant blueberry producing counties in Georgia. It includes both the input data and the tools that were used in the analysis along with the produced output data. Several other submodels were developed to produce the five reclassified rasters that were discussed previously. Finally, all these submodels were integrated into one single

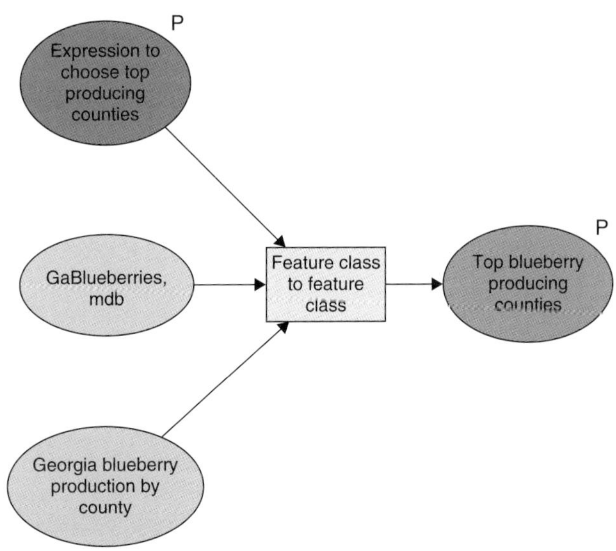

Figure 14.1. Submodel developed in ArcGIS ModelBuilder to obtain top blueberry producing counties in Georgia.

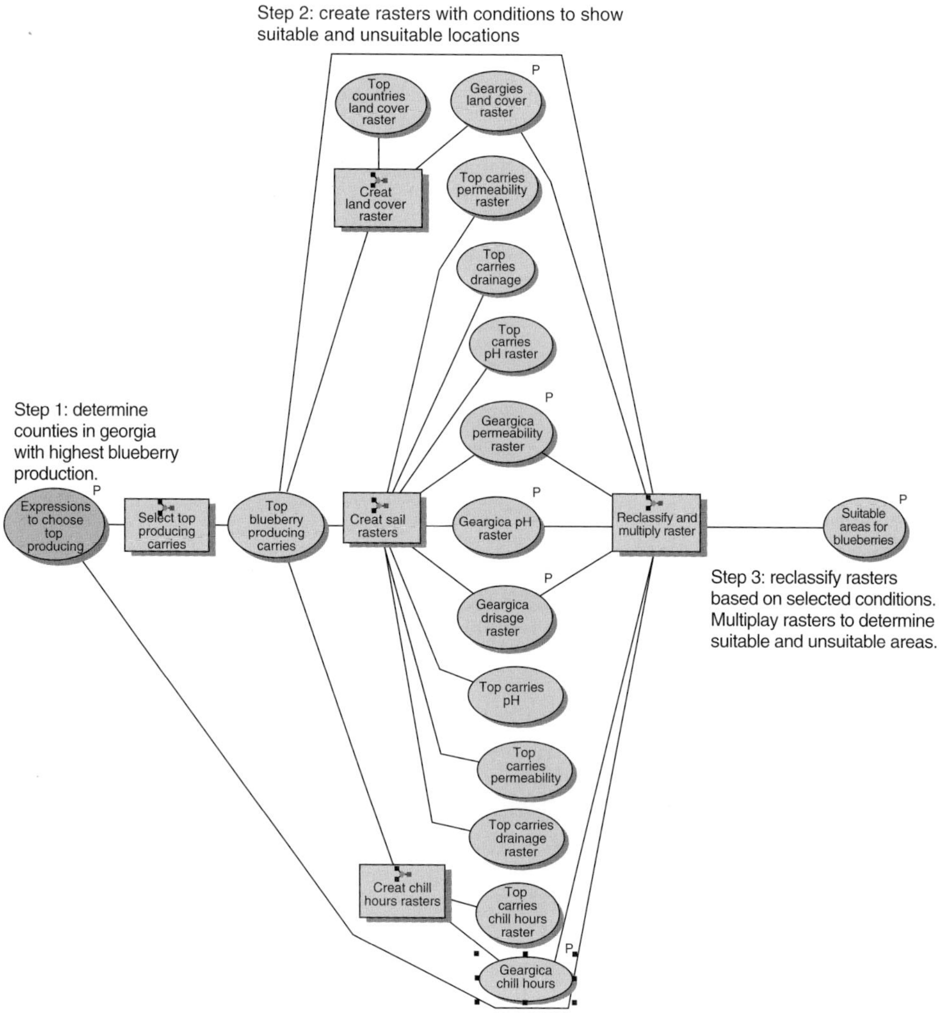

Figure 14.2. Comprehensive geospatial model developed in ArcGIS ModelBuilder for blueberry production suitability analysis in Georgia.

model that performed the map algebra function (multiplication) to combine all five rasters, and resulted in a blueberry production suitability raster. The single automated geospatial model that produced the final map showing the suitable blueberry production locations in Georgia is shown in Figure 14.2.

Figure 14.3 shows the climatologic-factor-based chilling hour map developed for blueberry crop production suitability analysis. Figure 14.4 is the LULC map of Georgia used in the blueberry crop production suitability analysis. Figures 14.5a–14.5c show the soil factor rasters (pH, permeability, and drainage) prepared for the study area for blueberry crop production suitability analysis. Figures 14.6a–14.6e are the spatial suitable raster map of Georgia for blueberry crop production. Figure 14.7 shows the spatial locations most suitable for blueberry

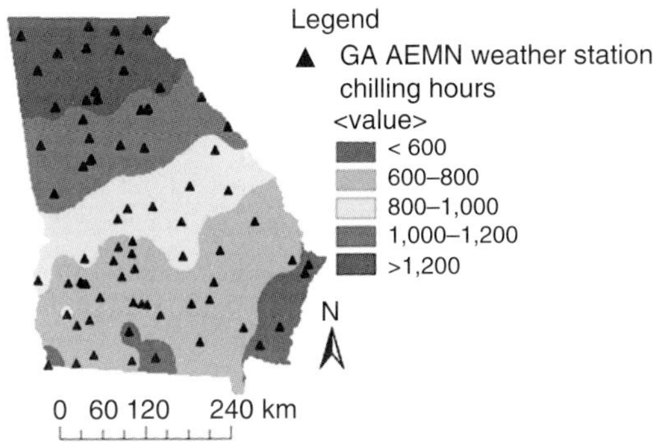

Figure 14.3. Chilling hour map for Georgia. For color detail, please see color plate section.

production in Georgia. Blueberry can be produced in other parts of Georgia, but it will not provide the optimum yield and is not sustainable, because cost of production may exceed the returns from the crop. Similar geospatial maps can be developed as part of the biofuel crop production suitability analysis.

Precision Agriculture for Higher Biomass Crop Yield

Crop growth observation, impact assessment, and timely strategic response to small variations in production define precision agriculture. It has been used in a wide range of agricultural activities including field crop production, horticulture, and forest management (Panda *et al.*, 2010b). Site-specific crop management (SSCM) is one facet of precision agriculture, which involves spatial referencing, crop and climate monitoring, attribute mapping, decision support systems, and differential action. SSCM is carried out with a greater degree of precision through the use of geospatial technologies. Geospatial technology is a combination of four essential tools: remote sensing, GIS, global positioning system (GPS), and information technology or data management (Otter-Nacke *et al.*, 1986; Casanova *et al.*, 1998; Panda,

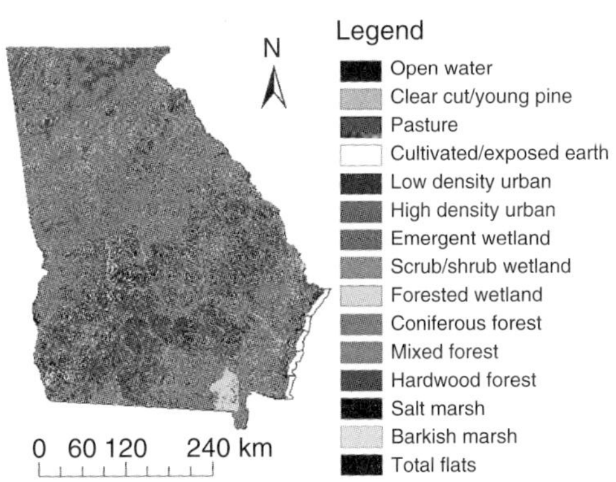

Figure 14.4. LULC analysis map of Georgia. For color detail, please see color plate section.

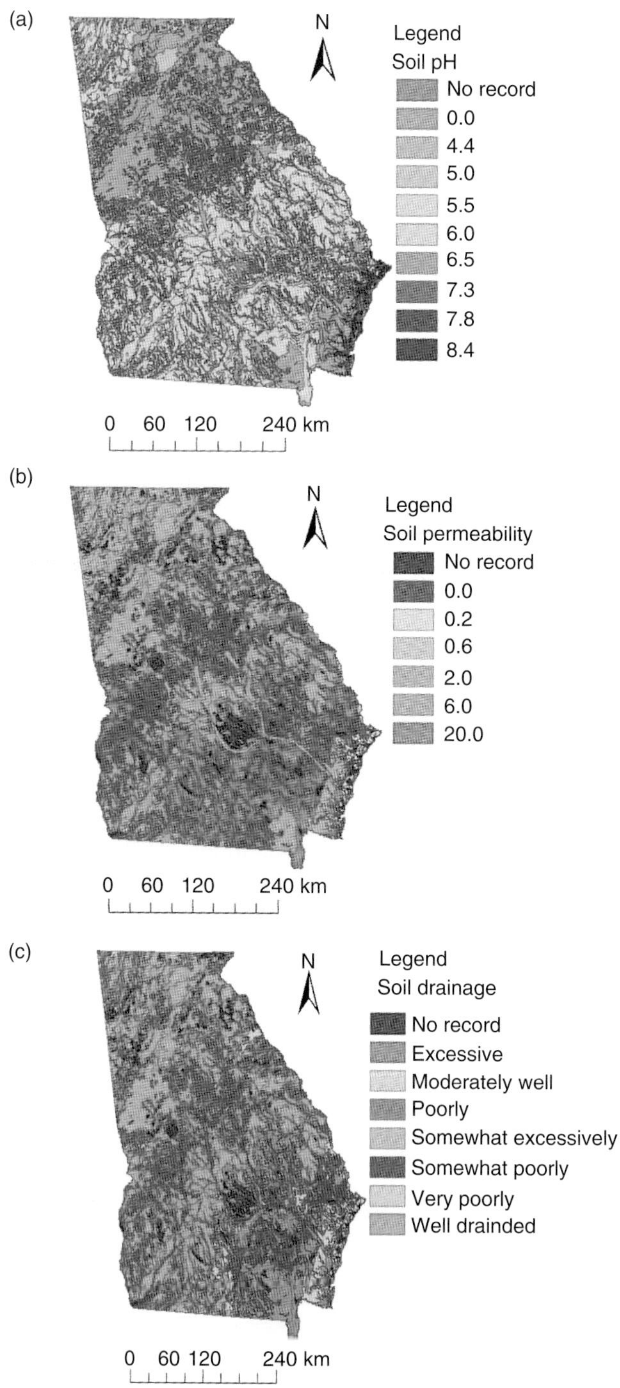

Figure 14.5. Maps showing the soil attributes that affect blueberry production, including (a) pH, (b) permeability, and (c) drainage. For color detail, please see color plate section.

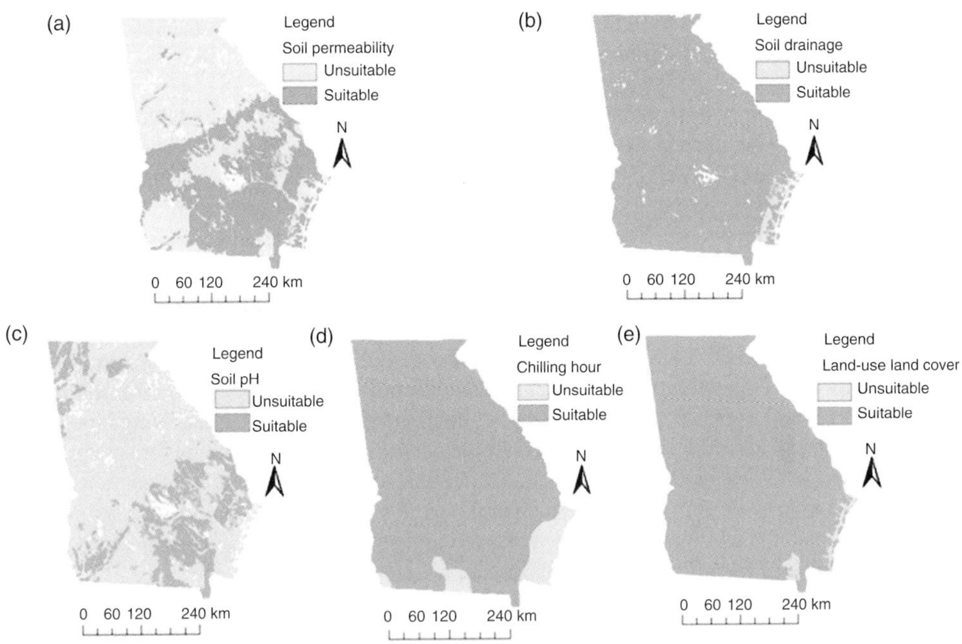

Figure 14.6. Maps showing the suitable areas in Georgia for various blueberry production parameters: (a) soil permeability, (b) soil drainage, (c) soil pH, (d) chilling hours, and (e) land use. For color detail, please see color plate section.

2003; Baez-Gonzalez *et al.*, 2005; Lobell *et al.*, 2005; Magri, 2005; Li *et al.*, 2009). SSCM has become very common in the management of field and row crops and is also being used for horticultural crops for higher yield with less investment (Panda *et al.*, 2009). Companies, such as Lanworth (http://www.lanworth.com/), which gather intelligence on natural resources, use geospatial technologies to map, analyze, and forecast agriculture and forest output. This

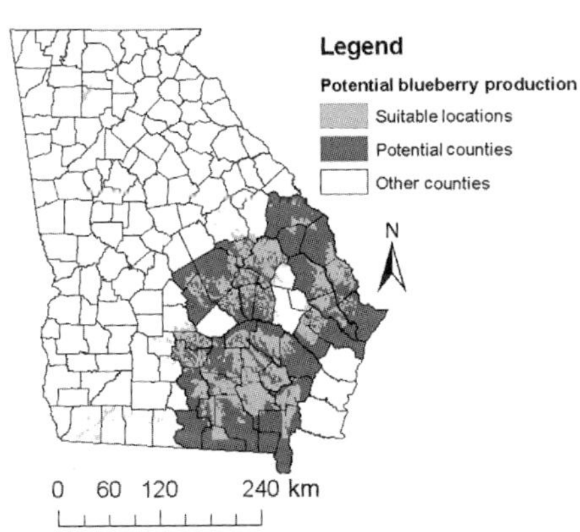

Figure 14.7. Blueberry production suitability map for Georgia. For color detail, please see color plate section.

is based on crop and forest growth parameters, which is a large amount of data, terabytes in size. It helps in the SSCM of agricultural and forest products, and increases the yield in a sustainable manner.

Biofuel crops like row crops, pastures, and forest plants can be sustainable with SSCM or precision agriculture management practices. Examples of SSCM application in the world are immense. It is unnecessary to refer to such thousands of articles. We have developed two models through his research on "data mining application for production management of crop." The first model was developed to predict maize crop yield from fields on a spatial basis using the mid-crop season aerial imagery (Panda *et al.*, 2010a, 2010c). The aerial imagery, individual band digital information along with indices developed with the bands were used to develop a yield prediction model using neural network techniques, such as backpropagation neural network (BPNN), radial basis function network (RBFN), self-organizing map (SOM), and learning vector quantization (LVQ). Farmers or end users will be able to use the model to predict the crop yield on a spatial basis from their field, with the aerial imagery in the mid-season. It will also enable them to identify low growth areas and reasons for such growth, allowing them to take remedial measures in time to achieve optimum yields. Such models can be developed for any biofuel crop to help enhance yield and to improve the crop sustainability. The second model was to predict crop water need and schedule irrigation for the crop during the season to minimize water wastage and provide water as required on a spatial basis to reduce plant stress and increase production. Below is the example of a BPNN-based neural network model, developed for corn crop yield prediction using several vegetation indices developed from a four-band aerial imagery acquired during mid-crop season.

Four widely used spectral indices were investigated in a study of irrigated corn crop yields in the Oakes Irrigation Test Area Research Site of North Dakota. These indices were: (a) red (R)- and near-infrared (NIR)-based normalized difference vegetation index (NDVI); (b) green (G)- and NIR-based green vegetation index (GVI); (c) R- and NIR-based soil-adjusted vegetation index (SAVI); and (d) R- and NIR-based perpendicular vegetation index (PVI). These four indices were investigated for maize yield during 3 years (1998, 1999, and 2001) and for the pooled data of these 3 years. Initially, BPNN models were developed, including 16 models (four indices × 4 years including the data from the pooled years) to test for the efficiency determination of those four vegetation indices in corn crop yield prediction.

The maize yield was best predicted using BPNN models that used the means and standard deviations of PVI grid images. In all 3 years, it provided higher prediction accuracies, coefficient of determination (r^2), and lower standard error of prediction than the models involving GVI, NDVI, and SAVI image information. The GVI, NDVI, and SAVI models for all 3 years provided average testing prediction accuracies of 24.26–94.85%, 19.36–95.04%, and 19.24–5.04%, respectively, while the PVI models for all 3 years provided average testing prediction accuracies of 83.50–96.04%. The PVI pool model provided better average testing prediction accuracy of 94% with respect to other vegetation models, for which it ranged from 89% to 93%. Similarly, the PVI pool model provided an r^2 value of 0.45 as compared to 0.31–0.37 for other index models. The $\log_{10}$ data transformation technique was used to enhance the prediction ability of the PVI models of the years 1998, 1999, and 2001 as it was chosen as the preferred index. Another model (Transformed PVI (Pool)) was developed using the $\log_{10}$ transformed PVI image information to show its global application. The transformed PVI models provided average maize yield prediction accuracies of 90%, 97%, and 98% for the years 1998, 1999, and 2001, respectively. The pool PVI transformed model provided an average testing accuracy of 93% along with an r^2 value of 0.72 and standard error of prediction of 0.05 tonnes ha^{-1}.

The equations used to develop the indices are as follows (Panda *et al.*, 2010b):

NDVI is determined using the R and NIR bands of a given image and is expressed as

$$\mathrm{NDVI} = \frac{\rho_{\mathrm{ir}} - \rho_{\mathrm{r}}}{\rho_{\mathrm{ir}} + \rho_{\mathrm{r}}} \tag{14.1}$$

where ρ_{r} and ρ_{ir} are spectral reflectance from the R- and NIR-band images, respectively. The GVI is determined using

$$\mathrm{GVI} = \frac{\rho_{\mathrm{ir}} - \rho_{\mathrm{g}}}{\rho_{\mathrm{ir}} + \rho_{\mathrm{g}}} \tag{14.2}$$

where ρ_{g}, and ρ_{ir} are spectral reflectance from the G- and NIR-band images.

SAVI tends to minimize soil brightness, a phenomenon that has been demonstrated by many researchers (Miura *et al.*, 2000; Jayaraman *et al.*, 2000). Huete (1988) introduced a soil calibration factor in the NDVI equation to account for the first order soil–vegetation optical interactions. SAVI is a compromise between NDVI and PVI and is defined as

$$\mathrm{SAVI} = \left[\frac{(\rho_{\mathrm{ir}} - \rho_{\mathrm{r}})}{(\rho_{\mathrm{ir}} + \rho_{\mathrm{r}} + L)}\right](1 + L) \tag{14.3}$$

where L is a constant that is a surrogate for the leaf area index (LAI). Huete (1988) defined the optimal adjustment factor of $L = 0.25$ to be considered for higher vegetation density in the field, $L = 0.5$ for intermediate vegetation density, and $L = 1$ for the low vegetation density. He suggested that SAVI ($L = 0.5$) successfully minimized the effect of soil variations in green vegetation compared to NDVI. The PVI equation is expressed as the function of the slope and intercept of the vegetation images of the R and NIR bands, and the soil images of the R and NIR bands (Thiam and Eastman, 1999).

$$\mathrm{PVI} = \sqrt{(\rho G_{\mathrm{ir,s}} - \rho P_{\mathrm{ir}})^2 + (\rho G_{\mathrm{r,s}} - \rho P_{\mathrm{r}})^2} \tag{14.4}$$

where $\rho_{\mathrm{r,s}}$ and $\rho_{\mathrm{ir,s}}$ are reflectance of soil background in the R and NIR bands, respectively; and ρ_{r} and ρ_{ir} = reflectance of vegetation in the R and NIR bands, respectively. PVI is determined using the distance between the intersection point ($\rho G_{\mathrm{ir,s}}$ and $\rho G_{\mathrm{r,s}}$) and the vegetation image pixel coordinate (ρP_{ir} and ρP_{r}) by the Pythagorean Theorem.

An example of a four-band study area image with 100 grids to extract 100 images from it is provided in Figure 14.8. Figure 14.9 provides the schema of the working procedure for imagery input parameter extraction for the BPNN model. Figure 14.10 shows the schema of BPNN architecture and input parameters are the indices developed from the aerial images using the equation above. Figure 14.11 provides a comparison of the actual spatial yield versus the model predicted yield using transformed PVI pool (randomly selected testing data from all the data used for the study) data. As observed in the figure, the model predicted very accurately with 72% r^2, 93.05% average accuracy, and 0.05 tonnes ha^{-1} of standard error of prediction.

This study can easily be replicated for any biofuel crop and advanced neural-network-based models or statistical models can be developed to predict crop or tree yield in mid-crop season. This will make it possible for farmers to take corrective measures in low-yield locations to enhance yield and make the biofuel crop yield sustainable.

Figure 14.8. Aerial image of the OITA NW29 quarter in July 30, 1998, showing grid image pattern. For color detail, please see color plate section.

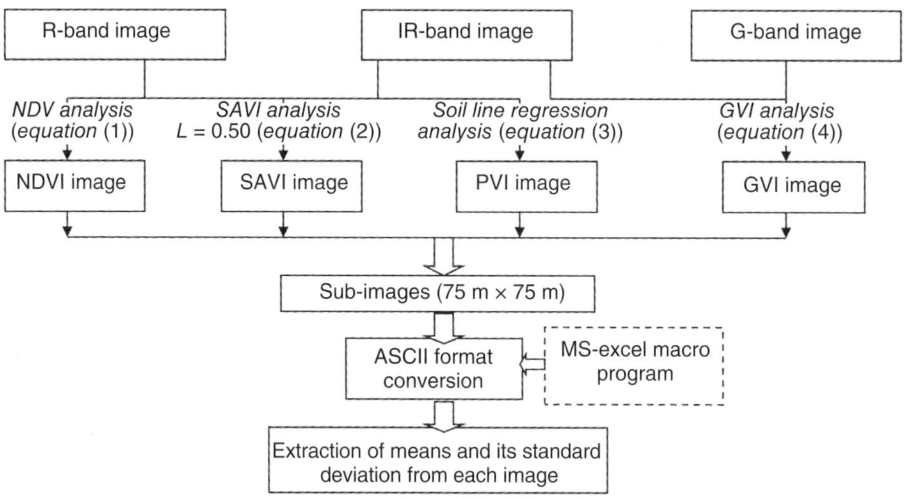

Figure 14.9. Schema of the procedure for imagery input parameter extraction for BPNN model.

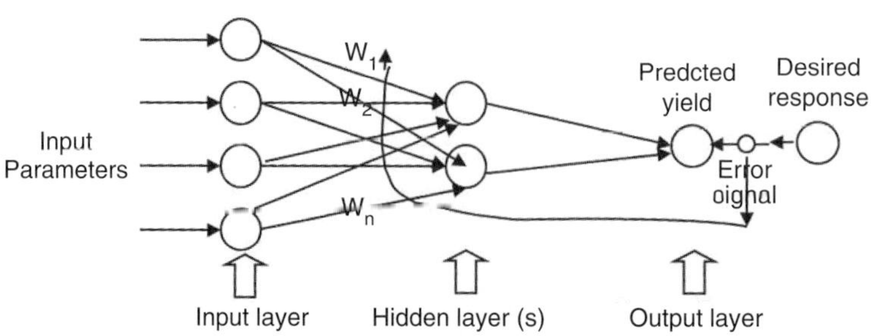

Figure 14.10. Back propagation neural network (BPNN) architecture.

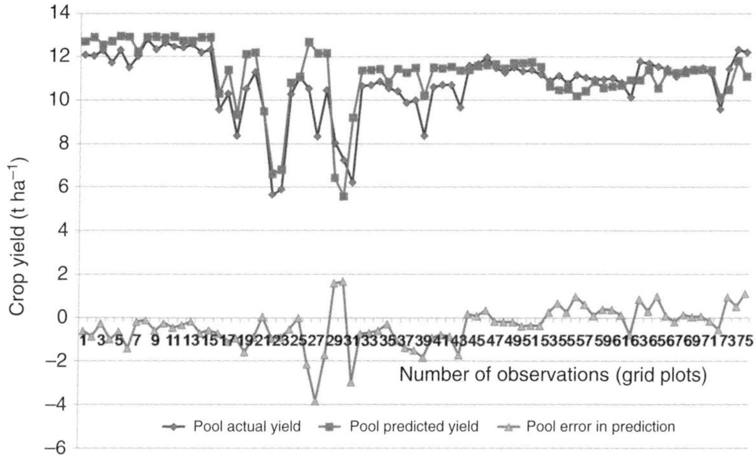

Figure 14.11. Comparison of actual versus predicted corn yield using testing dataset selected from all 3 years (1998, 1999, and 2001) pool data (Transformed PVI (Pool)).

Model or Procedure Development for Forest Quality Analysis

Remote sensing imageries of forest cover with advance image processing approach can help determine the quality of the forest as well as classify the species in the forest. Forest plant speciation is very tricky task. However, by combining high-resolution imagery and advanced image processing with ground truthing, it is possible to accurately classify the forest trees. We have developed a tool (program model) in ArcObjects environment that would query a Feature Layer containing four data fields representing the four mean reflectance values (R, G, B, and NIR), create mean vegetation species index (MVSI) fields calculated from them, and finally create a field containing tree species classification according to a predetermined protocol. The MVSI used the equations, NIR^2/BG, NIR^2/BR, and NIR^2/BR, where B, G, R, and NIR represent the Blue, Green, Red, and Near-Infrared bands of false color Multi-spectral Scanner (MSS) imagery. The protocol for assigning MVSI values to each plant species was developed based on field ground truthing. All four bands of QuickBird imagery of Elachee Nature Science Center, Gainesville, GA was segmented in the object-based image analysis (OBIA) software, eCognition Developer 8, based on mean band values and exported as the shapefile, to be classified by the tool developed in ArcObjects. The index values representing different species of plants in the study area forest were determined through an intense rule set (Figure 14.12) developed with OBIA software, eCognition Developer 8 using MVSI imagery of the study area. Figure 14.13 shows the final classified imagery of the study area. This advanced process would aid end users to obtain forest growth information to aid sustainable biofuel production.

Environmental (Soil and Hydrologic) Impact Analysis with Geospatial Technology

It is essential to study the hydrologic, soil, and environmental impact due to production of these renewable-energy-producing crops. The use of geospatial technology for such determinations is described below.

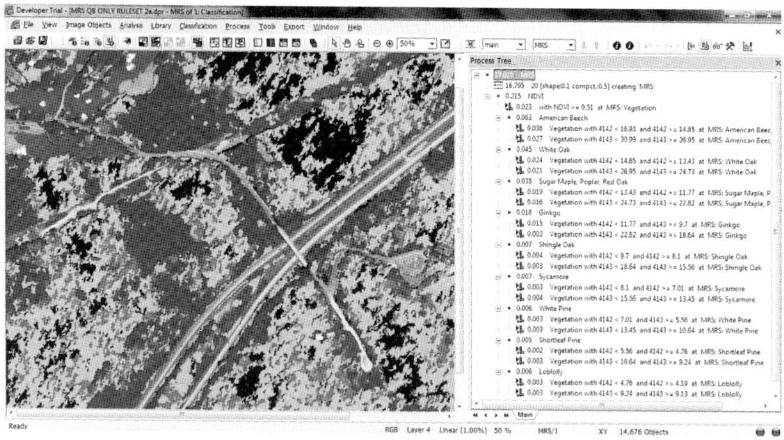

Figure 14.12. Advanced rule set (right) developed with OBIA software eCognition Developer 8 using MVSI imagery of the study area. For color detail, please see color plate section.

According to Panda *et al.* (2012), it is difficult to estimate the hydrologic parameters of vast forest cover and switchgrass through field measurements. Remote sensing has the proven ability for measuring some of those parameters in a quick and cost-effective manner. Thus, remotely sensed imageries are being used to estimate evapotranspiration (ET) amount through modeling. ET has a direct correlation with plant stomatal conductance (g_s), LAI, and canopy temperature, albedo, and soil moisture. Different forest plant species along with switchgrass demonstrate different values for these hydrologic parameters that ultimately help estimate the ET. Remotely sensed images including Landsat, SPOT, NAIP orthoimagery can help predict plant stomatal conductance, surface temperature, vegetation stress, and soil moisture amount (Lillesand and Kiefer, 1994). Vegetation indices like NDVI or SAVI, developed with NIR

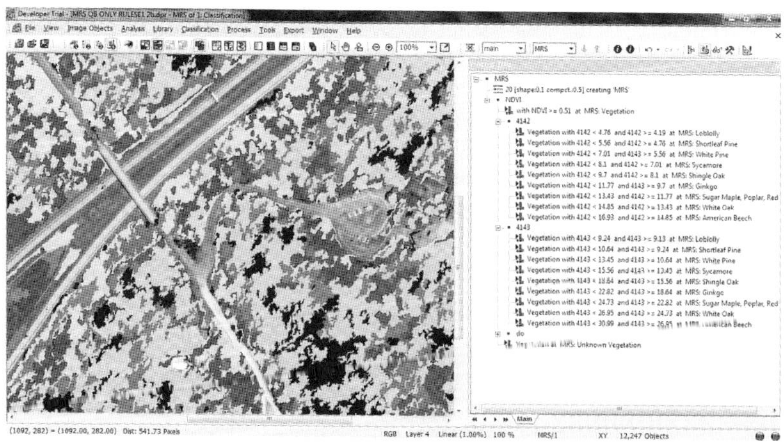

Figure 14.13. Final classified imagery of the study area to differentiate individual plant species. For color detail, please see color plate section.

Figure 14.14. LAI raster spatial map of a blueberry orchard in Southeast Georgia developed using geospatial model. For color detail, please see color plate section.

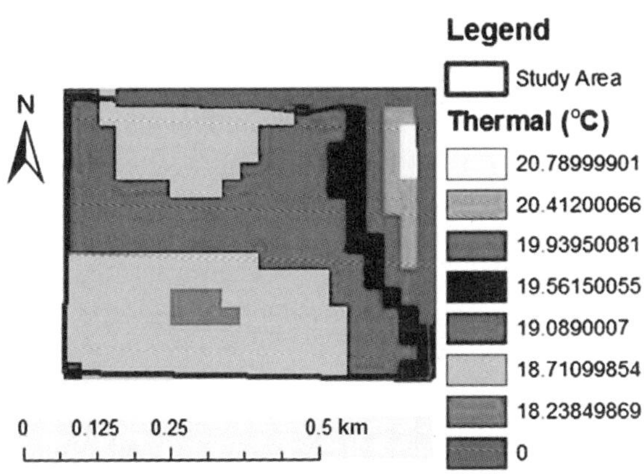

Figure 14.15. Thermal raster of showing the surface temperature variation of pine tree plantation in a study plot at Carteret, NC. For color detail, please see color plate section.

and R bands, are able to remotely determine such forest hydrologic parameters including LAI (Lillesand and Kiefer, 1994). Casanova *et al.* (1998) have previously used remote sensing to estimate the biomass and LAI of a row crop like rice. Panda *et al.* (2011) and Amatya *et al.* (2011) have shown that remotely sensed data are efficient in estimating stomatal conductance, canopy temperature, and LAI. Figures 14.14 and 14.15 show the spatial LAI and plant canopy temperature raster prepared for two study areas using the geospatial models developed for this purpose.

From the discussion above, it can be safely concluded that remote sensing, GIS, and GPS, along with model building are useful tools that can be applied advantageously in the identification of suitable climatic and edaphic regions for the cultivation of important biofuel crops, devising precise cultivation methods, and in analyzing, estimating, or monitoring the environmental impact of biofuel crops.

References

Amatya, D., Panda, S.S., Cheschair, G., Nettles J., Appleboom, T., and Skaggs, W. (2011) Evaluating evapotranspiration and stomatal conductance of matured pine using geospatial technology. American Geophysical Union Conference 2011, December 5–9, 2011.

Anderson, P.C., Crocker, T.E., and Lyrene, P.M. (1991) Vegetative and reproductive characteristics of hexaploid and tetraploid blueberry cultivars in north Florida. *Proceedings of Florida State Horticulture Society* 104: 12–15.

Baez-Gonzalez, A.D., Kiniry, J.R., Maas, S.J., *et al.* (2005) Large-area maize yield forecasting using leaf area index based yield model. *Agronomy Journal* 97: 418–425.

Biello, D. (2012) Grass makes better ethanol than corn does. Scientific American. Available at: http://www.scientificamerican.com/article.cfm?id=grass-makes-better-ethanol-than-corn (accessed 5 September 2012).

Bowen, C.R. and Hollinger, S.E. (2004) Geographic screening of potential alternative crops. *Renewable and Agriculture Food Systems* 19: 141–151.

Braswell, J., Stringer, S.J., Sampson, B.J., and Ingrahm, D. (2009) Establishment and maintenance of blueberries. USDA-ARS Extension Publication pp. 1–20.

Braswell, J.H., Spiers, J.M., and Hegwood, C.P. Jr. (1991) Establishment and maintenance of blueberries. Mississippi State University, Publication 1758.

Carr, M.H. and Zwick, P.D. (2007) *Smart Land-Use Analysis, The LUCIS Model: Land-use Conflict Identification Strategy*, ESRI Press, Redlands, CA.

Casanova, D., Epema, G.F., and Goudriaan, J. (1998) Monitoring rice reflectance at field level for estimating biomass and LAI. *Field Crops Research* 55: 83–92.

Erwig, M. and Schneider, M. (2000) Formalization of advanced map operations. 9th International Symposium on Spatial Data Handling, Beijing, IGU.

ESRI (Environmental Systems Research Institute) (2012). An overview of ModelBuilder. http://webhelp.esri.com/arcgisdesktop/9.2/index.cfm?TopicName=An_overview_of_ModelBuilder (accessed 25 July 2012).

Farajzadeh, M., Bayanti, R.M., Rahimi, M., Modares, T., and Amiri, R. (2007) Preparation of saffron cultivation suitability map based on the comparison of different weighting methods in GIS environment. Map Asia 2007. Available at: http://www.gisdevelopment.net/application/agriculture/cropping/ma07296.htm (accessed 25 July 2012).

Hewson, J. and Steininger, M. (2012) Using geospatial technology to map potential biofeedstock crop cultivation zones and identify potential conflict with areas of high biodiversity or ecosystem service value. Conservation International. Available at: http://www1.eere.energy.gov/biomass/pdfs/Biomass_2009_Sustainability_II_Hewson.pdf (accessed 25 July 2012).

Hoogenboom, G. (2012) Georgia automated environmental monitoring network. Available at: http://www.griffin.uga.edu/aemn/cgi-bin/AEMN.pl?site=AAAA&report=bg (accessed 25 July 2012).

Huete, A.R. (1988) A soil adjusted vegetation index (SAVI). *Remote Sensing Environment* 25: 295–309.

Jayaraman, V., Srivastava, S.K., Raju, D.K., and Rao, U.R. (2000) Total solution approach using IRS-1C and IRS-P3 data. *IEEE Transactions on Geoscience and Remote Sensing* 38: 587–604.

Krewer, G. and Meyers, S. (1989) Commercial blueberry culture. University of Georgia, Cooperative Extension Service Circulation 713, Athens, GA.

Krewer, G. and NeSmith, D.S. (2006) Blueberry cultivars for Georgia. Available at: http://www.smallfruits.org/Blueberries/production/06bbcvproc_Nov0206.pdf (accessed 25 July 2012).

Lance, L. and Panda, S.S. (2008) Delineation of cougar habitat in Appalachian mountain range in Georgia with geospatial modeling. Georgia and Florida Academy of Science Conference 2008, Jacksonville, FL.

Li, T., Feng, Y., and Li, X. (2009) Predicting crop growth under different cropping and fertilizer management practices. *Agricultural and Forest Meteorology* 149: 985–998.

Lillesand, T.M. and Kiefer, R.W. (1994) *Remote Sensing and Image Interpretation*, 3rd edn, John Wiley and Sons, Inc., New York, p. 376.

Lobell, D.B., Ortiz-Monasterio, J.I., Asner, G.P., Naylor, R.L., and Falcon, W.P. (2005) Combining field surveys, remote sensing, and regression trees to understand yield variations in an irrigated wheat landscape. *Agronomy Journal* 97: 241–249.

Magri, A. (2005) Soil test, aerial image and yield data as inputs for site-specific fertility and hybrid management under maize. *Precision Agriculture* 6: 87–110.

Miura, T., Huete, A.R., and Yoshioka, H. (2000) Evaluation of sensor calibration uncertainties on vegetation indices for MODIS. *IEEE Transactions on Geoscience and Remote Sensing* 38: 1399–1409.

Otter-Nacke, S., Godwin, D.C., and Ritchie, J.T. (1986) Testing and Validating the CERES-wheat Model in Diverse Environments, AGGRISTARS YM-15-00407, Johnson Space Center, Houston, TX, USA.

Panda, S.S. (2003) Data Mining application in production management of crop (Paper 1). Ph.D. Dissertation, North Dakota State University, Fargo, ND, USA.

Panda, S.S., Ames, D.P., and Panigrahi, S. (2010a) Application of vegetation indices for agricultural crop yield prediction using neural network. *Remote Sensing* 2: 673–696.

Panda, S.S., Hoogenboom, G., and Paz, J. (2009) Distinguishing blueberry bushes from mixed vegetation land use using high resolution satellite imagery and geospatial techniques. *Computers and Electronics in Agriculture* 67: 51–59.

Panda, S.S., Hoogenboom, G., and Paz, J. (2010b) Remote sensing and geospatial technological applications for site-specific management of fruit and nut crops: a review. *Remote Sensing* 2: 1973–1997.

Panda, S.S., Nolan, J., Amatya, D., *et al.* (2012) Stomatal conductance and leaf area index estimation using remotely sensed information and forest speciation. Poster presented in the 3rd International Conference on Forests and Water in a Changing Environment, 18–20 September, 2012, Fukuoka, Japan.

Panda, S.S., Panigrahi, S., and Ames, D.P. (2010c) Crop yield forecasting from remotely sensed aerial images with self-organizing maps. *Transactions of ASABE* 53: 323–338.

Panda, S.S., Peters, P., Harris, R., and Skarda, R.J. (2011) Remote measurement of potential water loss through evapo-transpiration of kudzu during the growing season. National Water Conference 2011, January 31–February 1, 2011, Washington, DC.

Reshmidevi, T.V., Eldho, T.I., and Jana, R. (2009) A GIS-integrated fuzzy rule-based inference system for land suitability evaluation in agricultural watersheds. *Agricultural Systems* 101: 101–109.

Thiam, A. and Eastman, J.R. (1999) Vegetation Indices, in *Guide to GIS and Image Processing,* V.2. Idrisi 32. Clark Labs, Clark University, Worcester, MA.

Williamson, J.G. (1994) Dooryard blueberry production in Florida. *Proceedings of Florida State Horticulture Society* 102: 399–403.

Williamson, J.G. and Lyrene, P.M. (2004). Blueberry varieties for Florida. HS967. University of Florida Extension. Institute of Food and Agricultural Sciences, Gainesville, FL.

Appendix I
Botanical Names

Common name	**Botanical name**
Alba mustard	*Sinapis alba*
Australian lemongrass	*Cymbopogon ambiguus*
Banana	*Musa acuminata*
Barley	*Hordeum vulgare*
Bethlehem	*Ipomoea quamoclit*
Birdsfoot trefoil	*Lotus corniculatus*
Blueberry	*Vaccinium* spp.
Camelina	*Camelina sativa*
Cardoon	*Cynara cardunculus*
Cassava	*Manihot esculenta*
Castor	*Ricinus communis*
Cucumber	*Cucumis* spp.
Convolvulus	*Ipomoea* spp.
Cotton	*Gossypium hirsutum*
Elephant grass	*Pennisetum purpureum*
Energy cane	*Saccharum* spp.
Ethiopian mustard	*Brassica carinata*
Eucalyptus	*Eucalyptus* spp.
Giant reed	*Arundo donax*
Hemp	*Cannabis sativa*
Jatropha	*Jatropha curcas*
Jerusalem artichoke	*Helianthus tuberosus*
Johnsongrass	*Sorghum halepense*
Linseed	*Linum usitatissimum*
Loblolly pine	*Pinus taeda*
Lupins	*Lupinus angustifolius*

Biofuel Crop Sustainability, First Edition. Edited by Bharat P. Singh.

Common name	**Botanical name**
Maize/Corn	*Zea mays*
Mexican fire plant	*Euphorbia heterophylla*
Morning glory	*Ipomoea* spp.
Mouse-ear cress	*Arabidopsis thaliana*
Miscanthus	*Miscanthus* spp.
Mustard	*Brassica juncea*
Oats	*Avena fatua*
Olive	*Elaeagnus* spp.
Oil palm	*Elaeis guineensis*
Passion vine	*Passiflora* spp.
Pearl millet	*Pennisetum glaucum*
Peas	*Pisum sativum*
Pineapple	*Ananas comosus*
Pine	*Pinus* spp.
Poplar	*Populus* spp.
Rapeseed	*Brassica napus*
Reed canarygrass	*Phalaris* spp.
Rice	*Oryza sativa*
Rye	*Lolium* spp.
Sorghum	*Sorghum bicolor*
Soybean	*Glycine max*
Signal grass	*Brachiaria decumbens*
Sunflower	*Helianthus annuus*
Sugarcane	*Saccharum* spp.
Sweet potato	*Ipomoea batatas*
Switchgrass	*Panicum virgatum*
Sugar beet	*Beta vulgaris*
Tomato	*Lycopersicon esculentum*
Triticale	*Triticum secale*
Weeping grass	*Microlaena stipoides*
Wheat	*Triticum aestivum*
Willow	*Salix* spp.

Index

Note: Page number followed by f and t indicates figure and table respectively.

Biofuel Crop Sustainability, First Edition. Edited by Bharat P. Singh.